ELECTRIC MOTOR DRIVE
Installation and Troubleshooting
second edition

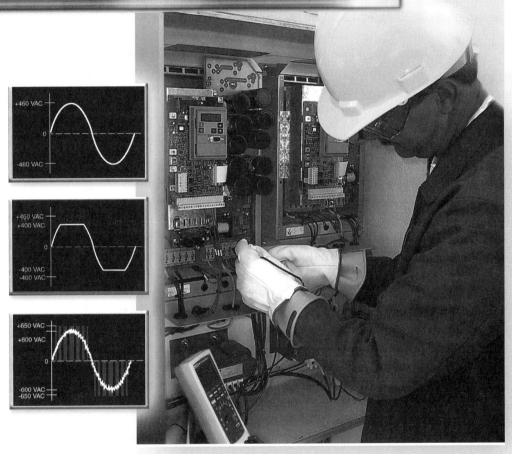

AMERICAN TECHNICAL PUBLISHERS, INC.
HOMEWOOD, ILLINOIS 60430-4600

Glen A. Mazur
William J. Weindorf

Electric Motor Drive Installation and Troubleshooting, 2nd Edition, contains procedures commonly practiced in industry and the trade. Specific procedures vary with each task and must be performed by a qualified person. For maximum safety, always refer to specific manufacturer recommendations, insurance regulations, specific job site and plant procedures, applicable federal, state, and local regulations, and any authority having jurisdiction. The material contained is intended to be an educational resource for the user. American Technical Publishers, Inc. assumes no responsibility or liability in connection with this material or its use by any individual or organization.

American Technical Publishers, Inc., Editorial Staff

Editor in Chief:
    Jonathan F. Gosse
Vice President—Production:
    Peter A. Zurlis
Art Manager:
    James M. Clarke
Technical Editors:
    Russell G. Burris
    Scott C. Bloom
Copy Editors:
    Catherine A. Mini
    Diane J. Weidner
Cover Design:
    James M. Clarke
Illustration/Layout:
    James M. Clarke
    William J. Sinclair
    Aimeé M. Gurski
    Nicole S. Polak
CD-ROM Development:
    Carl R. Hansen
    Christopher J. Bell
    Peter J. Jurek

2 3 4 5 6 7 8 9 – 08 – 9 8 7 6 5 4 3 2

Printed in the United States of America

ISBN 978-0-8269-1252-7

 This book is printed on recycled paper.

Acknowledgments

The authors and publisher are grateful to the following companies, organizations, and individuals for providing photographs, information, and technical assistnace:

ABB Inc., Drives & Power Electronics

Advanced Assembly Automation Inc.

Advanced Test Products

ASI Robicon

Atlas Technologies, Inc.

Baldor Electric Company

Baldor Motor and Drives

Browning; Emerson Power Transmission

Cincinnati Machine, a UNOVA Company

Cleveland Motion Controls

Coleman Cable Systems, Inc.

Control Techniques

Curtis Instruments, Inc.

Danfoss Drives

DoAll Company

Electrical Apparatus Service Association, Inc.

Extech Instruments

Fluke Corporation

GE Motors & Industrial Systems

International Rectifier

Kebco Power Transmission

Lab Safety Supply, Inc.

Leeson Electric Corporation

Lovejoy, Inc.

March Manufacturing Inc.

Motortronics

MTE Corporation

Omron IDM Controls

Rockwell Automation, Allen-Bradley Company, Inc.

Rockwell Automation/Reliance Electric

Saftronics Inc.

Siemens

Square D Company

Square D – Schneider Electric

Thermik Corporation

The Stanley Works

Unico, Inc.

Contents

CD-ROM Contents

Using the CD-ROM

Quick Quizzes®

Illustrated Glossary

Electrical Power Data Sheets

Energy Savings Spreadsheet

Test Tool Connections

Media Clips

ATPeResources.com

Introduction

Electric Motor Drive Installation and Troubleshooting, 2nd Edition, is a comprehensive reference for use in the electrical industry, the maintenance industry, electrical training programs, and related fields. Each chapter includes review questions and activities that reinforce the concepts presented. Topics covered range from motor types and controls to installing and troubleshooting electric motor drives. Large, detailed illustrations provide a visual, systematic approach to installing and troubleshooting electric motor drives.

This edition has been reorganized to provide coverage of an overview of electric motor drives at the beginning of the text. Electric motor drive selection has been expanded to include content on electric motor drive retrofits. The *Electric Motor Drive Installation and Troubleshooting CD-ROM* in the back of the book includes Quick Quizzes®, Illustrated Glossary, Electrical Power Data Sheets, Energy Savings Spreadsheet, Test Tool Connections, Media Clips, and ATPeResources.com. The Quick Quizzes® provide an interactive review of the major topics covered in each chapter. The Illustrated Glossary provides a helpful reference to key terms commonly used in the field. Selected terms are linked to illustrations and animated clips that supplement the definitions. The Electrical Power Data Sheets provide US Department of Energy statistics on electrical power generation and consumption for the US and individual states. The Energy Savings Spreadsheet provides a Microsoft® Excel® worksheet that calculates the energy savings for electric motor drives based on motor characteristics, operating hours, cost of electricity, percent of time at a specified flow, and investment cost. The Test Tool Connections provide common test tool connections used for troubleshooting electric motor drives systems, circuits, and components. The Media Clips provide a convenient link to selected video clips and animated graphics, which illustrate electric motor drive principles. ATPeResources.com provides a comprehensive array of instructional resources including Internet links to manufacturers, associations, and ATP resources. Instructions for using the CD-ROM are listed on the last page of this book.

Electric Motor Drive Installation and Troubleshooting, 2nd Edition, is one of several products published by American Technical Publishers, Inc. To obtain information on related educational products, please visit the American Tech web site at www.go2atp.com.

The electrical circuits in this book are drawn using electron theory concepts. The appendix contains the electrical circuits drawn using conventional theory concepts.

The Publisher

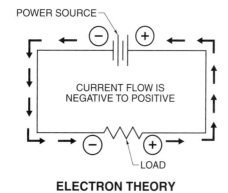

ELECTRON THEORY

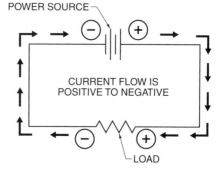

CONVENTIONAL THEORY
(⊕ TO ⊖ DIAGRAM)

Electric Motor Drive
Installation and Troubleshooting

Features

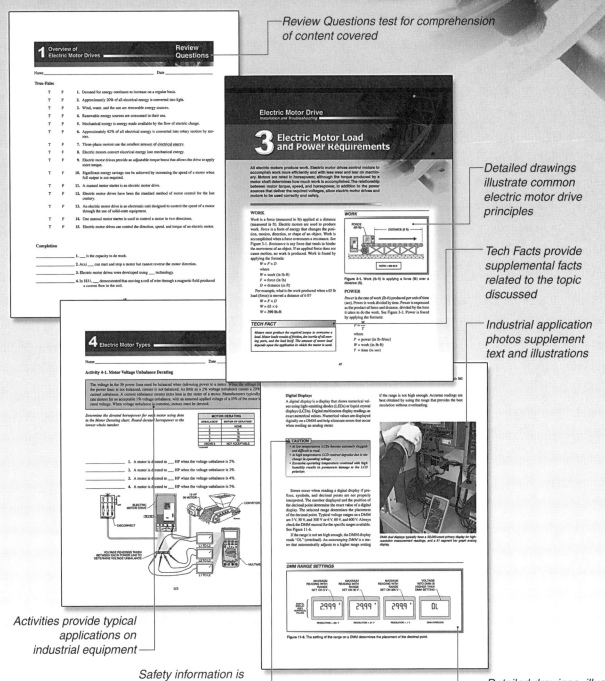

Review Questions test for comprehension of content covered

Detailed drawings illustrate common electric motor drive principles

Tech Facts provide supplemental facts related to the topic discussed

Industrial application photos supplement text and illustrations

Activities provide typical applications on industrial equipment

Safety information is Included throughout the book

Detailed drawings illustrate common electric motor drive principles

Electric Motor Drive
Installation and Troubleshooting

CD-ROM Features

Quick Quizzes® reinforce fundamental concepts with 10 questions per chapter

Using This CD-ROM provides information about components included on the CD-ROM

Illustrated Glossary provides helpful reference to electric motor drive terms

Electric Power Data Sheets provide statistics on electrical power generation and consumption

Energy Savings Spreadsheet calculates energy savings for electric motor drives

Test Tool Connections provide common test tool connections used for troubleshooting electric motor drive systems

Using This CD-ROM

Quick Quizzes®

Illustrated Glossary

Electrical Power Data Sheets

Energy Savings Spreadsheet

Test Tool Connections

Media Clips

www.ATPeResources.com

Visit www.go2atp.com

EXIT

Media Clips provide selected video clips and animated graphics that illustrate electric motor drive principles

www.go2atp.com provides a direct link to the American Tech web site

www.ATPeResources.com provides a comprehensive array of instructional resources

Electric Motor Drive
Installation and Troubleshooting

1 Overview of Electric Motor Drives

We live in a world that runs on energy and our demand for energy continues to increase on a regular basis. Electronic devices have become a part of our life, whether for comfort, safety, communication, or entertainment. Devices are added to homes, schools, workplaces, stores, and elsewhere to improve the quality of our lives. Government agencies and corporations are constantly creating new devices or improving old devices to help and protect us. These devices require energy, and the number of devices relied upon is constantly increasing. Over the next 30 years total energy consumption is expected to double and electrical energy consumption is expected to triple.

ENERGY

Energy is the capacity to do work. The primary sources of energy are coal, oil, gas, and nuclear energy. These sources of energy cannot be replaced since they are consumed in their use. Renewable energy sources, such as solar, wind, water, and thermal energy are not consumed in their use. These energy sources are used to produce work when converted to electricity, steam, heat, and mechanical force. See Figure 1-1.

Because energy continues to increase in cost and non-renewable energy sources are limited in supply, devices like motor drives that can save energy while performing the required work are in high demand and continue to increase in usage. Electric motor drives are used to control motors, which are the largest consumers of electrical energy of all the electrical devices used today.

Electrical output devices produce motion, light, heat, and sound. No usable work is performed without a rotating or nonrotating final output device. Electric motors are rotating output devices. Nonrotating output devices include lamps, heaters, linear motion devices such as solenoids, audible devices such as alarms and horns, and visual devices such as indicating lamps and instrument displays.

Electrical Energy

Electrical energy is energy made available by the flow of electric charge. All electrical output devices, whether rotating or nonrotating, convert electrical energy into another form of energy (mechanical, heat, light, sound, etc.). This changed energy form is then used to produce work. See Figure 1-2.

Electrical energy is converted into motion, light, heat, sound, and visual information. Approximately 62% of all electrical energy is converted into rotary motion by motors. These motors, which are used in large numbers, are available in sizes exceeding 500 HP. Even small motors use large amounts of electrical energy as compared to other types of output devices. For example, a ½ HP, 115 V, 1φ motor uses about the same amount of energy as eight 19″ color televisions, a 25 W energy-efficient lamp provides approximately the same minimum light output as a 75 W incandescent lamp. Typical residential uses for a ½ HP, 115 V, 1φ motor include a large residential refrigerator, a sump pump, or a garage door opener.

TECH FACT

Renewable energy is a class of energy resources that are replaced rapidly by natural processes.

ENERGY SOURCES

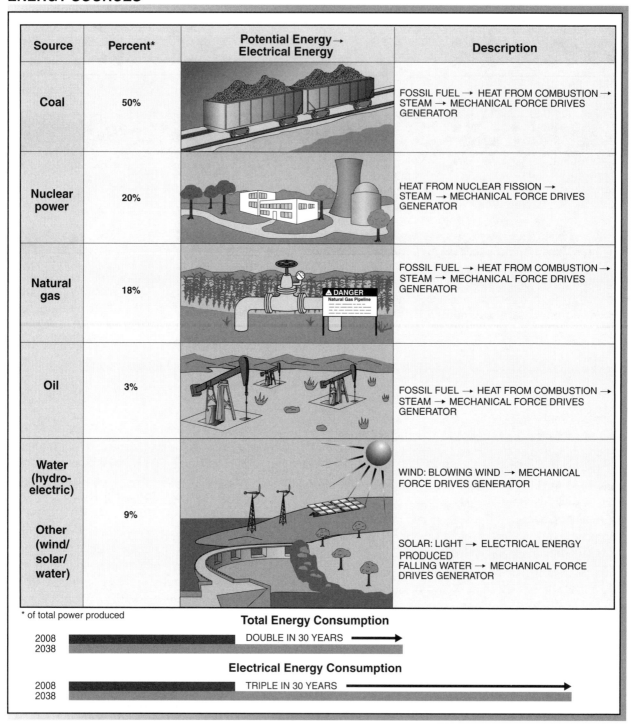

Source	Percent*	Potential Energy → Electrical Energy	Description
Coal	50%		FOSSIL FUEL → HEAT FROM COMBUSTION → STEAM → MECHANICAL FORCE DRIVES GENERATOR
Nuclear power	20%		HEAT FROM NUCLEAR FISSION → STEAM → MECHANICAL FORCE DRIVES GENERATOR
Natural gas	18%		FOSSIL FUEL → HEAT FROM COMBUSTION → STEAM → MECHANICAL FORCE DRIVES GENERATOR
Oil	3%		FOSSIL FUEL → HEAT FROM COMBUSTION → STEAM → MECHANICAL FORCE DRIVES GENERATOR
Water (hydro-electric) Other (wind/ solar/ water)	9%		WIND: BLOWING WIND → MECHANICAL FORCE DRIVES GENERATOR — SOLAR: LIGHT → ELECTRICAL ENERGY PRODUCED FALLING WATER → MECHANICAL FORCE DRIVES GENERATOR

* of total power produced

Total Energy Consumption

2008
2038   DOUBLE IN 30 YEARS ⟶

Electrical Energy Consumption

2008
2038   TRIPLE IN 30 YEARS ⟶

Figure 1-1. Energy sources include coal, nuclear power, natural gas, oil, water (hydroelectric or mechanical power), wind, and the sun.

Three-phase motors use the largest amount of electrical energy in commercial and industrial applications. Three-phase motors are the most commonly used motors in commercial and industrial applications because they are the most energy efficient. Even though they are the most energy-efficient motor type,

3φ motors use approximately 85% of all energy consumed by motors and approximately 52.7% of the total electrical energy consumed. Most motor manufacturers offer "energy-efficient" motors in addition to their standard line because of the amount of energy standard electric motors consume. Standard motors have an average efficiency of 89%. It is usually worth paying the 20% higher cost of purchasing an energy-efficient motor if a motor is to be operated long term.

Light accounts for the second-largest use of electrical energy. Approximately 20% of all electrical energy is converted into light. Light is produced by lamps (bulbs). The most common lamp used in residential lighting is the incandescent lamp. The most common lamps used in commercial and industrial lighting are fluorescent lamps for office installations, and high-intensity discharge (HID) lamps for warehouse and factory installations. Types of HID lamps include low-pressure sodium, mercury vapor, metal halide, and high-pressure sodium. HID lamps are also the most common lamps used for exterior lighting.

⚠ CAUTION

Even when the control signal is 12 VDC, control-signal wiring rated for 600 V should be used because drives can contain voltages of 600 V or more, which can result in electrical shock and is a potential fire hazard.

ELECTRICAL ENERGY CONSUMPTION . . .

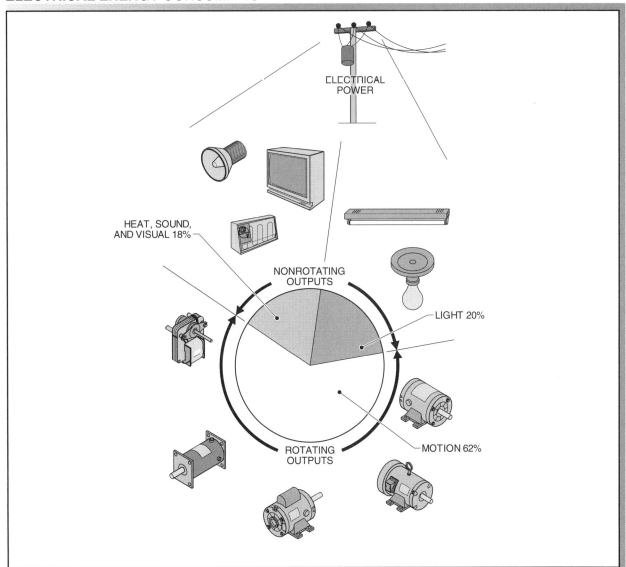

Figure 1-2 . . .

... ELECTRICAL ENERGY CONSUMPTION

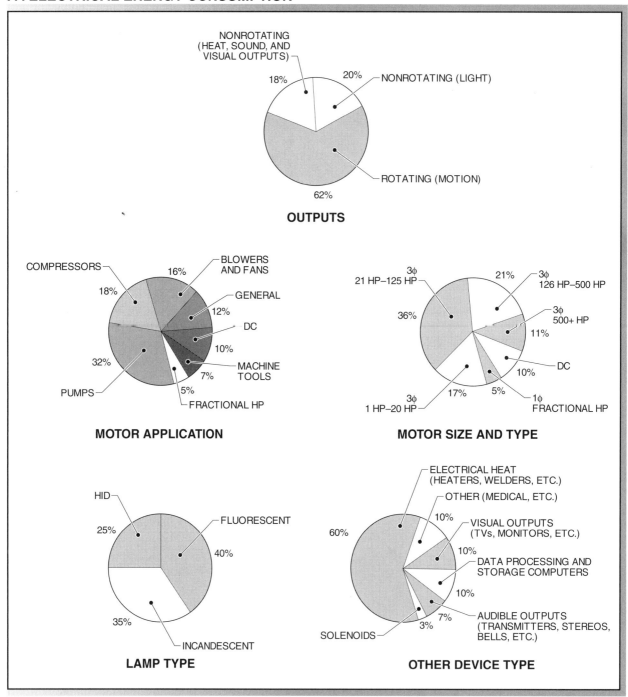

Figure 1-2. Electrical energy consumption varies for different types of devices and motors.

Electrical energy is also used to produce heat, linear motion, audible signals, and visual information. Approximately 18% of all electrical energy is converted into heat, linear motion, audible signals, and visual information. This group includes a large number of output devices that consume very little energy as compared to rotating output devices. Electric heating is the exception in this group because producing heat from electricity is clean, safe, and convenient but requires large amounts of energy.

ELECTRIC MOTORS

An *electric motor* is a rotating output device that converts electrical energy into rotating mechanical energy. Electric motors may use 1φ AC, 3φ AC, or DC power. The type of motor used depends on the application and involves factors such as the type of load the motor is operating, the required starting and running torque, environmental conditions, and available energy. See Figure 1-3. For example, DC motors operated from a battery are used for applications where AC power is not easily available. In addition, 1φ AC motors are used in all residential applications because 3φ power is not available for residential use, even though 3φ motors would save energy and cost. Another major consideration is the starting torque capability of the motor type since some loads require a higher starting torque to get the load turning.

However, by using electric motor drives, use of a specific motor type that is best for a given application is no longer necessary. Electric motor drives are available in sizes as small as fractional horsepower that can be powered from a single-phase 120/240 V power supply and deliver a 230 V, 3φ output. For this reason, 3φ motors are becoming the standard for residential HVAC units, washing machines, and other applications where once only 1φ motors were used. Also, with the use of inverters converting DC to AC, a DC power source can be connected to an inverter, the inverter to a motor drive, and the drive to an AC motor for speed control without the high maintenance requirements that DC motors have because of brush wear.

ELECTRIC MOTORS

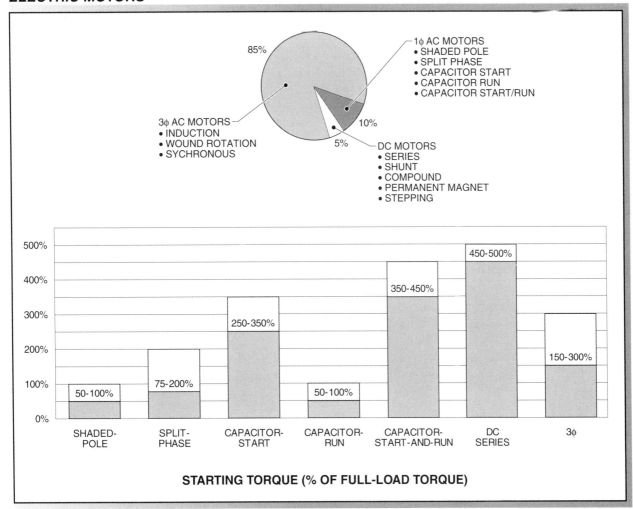

Figure 1-3. Electric motors may use 1φ AC, 3φ AC, or DC power.

Electric motor drives also provide an adjustable torque boost that allows the drive to apply more torque when trying to start heavy loads. This allows drives to offer the same high torque features found with DC and capacitor-start motors.

Even when the best motor type is used for application, almost all motors waste some energy because most motors in operation run at full speed, using maximum energy, even when only a fraction of the motor's total energy output is needed to perform the required work. When motors are used to operate pumps and fans, running the motor at half maximum speed consumes only one-eighth of the energy. Significant energy savings can be achieved by reducing the speed of the motor when full output is not required. The technology to operate motors at reduced speeds was not available in a reliable and cost-effective package until electric motor drives that could control motors of sizes ranging from fractional horsepower to thousands of horsepower became available.

HISTORY OF MOTOR CONTROL

Electric motor development started in the early 1800s when several individuals contributed to the development of electromagnetic field theories. In 1831, Michael Faraday demonstrated that moving a coil of wire through a magnetic field produced a current flow in the coil. Faraday's theory and other such theories lead to the development of the electric generator. After the development of the electric generator, the electric motor was developed. In 1873, the first commercial DC motor was demonstrated in Vienna. In 1888, Nikola Tesla invented the AC motor. By the start of the 1900s, DC, AC, and universal (AC/DC) motors were being used in an increasing number of applications. Today, it is uncommon to go through a day without using a motor.

The first motors were controlled by simple knife switches that started and stopped the motor by applying or removing the source of electric energy from the motor. The knife switches had exposed metal parts that carried the electricity and, because of the slow manual operation of the opening of the contacts, sparking would occur. Although the knife switches controlled the motor, they produced an electrical shock and fire hazard. See Figure 1-4.

As motor use increased, so did the need for a better and safer method of controlling them. Manual motor starters were developed from knife switches to increase safety and add motor overload protection. A *manual motor starter* is a control device used to control a motor by having technicians or operators control the motor directly at the location of the starter. Manual motor starters have

been used to control motors for over 100 years and are still used in some motor control applications that require only the manual turning ON and OFF of smaller motors (less than 1 HP).

KNIFE SWITCHES

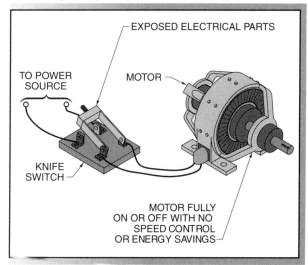

Figure 1-4. Knife switches were the first devices used to start and stop electric motors.

Magnetic motor starters were developed from manual starters to provide additional control functions and allow large motors to be controlled. A *magnetic motor starter* is a starter that has an electrically operated switch (contactor) and includes motor overload protection. Magnetic motor starters have been the standard method of motor control for most of the last century. Magnetic motor starters are still used for controlling motors in applications in which the only motor control function is to start and stop the motor and saving energy is not a major consideration.

An *electric motor drive* is an electronic device that controls the direction, speed, torque, and other operating functions of an electric motor in addition to providing motor protection and monitoring functions. Electric motor drives were developed from magnetic motor starters. Electric motor drives provide numerous motor control functions and can save energy in most motor applications. Magnetic motor starters are being replaced in ever-increasing numbers by electric motor drives. Most new motor control applications use motor drives. Motor drives are being used by the millions each year in various applications that include controlling the airflow in automobile, home, and office HVAC systems; controlling the speed of washing machines, conveyors, processing/production machines; and delivering a space shuttle to a launch pad.

MOTOR CONTROL METHODS

Today, motor control involves much more than just turning a motor ON and OFF and providing basic overload protection to prevent motor/circuit damage when the motor is overloaded. Motor control must include the safest and best method for controlling the motor for a given application. Today, motors can be controlled by means ranging from a simple manual ON/OFF switch to the use of a computer and communication network from any location on Earth.

With the increasing cost of electrical energy, motor control also means controlling the cost of operating the motor. The fact that electric motors cost so much to operate has become a major concern. With today's electric rates, it costs more to operate most motors for 8 hours a day for 6 months than it costs to purchase the motor. This is equivalent to purchasing an automobile that costs more to run for 2 hours a day for 4 years than it costs to purchase the vehicle. Whether it is the cost of electrical energy (or gas for a vehicle), any device that can save energy and still deliver the required work will continue to be used and improved.

Manual Motor Starters

Manual motor starters were the first major devices used to control motors. Manual motor starters include only power contacts. Power contacts are the electrical contacts that directly start and stop the flow of current to the motor. One power contact is required for every ungrounded (hot) conductor going to the motor. For DC and 1φ, 120 VAC (or less) motors, one power contact is required. For single-phase, 230 VAC (or more) motors, two power contacts are required. For all 3φ motors, three power contacts are required. See Figure 1-5.

MANUAL MOTOR STARTERS

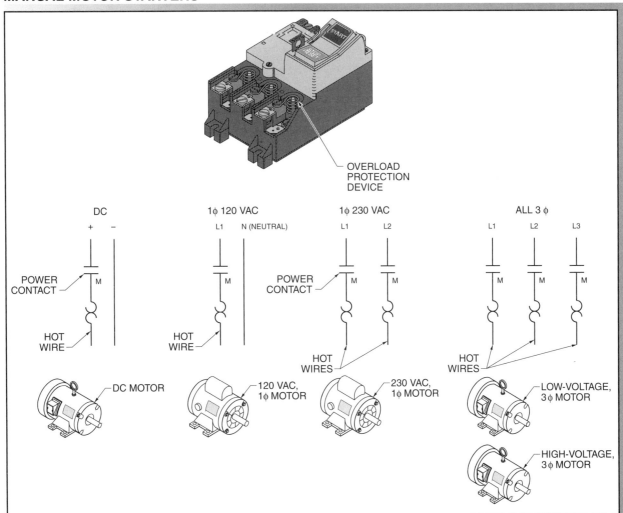

Figure 1-5. Manual motor starters can be used to control various types of motors.

Manual motor starter functions include the following:
- starting a motor when switched to the ON position
- stopping a motor when switched to the OFF position
- providing motor overload protection by monitoring the motor's current draw through overload devices (called heaters) and automatically turning off the motor if the motor is overloaded for too long

Note: The amount of time before the motor is turned off is determined by the amount of overload. The greater the overload current, the shorter the trip time. The time delay period is not adjustable.

A manual motor starter can start and stop a motor but cannot reverse the motor direction. If a motor application requires a motor to be controlled in both directions, two manual starters must be used. Commercially available reversing manual starters include two starters combined into one unit. See Figure 1-6.

MANUAL MOTOR STARTERS (FORWARD AND REVERSING)

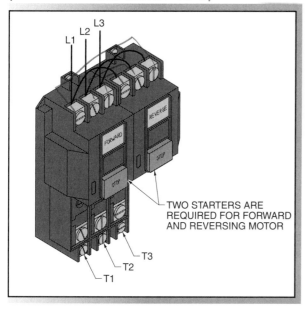

Figure 1-6. Two manual starters are required for forward and reverse operation of motors.

Magnetic Motor Starters

A magnetic motor starter is an electrically operated switch (contactor) that includes motor overload protection. Magnetic motor starters were developed to add control functions such as controlling a motor from any location and to allow a low control voltage to control a high motor voltage. Typical control circuits of 24 V are used to control 120 V motors and 120 V control circuits are used to control 230 V, 460 V, and higher-voltage motors. See Figure 1-7.

MAGNETIC MOTOR STARTERS

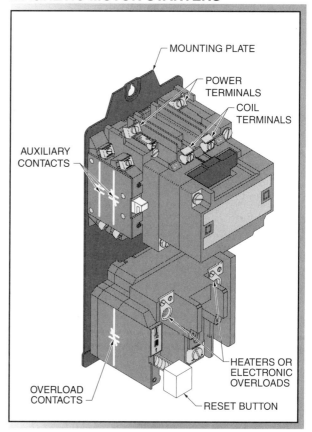

Figure 1-7. Magnetic motor starters are used to control most types of motors. Magnetic motor starters also feature overload protection.

Magnetic motor starters perform the following functions:
- Start a motor from any location when the starter coil is energized. The starter coil closes and opens the power contacts controlling the motor.
- Stop a motor from any location by de-energizing the starter coil.
- Provide motor overload protection by monitoring the motor current draw through overload devices (called heaters) and automatically turning the motor to OFF if the motor is overloaded for too long. The amount of time before turning off the motor is determined by the amount of overload. The greater the overload current, the shorter the trip time. The time-delay period is not easily adjustable. Some magnetic motor starters provide a means to allow some adjustment to the trip time—typically 10%. The most common method used is to select one of four different classes of overloads (class 10, 15, 20, or 30). The smaller the trip class number, the shorter the trip time.

Electric Motor Drives

Electric motor drives control the direction, speed, torque, and other operating functions of an electric motor, in addition to providing motor protection and monitoring functions. Electric motor drives are the latest major devices developed to control motors. Electric motor drives were developed using solid-state technology with not only standard motor control uses in mind, but also the need for solving newer and more challenging motor application problems and the need for saving energy. Because electric motor drives include the latest computer technology, they can perform all the functions of a magnetic motor starter, and they can also perform dozens of additional control functions, such as controlling motor speed, motor acceleration and deceleration times, applying a braking force to the motor/load, monitoring and displaying the motor's voltage, current, and power, and many other control functions. Although an electric motor drive can perform many more control functions than a magnetic motor starter, it typically costs less than a magnetic motor starter rated for the same horsepower or power rating. See Figure 1-8.

ELECTRIC MOTOR DRIVES

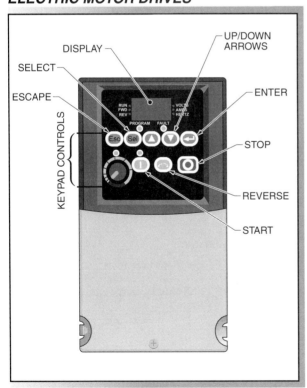

Figure 1-8. Electric motor drives offer many motor control functions, including reduced motor speed and controlled acceleration/deceleration.

Electric motor drives perform the following functions:
- Start a motor from any location. Since the drive automatically provides a safe low voltage (12 VDC – 24 VDC) to the drive's control circuit terminals, added safety and ease of wiring the control circuit are included features. There is no need for a step-down control transformer as there is with a magnetic motor starter control circuit.
- Stop a motor from any location. The ability to stop a motor at the drive (as with a manual starter) and away from the drive (as with a magnetic motor starter) provides additional control and safety to the circuit.
- Provide motor overload protection through the electric motor drive's direct monitoring of the motor's current draw. The electric motor drive turns off the motor when there is an overload. When the drive detects an overload, the drive records the fault and displays a fault code. Since the amount of current is adjustable, there is no need to stock dozens of various heater sizes for a given starter size, as is necessary with magnetic motor starters. Also, the trip time is fully adjustable, which allows each drive/motor to be customized to the application requirements.
- Provide a full range of speed control from 0 rpm to full motor speed.
- Control motor acceleration time, allowing the motor to start any load without product damage.
- Control motor deceleration time and apply a braking force when needed. This may eliminate the need for a separate braking circuit (dynamic braking, mechanical shoe brakes, etc.) as required with magnetic motor starter circuits.
- Set minimum and maximum speed limits on the motor so the motor operates within a certain speed range.
- Connect and interface with PCs, which allow drive parameters and drive faults to be displayed or output to a printer.
- Interface with PLCs and computer controls.

ELECTRIC MOTOR DRIVE FEATURES

Electric motor drives include many features that make them versatile, cost effective, and energy efficient. Most electric motor drives include the following features:

- a voltmeter that displays the voltage applied to the motor and the drive's internal DC bus voltage
- an ammeter that displays the current drawn by the motor
- a wattmeter that displays the drive's output power to the motor
- a frequency meter that displays the frequency applied to the motor
- a temperature meter that displays the drive's internal temperature
- output switches to be used for additional control functions or circuit feedback

Electric motor drives also include circuit and system monitoring of vital motor and circuit data so that the drive can automatically turn the motor OFF and display the fault that took place (e.g., short, ground, high/low voltage, high current, or high temperature). In addition, an electric motor drive also includes components that determine how a motor should stop (coast, ramp, or fast break), and allow the motor control circuit to use digital (ON/OFF) or analog control inputs.

Built-In Voltage, Current, Power, Frequency, and Temperature Meters

In electrical systems, electrical measurements are required to determine circuit and load operation and to troubleshoot potential problems. One major advantage of using a motor drive is that drives have built-in meters that measure and display voltage, current, power, frequency, and temperature. The use of external test instruments is reduced because motor drives include a display that indicates the voltage, current, power, frequency, and temperature of the motor. Also, a display of the drives' internal DC bus voltage and temperature aid in troubleshooting potential system problems. See Figure 1-9.

DRIVE METERING DISPLAY

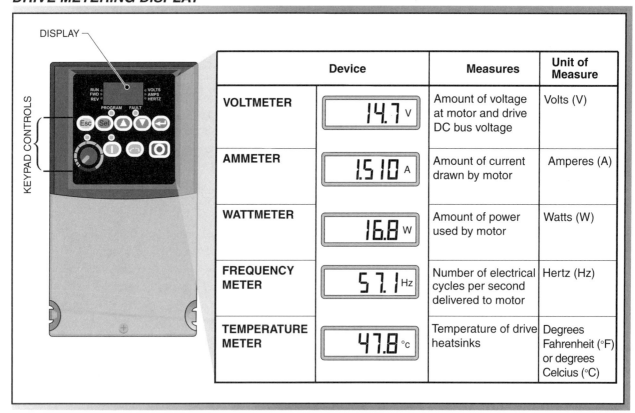

Device		Measures	Unit of Measure
VOLTMETER	14.7 v	Amount of voltage at motor and drive DC bus voltage	Volts (V)
AMMETER	1.510 A	Amount of current drawn by motor	Amperes (A)
WATTMETER	16.8 w	Amount of power used by motor	Watts (W)
FREQUENCY METER	57.1 Hz	Number of electrical cycles per second delivered to motor	Hertz (Hz)
TEMPERATURE METER	47.8 °C	Temperature of drive heatsinks	Degrees Fahrenheit (°F) or degrees Celcius (°C)

Figure 1-9. Electric motor drives include built-in meters that measure and display voltage, current, power, frequency, and temperature.

Motor and System Safety Features

Troubleshooting is required whenever there is a problem with a motor or any other part of the system. *Troubleshooting* is the systematic elimination of various parts of a system to locate a malfunctioning part. Troubleshooting requires measurements to be taken to determine the problem in a system. Problems may include shorts and ground faults. Ground faults can cause electrical shock. In addition, shorts, overloads, and excessively high temperatures can cause fires. Electric motor drives include built-in monitoring features to track system operation, automatically shut the system down if a problem is detected, and display a fault code and/or information. Because most drives can detect problems such as excessively high temperatures, overvoltages and overcurrents, shorts, and ground faults, they are much safer to use than magnetic motor control circuits. See Figure 1-10.

Motor Acceleration, Deceleration, and Speed Control

One of the main advantages of using an electric motor drive is that the drive can be used to adjust the motor speed from 0 rpm to full motor speed. In addition, the drive can also be programmed to allow the motor to operate within a set minimum and maximum speed range. This allows the drive to be programmed to prevent damage to the motor, the load, and the product being produced or moved. See Figure 1-11.

Heidelberg Harris, Inc.
Printing presses require that torque, horsepower, and speed be precisely controlled.

FAULTS

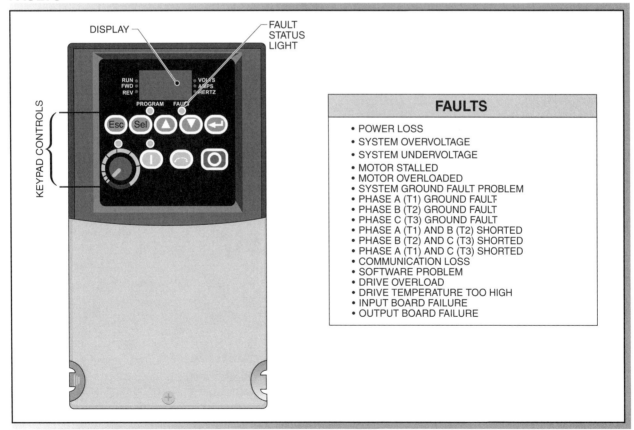

Figure 1-10. Motor drives include built-in monitoring and display of faults.

MOTOR SPEED AND SPEED RANGE

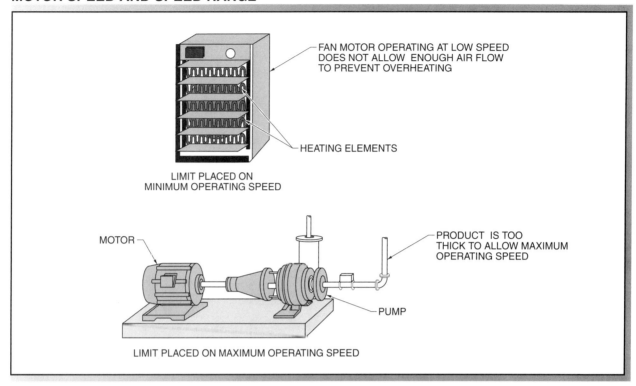

Figure 1-11. In addition to controlling the motor speed, drives can be programmed for certain speed ranges.

Motor Stopping and Braking Control

When a manual or magnetic motor starter turns the power to a motor OFF, the time it takes the motor to come to a full stop depends on the load connected to the motor. For some applications, a coast-to-stop works best. For other applications, a controlled stop works best. Electric motor drives can be programmed to allow several different stopping methods to be applied to a motor. This allows the motor to match the needs of the application. See Figure 1-12.

Energy Savings

Electric motor drives can reduce energy consumption and cost of operating almost up to 70% as compared to a standard motor control circuit. The amount saved depends on the driven load, application, operating conditions, and requirements. An electric motor drive saves energy and money by not using full power at all times. Applications that do not use an electric motor drive to reduce motor speed, use gears and pulleys to obtain a slower speed than the motor shaft rotation (nameplate-rated revolutions per minute). The time in which the work is to be accomplished must also be considered. For example,

energy and cost can be saved by operating a motor at a slower speed. In this case, the same amount of work is accomplished while only the amount of time is different. However, more energy is used to operate a motor at full speed than at a reduced speed.

Most motor applications are designed using the largest motor size possible for the worst-case operating condition. This means that the motor drives the load when the motor is loaded to the fullest capability of the driven load, but it also means that whenever the motor is less than fully loaded, it wastes energy and costs more to operate. In fact, for most motor applications, the motor speed can be reduced by at least 10% to 20% with no negative operating effects. See Figure 1-13.

Common Drive Usages

Electric motor drives are now used in almost every motor control application. The total number of drives used is increasing every year and the percentage of drives that are being used in place of magnetic motor starters is growing at a faster rate every year.

MOTOR STOPPING

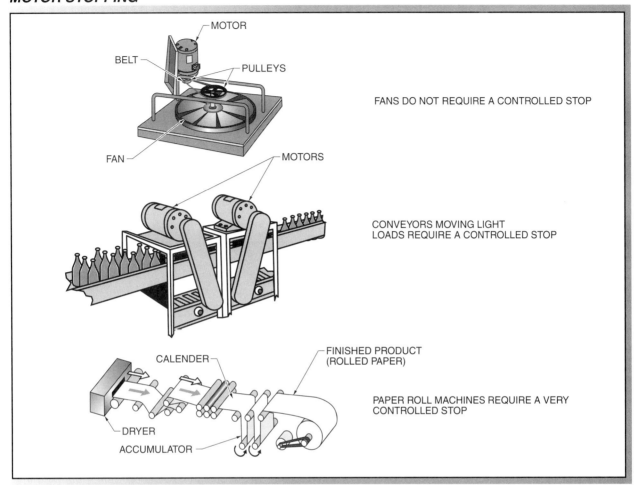

Figure 1-12. Drives can be programmed to allow a motor to coast to a stop or they can be programmed to apply a controlled braking force.

ELECTRIC MOTOR DRIVES AND MANUAL AND MAGNETIC MOTOR STARTERS

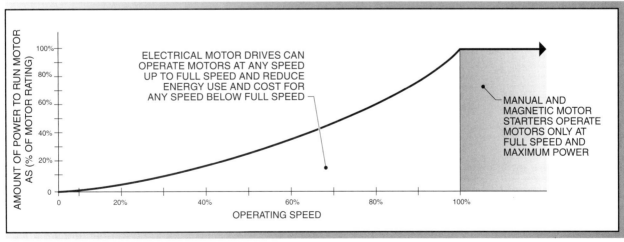

Figure 1-13. Using an electric motor drive may save up to 70% in energy use and cost.

Name_____ Date _____

True-False

T F   **1.** Demand for energy continues to increase on a regular basis.

T F   **2.** Approximately 20% of all electrical energy is converted into light.

T F   **3.** Wind, water, and the sun are renewable energy sources.

T F   **4.** Renewable energy sources are consumed in their use.

T F   **5.** Mechanical energy is energy made available by the flow of electric charge.

T F   **6.** Approximately 62% of all electrical energy is converted into rotary motion by motors.

T F   **7.** Three-phase motors use the smallest amount of electrical energy.

T F   **8.** Electric motors convert electrical energy into mechanical energy.

T F   **9.** Electric motor drives provide an adjustable torque boost that allows the drive to apply more torque.

T F   **10.** Significant energy savings can be achieved by increasing the speed of a motor when full output is not required.

T F   **11.** A manual motor starter is an electric motor drive.

T F   **12.** Electric motor drives have been the standard method of motor control for the last century.

T F   **13.** An electric motor drive is an electronic unit designed to control the speed of a motor through the use of solid-state equipment.

T F   **14.** One manual motor starter is used to control a motor in two directions.

T F   **15.** Electric motor drives can control the direction, speed, and torque of an electric motor.

Completion

Energy   **1.** ___ is the capacity to do work.

Manual motor starter   **2.** A(n) ___ can start and stop a motor but cannot reverse the motor direction.

Solid-state   **3.** Electric motor drives were developed using ___ technology.

Michael Faraday   **4.** In 1831, ___ demonstrated that moving a coil of wire through a magnetic field produced a current flow in the coil.

electric motor Drive **5.** A(n) ___ is an electronic device that controls operating functions of an electric motor and provides motor protection and monitoring functions. _pg 6_

magnetic Motor Starter **6.** A(n) ___ is an electrically operated switch (contactor) that includes motor overload protection. _pg 8_

motor Drive **7.** ___ is used to produce heat, linear motion, audible signals, and visual information.

Nikola Tesla **8.** ___ invented the AC motor. _pg 6_

_O_ **9.** Electric motor drives provide a full range of speed control from ___ rpm to full motor speed.

Electric motor Drive **10.** ___ have built-in meters that measure and display voltage, current, power, frequency, and temperature. _pg 10_

Multiple Choice

_B_ **1.** A(n) ___ is an electronic device that controls operating functions of an electric motor and provides motor protection and monitoring functions.
A. magnetic motor starter
B. electric motor drive
C. manual motor starter
D. transformer

_D_ **2.** A(n) ___ is a control device used to control a motor by having technicians or operators control the motor directly at the location of the starter.
A. magnetic motor starter
B. electric motor drive
C. transformer
D. manual motor starter

_C_ **3.** A(n) ___ is a rotating output device that converts electrical energy into rotating mechanical energy.
A. motor drive _pg 5_
B. inductor
C. electric motor
D. manual motor starter

_A_ **4.** ___ energy is energy made available by the flow of electric charge.
A. Electrical _pg 2_
B. Mechanical
C. Renewable
D. Static

_B_ **5.** ___ is the capacity to do work.
A. Power
B. Energy _pg 1_
C. Magnetism
D. Impedance

Name_____ Date_____

Activity 1-1. Electric Motor Drive Energy Savings

> The United States government tracks the amount of electrical power generated in the United States. The information provided is based on fuel types such as coal, hydroelectric, solar, etc.

Click on the Electrical Power Data Sheets button on the CD-ROM. Next, click on the Total Electric Power Industry Summary Statistics button to answer questions 1 through 7.

_____ 1. The total electrical power generated was ___ thousand MWh.

_____ 2. The total electrical power generated was ___ MWh.

_____ 3. The total electrical power generated was ___ kWh.

_____ 4. Assuming motors consume approximately 62% of electrical power, motors consumed ___ MWh.

_____ 5. Assuming three-phase motors consume approximately 85% of all electrical power used by all motor types, three-phase motors consumed ___ MWh.

_____ 6. If electric motor drives were used to control approximately 10% of all three-phase motors and the average savings by using an electric motor drive over a magnetic motor starter was 10%, the total energy savings would have been ___ MWh.

_____ 7. If electric motor drives were used to control 25% of all three-phase motors, and the average savings by using an electric motor drive over a magnetic motor starter was 20%, the total energy savings would have been ___ MWh.

Click on the Electrical Power Data Sheets button on the CD-ROM. Next, click on the Net Generation by State by Sector button to answer questions 8 through 14.

_____ 8. The total electrical power generated in your state was ___ thousand MWh.

_____ 9. The total electrical power generated in your state was ___ MWh.

_____ 10. The total electrical power generated in your state was ___ kWh.

_____ 11. Motors consume approximately 62% of electrical power. For the four-month period motors consumed ___ MWh in your state.

_____ 12. If the remainder of 2007 had the same monthly average of power consumption in your state, motors would have consumed ___ MWh in your state for all of 2007.

_____ **13.** If three-phase motors consumed 85% of all electrical power used by all motor types, three-phase motors consumed ___ MWh in your state for all of 2007.

_____ **14.** If electric motor drives were used to control 20% of all three-phase motors, and the average energy savings from using an electric motor drive instead of a magnetic motor starter was 15%, the total energy savings would have been ___ MWh in your state for all of 2007.

Activity 1-2. History of Motors and Motor Controls

> Electric motor development started in the early 1800s when several individuals contributed to the development of electromagnetic field theories.

Match the following events with the year that they took place. Note: _Not all answers are found in the textbook._

_____ **1.** Temperature records started to be recorded.

_____ **2.** Nikola Tesla invented the AC motor.

_____ **3.** New York City experienced a blackout due to equipment failure on an overloaded distribution network.

_____ **4.** Michael Faraday demonstrated how a moving coil of wire through a magnetic field produced a current flow in the coil.

_____ **5.** The world's first nuclear power plant, located in Russia, started generating electricity.

_____ **6.** The first commercial DC motor was demonstrated in Vienna.

_____ **7.** Thales (a Greek philosopher) found that when amber was rubbed with silk it became electrically charged and attracted objects.

_____ **8.** The Westinghouse Electric Company used an alternating current (AC) system to light the Chicago World's Fair.

_____ **9.** Reliance Electric was founded based on a new type of adjustable-speed direct-current motor. DC was the primary means of electrification at the time because AC was considered dangerous and unpredictable.

A. 600 B.C.
B. 1831
C. 1850
D. 1873
E. 1888
F. 1893
G. 1904
H. 1954
I. 1977

Activity 1-3. Drive Temperature Display

All drives have an ambient operating temperature rating and an allowable heat sink temperature before the drive automatically turns off the motor. The ambient temperature rating for most drives is 40°C. Some drives have a 50°C rating if the drive is not placed in an enclosure or if the enclosure has forced-air cooling. Since stated ambient operating temperatures are often given in degrees Celsius, converting to degrees Fahrenheit may be necessary.

CELSIUS TO FAHRENHEIT CONVERSION

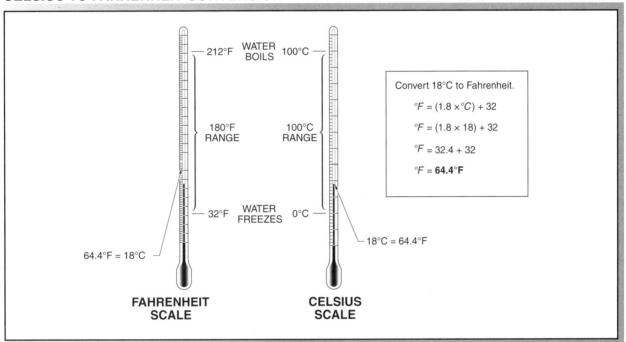

Convert 18°C to Fahrenheit.

$°F = (1.8 \times °C) + 32$

$°F = (1.8 \times 18) + 32$

$°F = 32.4 + 32$

$°F = \textbf{64.4°F}$

Convert temperatures from degrees Celsius to degrees Fahrenheit.

_____ **1.** 38°C = ___°F

_____ **2.** 56°C = ___°F

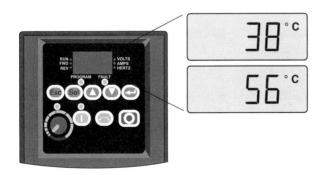

_____143.6_____ **3.** 62°C = ___°F

_____84.6_____ **4.** 29°C = ___°F

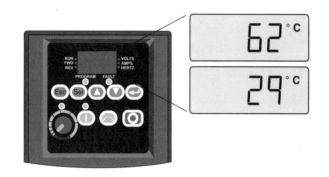

_____89.6_____ **5.** 32°C = ___°F

_____152.6_____ **6.** 67°C = ___°F

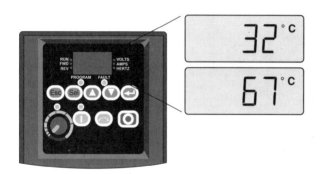

_____158_____ **7.** 70°C = ___°F

_____82.4_____ **8.** 28°C = ___°F

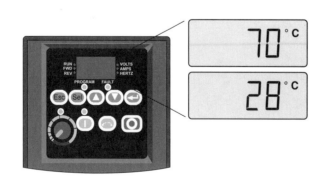

Activity 1-4. Drive Power Display

Most drives display the actual power the motor is using in watts (W) or kilowatts (kW). Thus, a conversion from power (in watts or kilowatts) to horsepower (HP) must be made for motors that have a horsepower rating.

Converting Watts to Horsepower

$HP = W \times .00134$

Where

HP = horsepower

W = Watts

$.00134$ = constant

Calculate the horsepower the motor is drawing to answer questions 1 through 4.

_____ 11.658

1. What is the horsepower equivalent of a drive displaying 8.7 kW?

_____ 3.082

2. What is the horsepower equivalent of a drive displaying 2.3 kW?

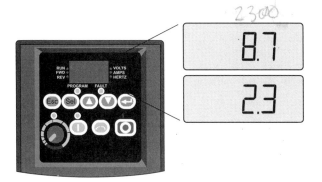

_____ .335

3. A drive displaying 0.25 kW of power is equivalent to ___ HP.

_____ 1.04118

4. A drive displaying 0.77 kW of power is equivalent to ___ HP.

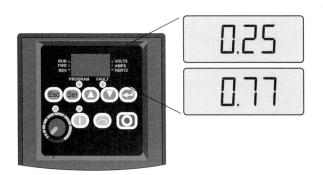

Calculate the percent the motor is actually working to answer questions 5 through 8.

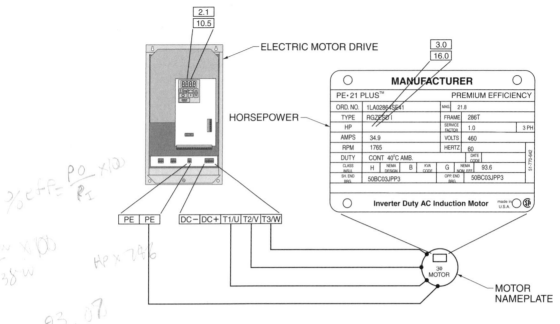

Handwritten notes (left margin):

$\%eff = \dfrac{P_O}{P_I} \times 100$

$\dfrac{2100\,w \times 100}{2238\,w}$

$HP \times 746$

93.07

_____ **5.** A motor with a nameplate rating of 3.0 HP connected to a drive that reads 2,083 kW
93.07   is working at ___%.

88.0278 **6.** A motor with a nameplate rating of 16.0 HP connected to a drive that reads 10.507 kW
_____   is working at ___%.

85.03   **7.** A motor connected to a
_____   drive that reads 3.806 kW is
      working at ___%.

MANUFACTURER			
PE · 5 PLUS™		PREMIUM EFFICIENCY	
ORD. NO.	1LA02864SE32	MAG. 21.8	
TYPE	3646L	FRAME	182TZ
HP	6.0	SERVICE FACTOR 1.0	1 PH
AMPS	34.9	VOLTS	230
RPM	3450	HERTZ	60
DUTY	CONT 40°C AMB.	DATE CODE	
CLASS INSUL	F NEMA DESIGN L KVA CODE	M NEMA NOM. EFF. 93.6	
SH. END BRG.	6206	OPP. END BRG.	6205
		Inverter Duty AC Induction Motor	made in U.S.A.

72.037   **8.** A motor connected to a
_____   drive that reads 5.374 kW is
      working at ___%.

MANUFACTURER			
PE · 10 PLUS™		PREMIUM EFFICIENCY	
ORD. NO.	1LA02864SE31	MAG. 21.8	
TYPE	0748M	FRAME	215T
HP	10.0	SERVICE FACTOR 1.5	3 PH
AMPS	12.5	VOLTS	230
RPM	1765	HERTZ	60
DUTY	CONT 40°C AMB.	DATE CODE	
CLASS INSUL	F NEMA DESIGN A KVA CODE	H NEMA NOM. EFF. 93.6	
SH. END BRG.	6307	OPP. END BRG.	6307
		Inverter Duty AC Induction Motor	made in U.S.A.

Electric Motor Drive
Installation and Troubleshooting

2 Electric Motor Drive Safety

Electric motor drives are used in electrical systems to control AC and DC motors. Working on electric motor drives, electric motors, and electrical systems requires following various NFPA, OSHA, and company safety rules, and the general principles of electrical safety. Safety rules include wearing approved protective clothing and using protective equipment. Additional personnel requirements include an understanding of the different danger, warning, and caution labels used with electric motor drives.

QUALIFIED PERSON

To prevent an accident, electrical shock, and damage to electrical equipment, electric motor drives must be installed and programmed by qualified persons. A *qualified person* is a person who is trained and has special knowledge of the construction and operation of electrical equipment or a specific task, and is trained to recognize and avoid electrical hazards that might be present with respect to the equipment or specific task. NFPA 70E Part II *Safety-Related Work Practices*, Chapter 1 *General*, Section 1-5.4.1 *Qualified Persons* provides additional information regarding the definition of a qualified person. A qualified person does the following:

- determines the voltage of energized electrical parts
- determines the degree and extent of hazards and uses the proper personal protective equipment and job planning to perform work safely on electrical equipment by following all OSHA, NFPA, equipment manufacturer, and company safety procedures and practices
- understands electric motor drive operation and follows all manufacturer procedures and approach distances specified by the NFPA
- performs the appropriate task required during an accident or emergency situation
- understands the operation of test equipment and follows all manufacturer procedures
- informs other technicians and operators of tasks being performed and maintains all records

SAFETY LABELS

A *safety label* is a label that indicates areas or tasks that can pose a hazard to personnel and/or equipment. Safety labels appear several ways on equipment and in equipment manuals. Safety labels use signal words to communicate the severity of a potential problem. The three most common signal words are danger, warning, and caution. See Figure 2-1.

WARNING

MOTOR MUST BE GROUNDED IN ACCORDANCE WITH THE NATIONAL ELECTRICAL CODE AND LOCAL CODES, BY TRAINED PERSONNEL TO PREVENT SERIOUS ELECTRICAL SHOCKS.

TO SERVICE MOTOR, DISCONNECT POWER SOURCE FROM MOTOR AND ANY ACCESSORY DEVICES AND ALLOW MOTOR TO COME TO A COMPLETE STAND STILL.

AVIS

IL EST NECESSAIRE DE METTRE LE MOTEUR A LA TERRE EN ACCORD AVEC LES NORMES DU CODE ELECTRIQUE NATIONAL ET LES NORMES LOCALES. CELUI-CI DOIT ETRE INSTALLE PAR UN PERSONNEL QUALIFIE AFIN DE PREVENIR TOUT INCIDENT ELECTRIQUE.

AVANT TOUTE INTERVENTION SUR LE MOTEUR ÉT SES ACCESSOIRES, IL EST OBLIGATOIRE DE DISCONNECTER L'ALIMENTATION ET D'ATTENDRE L'ARRRET COMPLET DU MOTEUR.

Baldor Electric Company
Electric motors have safety labels warning of hazards or stating important precautions such as ensuring that the motor is grounded according to NEC® and local codes, or, when using the motor with electric motor drives, being certain that the maximum speed rating (on nameplate) is not exceeded.

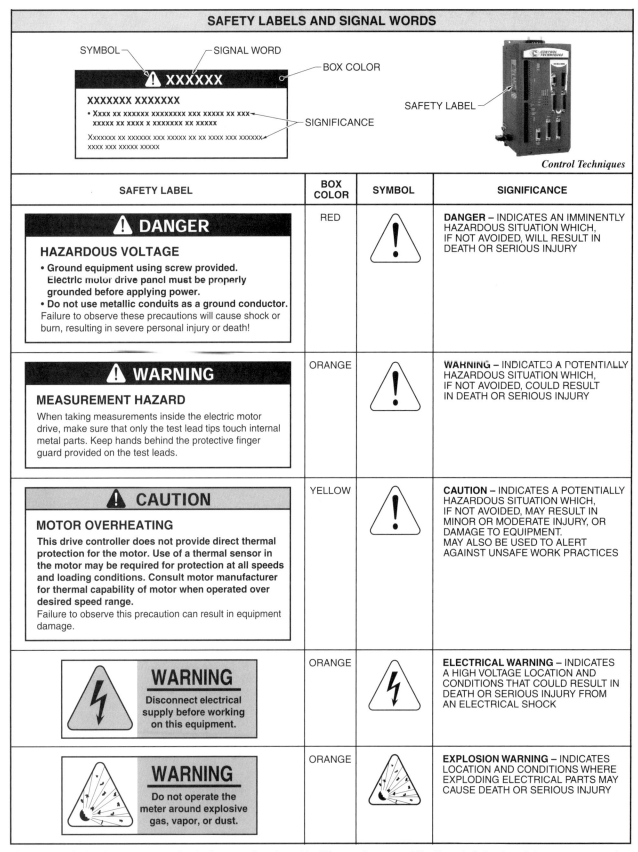

Figure 2-1. Safety labels are used to indicate a situation with different degrees of likelihood of death or injury to personnel.

Danger Signal Word

Danger signal word is a word used to indicate an imminently hazardous situation, which if not avoided, results in death or serious injury. The information indicated by a danger signal word indicates the most extreme type of potential situation, and must be followed. The danger symbol is an exclamation mark enclosed in a triangle followed by the word danger written boldly in a red box. See Figure 2-2.

Figure 2-2. Danger signal words indicate an imminently hazardous situation, which if not avoided, results in death or serious injury.

⚠ DANGER

Do not touch any internal metal parts of an electric motor drive. Live metal parts can cause an electrical shock, and charged capacitors can present a hazard even if the source of power is removed. Make sure capacitors are discharged by measuring the DC voltage between the DC + and DC – bus terminals.

Warning Signal Word

Warning signal word is a word used to indicate a potentially hazardous situation which, if not avoided, could result in death or serious injury. The information indicated by a warning signal word indicates a potentially hazardous situation and must be followed. The warning symbol is an exclamation mark enclosed in a triangle followed by the word warning written boldly in an orange box.

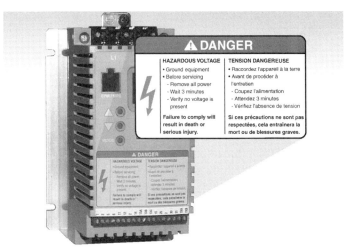

Square D – Schneider Electric
Electric motor drives have safety labels warning of dangers to be avoided, for example, ensuring that the drive is properly grounded; or, before performing any maintenance, removing all power, waiting three minutes, then verifying that there is no voltage present.

⚠ WARNING

When taking measurements inside an electric motor drive, make sure that only the test lead tips touch internal metal parts. Keep hands behind the protective finger guard provided on the test leads.

Caution Signal Word

Caution signal word is a word used to indicate a potentially hazardous situation which, if not avoided, may result in minor or moderate injury. The information indicated by a caution signal word indicates a potential situation that may cause a problem to people and/or equipment. A caution signal word also warns of problems due to unsafe work practices. The caution symbol is an exclamation mark enclosed in a triangle followed by the word caution written boldly in a yellow box.

Other signal words may also appear with the danger, warning, and caution signal words used by manufacturers. ANSI Z535.4, *Product Safety Signs and Labels*, provides additional information concerning safety labels. Additional signal words may be used alone or in combination on safety labels. Additional signal words used by electric motor drive manufacturers are electrical warning and explosion warning.

Electrical Warning Signal Word

Electrical warning signal word is a word used to indicate a high-voltage location and conditions that could result in death or serious personal injury from an electrical shock if proper precautions are not taken.

The electrical warning safety label is usually placed where there is a potential for coming in contact with live electrical wires, terminals, or parts. The electrical warning symbol is a lightning bolt enclosed in a triangle. The safety label may be shown with no words or may be preceded by the word warning written boldly.

Explosion Warning Signal Word

Explosion warning signal word is a word used to indicate locations and conditions where exploding parts may cause death or serious personal injury if proper precautions and procedures are not followed. The explosion warning symbol is an explosion enclosed in a triangle. The safety label may be shown with no words or may be preceded by the word warning written boldly.

PERSONAL PROTECTIVE EQUIPMENT

Personal protective equipment (PPE) is gear worn by a technician to reduce the possibility of injury in the work area. Personal protective equipment should be worn when installing or maintaining any electrical system, electric motor drive, or motor. All personal protective equipment must meet OSHA Standard Part 1910 Subpart I – *Personal Protective Equipment* (1910.132 through 1910.138), applicable ANSI standards, and other safety mandates. See Figure 2-3. Personal protective equipment includes protective clothing, head protection, eye protection, ear protection, hand and foot protection, back protection, knee protection, and rubber insulating matting.

PERSONAL PROTECTIVE EQUIPMENT

Figure 2-3. Personal protective equipment is used when taking electrical measurements to reduce the possibility of an injury.

Protective Clothing

Protective clothing is clothing that provides protection from contact with sharp objects, hot equipment, and harmful materials. Protective clothing made of durable material such as denim should be snug, yet allow ample movement. Pockets should allow convenient access, but should not snag on tools or equipment. Soiled protective clothing should be washed to reduce the flammability hazard.

Arc-resistant clothing must be used when working with live high-voltage electrical circuits. Arc-resistant clothing is made of materials such as Nomex®, Basofil®, and/or Kevlar® fibers. The arc-resistant fibers can be coated with PVC to offer weather resistance and to increase arc resistance. Arc-resistant clothing must meet three requirements:

- Clothing must not ignite and continue to burn.
- Clothing must provide an insulating value to the wearer to dissipate heat throughout the clothing and away from the skin.
- Clothing must provide resistance to the break-open forces generated by the shock wave of an arc.

The National Fire Protection Association (NFPA) specifies boundary distances where arc protection is required. All personnel working within specified boundary distances require arc-resistant clothing and equipment. Boundary distances vary depending on the voltage involved.

Loose-fitting clothing and long hair must be secured and jewelry removed to prevent them from getting caught in rotating equipment. Electrical shock results if jewelry makes contact with energized electrical circuits.

Head Protection

Head protection requires using a protective helmet. A *protective helmet* is a hard hat that is used in the workplace to prevent injury from the impact of falling and flying objects, and from electrical shock. Protective helmets resist penetration and absorb impact force. Protective helmet shells are made of durable, lightweight materials. A shock-absorbing lining consists of crown straps and a headband that keeps the shell away from the head to provide ventilation.

Protective helmets are identified by class of protection against specific hazardous conditions. Class A, B, and C helmets are used for construction and industrial applications. Class A protective helmets protect against low-voltage shock and burns, impact hazards, and are commonly used in construction and manufacturing facilities. Class B protective helmets protect against high-voltage shock and burns, impact hazards, and penetration by falling or flying objects. Class C protective helmets are manufactured with lighter materials yet provide adequate impact protection. See Figure 2-4.

PROTECTIVE HELMETS

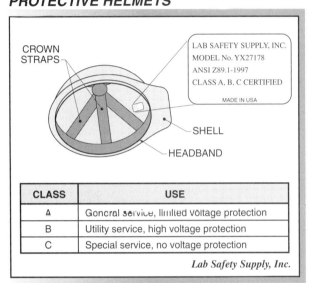

CLASS	USE
A	General service, limited voltage protection
B	Utility service, high voltage protection
C	Special service, no voltage protection

Lab Safety Supply, Inc.

Figure 2-4. Protective helmets are identified by class of protection against hazardous conditions.

Eye Protection

Eye protection must be worn to prevent eye or face injuries caused by flying particles, contact arcing, and radiant energy. Eye protection must comply with OSHA 29 CFR 1910.133, *Eye and Face Protection.* Eye protection standards are specified in ANSI Z87.1, *Occupational and Educational Eye and Face Protection.* Eye protection includes safety glasses, face shields, and goggles. See Figure 2-5.

Safety glasses are an eye protection device with special impact-resistant glass or plastic lenses, reinforced frames, and side shields. Plastic frames are designed to keep the lenses secured in the frame if an impact occurs and minimize the shock hazard when working with electrical equipment. Side shields provide additional protection from flying objects. Tinted-lens safety glasses protect against low-voltage arc hazards.

A *face shield* is an eye and face protection device that covers the entire face with a plastic shield, and is used for protection from flying objects. Tinted face shields protect against low-voltage arc hazards. *Goggles* are

an eye protection device with a flexible frame that is secured on the face with an elastic headband. Goggles fit snugly against the face to seal the areas around the eyes, and may be used over prescription glasses. Goggles with clear lenses protect against small flying particles or splashing liquids. Tinted goggles are sometimes used to protect against low-voltage arc hazards.

EYE PROTECTION

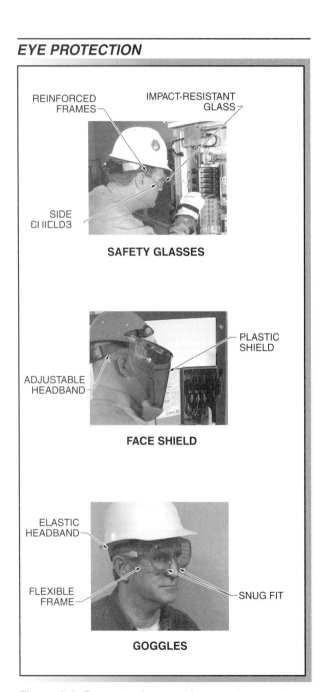

Figure 2-5. Eye protection must be worn to prevent eye or face injuries caused by flying particles, contact arcing, and radiant energy.

Safety glasses, face shields, and goggle lenses must be properly maintained to provide protection and clear visibility. Lens cleaners are available that clean without risk of lens damage. Pitted or scratched lenses reduce vision and may cause lenses to fail on impact.

Ear Protection

Ear protection are devices worn to limit the noise entering an ear and includes earplugs and earmuffs. An *earplug* is an ear protection device made of moldable rubber, foam, or plastic, and inserted into the ear canal. An *earmuff* is an ear protection device worn over the ears. A tight seal around an earmuff is required for proper protection.

Power tools and equipment can produce excessive noise levels. Technicians subjected to excessive noise levels may develop hearing loss over a period of time. The severity of hearing loss depends on the intensity and duration of exposure. Noise intensity is expressed in decibels. A *decibel (dB)* is a unit of measure used to express the relative intensity of sound. See Figure 2-6. Ear protection is worn to prevent hearing loss.

Ear protection devices are assigned a noise reduction rating (NRR) number based on the noise level reduced. For example, an NRR of 27 means that the noise level is reduced by 27 dB when tested at the factory. To determine approximate noise reduction in the field, 7 dB is subtracted from the NRR. For example, an NRR of 27 provides a noise reduction of approximately 20 dB in the field.

Hand and Foot Protection

Hand protection are gloves worn to prevent injuries to hands caused by cuts or electrical shock. The appropriate hand protection is determined by the duration, frequency, and degree of the hazard to the hands. *Electrical gloves* are gloves made of latex that are used to provide maximum insulation from electrical shock. Latex gloves are stamped with a working voltage range such as 500 V–26,500 V. See Figure 2-7. *Cover gloves* are gloves worn over latex electrical gloves to prevent penetration of the latex gloves and provide added protection against electrical shock.

Foot protection are shoes worn to prevent foot injuries that are typically caused by objects falling less than 4′ and having an average weight less than 65 lb. Safety shoes with reinforced steel toes protect against injuries caused by compression and impact. Insulated rubber-soled shoes are commonly worn during electrical work to prevent electrical shock. Protective footwear must comply with ANSI Z41, *Personal Protection—Protective Footwear*.

SOUND LEVELS			
AVERAGE DECIBEL (dB)	**LOUDNESS**	**EXAMPLES**	**EXPOSURE DURATION**
140	Deafening	Jet airplane taking off, air raid siren, locomotive horn	
130	Pain threshold		2 min
120	Feeling threshold		7 min
110	Uncomfortable		30 min
100	Very loud	Chain saw	2 hr
90	Noisy	Shouting, auto horn	4 hr
85			8 hr
80	Moderately loud	Vacuum cleaner	25.5 hr
70	Loud	Telephone ringing, loud talking	——
60	Moderate	Normal conversation	——
50	Quiet	Hair dryer	——
40	Moderately quiet	Refrigerator running	——
30	Very quiet	Quiet conversation, broadcast studio	——
20	Faint	Whispering	——
10	Barely audible	Rustling leaves, soundproof room, human breathing	——
0	Hearing threshold	Intolerably quiet	——

Figure 2-6. A decibel is a unit of measure used to express the relative intensity of sound.

HAND AND FOOT PROTECTION

ELECTRICAL SAFETY GLOVES		
LATEX THICKNESS*	TEST VOLTAGE	WORKING VOLTAGE†
0.5	2500	500
1.0	5000	1000
1.5	10,000	7500
2.3	20,000	17,000
2.9	30,000	26,500

* in mm
† maximum

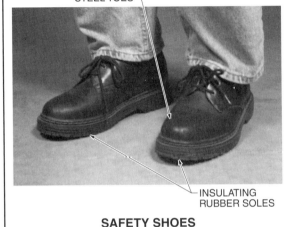

STEEL TOES

INSULATING RUBBER SOLES

SAFETY SHOES

Figure 2-7. Rubber insulated gloves and safety shoes provide insulation from electrical shock.

Back Protection

A back injury is one of the most common injuries resulting in lost time in the workplace. Back injuries are the result of improper lifting procedures. Back injuries are prevented through proper planning and work procedures. Assistance should be sought when moving heavy objects. When lifting objects from the ground, ensure the path is clear of obstacles and free of hazards. Bend the knees and grasp the object firmly. Next, lift the object, straightening the legs and keeping the back as straight as possible. Finally, move forward after the whole body is in the vertical position. Keep the load close to the body and keep the load steady. See Figure 2-8.

TECH FACT

The UV rays of sunlight break down the materials of protective helmets, requiring that protective helmets be replaced at least every 10 years.

PROPER LIFTING

1. BEND KNEES AND GRASP OBJECT FIRMLY

KEEP BACK STRAIGHT

2. LIFT OBJECT BY STRAIGHTENING LEGS

3. MOVE FORWARD AFTER WHOLE BODY IS IN VERTICAL POSITION

Figure 2-8. Lifting an object with the legs reduces the possibility of a back injury.

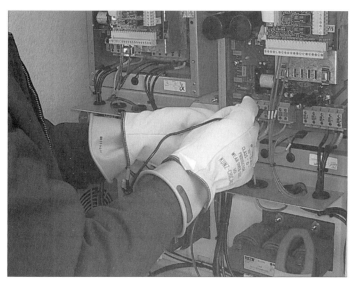

Cover gloves must be worn over latex electrical gloves to prevent electrical glove penetration.

Long objects such as conduit may not be heavy, but the weight might not be balanced; therefore, it should be carried by two or more people. When carried on the shoulder by one person, conduit should be transported with the front end pointing downward to minimize the possibility of injury to others when walking around corners or through doorways. See Figure 2-9.

CARRYING LOADS ON SHOULDER

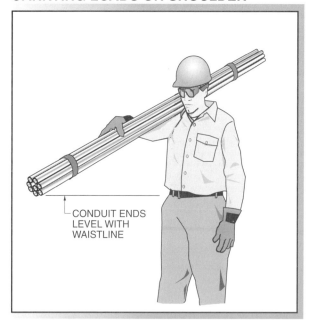

CONDUIT ENDS LEVEL WITH WAISTLINE

Figure 2-9. When carried on the shoulder by one person, conduit should be transported with the front end down.

Knee Protection

A *knee pad* is a rubber, leather, or plastic pad strapped onto the knees for protection. A knee pad is worn by technicians who spend considerable time working on their knees or who work in close areas and must kneel for proper access to motors and electric motor drives. Buckle straps or Velcro™ closures secure knee pads in position. See Figure 2-10.

KNEE PROTECTION

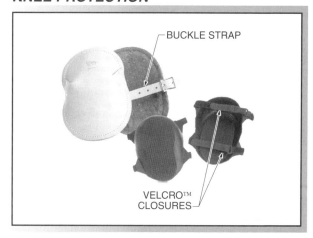

BUCKLE STRAP

VELCRO™ CLOSURES

The Stanley Works

Figure 2-10. Knee pads are used to provide protection and comfort to technicians who spend considerable time on their knees.

Rubber Insulating Matting

Rubber insulating matting is a floor covering that provides technicians protection from electrical shock when working on live electrical circuits. Dielectric black fluted rubber matting is specifically designed for use in front of open cabinets or high-voltage equipment. Matting is used to protect technicians when voltages are over 50 V. Two types of matting that differ in chemical and physical characteristics are designated as Type I natural rubber and Type II elastomeric compound matting. See Figure 2-11.

ELECTRICAL SHOCK

An *electric shock* is a shock that results any time a body becomes part of an electrical circuit. Electrical shock effects vary from a mild sensation, to paralysis, to death. The severity of an electric shock depends on the amount of electric current in milliamps (mA) that flows through the body, the length of time the body is exposed to the current flow, the path the current takes through the body, and the physical size and condition of the body through which the current passes. See Figure 2-12.

RUBBER INSULATING MATTING RATINGS					
SAFETY STANDARD	MATERIAL THICKNESS		MATERIAL WIDTH (in.)	TEST VOLTAGE	MAXIMUM WORKING VOLTAGE
	Inches	Millimeters			
BS921*	.236	6	36	11,000	450
BS921*	.236	6	48	11,000	450
BS921*	.354	9	36	15,000	650
BS921*	.354	9	48	15,000	650
VDE0680[†]	.118	3	39	10,000	1000
ASTM D178[‡]	.236	6	24	25,000	17,000
ASTM D178[‡]	.236	6	30	25,000	17,000
ASTM D178[‡]	.236	6	36	25,000	17,000
ASTM D178[‡]	.236	6	48	25,000	17,000

* BSI–British Standards Institiute
[†] VDE–Verband Deutscher Elektrotechniker Testing and Certification Institute
[‡] ASTM International–American Society for Testing and Materials

Figure 2-11. Rubber insulating matting provides protection from electrical shock when working on live electrical circuits.

ELECTRICAL SHOCK EFFECTS	
APPROXIMATE CURRENT*	EFFECT ON BODY[†]
over 20	Causes severe muscular contractions, paralysis of breathing, heart convulsions
15-20	Painful shock May be frozen or locked to point of electrical contact until circuit is de-energized
8-15	Painful shock Removal from contact point by natural reflexes
8 or less	Sensation of shock but probably not painful

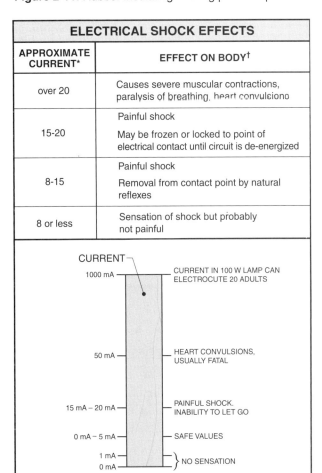

* in mA
[†] effects vary depending on time, path, amount of exposure, and condition of body

Figure 2-12. Electrical shock is a condition that results any time a body becomes part of an electrical circuit.

During an electrical shock, the body of a person becomes part of an electrical circuit. The resistance a body of a person offers to the flow of current varies. Sweaty hands have less resistance than dry hands. A wet floor has less resistance than a dry floor. The lower the resistance, the greater the current flow. As the current flow increases, the severity of an electrical shock increases. Safe working habits with proper protective equipment and instrument usage are required to prevent electrical shock when troubleshooting exposed electrical devices that are normally enclosed.

Grounding

Grounding is the connection of all exposed non-current-carrying metal parts to the earth. Grounding provides a direct path for unwanted fault current to the earth without causing harm to persons or equipment. Electrical circuits are grounded to safeguard equipment and personnel against the hazards of electrical shock and electrostatic discharge. Proper grounding of electrical tools, motors, equipment, enclosures, electric motor drives, and other control circuitry helps prevent hazardous conditions. On the other hand, improper electrical wiring or misuse of electricity causes destruction of equipment and fire damage to property as well as personal injury.

In addition to helping prevent electrical shock, grounding an electric motor drive also helps reduce electrical noise that can cause problems within the electrical system or drive. Electric motor drives should not be operated unless they are properly grounded. Electric motor drives are grounded as per NEC® Article 250 and applicable state and local codes. Ground the electric motor drive by connecting the ground terminal to the electrical system ground. The ground screw is green and marked with the symbol for ground (⏚) and/or the abbreviation for ground (GND). See Figure 2-13.

ELECTRIC MOTOR DRIVE GROUNDING

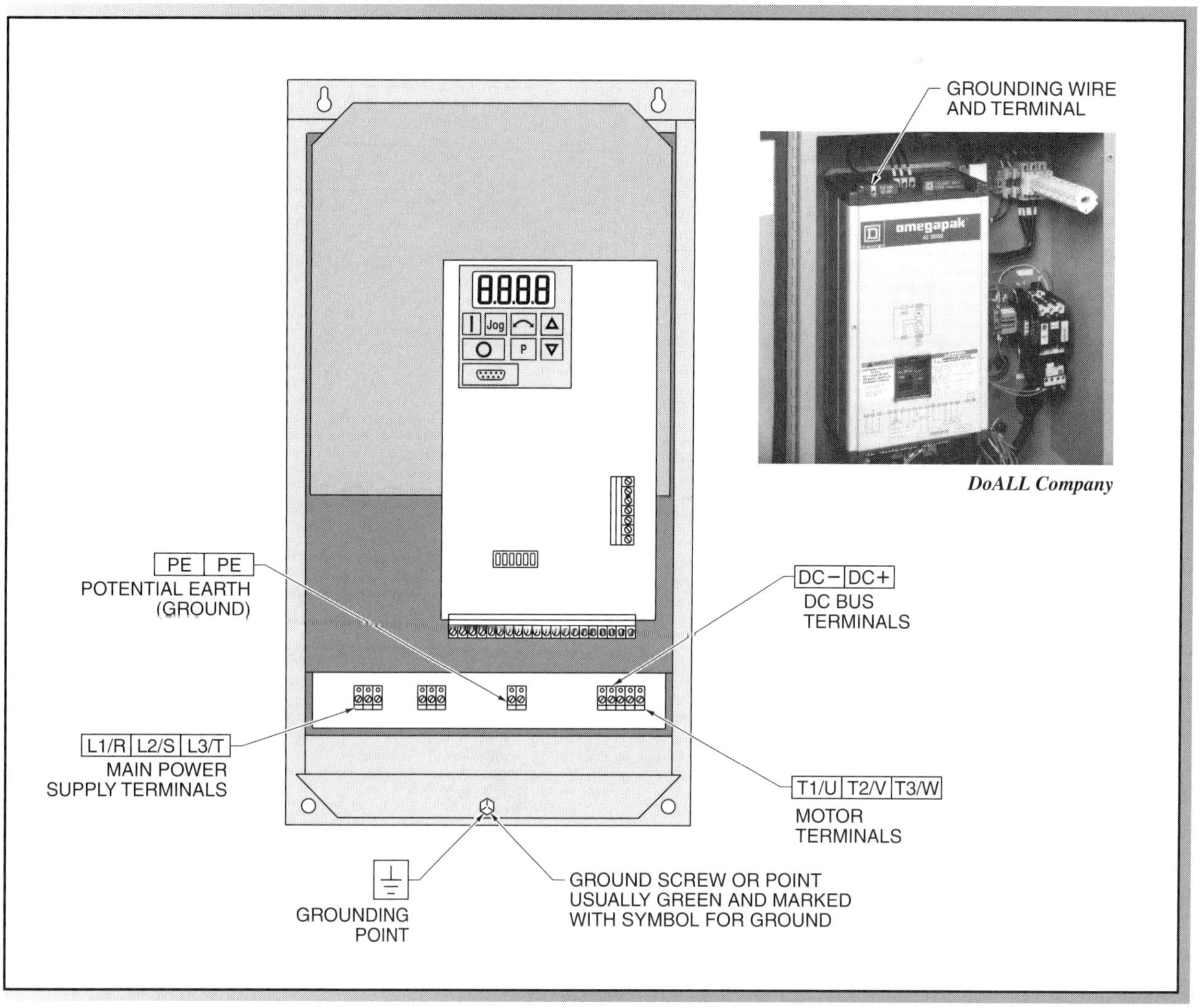

Figure 2-13. Proper grounding of an electric motor drive helps prevent electrical shock and reduces electrical noise that can cause problems within an electrical system.

To prevent problems, a grounding path must:
• be as short as possible and of sufficient size recommended by the manufacturer (minimum 14 AWG copper)
• never be fused or switched
• be a permanent part of the electrical circuit
• be continuous and uninterrupted from the electrical circuit to earth

Ground Fault Circuit Interrupters

A *ground fault circuit interrupter (GFCI)* is a device that protects against electrical shock by detecting an imbalance of current in the normal conductor pathways and opening the circuit. When current in the two conductors of an electrical circuit varies by more than 5 mA, a GFCI opens

the circuit. A GFCI is rated to trip quickly enough (1/40 of a second) to prevent electrocution. See Figure 2-14.

A GFCI protects against the most common form of electrical shock hazard, the ground fault. A GFCI does not protect against line-to-line contact hazards, such as a technician holding two hot wires or a hot and a neutral wire in each hand. GFCI protection is required in addition to OSHA grounding requirements.

Wet plugs, receptacles, and tools may cause tripping or interruption of current flow of a GFCI. Limit exposure of tool connectors to excessive moisture by using watertight connectors. Providing more GFCIs or shorter circuits can also prevent tripping caused by the cumulative leakage from several tools or by leakage from extremely long circuits.

GROUND FAULT CIRCUIT INTERRUPTER

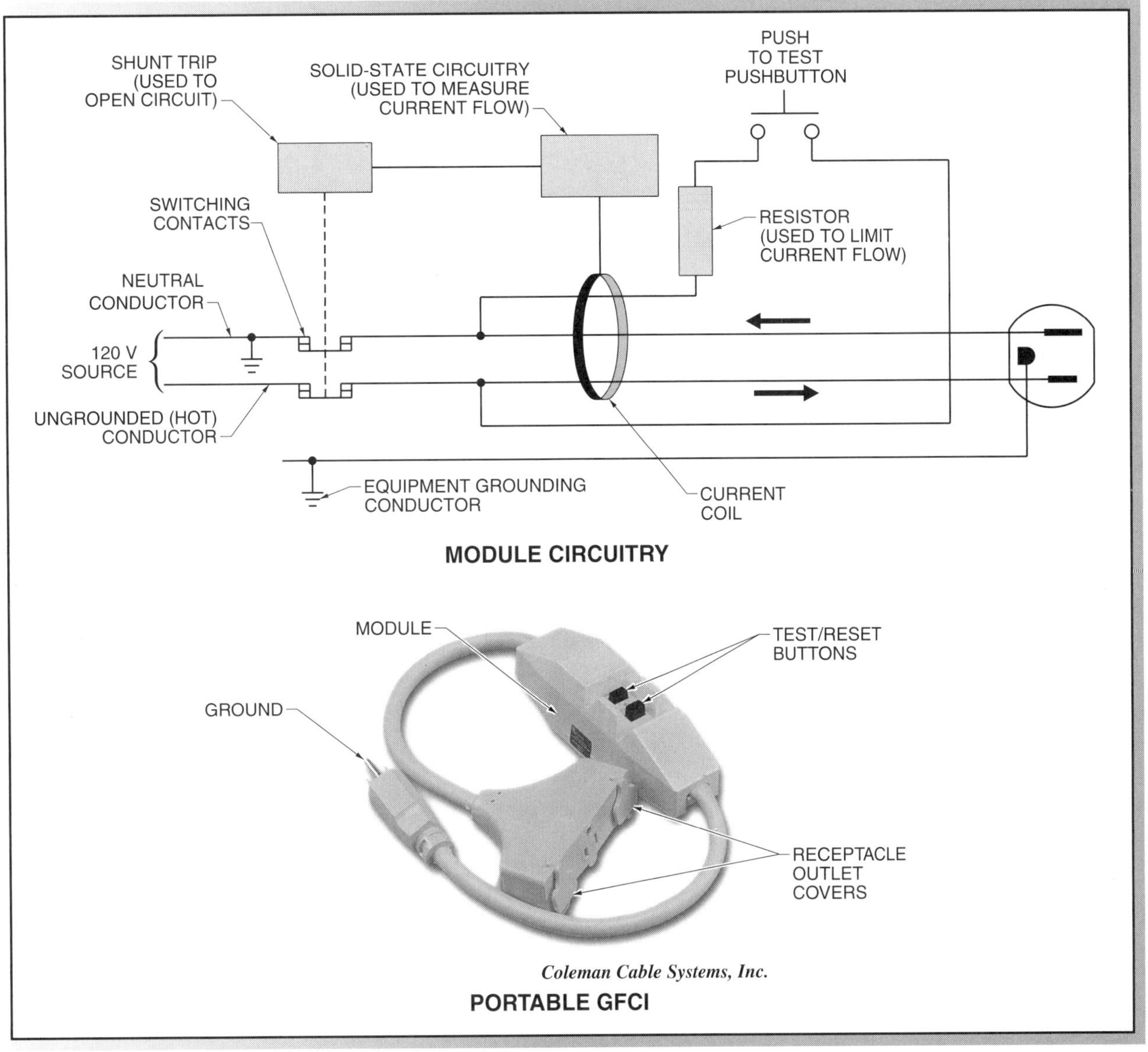

Figure 2-14. A portable GFCI compares the amount of current in the hot or ungrounded conductor with the amount of current in the common or grounded conductor. The GFCI immediately breaks the circuit if the current difference exceeds 5 mA.

Portable GFCIs are designed to be easily moved from one location to another. Portable GFCIs commonly contain more than one receptacle outlet protected by the electronic circuit module. GFCIs incorporate a no-voltage release device that disconnects power to the outlets if any current imbalance exists between the two circuit conductors. Portable GFCIs should be inspected and tested before each use. GFCIs have a built-in test circuit to ensure that the ground fault protection is operational.

LOCKOUT/TAGOUT

Lockout is the process of removing the source of electrical power and installing a lock, which prevents the power from being turned ON. To ensure the safety of personnel working with equipment, all electrical, pneumatic, and hydraulic power is removed and the equipment must be locked out and tagged out. *Tagout* is the process of placing a danger tag on the source of electrical power, which indicates that the equipment may not be operated until the danger tag is removed.

Per OSHA standards, equipment is locked out and tagged out before any installation or preventive maintenance is performed. See Figure 2-15.

LOCKOUT/TAGOUT

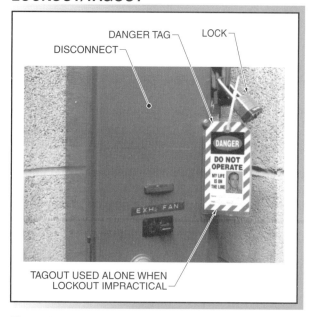

Figure 2-15. Lockouts and tagouts are applied to equipment to prevent injury from energized circuits and equipment operation during installation and maintenance.

A danger tag has the same importance and purpose as a lock and is used alone only when a lock does not fit the disconnect device. The danger tag shall be attached at the disconnect device with a tag tie or equivalent and shall have space for the technician's name, craft, and other company-required information. A danger tag must withstand the elements and expected atmosphere for the maximum period of time that exposure is expected. Lockout/tagout is used when:

- power is not required to be on to a piece of equipment to perform a task
- machine guards or other safety devices are removed or bypassed
- the possibility exists of being injured or caught in moving machinery
- jammed equipment is being cleared
- the danger exists of being injured if equipment power is turned ON

Lockout and tagouts do not by themselves remove power from a machine or its circuitry. OSHA provides a standard procedure for equipment lockout/tagout. Lockout is performed and tagouts are attached only after the equipment is turned OFF and tested. Lockout/Tagout procedures are as follows:

1. Notify all affected persons that a lockout/tagout is required. Notification should include the reason for the lockout/tagout, and the expected duration.
2. If the equipment is operating, shut it down using the normal procedures.
3. Operate the energy isolating device(s) so that the equipment is isolated from all energy sources. Stored energy, such as in springs, elevated machine members, capacitors, etc., must be dissipated or restrained by blocking, discharging, or other appropriate methods.
4. Lockout and/or tagout the energy-isolating devices with assigned lock(s) and/or danger tag(s).
5. After ensuring that no personnel are exposed, operate the normal operating controls, verifying that the equipment is inoperable and that all energy sources have been isolated.
6. Inspect and test the equipment with appropriate test instruments to verify that all energy sources are disconnected. Multi-phase electrical power requires that each phase be tested. The equipment is now locked out and/or tagged out.

A lockout/tagout must not be removed by any person other than the authorized person who installed the lockout/tagout, except in an emergency. In an emergency, only supervisory personnel may remove a lockout/tagout, and only upon notification of the authorized person. A list of company rules and procedures is given to authorized personnel and any person who may be affected by a lockout/tagout.

When more than one technician is required to perform a task on a piece of equipment, each technician shall place a lockout and/or tagout on the energy-isolating device(s). A multiple lockout/tagout device (hasp) must be used because energy-isolating devices typically cannot accept more than one lockout/tagout. A *hasp* is a multiple lockout/tagout device.

> ⚠ **CAUTION**
>
> *Live electrical circuits in an electric motor drive may emit an arc at any time.*

ELECTRICAL SAFETY

Safety rules must be followed when working with electrical equipment and electric motor drive systems. Following electrical safety rules helps prevent injuries from electrical energy sources. Electrical safety rules include the following:

- Always comply with the NEC®, state, and local codes.
- Use UL® approved equipment, components, and test equipment.
- Before removing any fuse from a circuit, be sure the switch for the circuit is open or disconnected. When removing fuses, use an approved fuse puller and break contact on the line side of the circuit first. When installing fuses, install the fuse first into the load side of the fuse clip, then into the line side.
- Inspect and test grounding systems for proper operation. Ground any conductive component or element that is not energized.
- Turn OFF, lockout, and tagout any circuit that is not required to be energized when maintenance is being performed.
- Always use PPE and safety equipment.
- Perform the appropriate task required during an emergency situation.
- Use only a Class C rated fire extinguisher on electrical equipment. A Class C fire extinguisher is identified by the color blue inside a circle. See Figure 2-16.
- Always work with another individual when working in a dangerous area, on dangerous equipment, or with high voltages.
- Do not work when tired or taking medication that causes drowsiness unless specifically authorized by a physician.
- Do not work in poorly lighted areas.
- Ensure there are no atmospheric hazards such as flammable dust or vapor in the area.
- Use one hand when working on a live circuit to reduce the chance of an electrical shock passing through the heart and lungs.
- Never bypass fuses, circuit breakers, or any other safety device.

TECH FACT

Fires or explosions may result if electric motor drives used in hazardous locations are not installed in the proper, approved enclosure. Hazardous locations are defined by their class, division, and group.

FIRE EXTINGUISHERS

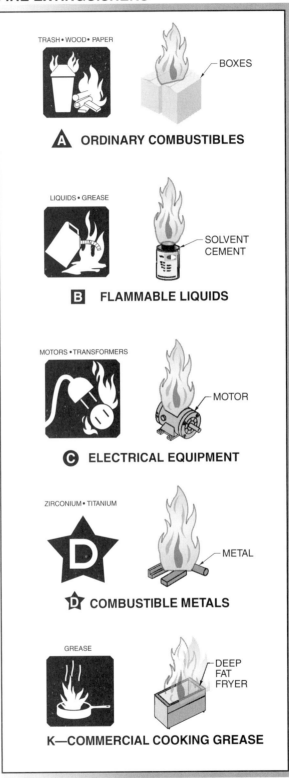

Figure 2-16. Fire extinguishers are classified for use by the specific combustible material to be extinguished.

Electrostatic Discharge

Electrostatic discharge (ESD) is the movement of electrons from a source to an object. *Static electricity* is an electrical charge at rest. The most common way to build up static electricity is by friction. Friction causes electrons from one source to flow to another material, causing an electron buildup (negative static charge). When a person (negative charged) contacts a positive charged or a grounded object, all the excess electrons flow (jump) to that object. Electrostatic discharge can be 35,000 V or more. People normally do not feel electrostatic discharges until the discharge reaches 3000 V. See Figure 2-17. Solid-state devices and circuits may be damaged or destroyed by a 10 V electrostatic discharge. Technicians should wear a wrist grounding strap or other type of grounding device to avoid damage to solid state devices and circuits from ESD.

ELECTROSTATIC DISCHARGE

Figure 2-17. An excess of electrons produces an arc during electrostatic discharge (ESD).

The safety rules that are effective in preventing static damage include the following:

- Always keep a work area clean and clear of unnecessary materials, especially common plastics.
- Never handle electronic devices by their leads.
- Test grounding devices daily to make sure that they have not become loose or intermittent.
- Never work on an ESD-sensitive object without a properly working grounding device.
- Always handle printed circuit boards (PC boards) by their outside corners.
- Always transport PC boards in anti-static trays.
- Never stack PC boards.
- When transporting single electronic devices, always keep them sealed in conductive static shielding.

ELECTRIC MOTOR SAFETY

Electric motor drives are used to control electric motors. Electric motors are part of all electric motor drive installations. Caution is required when working in the electric motor hazard areas. The electric motor hazard areas are the electrical circuit and rotating shaft. See Figure 2-18. Electric motor safety rules include the following:

- Always apply lockout/tagouts during services that do not require that power be applied to the motor or electric motor drive.
- Program the electric motor drive with the correct overload and overcurrent protection information to protect the motor during all operating conditions.
- Ensure the motor frame is connected to the proper grounding system.
- Ensure that rotating parts of a motor have a guard or housing.
- Use the correct motor type for the location.
- Connect the motor to the correct electric motor drive type, size, and drive output terminals.

ELECTRIC MOTOR DRIVE SAFETY

An *electric motor drive* is an electronic device that controls the direction, speed, torque, and other operating functions of an electric motor in addition to providing motor protection and monitoring functions. An electric motor drive may also be called by the specific type of drive design, such as adjustable-speed DC drive, inverter drive, adjustable-speed AC drive, variable-frequency drive, direct-torque drive, vector drive, or regenerative drive.

Electric motor drives are used to control high-voltage motors that typically operate at 115, 208, 240, 460, 480 VAC, or higher. Any electric motor drive installation or troubleshooting task can present a dangerous situation that can lead to an accident or an electrical shock. Improper electric motor drive programming can also lead to damage of equipment and produce a danger to operators and maintenance personnel.

⚠ CAUTION

High ambient temperature affects the life of an electric motor drive and can affect the performance of the drive. Do not install an electric motor drive in any location that exceeds the rated ambient temperature limit.

ELECTRIC MOTOR HAZARD AREAS

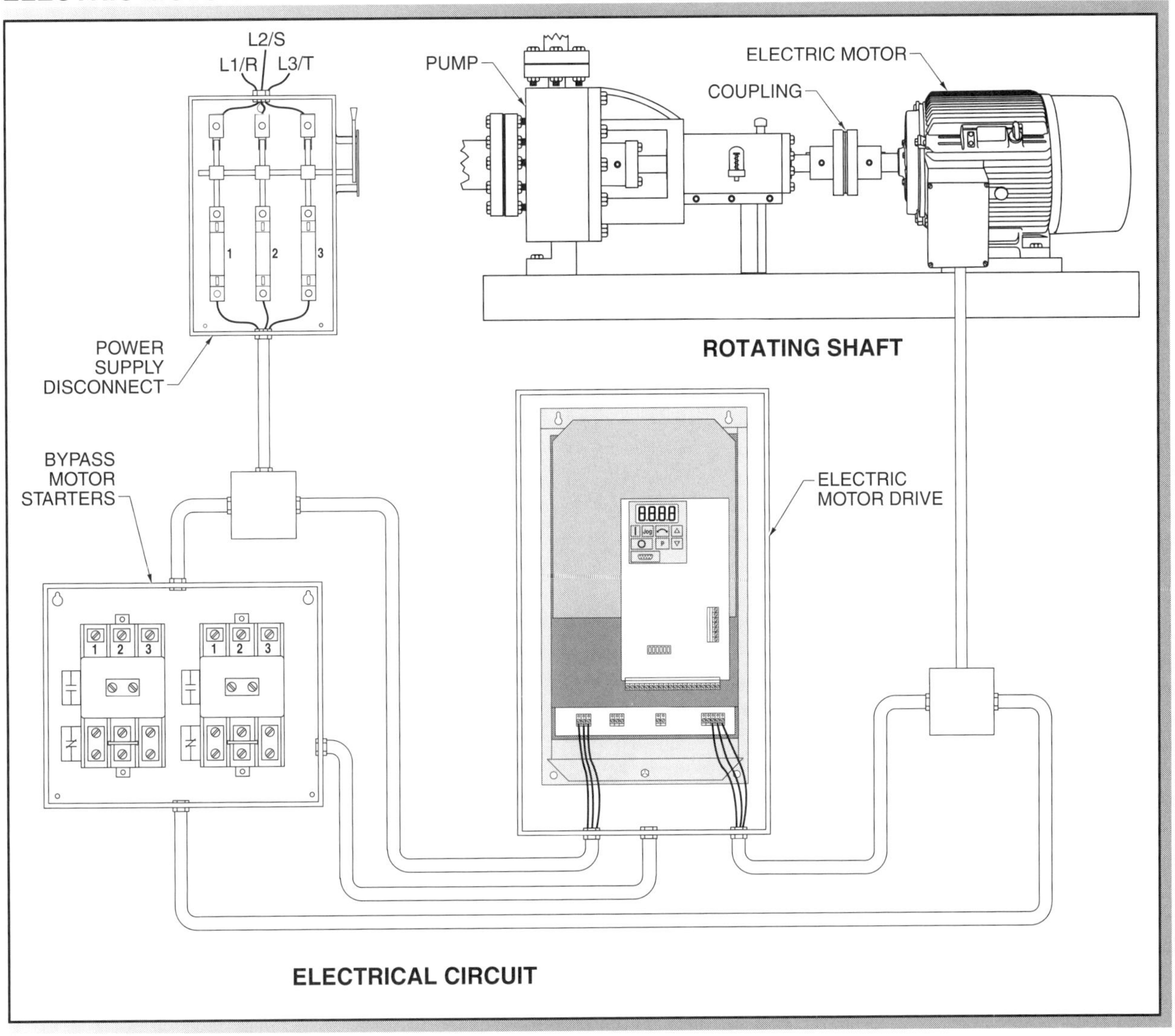

Figure 2-18. Electric motor installation is part of all electric motor drive installations.

An electric motor drive is part of an electrical system. An electric motor drive is connected to a power supply and delivers modified power to a motor. When including an electric motor drive in a system, the following rules should be applied:

• Do not apply power until the entire operating manual of the electric motor drive is understood.
• Only qualified persons should have access to the electric motor drive adjustments. Changing an electric motor drive setting can affect the entire system.
• Never touch any internal part of the electric motor drive when power is applied.

• Do not set the electric motor drive output frequency too high. Operating a motor above its rated speed can decrease the motor's torque and may result in damage to the motor and/or the motor's driven equipment.
• Do not set the electric motor drive acceleration and deceleration times too short. Short times can cause mechanical stress on the driven load and produce electric motor drive tripping.
• Incoming power must be connected to the electric motor drive's input terminals L1 (R), L2 (S), and L3 (T). Do not apply incoming power to the output terminals T1 (U), T2 (V), or T3 (W). This can damage the electric motor drive and produce a hazardous situation.

• Use separate metal conduits for routing the input power conductors, output power conductors, and control circuit conductors. See Figure 2-19.

MACHINE SAFETY

Installing or performing maintenance tasks on an electric motor drive requires that the drive be properly programmed to the function of the machine or piece of equipment. Electric motor drives are used to control electric motors connected to conveyors, pumps, blowers and fans, production and packing machinery, and other assorted loads. An electric motor drive must be designed and programmed with the safety of the entire system in mind. Any system that includes an electric motor drive must be checked and double-checked to ensure the system operates safely during all possible control functions.

ELECTRIC MOTOR DRIVE CONDUIT USAGE

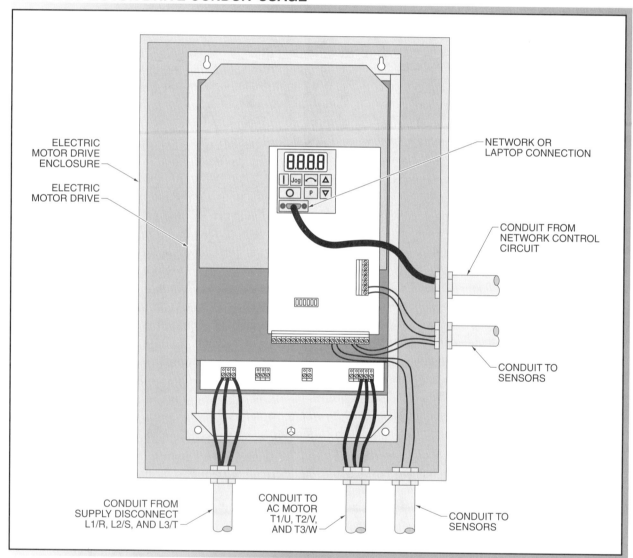

Figure 2-19. Use separate metal conduits for routing the input power conductors, output power conductors, and control circuit conductors.

Name _____ Date _____

True-False

T　F　**1.** A qualified person is a person who has special knowledge, training, and experience in the installation, programming, maintenance, and troubleshooting of electrical equipment.

T　F　**2.** Safety labels are used to indicate a situation with different degrees of likelihood of death or injury to personnel.

T　F　**3.** The information indicated by a danger signal word indicates the least extreme type of potential situation.

T　F　**4.** A protective helmet is a hard hat that is used in the workplace to prevent injury from the impact of falling and flying objects and from electrical shock.

T　F　**5.** Cover gloves are gloves worn under latex electrical gloves to prevent penetration of the latex gloves and provide added protection against electrical shock.

T　F　**6.** An electric shock results any time a body becomes part of an electrical circuit.

T　F　**7.** An electric shock with a current of 50 mA is rarely fatal.

T　F　**8.** A lockout/tagout must not be removed by any person other than the authorized person who installed the lockout/tagout, except in an emergency.

T　F　**9.** Solid-state devices and circuits cannot be damaged or destroyed by a 10 V electrostatic discharge.

T　F　**10.** Always handle printed circuit boards by their outside corners to prevent static damage.

T　F　**11.** Electric motor drives are grounded as per NEC® Article 250 and applicable state and local codes.

T　F　**12.** A warning signal word is a word used to indicate a potentially hazardous situation which, if not avoided, may result in minor or moderate injury.

T　F　**13.** Personal protective equipment is gear worn by a technician to reduce the possibility of injury in the work area.

T　F　**14.** A decibel is a unit of measure used to express the relative intensity of sound.

T　F　**15.** When carried on the shoulder by one person, conduit should be transported with the front end pointing upward to minimize the possibility of injury to others when walking around corners or through doorways.

Completion

_AC & DC_   1. Electric motor drives are used in electrical systems to control ___ and ___ motors.

warning signal   2. A(n) ___ signal word is a word used to indicate a potentially hazardous situation which, if not avoided, could result in death or serious injury.

_____ 3. The electrical warning safety label is usually placed where there is a potential for coming in contact with ___ electrical wires, terminals, or parts.

face shield   4. ___ must be worn to prevent eye or face injuries caused by flying particles, contact arcing, and radiant energy.

_____ 5. ___ is the connection of all exposed non-current-carrying metal parts to the earth.

GFCI   6. A(n) ___ is a device that protects against electrical shock by detecting an imbalance of current in the normal conductor pathways and opening the circuit.

lockout   7. ___ is the process of removing the source of electrical power and installing a lock, which prevents the power from being turned ON.

_____ 8. When removing fuses, use an approved fuse puller and break contact on the ___ side of the circuit first.

electric motor drive   9. A(n) ___ is an electronic device that controls the direction, speed, torque, and other operating functions of an electric motor, in addition to providing motor protection and monitoring functions.

input   10. Incoming power must be connected to the electric motor drive's ___ terminals L1 (R), L2 (S), and L3 (T).

Multiple Choice

_B_   1. The three most common signal words on safety labels are ___, warning, and caution.
     A. attention   C. explosion
     B. danger   D. safety

_C_   2. A(n) ___ injury is one of the most common injuries resulting in lost time in the workplace.
     A. ankle   C. back
     B. leg   D. hand

_D_   3. A GFCI protects against the most common form of electrical shock hazard, the ___.
     A. line-to-line short circuit   C. overload
     B. static discharge   D. ground fault

_C_   4. Use only a Class ___ rated fire extinguisher on electrical equipment.
     A. A   C. C
     B. B   D. K

_B_   5. ___ discharge is the movement of electrons from a source to an object.
     A. Current   C. Ion
     B. Electrostatic   D. Voltage

Name _____ Date _____

Activity 2-1. Safety Precautions Checklist

Before working on an electric motor drive and motor circuit, technicians and qualified persons must understand the electrical system and test equipment used.

1. Complete the following checklist to prepare for working on an electric motor drive and motor application.

SAFETY PRECAUTIONS CHECKLIST
General Information
Where is the nearest fire extinguisher located?
Are the procedures to operate a fire extinguisher understood?
Is the fire extinguisher approved for electrical fires?
Has the proper protective equipment (eye protection, etc.) been supplied for working in the electric motor drive and motor area?
Where is the nearest telephone?
What is the number to be dialed in case of an emergency?
Have the company or school safety rules and policies been given to all personnel?
Have the company or school safety rules and policies been read and understood?

Electric Motor Drive Information
What is the model number of the electric motor drive?
What is the rated input voltage and rated output voltage of the electric motor drive?   Input   Output
What is the current rating or horsepower rating of the electric motor drive?
What is the supply voltage to the electric motor drive?
Are the service manuals of the electric motor drive available?

Test Equipment Information
What type of test equipment (voltmeter, digital multimeter, etc.) is available to take measurements?
What type of additional test equipment (megohmmeter, scopemeter, etc.) is available to take measurements when troubleshooting electric motor drive and motor problems?
Has the test equipment operation and proper usage been explained?
Have the methods of checking test meters been explained?

Activity 2-2. Lockout/Tagout Compliance Checklist

Lockout/tagout is applied when equipment to be serviced does not require power. In order to ensure that there is a lockout/tagout procedure in place and that this procedure is followed, a lockout/tagout checklist should be

1. Complete the following lockout/tagout compliance checklist for each electric motor drive and motor application that requires locking out.

LOCKOUT/TAGOUT COMPLIANCE CHECKLIST		
Operation	**Date & Time Begun**	**Date & Time Completed**
Obtain a work order to service an electric motor drive, motor, and machine.		
Obtain all necessary electric motor drive, motor, and machine manufacturer manuals.		
Prepare electric motor drive, motor, and machine for shutdown.		
Notify all employees who might be affected before removing power from an electric motor drive and applying a lockout/tagout.		
Remove power and measure all power sources to verify that power is OFF. Also verify that all DC bus capacitors are fully discharged.		
Apply appropriate lockout/tagout devices per company or school policy.		
Service electric motor drive, motor, and machine using manufacturer and company or school procedures.		
Notify all employees who might be affected that power will be turned ON.		
Turn power ON and measure to verify that power is at correct level.		
Verify that electric motor drive, motor, and machine are operating properly.		
Complete all required documentation.		

Activity 2-3. Insulation Life and Temperature Chart

Manufacturers use charts, such as the Insulation Life and Temperature Chart, to determine technical information required for safe electric motor drive and motor installation. All insulation deteriorates over time. Temperature affects motor insulation more than any other variable. Typical insulation is rated for an operating temperature of 77°F (25°C) for 100% insulation life. Insulation may last longer than rated life when the temperature remains below the rated level. Insulation life decreases as the temperature rises above the rated level.

Use the Insulation Life and Temperature Chart to determine expected motor insulation operating life.

_____ **1.** At 77°F, the rated insulation life is ___%.

_____ **2.** At 50°C, the rated insulation life is ___%.

_____ **3.** At 38°F, the rated insulation life is ___%.

_____ **4.** At 75°C, the rated insulation life is ___%.

_____ **5.** At 302°F, the rated insulation life is ___%.

TEMPERATURE CONVERSION CHART	
DEGREES F	DEGREES C
77	25
122	50
167	75
212	100
257	125
302	150
347	175
392	200

INSULATION LIFE AND TEMPERATURE CHART

Activity 2-4. Cable Length, Carrier Frequency, and Insulation Life

As the carrier frequency of an electric motor drive increases, motor insulation life decreases. The length of the motor leads from the electric motor drive to the motor (cable length) also affects insulation life.

Use the Cable Length, Carrier Frequency, and Insulation Chart to determine expected motor insulation life.

_____ **1.** When an electric motor drive uses a carrier frequency of 3 kHz and has a cable length of 300′, the expected insulation life is ___ hr.

_____ **2.** When an electric motor drive uses a carrier frequency of 6 kHz and has a cable length of 200′, the expected insulation life is ___ hr.

_____ **3.** When an electric motor drive uses a carrier frequency of 9 kHz and has a cable length of 200′, the expected insulation life is ___ hr.

_____ **4.** When an electric motor drive uses a carrier frequency of 12 kHz and has a cable length of 250′, the expected insulation life is ___ hr.

_____ **5.** When an electric motor drive uses a carrier frequency of 6 kHz and has a cable length of 100′, the expected insulation life is ___ hr.

_____ **6.** When an electric motor drive uses a carrier frequency of 6 kHz and has a cable length of 500′, the expected insulation life is ___ hr.

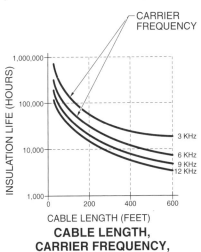

CABLE LENGTH, CARRIER FREQUENCY, AND INSULATION CHART

Activity 2-5. Selecting Fuse and Wire Size

Electric motor drive manufacturers provide charts to help select fuses and wire sizes for drive applications. Fuse and wire charts are used for general selection, and actual sizes can change based on cable lengths, operating temperatures, and other variables.

Use the Manufacturer Chart to determine the fuse size and wire size required for 460 VAC rated drives.

_____ **1.** A 7.5 HP rated drive requires a(n) ___ AWG wire size.

_____ **2.** A 10 HP rated drive requires a(n) ___ A rated time delay fuse.

_____ **3.** A 27 A rated drive requires a(n) ___ AWG wire size.

_____ **4.** A 40 A rated drive requires a(n) ___ A rated time delay fuse.

_____ **5.** A 50 HP rated drive requires a(n) ___ A rated fast acting fuse.

_____ **6.** A 125 A rated drive requires a(n) ___ A rated fast acting fuse.

_____ **7.** A #14 AWG copper wire is used to supply power to 460 VAC rated drives up to ___ HP.

_____ **8.** Single conductor wires are used up to ___ HP.

CATALOG NUMBER	CONTROL RATING		INPUT BREAKER*	INPUT FUSE*		WIRE GAUGE†‡
	Amps	HP		Fast Acting	Time Delay	
1 – DA – 001	2	0.75	3	2	2	14
2 – DA – 002	2	1	3	3	2.5	14
3 – DA – 003	4	2	7	5	4.5	14
4 – DA – 004	5	3	7	8	6.3	14
5 – DA – 005	8	5	15	12	10	14
6 – DA – 006	11	7.5	15	17.5	15	14
7 – DA – 007	14	10	20	20	17.5	12
8 – DA – 008	21	15	30	30	25	10
9 – DA – 009	27	20	40	40	35	10
10 – DA – 010	34	25	50	50	45	8
11 – DA – 011	40	30	50	60	50	8
12 – DA – 012	52	40	70	80	70	6
13 – DA – 013	65	50	90	100	90	4
14 – DA – 014	77	60	100	125	100	3
15 – DA – 015	96	75	125	150	125	2
16 – DA – 016	125	100	175	200	175	1/0
17 – DA – 017	160	125	200	250	200	2/0
18 – DA – 018	180	150	225	300	250	3/0
19 – DA – 019	240	200	300	350	300	(2) 2/0
20 – DA – 020	300	250	400	450	400	(2) 4/0
21 – DA – 021	360	300	450	600	450	(3) 2/0

460 VAC CONTROLS (3φ) WIRE SIZE AND PROTECTION DEVICES

* In amps
† AWG
‡ All sizes based on 167°F (75°C) rated copper wire and 104°F (40°C) ambient temperature

MANUFACTURER CHART

Activity 2-6. Lockout/Tagout

Determine if a lockout/tagout is required for each service call.

_____ **1.** Is lockout/tagout required for service call 1?

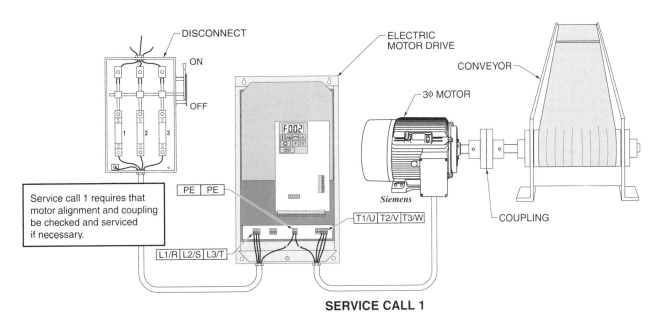

Service call 1 requires that motor alignment and coupling be checked and serviced if necessary.

SERVICE CALL 1

_____ **2.** Is lockout/tagout required for service call 2?

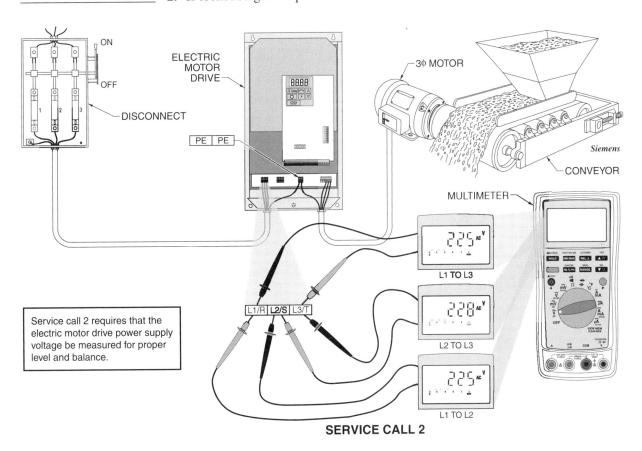

Service call 2 requires that the electric motor drive power supply voltage be measured for proper level and balance.

SERVICE CALL 2

_____   **3.** Is lockout/tagout required for service call 3?

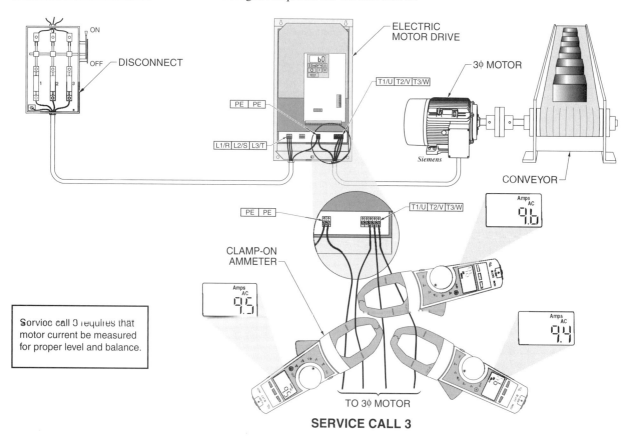

Sorvioc call 3 requires that motor current be measured for proper level and balance.

SERVICE CALL 3

_____   **4.** Is lockout/tagout required for service call 4?

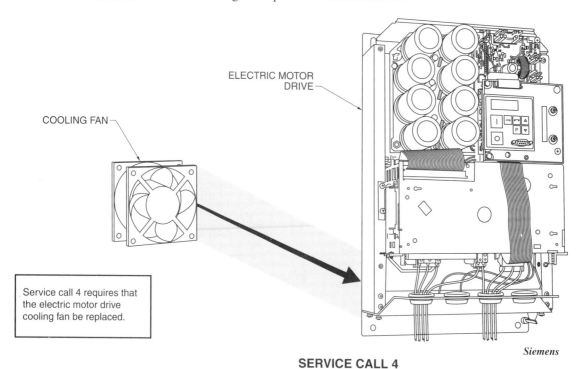

Service call 4 requires that the electric motor drive cooling fan be replaced.

SERVICE CALL 4

Electric Motor Drive
Installation and Troubleshooting

3 Electric Motor Load and Power Requirements

All electric motors produce work. Electric motor drives control motors to accomplish work more efficiently and with less wear and tear on machinery. Motors are rated in horsepower, although the torque produced by a motor shaft determines how much work is accomplished. The relationship between motor torque, speed, and horsepower, in addition to the power sources that deliver the required voltages, allow electric motor drives and motors to be used correctly and safely.

WORK

Work is a force (measured in lb) applied at a distance (measured in ft). Electric motors are used to produce work. *Force* is a form of energy that changes the position, motion, direction, or shape of an object. Work is accomplished when a force overcomes a resistance. See Figure 3-1. *Resistance* is any force that tends to hinder the movement of an object. If an applied force does not cause motion, no work is produced. Work is found by applying the formula:

$$W = F \times D$$

where

W = work (in lb-ft)

F = force (in lb)

D = distance (in ft)

For example, what is the work produced when a 65 lb load (force) is moved a distance of 6 ft?

$$W = F \times D$$
$$W = 65 \times 6$$
$$W = \textbf{390 lb-ft}$$

TECH FACT

Motors must produce the required torque to overcome a load. Motor loads consist of friction, the inertia of all moving parts, and the load itself. The amount of motor load depends upon the application in which the motor is used.

WORK

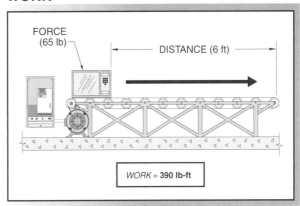

Figure 3-1. Work (lb-ft) is applying a force (lb) over a distance (ft).

POWER

Power is the rate of work (lb-ft) produced per unit of time (sec). Power is work divided by time. Power is expressed as the product of force and distance, divided by the time it takes to do the work. See Figure 3-2. Power is found by applying the formula:

$$P = \frac{W}{T}$$

where

P = power (in lb-ft/sec)

W = work (in lb-ft)

T = time (in sec)

For example, what is the power produced when a 65 lb load (force) is moved a distance of 6 ft in 5 sec?

$$P = \frac{W}{T}$$

$$P = \frac{390}{5}$$

$$P = \textbf{78 lb-ft/sec}$$

For example, what is the horsepower produced when a 65 lb force is moved a distance of 6 ft in 5 sec?

$$HP = \frac{P}{550}$$

$$HP = \frac{78}{550}$$

$$HP = \textbf{.14 HP}$$

POWER

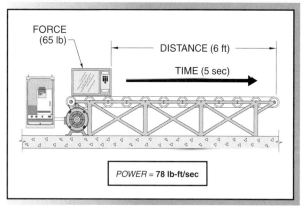

Figure 3-2. Power (lb-ft/sec) is the rate of work (lb-ft) produced per unit of time (sec).

HORSEPOWER

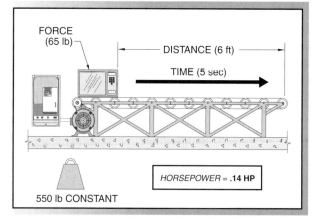

Figure 3-3. Horsepower is a unit for measuring power that equals 550 lb-ft/sec.

TECH FACT

Electrical energy supplied to the input of an electric motor drive is converted to mechanical energy at the motor shaft. No electric motor drive and motor system converts the energy at 100% efficiency. The total efficiency of the electric motor drive and motor system depends on energy losses (heat loss) within the drive and motor.

HORSEPOWER

Horsepower is a unit for measuring power. Horsepower equals 550 lb-ft/sec or 746 W electrically. See Figure 3-3. Horsepower is found by applying the formula:

$$HP = \frac{P}{550}$$

where

HP = power (in lb-ft/sec)

P = power (in lb-ft/sec)

550 = constant

TORQUE

Torque is the force that produces rotation. Torque causes an object attached to a motor shaft to rotate. Torque (lb-ft) consists of a force (lb) acting at a distance (ft). See Figure 3-4. Torque is measured in pound-feet (lb-ft). Torque, unlike work, may exist even though no movement occurs. Torque is found by applying the formula:

$$T = F \times r$$

where

T = torque (in lb-ft)

F = force (in lb)

r = radius (in ft)

For example, what is the torque produced by a 60 lb force pushing on a 3 ft lever arm?

$$T = F \times r$$

$$T = 60 \times 3$$

$$T = \textbf{180 lb-ft}$$

Work is accomplished when the amount of torque produced is large enough to cause movement. Torque increases as the force becomes larger, as the radius becomes longer, or a combination of the two.

TORQUE

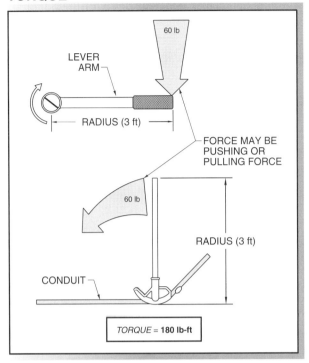

Figure 3-4. Torque (lb-ft) is the force that produces rotation.

Motors are sized according to horsepower, and the turning force of machinery shafts is rated by torque. Motor shafts produce torque when the motor is connected to a proper power supply. The amount of torque produced by a motor depends upon the motor type and the control device used. An electric motor drive used to control a motor not only controls the speed of the motor, but also determines the amount of torque the motor produces. Understanding torque is essential for the proper selection of any electric motor drive product.

MOTOR TORQUE

Motor torque is the force that produces or tends to produce rotation of a motor shaft. A motor must produce enough torque to start and maintain load movement. Motors produce four types of torque: locked rotor torque, pull-up torque, breakdown torque, and full-load torque. See Figure 3-5.

Locked Rotor Torque

Locked rotor torque (LRT) is the torque a motor produces when the shaft (rotor) is stationary and full power is applied to the motor. Locked rotor torque can also be the torque required to start a motor turning. Locked rotor torque or starting torque is usually expressed as a percentage of full-load torque. Many loads require a higher torque to start movement than to maintain movement. Motors can safely produce a higher torque output than the rated full-load torque for short periods of time, allowing a higher torque to be produced for starting the load.

Pull-Up Torque

Pull-up torque (PUT) is the torque required to bring a load up to its rated speed. Pull-up torque is also referred to as acceleration torque. If a motor is properly sized for the load, pull-up torque is needed only briefly. A motor that does not have sufficient pull-up torque may have adequate locked rotor torque to start the load turning but cannot bring the load up to rated speed (rpm). Once the motor is up to its rated speed, full-load torque keeps the load turning.

Breakdown Torque

Breakdown torque (BDT) is the maximum torque a motor can provide without an abrupt reduction in motor speed. An increasing load on a motor shaft requires that the motor produce more torque. As the load continues to increase, the point at which the motor stalls is reached. This point is the breakdown torque.

Atlas Technologies, Inc.
The breakdown torque of an electric motor must be greater than the torque required to rotate a hydraulic pump operating at maximum pressure.

MOTOR TORQUE

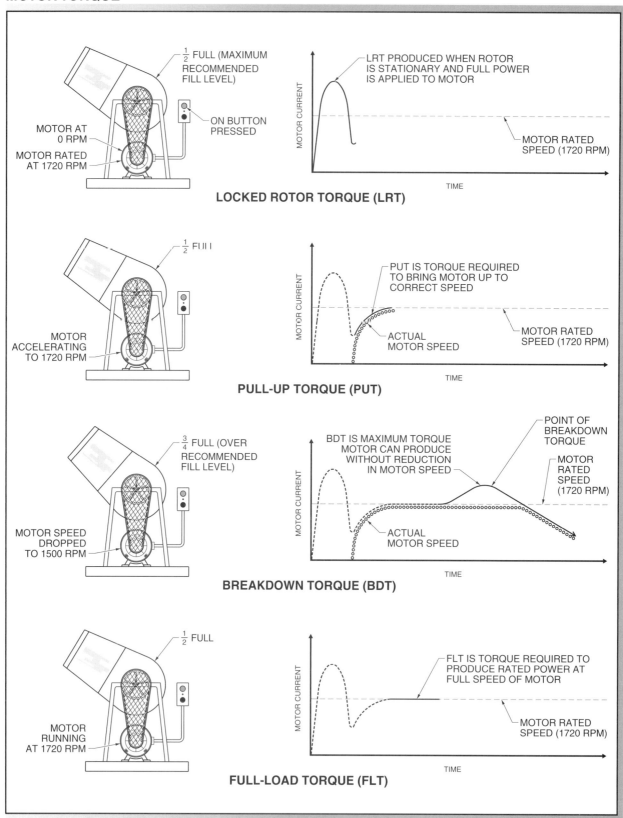

Figure 3-5. Motors must produce enough torque to start and maintain load movement.

Full-Load Torque

Full-load torque (FLT) is the torque required to produce the rated power at full speed of the motor. The amount of torque a motor produces at rated power and full speed (full-load torque) is found by using a horsepower to torque conversion chart, or applying a formula. See Figure 3-6.

HORSEPOWER TO TORQUE CONVERSION

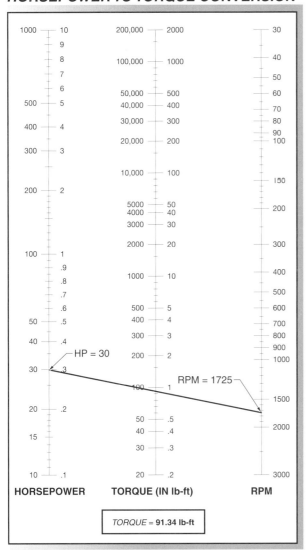

TORQUE = 91.34 lb-ft

Figure 3-6. The amount of torque (lb-ft) a motor produces at full speed (rpm) and rated power (HP) can be found by using a conversion chart.

To use the conversion chart, a straightedge is placed along the two known quantities and the unknown quantity is read on the third line. To determine torque when horsepower and speed are known, a straight line is drawn from the horsepower to speed (rpm). The point at which the line crosses the torque value is the full-load torque of the motor. Motor full-load torque can also be found by applying the formula:

$$T = \frac{HP \times 5252}{rpm}$$

where

T = torque (in lb-ft)

HP = horsepower

5252 = constant

rpm = revolutions per minute

For example, what is the FLT of a 30 HP motor operating at 1725 rpm?

$$T = \frac{HP \times 5252}{rpm}$$

$$T = \frac{30 \times 5252}{1725}$$

$$T = \frac{157,560}{1725}$$

T = **91.34 lb-ft**

A motor that is fully loaded produces full-load torque. A motor that is underloaded produces less than full-load torque. A motor that is overloaded must produce more than full-load torque to keep the load operating at rated speed but cannot exceed breakdown torque. See Figure 3-7.

For example, a 30 HP motor operating at 1725 rpm develops 91.34 lb-ft of torque. The 30 HP motor produces an output of 30 HP to develop 91.34 lb-ft of torque at 1725 rpm. The 30 HP motor produces an output of 15 HP to develop 45.67 lb-ft of torque at 1725 rpm, drawing less current (less power) from the power lines and operating at lower temperatures.

Connecting the 30 HP motor to a load that requires twice as much torque (182.68 lb-ft) at 1725 rpm forces the motor to produce an output of 60 HP. The 30 HP motor draws more current (more power) from the power lines and operates at higher temperatures. The overload protection devices automatically disconnect the motor from the power lines before any permanent damage is done.

TECH FACT

In the United States, the standard for rating power is horsepower (HP). The International Electromechanical Commission standard for rating power is the kilowatt (kW). One horsepower equals 746 W or .75 kW.

MOTOR TORQUE, SPEED, AND HORSEPOWER CHARACTERISTICS

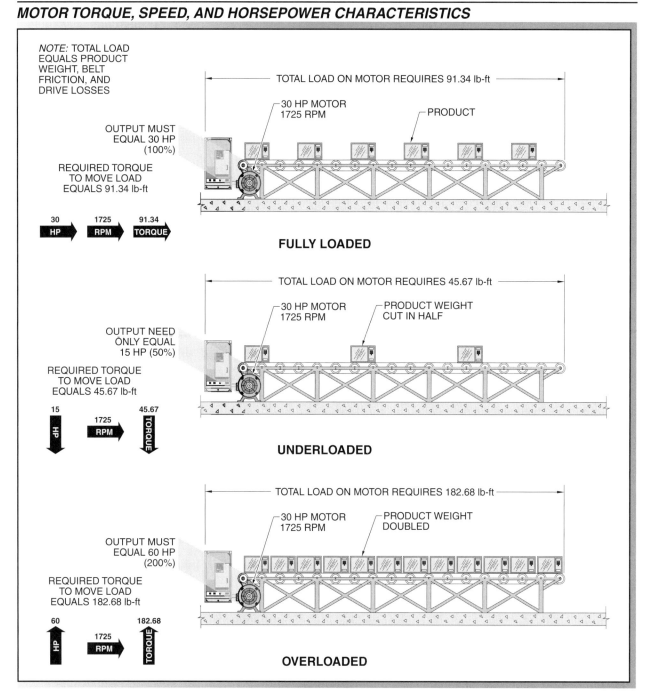

Figure 3-7. Motors may be fully loaded, underloaded, or overloaded.

MOTOR TORQUE CLASSIFICATIONS

Motors produce torque but not all motors produce the same torque characteristics. The National Electrical Manufacturers Association (NEMA) classifies motors according to electrical characteristics. Torque characteristics of a motor vary with the classification of the motor.

Motors are classified as Class A through E. Classes B, C, and D are most common, but manufacturer catalogs also list Class A.

Class B motors are the most widely used motor type. Class B motors increase the starting torque of the motor to 150% of the motor's full-load torque. As a Class B motor accelerates, pull-up torque can reach 240% of the motor's full-load torque. A Class C motor increases the starting torque of a motor by 225% and can have a pull-up torque as high as 200%. A Class D motor increases the starting torque of a motor by 275%. Although a Class D motor increases the starting torque the most, a Class D pull-up torque never increases above the starting torque value. See Figure 3-8.

MOTOR TORQUE CLASSIFICATION

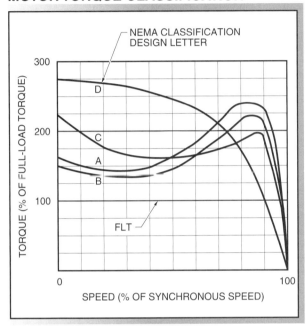

Figure 3-8. Motors are classified according to electrical characteristics.

MOTOR POWER

Motor power is the strength of a motor and is measured in watts or horsepower. A *watt (W)* is an electrical measurement equal to the power produced by 1 A of current across a potential difference of 1 V. A watt is ¹⁄₇₄₆ of 1 HP and is the base unit of electrical power. A *horsepower (HP)* is a unit of power equal to 746 W, 550 lb-ft/sec, or 33,000 lb-ft/min. See Figure 3-9.

Motor manufacturers use horsepower to measure the energy produced by a working motor. The horsepower of a motor, when current, efficiency, and voltage are known, is found by applying the formula:

$$HP = \frac{V \times I \times E_{ff}}{746}$$

where

HP = horsepower

V = voltage (in V)

I = current (in A)

E_{ff} = efficiency

746 = constant

For example, what is the horsepower of a 230 V motor pulling 4 A and having 82% efficiency?

$$HP = \frac{V \times I \times E_{ff}}{746}$$

$$HP = \frac{230 \times 4 \times 0.82}{746}$$

$$HP = \frac{754.4}{746}$$

$$HP = \textbf{1 HP}$$

The horsepower of a motor determines what size load a motor can operate and how fast the load rotates. The horsepower of a motor, when the speed and full-load torque are known, is found by applying the formula:

$$HP = \frac{rpm \times FLT}{5252}$$

where

HP = horsepower

rpm = revolutions per minute

FLT = full-load torque (in lb-ft)

5252 = constant

For example, what is the horsepower of a 1725 rpm motor with an FLT of 3.1 lb-ft?

$$HP = \frac{rpm \times FLT}{5252}$$

$$HP = \frac{1725 \times 3.1}{5252}$$

$$HP = \frac{5347.5}{5252}$$

$$HP = \textbf{1 HP}$$

Saftronics Inc.
Overhead cranes use 75 HP–125 HP motors for hoisting, 20 HP–25 HP brake motors for carriage movement, and 10 HP–25 HP gear motors for traveling.

MOTOR POWER

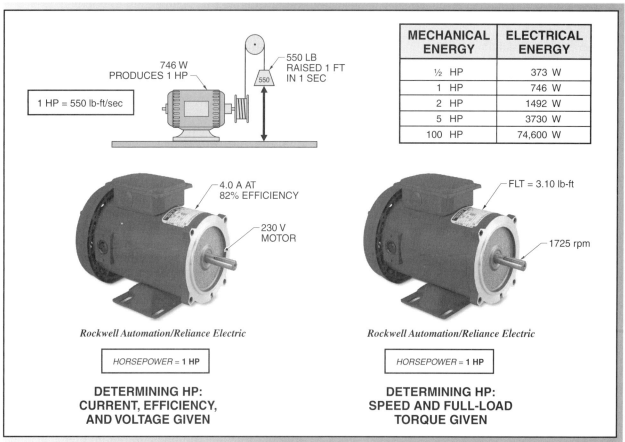

MECHANICAL ENERGY	ELECTRICAL ENERGY
½ HP	373 W
1 HP	746 W
2 HP	1492 W
5 HP	3730 W
100 HP	74,600 W

1 HP = 550 lb-ft/sec

746 W PRODUCES 1 HP

550 LB RAISED 1 FT IN 1 SEC

4.0 A AT 82% EFFICIENCY

230 V MOTOR

Rockwell Automation/Reliance Electric

HORSEPOWER = **1 HP**

DETERMINING HP: CURRENT, EFFICIENCY, AND VOLTAGE GIVEN

FLT = 3.10 lb-ft

1725 rpm

Rockwell Automation/Reliance Electric

HORSEPOWER = **1 HP**

DETERMINING HP: SPEED AND FULL-LOAD TORQUE GIVEN

Figure 3-9. Motor power is rated in watts or horsepower.

Torque and horsepower formulas are used to determine theoretical values. Motor loads that are difficult to start require higher ratings. Formulas applied to specific applications require an additional 15% to 40% capability to start the given load. To increase the rating, multiply the calculated theoretical value by 1.15 (for 15%) to 1.40 (for 40%).

$$HP = \frac{rpm \times FLT}{5252} \times percentage$$

where

HP = horsepower

rpm = revolutions per minute

FLT = full-load torque (in lb-ft)

$percentage$ = additional capacity

5252 = constant

For example, what is the horsepower of a 1725 rpm motor with an FLT of 3.1 lb-ft with an added 25% output capability?

$$HP = \frac{rpm \times FLT}{5252} \times percentage$$

$$HP = \frac{1725 \times 3.1}{5252} \times 1.25$$

$$HP = \frac{5347.5}{5252} \times 1.25$$

$$HP = 1.02 \times 1.25$$

$$HP = \textbf{1.28 HP}$$

RELATIONSHIP BETWEEN SPEED, TORQUE, AND HORSEPOWER

The operating speed, torque, and horsepower ratings determine the work a motor produces. The three factors are interrelated when applied to driving a load. See Figure 3-10. If the torque remains constant, speed and horsepower are proportional. Example A shows that, as speed increases, horsepower must increase to maintain a constant torque. Example B shows that, as speed decreases, the horsepower must decrease to maintain a constant torque.

MOTOR SPEED, TORQUE, AND HORSEPOWER RELATIONSHIP

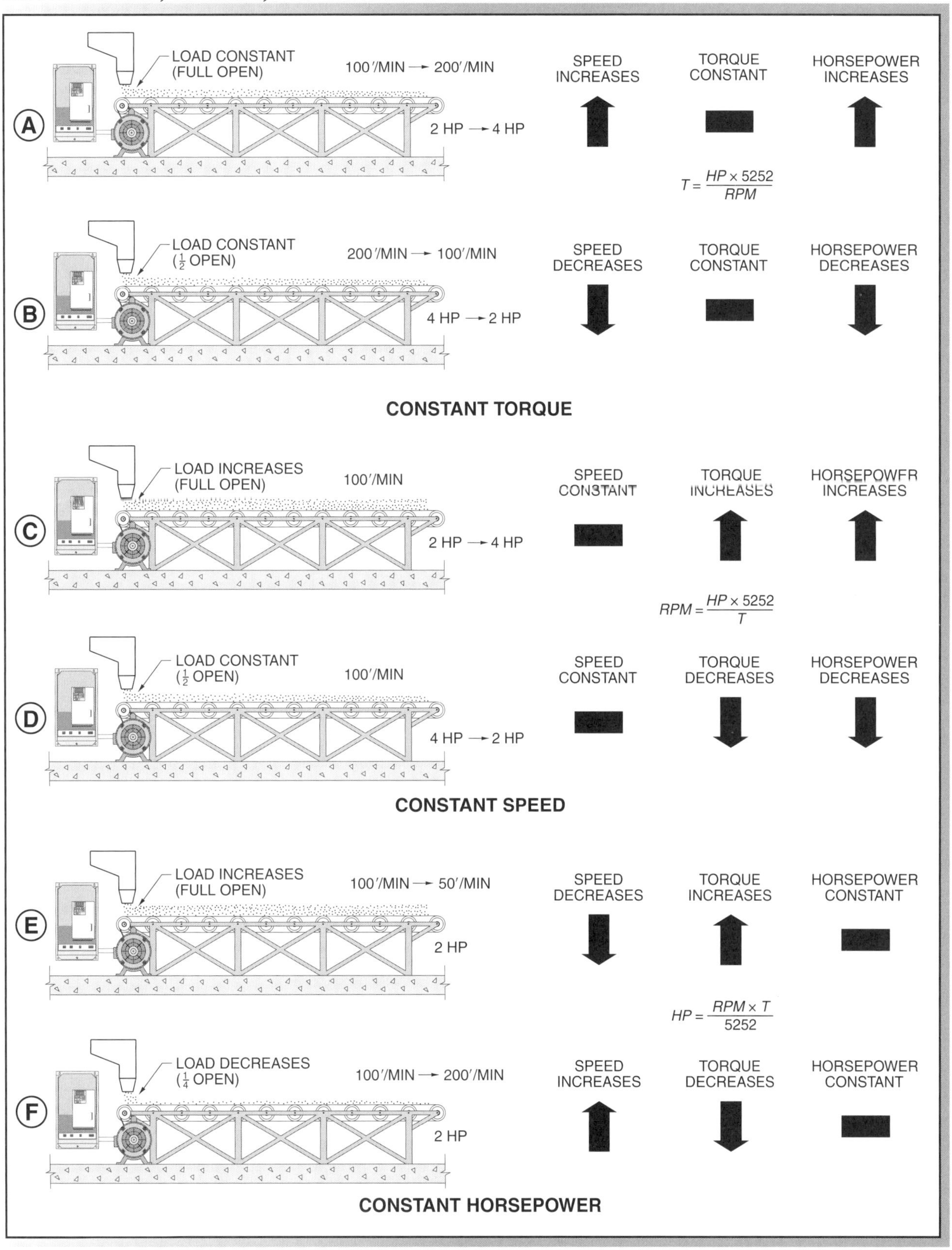

Figure 3-10. Operating speed, torque, and horsepower ratings determine the work a motor produces.

When speed remains constant, torque and horsepower are proportional. Example C shows that, as torque increases, the horsepower must increase to maintain a constant speed. Example D shows that, as torque decreases, the horsepower must decrease to maintain a constant speed.

When torque and speed vary simultaneously but in opposite directions, the horsepower remains constant. Example E shows that, as torque is increased and speed is reduced, horsepower remains constant. Example F shows that, as torque is reduced and speed is increased, horsepower remains constant.

MOTOR LOAD TYPES

Motors are used to move a variety of loads. A load may require constant torque, variable torque, or constant horsepower when operating at varying speeds. Each motor type has its own ability to control loads at varying speeds. The best type of motor to use for a given application depends upon the type of load the motor must move. Loads are classified as constant torque, constant horsepower, or variable torque. See Figure 3-11.

Constant Torque Load

A *constant torque (CT) load* is a load that requires the torque to remain constant. A constant torque load is sometimes called a constant torque/variable horsepower (CT/VH) load. A change in operating speed requires a change in horsepower. See Figure 3-12. Constant torque loads are the most common of all load types and include

MOTOR LOADS				
LOAD	MOTOR TORQUE*		CLASSIFICATION	NEMA MOTOR DESIGN
	LRT	PUT		
Ball mill (mining)	125–150	175–200	CT	C–D
Band saws Production Small	50–80 40	175–225 150	CT CT	C B
Car pullers Automobile Railroad	150 175	200–225 250–300	CH CH	C D
Chipper	60	225	CT	B
Compressor (air)	60	150	VT	B
Conveyors Unloaded at start Loaded at start Screw	50 125–175 100–125	125–150 200–250 50–175	CT CT CT	B C C
Crushers Unloaded at start With flywheel	75–100 125–150	150–175 175–200	CT CT	B D
Dryer, industrial (loaded rotary drum)	150–175	175–225	CT	D
Fan and blower	40	150	VT	B
Machine tools Drilling Lathe	40 75	150 150	CH CH	B B
Press (with flywheel)	50–100	250–350	CH	D
Pumps Centrifugal Positive displacement Propeller Vacuum pumps	50 60 50 60	150 175 150 150	VT CT VT CT	B B B B–C

* in % of FLT
CT = constant torque
CH = constant horsepower
VT = variable torque

Figure 3-11. Loads are classified as constant torque (CT), constant horsepower (CH), or variable torque (VT).

loads that produce friction. Examples of constant torque loads include conveyors, positive-displacement pumps, metal-cutting saws, load-lifting equipment, packaging machines, wire-drawing machines, saws, and other loads that operate at varying speeds.

When operating speeds vary, a constant torque load requires the same torque at low speeds as at high speeds. Since the torque requirement remains constant, an increase in speed requires an increase in horsepower.

CONSTANT TORQUE LOAD

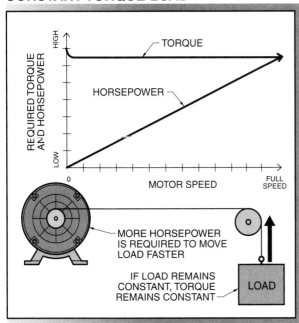

Figure 3-12. Constant torque (CT) loads are loads in which the torque remains constant.

Constant Horsepower Load

A *constant horsepower (CH) load* is a load that requires high torque at low speeds and low torque at high speeds. A constant horsepower load is sometimes called a constant horsepower/variable torque (CH/VT) load. As speed increases, horsepower remains constant and torque requirements decrease. Speed and torque are inversely proportional in constant horsepower loads. Examples of constant horsepower loads include drill presses, lathes, milling machines, and other types of machinery such as center-driven winders used to roll and unroll paper or metal materials. See Figure 3-13. Work performed on a varying diameter with constant tension and linear speed requires that horsepower be constant. Motor speed must vary to maintain a constant material linear speed. The diameter of the material driven by

the motor is constantly changing as material is added to the roll. At the start, the motor runs at high speed to maintain the correct material speed while torque is kept at a minimum. As the material is added, the motor must deliver more torque at a lower speed to provide constant horsepower.

CONSTANT HORSEPOWER LOAD

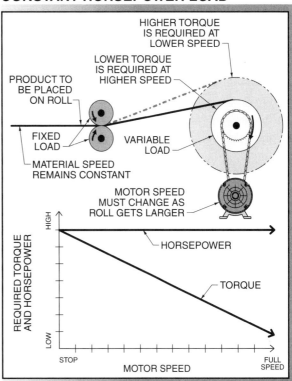

Figure 3-13. Constant horsepower (CH) loads are loads that require high torque at low speeds and low torque at high speeds.

Cincinnati Machine, a UNOVA Company
Constant horsepower is required to cut metal, whether the cut is completed in 30 sec or 5 min.

Variable Torque Load

A *variable torque (VT) load* is a load that requires varying torque and horsepower at different speeds. A variable torque load is often referred to as a variable torque/variable horsepower (VT/VH) load. See Figure 3-14. Variable torque loads require the motor to work harder to deliver more output at a faster speed. Both torque and horsepower are increased with increased speed. Examples of variable torque loads include fans, blowers, centrifugal pumps, mixers, and agitators.

VARIABLE TORQUE LOADS

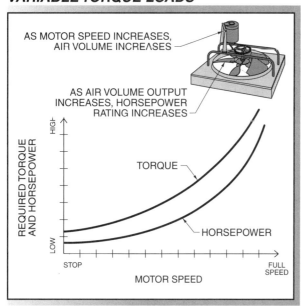

AS MOTOR SPEED INCREASES, AIR VOLUME INCREASES

AS AIR VOLUME OUTPUT INCREASES, HORSEPOWER RATING INCREASES

TORQUE

HORSEPOWER

REQUIRED TORQUE AND HORSEPOWER

HIGH

LOW

STOP

FULL SPEED

MOTOR SPEED

Figure 3-14. Variable torque (VT) loads are loads that require varying torque and horsepower at different speeds.

Saftronics Inc.
Centrifugal pumps are variable torque/variable horsepower loads that require heavy-duty, high-horsepower motors.

Variable torque loads change from a low torque at low speeds to a very high torque at high speeds. Doubling the speed of a fan or centrifugal pump increases the torque requirement by four times and the horsepower requirement by eight times. The Affinity Law that applies to fans and pumps states that as the speed of a centrifugal load increases, the horsepower requirement increases with the cube of the speed.

ELECTRICAL LOADS

An electrical load is any device that converts electrical energy into some other form of energy. The three basic types of electrical loads are resistive, inductive, and capacitive. Each type affects an electrical system in a different way. Electric motor drive and motor systems include a combination of all three types. Understanding each load type is required in order to understand an electrical system that includes electric motor drives and motors.

Resistive Loads

A *resistive load* is a load that contains only electrical resistance. *Resistance (R)* is the opposition to the flow of electrons. Resistance is measured in ohms and is represented by the Greek letter omega (Ω). Resistive loads are the simplest type of electrical loads. See Figure 3-15. The electrical characteristics of a resistive load circuit are the same regardless of whether an AC or DC voltage source is used. Voltage and current are in phase when a resistive load is connected to an alternating current (AC) power supply. *In phase* is the state when voltage and current reach their maximum amplitude and zero level simultaneously. Examples of resistive loads include heating elements and incandescent lamps.

Inductive Loads

An *inductive load* is a load that contains only electrical inductance. *Inductance (L)* is the property of a circuit that causes it to oppose a change in current due to energy stored in a magnetic field of a coil. The opposition to current change is the result of the current energy stored in the magnetic field of a coil. All coils (motor windings, transformers, solenoids, etc.) create inductance in an electrical circuit and oppose a change in current. A phase shift exists between alternating current and voltage in an inductive load. An inductive load circuit is a circuit in which current lags voltage. The greater the inductance of the load, the larger the phase shift. The unit of inductance is the henry (H). See Figure 3-16.

RESISTIVE CIRCUIT

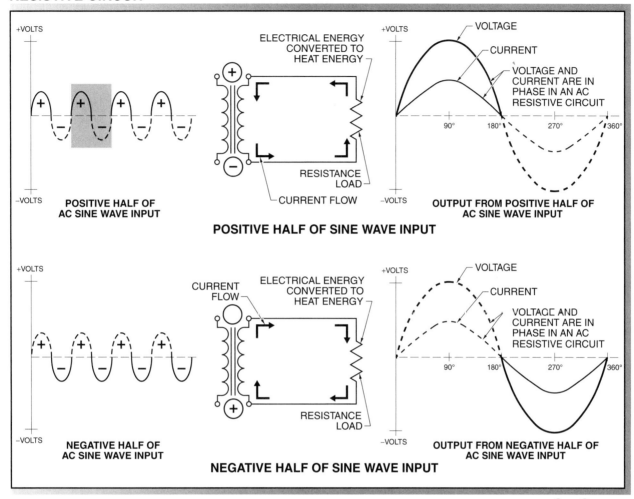

Figure 3-15. In a resistive load circuit, voltage and current are in phase.

SEE APPENDIX FOR ⊕ TO ⊖ DIAGRAM

Inductive Reactance. *Inductive reactance (X_L)* is an inductor's opposition to alternating current. Similar to resistance, inductive reactance is measured in ohms (Ω). The amount of inductive reactance in a circuit depends on:
• The amount of inductance of the inductor (coil). Inductance is normally a fixed amount for any given load.
• The frequency of the current. Frequency may be a fixed or variable amount. The higher the frequency or greater the inductance, the higher the inductive reactance.

Capacitive Loads

A *capacitive load* is a load that contains only electrical capacitance. *Capacitance (C)* is the ability of a component or circuit to store energy in the form of an electrical charge. Capacitors are a capacitive load and create capacitance in an electrical circuit. A *capacitor* is an

electric device specifically designed to store a voltage charge of energy. The unit of capacitance is the farad (F). A farad is a unit that is too large for most capacitance applications. Capacitance is typically stated in microfarads (μF) or picofarads (pF). A microfarad is one millionth of a farad and a picofarad is one trillionth of a farad.

TECH FACT

Electric motor drives may include capacitors connected between the main circuit and the frame (ground) of the drive. These capacitors increase the ground leakage current through the ground connection from the power supply and may cause ground fault circuit breakers to trip.

INDUCTIVE CIRCUIT

+VOLTS

+

+ + +

−

−VOLTS

**POSITIVE HALF OF
AC SINE WAVE INPUT**

ELECTRICAL ENERGY CONVERTED TO
HEAT ENERGY AND A MAGNETIC
FIELD PRODUCED

+

−

INDUCTIVE LOAD
CURRENT FLOW
NOTE: INDUCTIVE LOAD OFFERS
OPPOSITION TO CHANGE IN
CURRENT FLOW

+VOLTS

VOLTAGE
CURRENT LAGS
VOLTAGE BY 90°
IN AN AC INDUCTIVE
CIRCUIT

90° 180° 270° 360°

CURRENT

−VOLTS

**OUTPUT FROM POSITIVE HALF OF
AC SINE WAVE INPUT**

POSITIVE HALF OF SINE WAVE INPUT

+VOLTS

+ + + +

−

−VOLTS

**NEGATIVE HALF OF
AC SINE WAVE INPUT**

ELECTRICAL ENERGY CONVERTED TO
HEAT ENERGY AND A MAGNETIC
FIELD PRODUCED

CURRENT FLOW
−

+

INDUCTIVE LOAD
NOTE: INDUCTIVE LOAD OFFERS
OPPOSITION TO CHANGE IN
CURRENT FLOW

+VOLTS

VOLTAGE
CURRENT LAGS
VOLTAGE BY 90°
IN AN AC INDUCTIVE
CIRCUIT

90° 180° 270° 360°

CURRENT

−VOLTS

**OUTPUT FROM NEGATIVE HALF OF
AC SINE WAVE INPUT**

NEGATIVE HALF OF SINE WAVE INPUT

SEE APPENDIX FOR ⊕ TO ⊖ DIAGRAM

Figure 3-16. In an inductive load circuit, current lags voltage because an inductor opposes a change in current.

Kebco Power Transmission
Capacitors of various sizes are used in electric motor drives for filtering and to block DC voltages.

A phase shift exists between voltage and current in a capacitive circuit. A capacitive circuit is a circuit in which current leads voltage. The greater the capacitance in a circuit, the larger the phase shift. See Figure 3-17.

Capacitive Reactance. *Capacitive reactance (X_C) is the opposition to current flow by a capacitor.* Along with inductive reactance (X_L), capacitive reactance is expressed in ohms (Ω). The amount of capacitive reactance in a circuit depends on:

• The amount of capacitance (farads) of the capacitor. Capacitance is typically a fixed amount for any given capacitor.
• The frequency of the current. Frequency may be a fixed or variable amount. The higher the frequency or greater the capacitance, the lower the capacitive reactance.

CAPACITIVE CIRCUIT

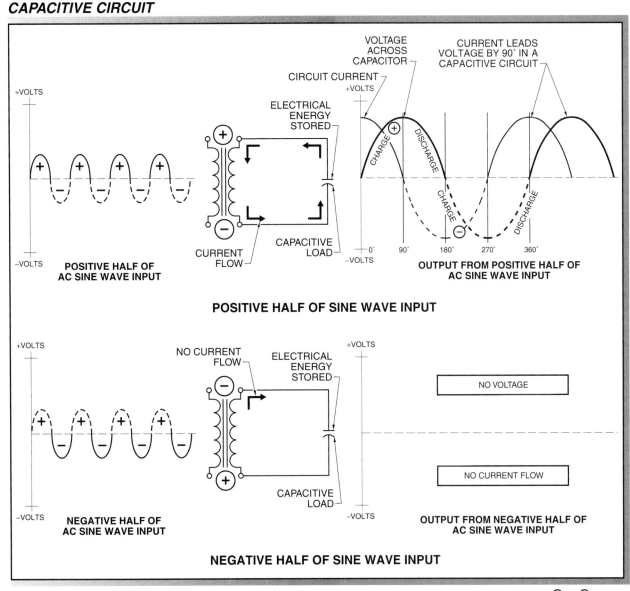

SEE APPENDIX FOR ⊕ *TO* ⊖ *DIAGRAM*

Figure 3-17. In a capacitive load circuit, current leads voltage because a capacitor opposes a change in voltage.

Impedance

Impedance (Z) is the total opposition of resistance, inductive reactance, and capacitive reactance offered to the flow of alternating current. Inductors and capacitors are electrical energy-storing devices used in AC and DC applications such as motors, electric motor drives, and filtering circuits. Inductors are used to store current charges. Capacitors are used to store voltage charges. When used together, inductors and capacitors each contribute to a circuits operation. Capacitors and inductors are used in filter circuits to smooth pulsating direct current. See Figure 3-18.

AC circuits use resistance (R), inductive reactance (X_L), and capacitive reactance (X_C) to limit current flow. Typically, AC circuits contain all three. The exact behavior of current in an AC circuit depends on the amount of resistance, inductive reactance, and capacitive reactance. The combined opposition to current flow in an AC circuit is impedance. Impedance, as with inductive reactance and capacitive reactance, is measured in ohms.

Ohm's law $(E = I \times R, I = E \div R, R = E \div I)$ is limited to circuits where electrical resistance is the only significant opposition to the flow of current. Ohm's law is used in DC circuits and AC circuits that do not contain a significant amount of inductance and/or capacitance. Circuits that contain impedance use Ohm's law with Z substituted for R in the formula. See Figure 3-19. See Appendix.

FILTER CIRCUITS

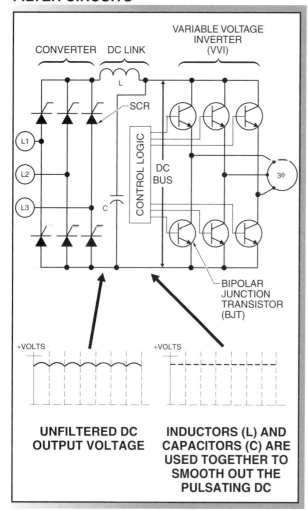

UNFILTERED DC OUTPUT VOLTAGE

INDUCTORS (L) AND CAPACITORS (C) ARE USED TOGETHER TO SMOOTH OUT THE PULSATING DC

Figure 3-18. Inductors are used to store a current charge and capacitors are used to store a voltage charge for filtering of the DC bus voltage.

POWER REACTORS

A *power reactor* is a device used to condition the supplied power to an electric motor drive and/or motor and protect the diodes, transistors, and other electronic components inside the drive from damage caused by transient voltages. Power reactors are installed in series with the power lines delivering voltage to the electric motor drive (line reactor) and/or between the drive output and the motor (load reactor). See Figure 3-20.

A power reactor is used to add impedance to a circuit. Adding impedance cleans and reduces the amount of current that causes damage when transient voltages occur. A transient voltage is a temporary, unwanted voltage in an electrical circuit. Power reactors are added to the electric motor drive supply circuit and motor circuit to limit transient voltages.

Electric motor drives and motors are designed to be supplied with power that is of the proper type (DC, 1ϕ AC, or 3ϕ AC), level (230 V, 460 V, etc.), amount (15 A, 50 A, etc.), and that is as clean as possible. Power delivered to an electric motor drive and motor that is not correct and clean affects performance and/or permanently damages the drive and motor.

Line and load reactors are used:
• When the power lines supplying the electric motor drive have power factor correction capacitors connected in the lines. Line reactors should be connected between the power factor correction capacitors and the electric motor drive.
• When the impedance of the power lines supplying the electric motor drive does not meet the drive manufacturer's minimum requirements for the drive. Line reactors should be connected to provide the needed circuit impedance. Electric motor drive manufacturer minimum required line impedance is 1% to 3%, depending upon the drive size.
• When the input power that is supplying power to the electric motor drive and motor exceeds 500 kVA, or is 10 times the kVA rating of the drive.
• To reduce the total harmonic distortion (THD) on the power supply lines.
• When there is a good chance of sudden voltage variations (sags and/or swells) in the supply voltage.
  Load reactors are used:
• To protect the electric motor drive from a short circuit at the motor.
• To limit the rate of rise of motor-produced surge currents.
• To slow the rate of change in the power that the electric motor drive delivers to the motor.
• To protect the motor insulation against electric motor drive short circuits.
• To reduce reflective wave damage and reduce harmonic distortion to the motor.

⚠ WARNING

Electric motor drive parameter settings and external control signals may allow a drive to automatically start a motor after an input power failure.

AC CIRCUIT FORMULAS

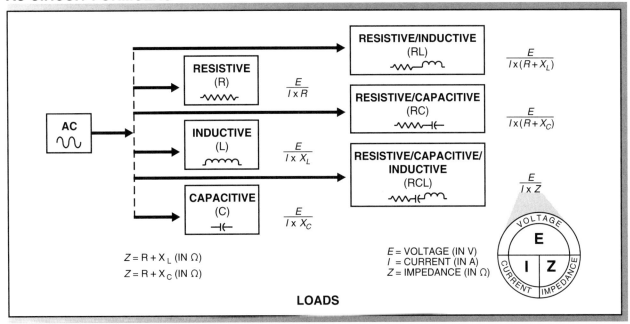

Figure 3-19. Ohm's law is used on circuits with impedance by substituting Z (impedance) for R (resistance) in the formula.

POWER REACTORS

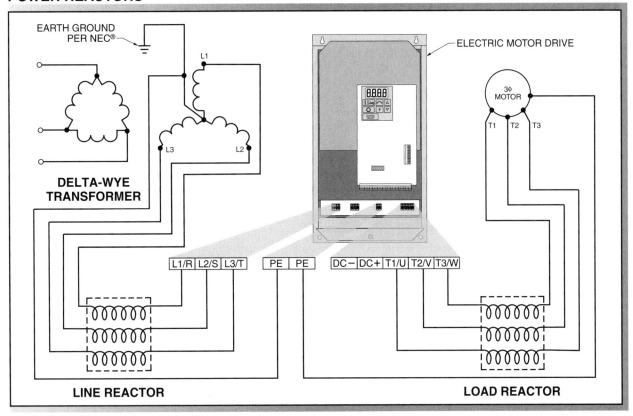

Figure 3-20. Reactors are used to condition the supply voltage (line reactor) to the electric motor drive and the load voltage (load reactor) to the motor.

The size of line reactors and load reactors is based on the electric motor drive rated nameplate current (full-load amps). Ensure the current rating taken from the motor's nameplate is the current listed for the voltage that is applied to the motor. Some motors have dual voltage ratings and dual current ratings. Line reactors and load reactors are installed as close to the electric motor drive as possible.

Reactors are relatively inexpensive when compared with the total cost of the electric motor drive and motor system, and are available from most drive manufacturers. AC line reactors are typically specified according to their voltage (230 V, 460 V, etc.), rated horsepower (1 HP, 50 HP, etc.), current (3.6 A, 130 A, etc.), inductance (3 mH, 0.10 mH, etc.), and percent impedance (2.94% Z, 3.54% Z). Check with the electric motor drive manufacturer for the best reactor to use with the drive and motor selected. See Figure 3-21.

The input impedance (Z) of the power lines can be determined by measuring the line-to-line voltage at no load (motor off) and at full load (motor full speed) and applying the formula:

$$Z_\% = \frac{V_{NL} - V_{FL}}{V_{NL}} \times 100$$

where

$Z_\%$ = input impedance of power lines (in %)

V_{NL} = measured no-load power line voltage when motor is off (in V)

V_{FL} = measured full-load power line voltage when motor is at full speed (in V)

100 = constant

Line and load reactors are used to reduce the effect of transient voltages and to clean the current to electric motor drives or motors.

AC REACTOR RATINGS				
AC REACTOR PART NUMBER	MOTOR RATED HP	CURRENT (A)	IMPEDANCE AS % OF Z	INDUCTANCE IN mH
200–230 V, 3ϕ LOW VOLTAGE				
1LV	1	3.6	2.94	3
2LV	3	9.6	3.26	1.25
3LV	5	15.2	3.31	0.8
4LV	10	28.0	3.05	0.4
5LV	25	68.0	3.70	0.2
6LV	40	104.0	4.24	0.15
7LV	50	130.0	3.54	0.10
8LV	60	154.0	3.14	0.08
9LV	75	192.0	2.87	0.06
400–460 V, 3ϕ HIGH VOLTAGE				
1HV	1	1.8	2.94	12
2HV	3	4.8	4.24	6.5
3HV	5	7.6	3.1	3.0
4HV	10	14.0	2.86	1.5
5HV	25	34.0	3.70	0.8
6HV	40	52.0	3.54	0.5
7HV	50	65.0	3.54	0.4
8HV	60	77.0	4.19	0.4
9HV	75	96.0	3.92	0.3

Figure 3-21. AC reactors are typically specified according to rated voltage, current, horsepower, impedance, and inductance.

For example, what is the input impedance (Z) of a power line in which the supply voltage was measured at 462 V with no load, and at 456 V under full load?

$$Z_\% = \frac{V_{NL} - V_{FL}}{V_{NL}} \times 100$$

$$Z_\% = \frac{462 - 456}{462} \times 100$$

$$Z_\% = \frac{6}{462} \times 100$$

$$Z_\% = .012987 \times 100$$

$$Z_\% = \textbf{1.3\%}$$

Line and load reactors should be ordered from the electric motor drive manufacturer to ensure proper selection. The reactor ordered is primarily based on the motor's rated nameplate current (full-load amps).

To determine the minimum inductance (L) required for a 3ϕ line and/or load reactor, apply the formula:

$$L = \frac{V_{LL} \times Z_{\%MIN}}{I \times 1.73 \times 377}$$

where

L = minimum inductance (in H)

V_{LL} = measured line-to-line input voltage

$Z_{\%MIN}$ = desired percent of input impedance [use 3% (.03) unless otherwise specified by manufacturer]

I = current rating of drive (in A)

1.73 = square root of 3 (constant used for 3ϕ power)

377 = constant used for 60 Hz power supplies (use 314 for 50 Hz power supplies)

For example, what is the minimum inductance (L) required of a 3ϕ reactor used with an electric motor drive that is rated 35 A at 480 V, with a measured line-to-line voltage of 462 V?

$$L = \frac{V_{LL} \times Z_{\%MIN}}{I \times 1.73 \times 377}$$

$$L = \frac{462 \times 0.03}{35 \times 1.73 \times 377}$$

$$L = \frac{13.86}{22,827.35}$$

$$L = \textbf{.0006071 H (.6071 mH or .60 mH)}$$

ELECTRICAL SYSTEMS AND ELECTRIC MOTOR DRIVES

An *electrical power system* is a system that generates, transmits, distributes, and/or delivers electrical power to satisfactorily operate electrical loads designed for connection to the system. An electrical system that includes electric motor drives may be small and simple or large and complex. Small electric motor drives are used to control the speed of fractional horsepower motors that control the flow of air in a heating and air conditioning (HVAC) system. Large electric motor drives are used to control the movement of products along an entire assembly line that includes numerous motors ranging in size of 100 HP or more.

An electric motor drive is part of a facility's electrical power system. The electric motor drive is connected to the power supply distribution system and reconditions the power for delivery to one or more motors. Electric motor drives affect the electrical power system both upstream and downstream from the drive. Understanding the electrical system and where an electric motor drive fits into the electrical system is a must for the drive to be installed and used correctly and without causing problems to the drive, motor, or other components within the system. See Figure 3-22.

Regardless of the size of the electrical power system, power must be supplied that allows the electric motor drives and motors to operate satisfactorily without causing additional problems within the system. Damage to electrical equipment occurs if power is not supplied that has the proper level (voltage), amount (current), type (DC, 1ϕ AC, or 3ϕ AC), and condition (purity).

Electric motor drives affect the power supply distribution system and must be installed and used correctly.

ELECTRICAL SYSTEMS AND ELECTRIC MOTOR DRIVES

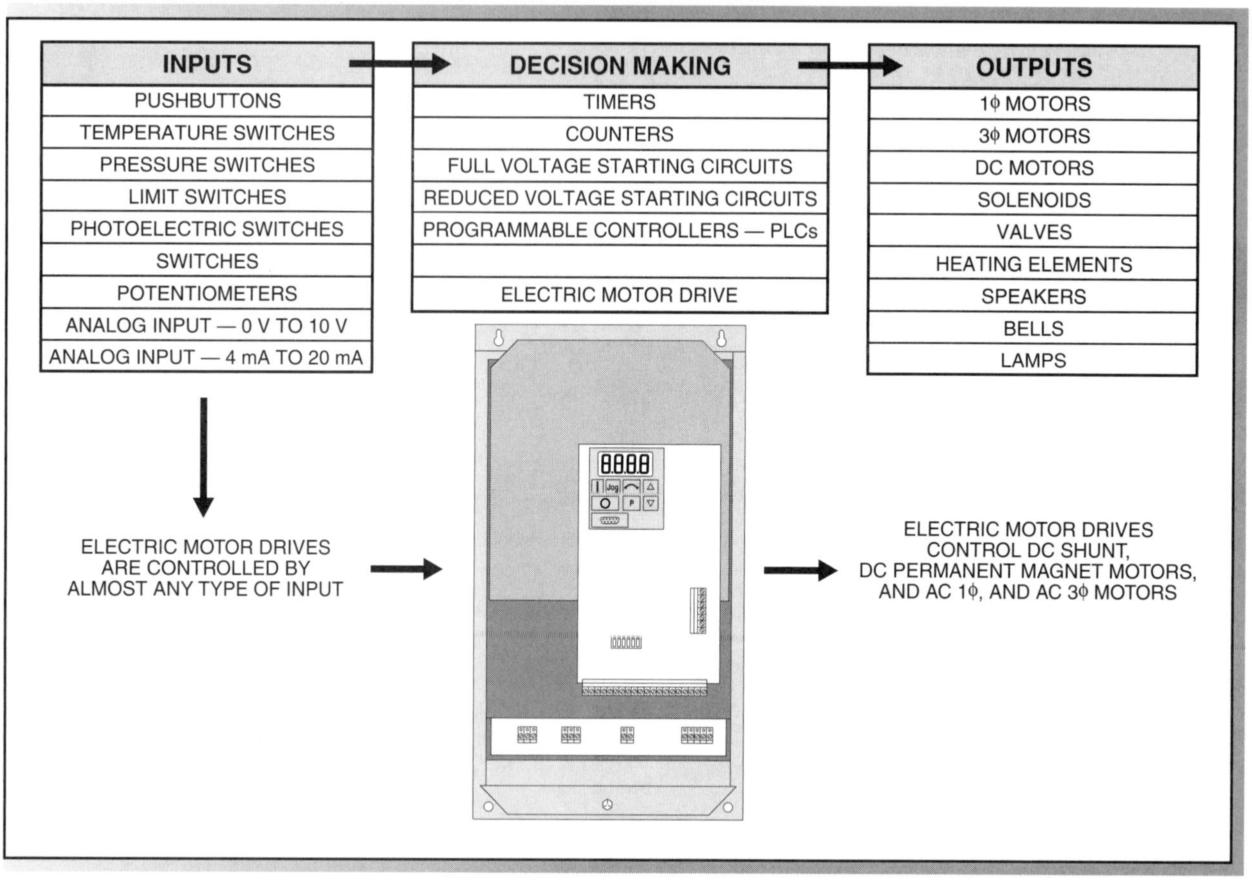

INPUTS	DECISION MAKING	OUTPUTS
PUSHBUTTONS	TIMERS	1φ MOTORS
TEMPERATURE SWITCHES	COUNTERS	3φ MOTORS
PRESSURE SWITCHES	FULL VOLTAGE STARTING CIRCUITS	DC MOTORS
LIMIT SWITCHES	REDUCED VOLTAGE STARTING CIRCUITS	SOLENOIDS
PHOTOELECTRIC SWITCHES	PROGRAMMABLE CONTROLLERS — PLCs	VALVES
SWITCHES		HEATING ELEMENTS
POTENTIOMETERS	ELECTRIC MOTOR DRIVE	SPEAKERS
ANALOG INPUT — 0 V TO 10 V		BELLS
ANALOG INPUT — 4 mA TO 20 mA		LAMPS

ELECTRIC MOTOR DRIVES ARE CONTROLLED BY ALMOST ANY TYPE OF INPUT

ELECTRIC MOTOR DRIVES CONTROL DC SHUNT, DC PERMANENT MAGNET MOTORS, AND AC 1φ, AND AC 3φ MOTORS

Figure 3-22. Electric motor drives are part of the electrical power system in a facility.

Power Quality Effects

A load (electric motor drive and motor) with proper electrical power supplied operates properly for its life expectancy unless there is a problem. A problem with the load results in the load requiring maintenance or replacement. A problem with a load is a safety hazard that results in production equipment damage and costly downtime. Performing maintenance or replacing the load is only a short-term solution when there is a problem with the incoming power.

Electric motor drives require correct incoming power and recondition that power for delivery to motors. The power into and out of an electric motor drive affects the electrical system, drive, and motor. An electric motor drive that is having maintenance performed or is being installed must have the quality of the incoming and outgoing power checked to ensure proper motor and system operation. Electric motor drives typically indicate a problem with the power supply, drive, or motor by removing power from the motor and displaying a fault message on the screen of the drive. See Figure 3-23. Fault messages that

are displayed when there is a power supply problem are unbalanced current in the 3φ motor leads, input phase loss (3φ), and power loss ride-through (500 ms).

Any time a power supply fault is detected and displayed on an electric motor drive, electrical measurements must be taken. Voltage and current measurements are the minimum measurements performed. Determining other problems requires that harmonic and transient voltage measurements be performed. *Harmonics* are frequencies that are whole-number multiples (second, third, fourth, fifth, etc.) of the fundamental frequency.

⚠ WARNING

When power is not required, lockout/tagout power and always measure voltage before working around exposed terminals, lugs, or parts of an electric motor drive because capacitors may still be charged. The drive power terminal strip may measure at high voltage levels when power is removed from an electric motor drive.

FAULT MESSAGES

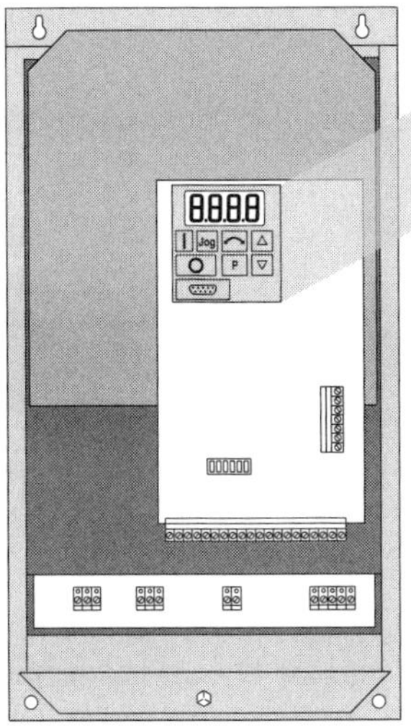

FAULT MESSAGE	POSSIBLE CAUSE	CORRECTIVE ACTION
GROUND FAULT	IMPROPER WIRING	1) DISCONNECT ALL POWER AND LOCKOUT/TAGOUT 2) CHECK FOR PROPER WIRING
	MOTOR WINDING SHORTED	1) LOOK FOR SIGNS OF SHORT
	POWER LINE SHORTED	1) TEST FOR SHORTS
OVERCURRENT TRIP		
OVERTEMPERATURE TRIP		
INPUT PHASE LOSS		
DC BUS HIGH		
DC BUS LOW	SEE TROUBLESHOOTING MATRICES	
LOST USER DATA		
MEMORY/ERROR		
OVERLOAD – 1 MIN		
OVERLOAD – 3 SEC		
OVERSPEED		

FAULT MESSAGE — ELECTRIC MOTOR DRIVES DISPLAY FAULT MESSAGES

OPERATION MANUAL LISTS CAUSE AND CORRECTIVE ACTIONS FOR FAULT MESSAGES

Figure 3-23. Electric motor drives display fault messages on the screen of the drive.

Phase Unbalance. *Phase unbalance (imbalance)* is the unbalance that occurs when the 3ϕ power lines are more than or less than 120° out of phase. Phase unbalance of a 3ϕ power system occurs when single-phase loads are applied, causing one or two of the lines to carry more of the load. Electricians balance the loads on the three phases of the power system during installation. An unbalance begins to occur as additional single-phase loads are added to the system. The unbalance causes the 3ϕ power lines to move out of phase so the lines are no longer 120° apart. See Figure 3-24.

Phase unbalance causes 3ϕ motors to operate at temperatures higher than the listed ratings. The greater the phase unbalance, the greater the temperature rise. High temperatures produce insulation breakdown and other related problems. A 3ϕ motor operating in an unbalanced circuit cannot deliver nameplate rated horsepower. A phase unbalance of 3% causes a motor to work at only 90% of its rated power so that it must be derated by 10%.

Voltage Unbalance. *Voltage unbalance* is the unbalance that occurs when the voltages at the three motor terminals or other 3ϕ loads are not equal. A motor that has voltage unbalance has one winding overheating, causing thermal deterioration of that winding. Voltage unbalance results in current unbalance. Line voltage must be checked for voltage unbalance periodically and during all maintenance calls. Voltage unbalance is typically not more than 1%. When there is a 2% or greater voltage unbalance, the following steps must be taken:

1. Check the surrounding power system for excessive loads connected to one supply line.
2. Adjust the load or motor rating by reducing the load on the motor or by oversizing the motor if the voltage unbalance cannot be corrected.
3. Notify the electrical utility company if the unbalance is at the main service entrance.

The primary source of voltage unbalance of less than 2% is single-phase loads on a 3ϕ circuit. Blown fuses can cause high-voltage unbalances in one phase of a 3ϕ capacitor bank. Voltage unbalance is found by applying the following procedure:

1. Measure the voltage between each incoming supply line. Take measurements from L1 to L2, L1 to L3, and L2 to L3. Add the voltages.
2. Find the average voltage by dividing by three.
3. Find the voltage deviation by subtracting the voltage average from the voltage with the largest deviation.
4. Find the voltage unbalance by applying the formula:

$$V_U = \frac{V_D}{V_A} \times 100$$

where

V_U = voltage unbalance (in %)
V_D = voltage deviation (in V)
V_A = voltage average (in V)
100 = constant

When a 3ϕ motor fails due to voltage unbalance, one or two of the stator windings become blackened. The darkest winding is the one with the largest voltage unbalance. See Figure 3-25.

Current Unbalance. *Current unbalance* is the unbalance that occurs when the current is not equal at the leads of a 3ϕ motor or other 3ϕ load. A small voltage unbalance causes a high current unbalance. A high current unbalance produces excessive heat which in turn can result in insulation breakdown. A voltage unbalance causes current unbalance at a rate of about 8:1. A 2% voltage unbalance results in a 16% current unbalance.

Current unbalances should not exceed 10%. A current unbalance that exceeds 10% requires checking for a voltage unbalance, and when there is a voltage unbalance of more than 1%, check for a current unbalance.

PHASE UNBALANCE

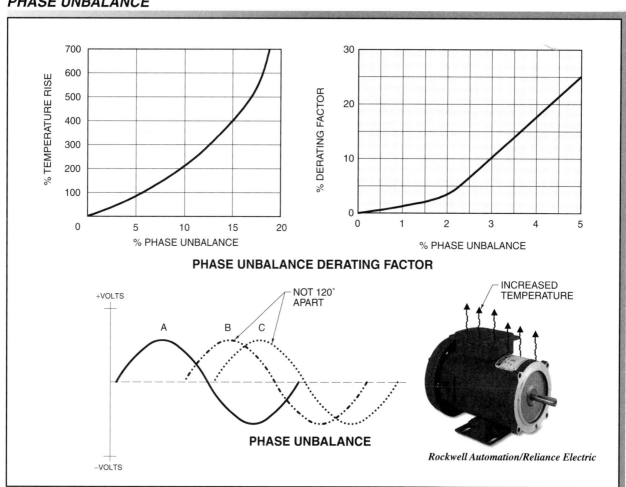

Figure 3-24. Phase unbalance is the unbalance that occurs when power lines are out of phase.

VOLTAGE UNBALANCE

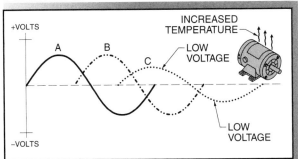

Figure 3-25. Voltage unbalance occurs when the voltage at the motor leads are not equal.

Current unbalance is found the same way as voltage unbalance is found, except current measurements are taken. Current unbalance is found by applying the following procedure:

1. Measure the current on each incoming power line. Take a current measurement on L1, L2, and L3.
2. Add the currents.
3. Find the current average by dividing by three.
4. Subtract the current average from the largest measurement to find the current deviation.
5. To find current unbalance, apply the formula:

$$I_U = \frac{I_D}{I_A} \times 100$$

where

I_U = current unbalance (in %)
I_D = current deviation (in A)
I_A = current average (in A)
100 = constant

Single-Phasing. *Single-phasing* is the operation of a 3ɸ load on two phases because one phase is lost. Single-phasing occurs when one of the 3ɸ lines to a 3ɸ motor does not deliver the required voltage. Single-phasing is the maximum condition of voltage unbalance. See Figure 3-26.

TECH FACT

Electrical unbalances raise amperage and temperatures of motors. Small-horsepower motors are more sensitive to electrical unbalances than high-horsepower motors. Standard efficiency motors are more sensitive to electrical unbalances than high-efficiency motors.

Electrical Apparatus Service Association, Inc.
Single-phasing causes severe burning and blackening of one phase of the stator windings.

SINGLE-PHASING

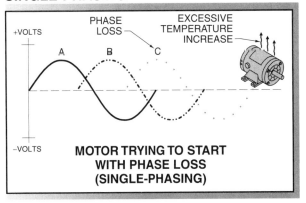

Figure 3-26. Single-phasing occurs when one of the three phase lines is lost.

Single-phasing occurs when one phase opens on either the primary or secondary power distribution system. Common causes of single-phasing include a blown fuse, a mechanical failure within the switching equipment, or lightning striking and knocking out one of the power lines. Single-phasing can go undetected on systems because a 3ɸ motor operating on two phases continues to operate in some applications. The motor will not start on two phases, but will continue to run if a phase is lost when the motor is already operating. A motor can operate until it burns out if not properly protected. A motor that is single-phasing draws all current from the two remaining supply lines.

Measuring voltage at a motor may not detect single-phasing. The open winding in the motor generates a voltage that is almost equal to the phase voltage that is lost. The open winding acts as the secondary of a transformer, and the two windings connected to the power source act as the primary.

Improper Phase Sequence. *Improper phase sequence* is the changing of the sequence of any two phases (phase reversal) in a 3φ motor control circuit. Improper phase sequence reverses motor rotation. Reversing motor rotation can damage driven machinery or injure personnel. See Figure 3-27.

POWER DISTRIBUTION SYSTEMS

A *power distribution system* is wires, wire ways, switchgear, and panels used to deliver the required type (DC, 1φ AC, or 3φ AC) and level (120 V, 230 V, 460 V, etc.) of electricity to loads connected to the system. Electric motor drives are supplied with power from branch circuits. A *branch circuit* is the portion of a distribution system between the final overcurrent protection device and the outlet (receptacle) or load. Motors are supplied with power from electric motor drives or motor starters.

The power lines supplying 3φ power to an electric motor drive are marked L1 (R), L2 (S), and L3 (T). The power lines supplying 1φ power to an electric motor drive are marked L and N. The power lines supplying DC power to an electric motor drive are marked DC+ and DC−. The 3φ power lines delivering power from an electric motor drive to the motor are marked T1 (U), T2 (V), and T3 (W). The DC power lines delivering power from an electric motor drive to the motor are marked A1+ and A2−. See Figure 3-28.

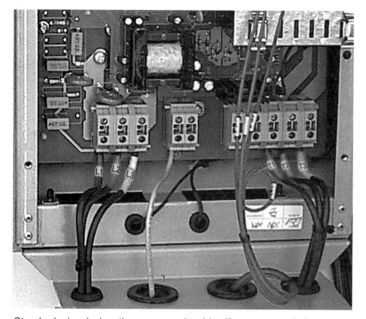

Standard wire designations are used to identify power supply lines to an electric motor drive and from the drive to the motor leads.

PHASE SEQUENCING

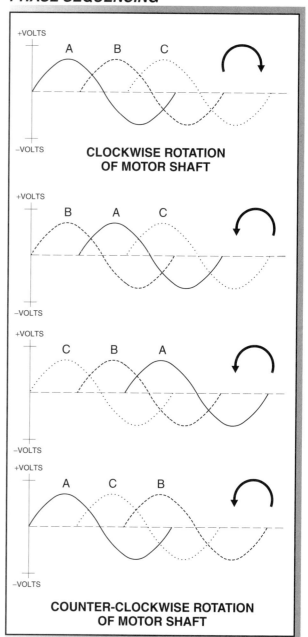

Figure 3-27. Improper phase sequence is the changing of the sequence of any two phases.

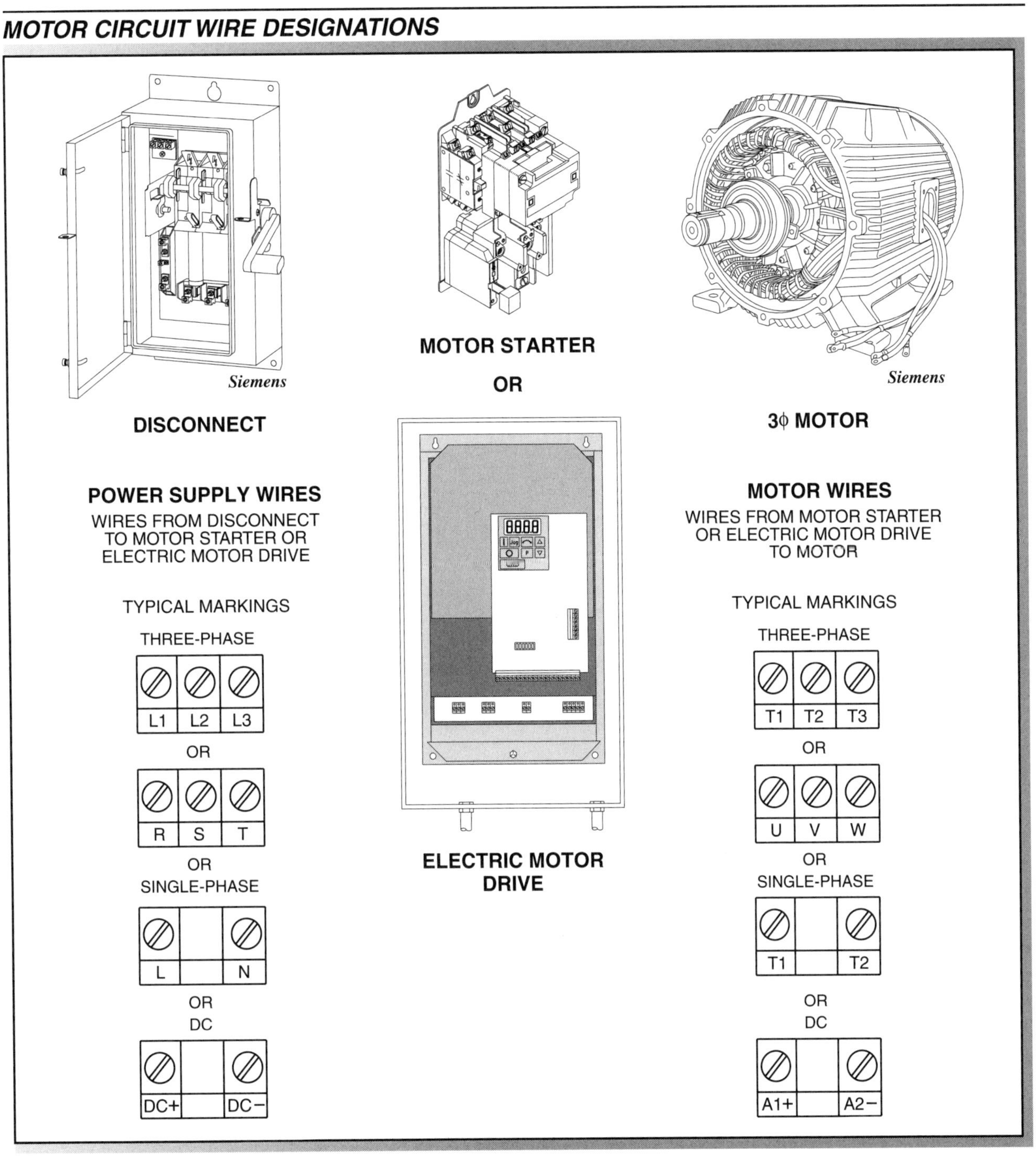

MOTOR CIRCUIT WIRE DESIGNATIONS

Figure 3-28. Electric motor drives and motor starters use a system of wire markings for the power supply wires and for the motor wires that distinguishes each wire.

Conductors are also color-coded. Conductor color-coding allows easier tracking of conductors, identification of types of service, load balancing among the different phases, and troubleshooting. Some colors have definite meaning. The color green always indicates a conductor used for grounding. A *grounding conductor* is a conductor that does not normally carry current, except during a fault (short circuit or ground fault). Other colors may have more than one meaning depending on the circuit. Some colors are required on conductors to meet NEC® requirements. NEC® Section 110.15 states in a 4-wire delta-connected secondary system, the higher voltage phase should be colored orange (or clearly marked) because it is too high for low-voltage single-phase power and too low for high-voltage single-phase power. See Figure 3-29.

TECH FACT

To achieve proper motor efficiency, a motor should be operated at 75% to 80% of its rating.

Use of 120/240 V, 1φ, 3-Wire Service

A 120/240 V, 1φ, 3-wire service is an electrical service used to supply power to customers that require large amounts of 120 V and 240 V, 1φ power. A 120/240 V, 1φ, 3-wire service includes two ungrounded (hot) wires and one neutral wire. The neutral wire is grounded and is not fused or switched at any point. See Figure 3-30.

CIRCUIT CONDUCTOR COLOR CODING

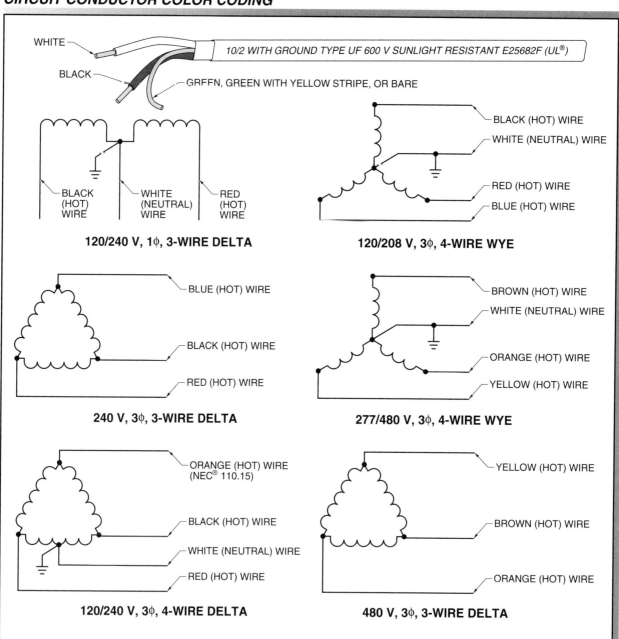

Figure 3-29. Conductors are color-coded for identification.

120/240 V, 1φ, 3-WIRE SERVICE

Figure 3-30. A 120/240 V, 1φ, 3-wire service is commonly used for wiring of lighting and small appliance circuits.

A 120/240 V, 1φ, 3-wire service is commonly used for interior wiring for lighting and small appliances. A 120/240 V, 1φ, 3-wire service is the primary service used to supply residential or small commercial applications. A 120/240 V, 1φ, 3-wire service is used to supply electric motor drives that operate HVAC systems. Electric motor drives are available that can be connected to either 120 V or 240 V 1φ power. The electric motor drive then converts the single-phase power into 3φ power for supplying 3φ motors.

NEC® Phase Arrangement

Three-phase circuits include three individual ungrounded (hot) power lines. The power lines are referred to as phases A (L1 or R), B (L2 or S), and C (L3 or T). Phases A, B, and C are connected to a switchboard or panelboard according to NEC® subsection 408.3(E). The phases must be arranged A, B, C from front to back, top to bottom, or left to right, as viewed from the front of the switchboard or panelboard. See Figure 3-31.

PHASE ARRANGEMENT AND HIGH PHASE MARKING

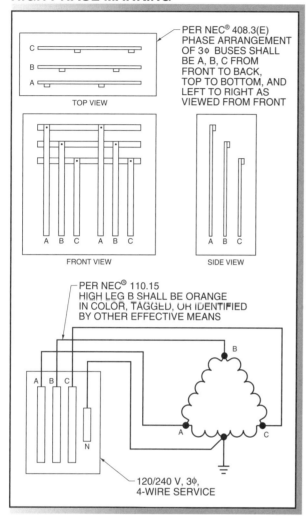

Figure 3-31. Phases inside a switchboard or panelboard are connected according to NEC® standards.

Phases are arranged in sequence from left to right: phase A, phase B, and phase C.

Use of 120/208 V, 3φ, 4-Wire Service

A 120/208 V, 3φ, 4-wire service is an electrical service used to supply customers that require a large amount of 120 V, 1φ power; 208 V, 1φ power; and low-voltage 3φ power. A 120/208 V, 3φ, 4-wire service includes three ungrounded (hot) lines and one grounded neutral line. Each hot line measures 120 V to ground when connected to the grounded neutral line. See Figure 3-32.

A 120/208 V, 3φ, 4-wire service is used to provide large amounts of low-voltage (120 V, 1φ) power. The 120 V circuits should be balanced to distribute the power equally among the three hot lines. This is accomplished by alternating connecting the 120 V circuits to the power panel so each phase (A to N, B to N, C to N) is divided among the loads. Heating elements and electric motor drive loads using 208 V, 1φ, must be balanced between phases (A to B, B to C, A to C).

⚠ WARNING

Ensure electric motor drives are properly grounded before applying power. An electric motor drive that is not properly grounded has dangerously high voltage levels between the conductive parts of the drive housing and ground due to high leakage current.

Use of 120/240 V, 3φ, 4-Wire Service

A 120/240 V, 3φ, 4-wire service is an electrical service used to supply customers that require large amounts of 3φ power by using all three transformers connected end-to-end. A 120/240 V, 3φ, 4-wire service works in commercial and industrial applications where the total amount of single-phase power used is small when compared to the total amount of 3φ power required. See Figure 3-33.

A 120/240 V, 3φ, 4-wire service supplies single-phase power by center-tapping one of the three transformers. Because only one transformer delivers all of the single-phase power, 120/240 V, 3φ, 4-wire service is used in applications that require mostly 3φ power or 240 V, 1φ power and some 120 V, 1φ power. Each transformer may be center-tapped if large amounts of 120 V, 1φ power are required.

120/208 V, 3φ, 4-WIRE SERVICE

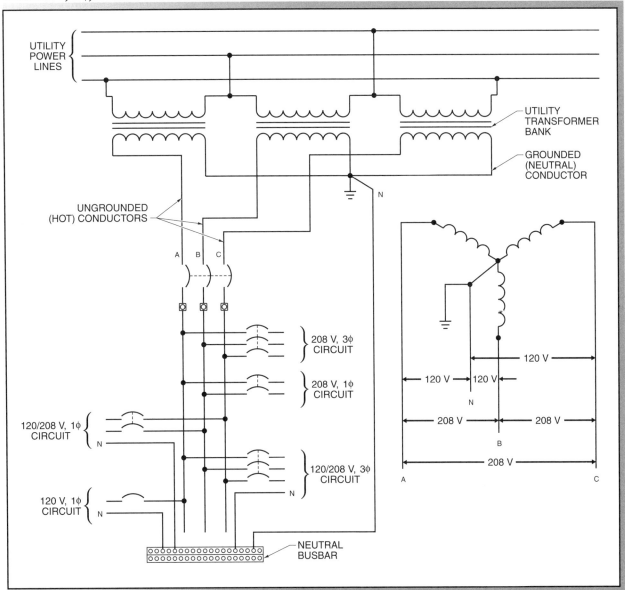

Figure 3-32. A 120/208 V, 3φ, 4-wire service is used to provide large amounts of low-voltage 120 V, 1φ power.

Use of 277/480 V, 3φ, 4-Wire Service

A 277/480 V, 3φ, 4-wire service is an electrical service that is the same as the 120/208 V, 3φ, 4-wire service except the voltage levels are higher. This service includes three ungrounded (hot) lines and one grounded neutral line. Each hot line has 277 V to ground when connected to the neutral or 480 V when connected between any two hot (A to B, B to C, or C to A) lines. See Figure 3-34.

A 277/480 V, 3φ, 4-wire service provides large amounts of power with 277 V, 1φ or 480 V 3φ power, but not 120 V, 1φ power. A 277/480 V, 3φ, 4-wire service cannot be used to supply 120 V, 1φ general lighting and

appliance circuits. A 277/480 V, 3φ, 4-wire service can be used to supply 277 V and 480 V, 1φ lighting circuits. High-voltage lighting circuits are used in commercial and industrial fluorescent and HID (high-intensity discharge) lighting circuits.

A system that cannot deliver 120 V, 1φ power appears to have limited use. Commercial applications (sport complexes, schools, offices, parking lots, etc.) have lighting as the major part of the electrical system. Large commercial applications include several sets of transformer banks that are connected to the 277/480 V, 3φ, 4-wire service to reduce the voltage to 120 V, 1φ.

120/240 V, 3φ, 4-WIRE SERVICE

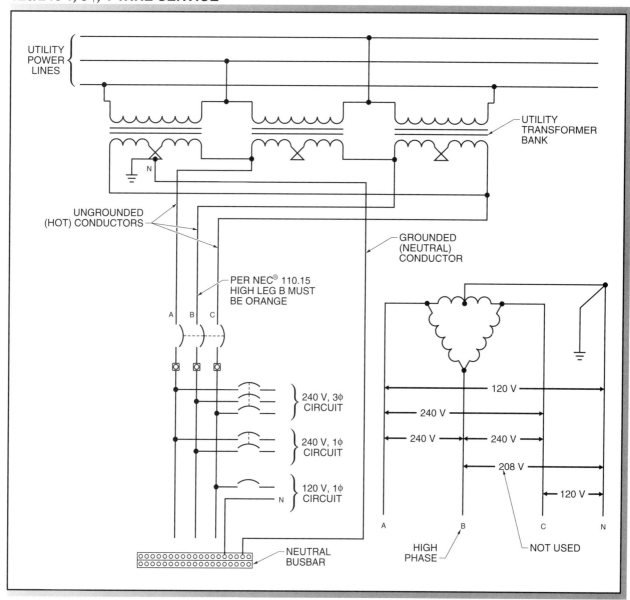

Figure 3-33. A 120/240 V, 3φ, 4-wire service is used to supply large amounts of 3φ power.

ASI Robicon

A 277/480 V, 3φ, 4-wire service supplies power for lighting, production machinery, and support equipment.

When electric motor drives are used for larger commercial HVAC systems or industrial applications, the drive is usually connected to 480 V, 3φ power lines. Connecting an electric motor drive to a 3φ power is always recommended, even if the drive can be connected to a single-phase supply.

TECH FACT

Energy-efficient motors are about 2.5% more efficient then standard motors, saving large amounts of power and money over time.

277/480 V, 3φ, 4-WIRE SERVICE

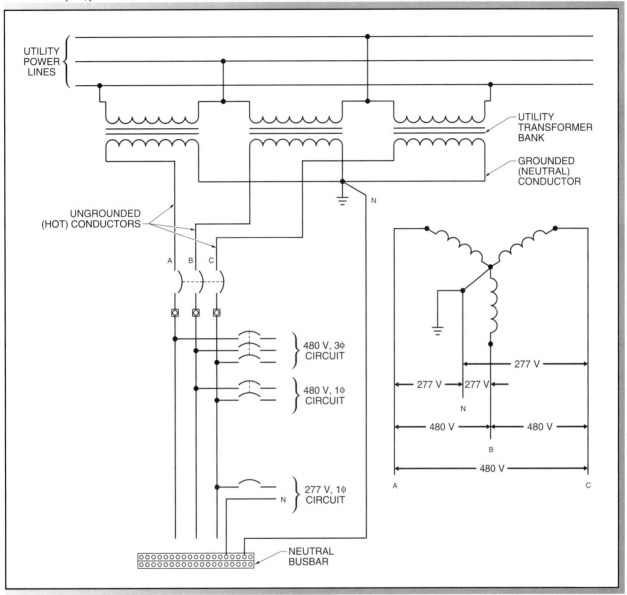

Figure 3-34. A 277/480 V, 3φ, 4-wire service is used to supply large amounts of 277 V, 1φ power for fluorescent and HID lighting.

Name _____ Date _____

True-False

T F **1.** Torque, unlike work, may exist even though no movement occurs.

T F **2.** Locked rotor torque is also referred to as acceleration torque.

T F **3.** The horsepower of a motor determines what size load a motor can operate and how fast the load rotates.

T F **4.** All coils (motor windings, transformers, solenoids, etc.) create capacitance in an electrical circuit and oppose a change in current.

T F **5.** Voltage unbalance is the unbalance that occurs when the 3ϕ power lines are more or less than 120° out of phase.

T F **6.** Single-phasing is the operation of a 3ϕ load on two phases because one phase is lost.

T F **7.** A grounding conductor is a conductor that normally carries current.

T F **8.** Electric motor drives are available that can be connected to either 120 V or 240 V 1ϕ power.

T F **9.** Connecting an electric motor drive to a 3ϕ power source is always recommended.

T F **10.** Current unbalance reverses motor rotation in a 3ϕ motor control circuit.

T F **11.** Motor torque is the force that produces or tends to produce motor rotation of a motor shaft.

T F **12.** A constant torque load is a load that requires high torque at low speeds and low torque at high speeds.

T F **13.** Capacitance is the total opposition of resistance, inductive reactance, and capacitive reactance offered to the flow of alternating current.

T F **14.** A branch circuit is the portion of a distribution system between the final overcurrent protection device and the outlet or load.

T F **15.** The National Electrical Manufacturers Association (NEMA) classifies motors according to electrical characteristics.

Completion

_____ **1.** ___ is accomplished when a force overcomes a resistance.

_____ **2.** ___ is the maximum torque a motor can provide without an abrupt reduction in motor speed.

_____Class B_____ *pg 52*

3. ___ motors increase the starting torque of the motor to 150% of the motor's full-load torque.

_____Constant torque load_____ *pg 57*

4. When operating speeds vary, a(n) ___ load requires the same torque at low speeds as at high speeds.

_____farad_____ *pg 59*

5. The unit of capacitance is the ___.

_____series_____ *pg 62*

6. Power reactors are installed in ___ with the power lines delivering voltage to the electric motor drive.

_____Harmonic distortion_____ *pg 62*

7. ___ are frequencies that are whole-number multiples of the fundamental frequency.

_____heat_____ *pg 68*

8. A high current unbalance produces excessive ___, which in turn can result in insulation breakdown.

_____green_____ *pg 71*

9. The color ___ always indicates a conductor used for grounding.

_____408.3 (E)_____ *pg 73*

10. Phases A, B, and C are connected to a switchboard or panelboard according to NEC® subsection ___.

Multiple Choice

_____D_____ **1.** ___ is/are the force that produces rotation.
A. Lumens   C. Speed
B. Resistance   D. Torque

_____B_____ **2.** ___ torque is the torque required to produce the rated power at full speed of the motor. *pg 48*
A. Breakdown   C. Locked rotor
B. Full-load   D. Pull-up

_____C_____ **3.** A ___ is an example of a constant horsepower load. *pg 51*
A. conveyor   C. lathe
B. fan   D. pump

_____D_____ **4.** ___ unbalance is the unbalance that occurs when the voltage at 3ϕ loads is not equal. *pg 57*
A. Current   C. Resistance
B. Phase   D. Voltage

_____B_____ **5.** The power lines supplying 3ϕ power to an electric motor drive can be marked ___. *pg 67*
A. DC+, DC–   C. T1, T2, T3
B. R, S, T   D. U, V, W

pg 70

Name _____ Date _____

Activity 3-1. Motor Data Sheet

To determine the type and size of electric motor drive a customer requires, drive manufacturers use motor data sheets to gather information. This information includes general information (customer, etc.) and technical information. To complete a data sheet to and use the sheet properly, technicians and qualified persons must understand the technical terms.

Identify the abbreviations and define the terms that follow the Motor Data Sheet.

MOTOR DATA SHEET		
Customer Information		
Customer:	Customer location:	
Application:		
Application conditions:		
Environment		
Duty: Standard ☐ Severe ☐	Max. motor ambient ☐ Max. drive ambient ☐	
Altitude: Up to 3300′ ☐ 3300′ and over ☐	Maximum number of starts per hour:	
Mounting: Horizontal ☐ Vertical ☐ Other ☐		
Explosionproof ☐ Class: Division: Group(s):		
Motor Parameters		
HP: kW: Volts: Hz: Frame:	Insulation class:	
Base rpm: Minimum speed: Maximum speed:	Service factor:	
Breakdown torque: lb-ft Full load torque: lb-ft	Load type: CT ☐ CH ☐ Special ☐	
Construction		
IEC ☐ NEMA ☐ Cast iron ☐ Aluminum ☐ Steel ☐		
Bearing type: Ball ☐ Sleeve ☐ Roller ☐	Enclosure: Open ☐ Enclosed ☐	
Connection to load: Coupling ☐ Belts ☐ Other ☐		
Electrical Feed		
☐ 1φ ☐ 3φY ☐ 3φΔ	Voltage rating: Amperage rating:	Fuse ☐ CBs ☐
Options		
Line reactor ☐ Load reactor ☐ Tachometer ☐ Encoder ☐		
Special paint ☐ Special lead terminations ☐ Other ☐		

1. What does the abbreviation kW stand for?

2. Define base rpm.

3. Define breakdown torque.

_____

4. Define full load torque.

_____

5. What does the abbreviation CT stand for?

_____

6. What does the abbreviation CH stand for?

_____

7. What does the abbreviation VT stand for?

_____

Activity 3-2. Motor Torque

Motor torque is the force that causes a motor shaft to rotate. The standard units of torque are pound-inches (lb-in) or pound-feet (lb-ft). The exact amount of torque a motor can produce depends on the motor design, applied voltage, and motor current. General motor torque rules can be applied to most motor applications.

Determine the motor torque using the General Motor Torque chart and motor nameplate.

_____ **1.** The developed motor torque for the motor is ____.

_____ **2.** If the motor had a nameplate rating of 1140 rpm, the developed motor torque would be ____.

_____ **3.** If the motor had a nameplate rating of 3450 rpm, the developed motor torque would be ____.

GENERAL MOTOR TORQUE		
NUMBER OF MOTOR POLES	MOTOR RATED SPEED*	MOTOR TORQUE†
2	3600 (3450)	1.5
4	1800 (1725)	3
6	1200 (1140)	4.5
8	900 (860)	6

* Synchronous (typical running)
† in lb-ft per HP

○ MANUFACTURER ○				
PE·21 PLUS™		PREMIUM EFFICIENCY		
ORD. NO.	1LA02864SE41	MAG.		
TYPE	RGZESDI	FRAME	286T	
H.P.	20.00	SERVICE FACTOR	1.0	3 PH
AMPS	24.0	VOLTS	460	
R.P.M.	1725	HERTZ	60	
DUTY	CONT 40°C AMB.	DATE CODE		
CLASS INSUL	H / NEMA DESIGN B / K.V.A. CODE G / NEMA NOM. EFF. 93.6			
SH. END BRG.	50BC03JPP3	OPP. END BRG.	50BC03JPP3	
○ Inverter Duty AC Induction Motor made in U.S.A. ○ ⓢⓟ				

Activity 3-3. Electric Motor Drive Power Derating

Electric motor drives have a power rating that indicates the maximum power a drive can deliver when operating under normal conditions (within drive ratings). An electric motor drive power rating is based on standard input voltages (230 V, 460 V, or 575 V). The power rating of the electric motor drive must be reduced when the drive is connected to a reduced input voltage. The amount of reduction is proportional to the voltage change. To determine the amount of power derating for an electric motor drive, apply the following formula:

$$D_{HP} = D_{RHP} \times \frac{V_{AP}}{V_R}$$

where

D_{HP} = electric motor drive derated horsepower (in HP)

D_{RHP} = electric motor drive rated horsepower (in HP)

V_{AP} = input voltage applied to electric motor drive (in V)

V_R = voltage rating of electric motor drive (in V)

Calculate the derated horsepower of the electric motor drives.

_____ **1.** The 230 V rated electric motor drive has a derated horsepower of ____ HP.

_____ **2.** The 460 V rated electric motor drive has a derated horsepower of ____ HP.

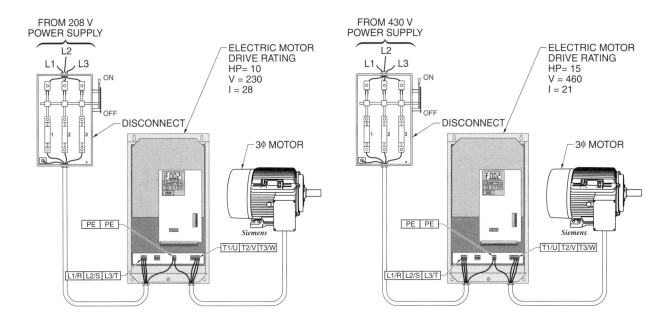

Activity 3-4. Connecting Load Reactor and Pushbuttons

1. Connect the disconnect (including ground) to the required terminals to supply power to the electric motor drive and motor circuit. Connect the reactor to the terminals so that it functions as a load reactor. Connect the electric motor drive so that it operates using external forward and reverse pushbuttons to control motor direction and an external potentiometer for speed control. Connect the motor terminals for low-voltage operation.

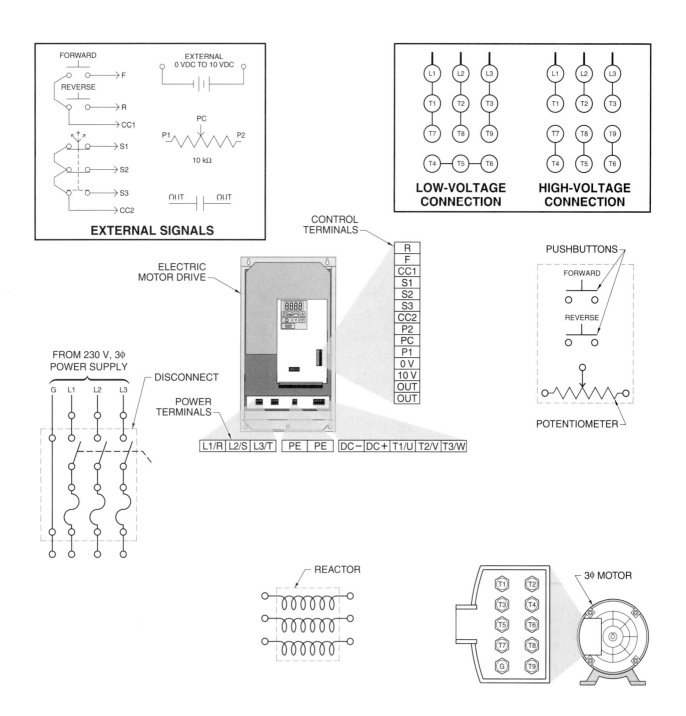

Activity 3-5. Connecting Line Reactor and Selector Switch

1. Connect the disconnect (including ground) to the required terminals to supply power to the electric motor drive and motor circuit. Connect the reactor to the terminals so that it functions as a line reactor. Connect the electric motor drive so that it operates using an external three-position selector switch to control motor direction and an external 0 VDC to 10 VDC power supply for speed control. Connect the motor terminals for high-voltage operation.

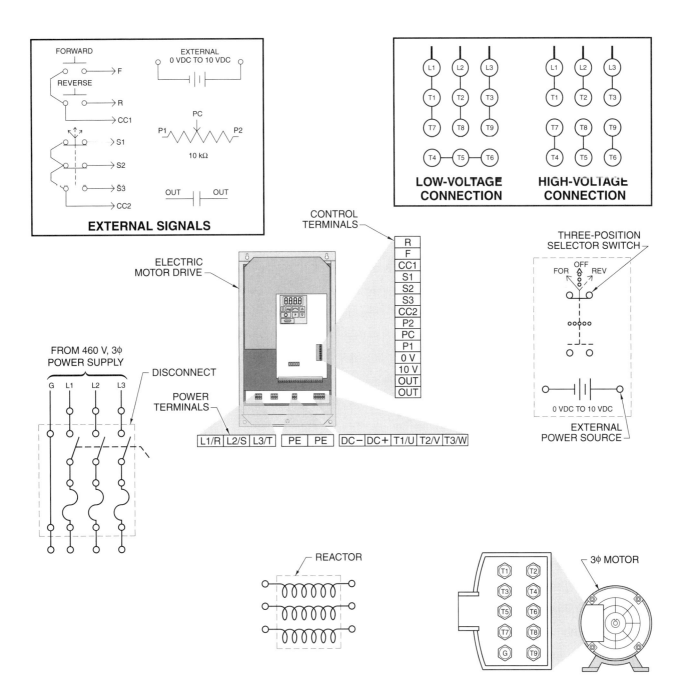

Activity 3-6. Connecting Line Reactor and Level Switch

1. Connect the disconnect (including ground) to the required terminals to supply power to the electric motor drive and motor circuit. Connect the reactor to the terminals so that it functions as a line reactor. Connect the electric motor drive so that it operates using an external level switch to control direction and a two-position selector switch that controls an input signal into S1 for high speed and S2 for low speed for speed control. Connect the motor terminals for high-voltage operation.

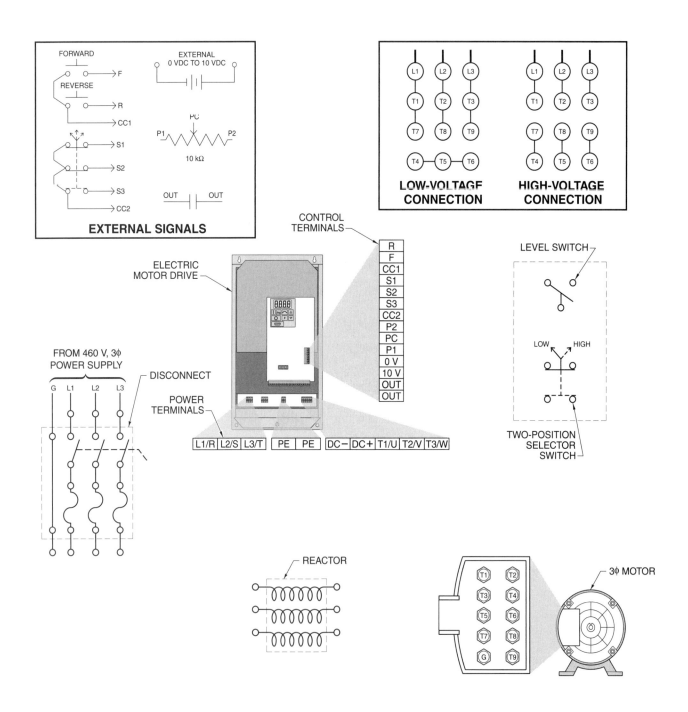

4 Electric Motor Types

Electric motors are available as DC motors or as 1φ or 3φ AC motors. Several types of DC motors and AC motors are used to meet various application requirements. Electric motor drive applications require that specifically rated motors be used. Understanding which type of AC motor can be used in an AC drive application and which type of DC motor can be used in a DC drive application creates dependable and efficient operating systems.

ELECTRIC MOTORS AND ELECTRIC MOTOR DRIVES

An *electric motor* is a rotating output device that converts electrical energy into rotating mechanical energy. General types of electric motors are AC motors, DC motors, and AC/DC (universal) motors. Various designs of electric motors allow motors to be used for numerous applications. The choice of an electric motor for a given application depends on the driven load, environmental conditions, available electric power, and efficiency requirements.

Electric motors are controlled by electric motor drives. Electric motor drives control motors and motor loads more efficiently than standard mechanical motor starter controls. The speed and torque control and high-efficiency of electric motor drives allows drives to save energy and money. The two categories of electric motor drives are direct current (DC) drives and alternating current (AC) drives. DC and AC electric motor drives are replacing manual motor starters, magnetic motor starters, and soft starters (solid-state starters) in most applications. See Figure 4-1.

DC drives are used to control DC series motors, DC shunt motors, DC compound motors, and DC permanent-magnet motors. DC drives are used to control DC motors by varying the voltage delivered to the motor. The speed (rpm) of a DC motor is proportional to the voltage applied to the motor. DC drives use electronic controls, such as silicon-controlled rectifiers (SCRs), for smooth motor speed control from 0 rpm to full rated speed.

Three-phase AC drives are used to control 3φ induction motors. Industry also uses 1φ power supplies to supply the power to electric motor drives to control 3φ motors. Single-phase AC drives are also used to control 1φ capacitor-run motors.

Rockwell Automation/Reliance Electric
Various types of DC and AC electric motors are used to accomplish work in industrial, commercial, and residential applications.

In applications where DC drives were once used, AC drives are replacing the DC drives. AC drives are used in a majority of new electric motor drive applications. AC drives are used in most new HVAC and industrial applications. Reasons for replacing DC drives with AC drives include the following:

• improved torque control available with newer AC drive designs

• easier installation, programming, and maintenance procedures

• easier interfacing with communication networks, other electric motor drives, and control devices

• significant improvements in electronic controls of AC drives, such as faster control, by using insulated gate bipolar transistors (IGBTs)

MOTOR TYPES

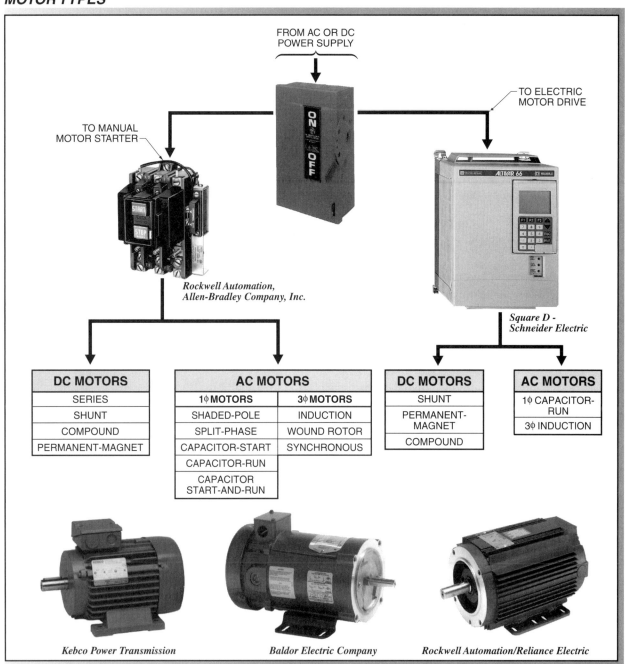

Figure 4-1. Electric motor drives are typically used to control DC shunt, DC permanent-magnet, DC compound, AC capacitor-run, and AC induction motors.

- ability to operate a motor above the nameplate rated speed without adding additional brush sparking as with a DC motor
- availability of smaller, more cost-effective AC drives that control AC motors from fractional horsepower to hundreds of horsepower
- lower maintenance costs of AC motors due to the lack of brushes that require replacement
- development of special inverter duty-rated AC motors designed to work in AC drive applications

DC drives continue to be used in some applications. Reasons for not replacing DC drives with AC drives include the following:
- DC drives have smaller packaging for the same rated horsepower than AC drives.
- DC motors produce high torque at low speeds.
- Industrial applications use large and expensive motors for many applications, and replacing large DC motors with AC motors may not be cost effective.

Consideration must be given to replacement motors that are already in stock at a facility and that trained maintenance personnel understand.

SINGLE-PHASE AC MOTORS

Single-phase AC motors are used in residential applications more than any other type of motor, and also in some industrial and commercial applications. Single-phase AC motor-driven appliances and devices used in residences and industrial and commercial facilities include furnaces, air conditioners, refrigerators, blowers, washing machines, dryers, ovens, clocks, and cooling fans for computers.

AC motors are used with single-phase electric motor drives, but industry typically uses 3ϕ electric motor drives to control 3ϕ motors. An AC drive converts the incoming 1ϕ power to 3ϕ power. The 3ϕ power is used to drive more efficient 3ϕ motors. See Figure 4-2.

Single-phase motors have two main parts, the stator and rotor. A *stator* is the stationary part of an AC motor

AC DRIVE 1ϕ TO 3ϕ CONVERSION

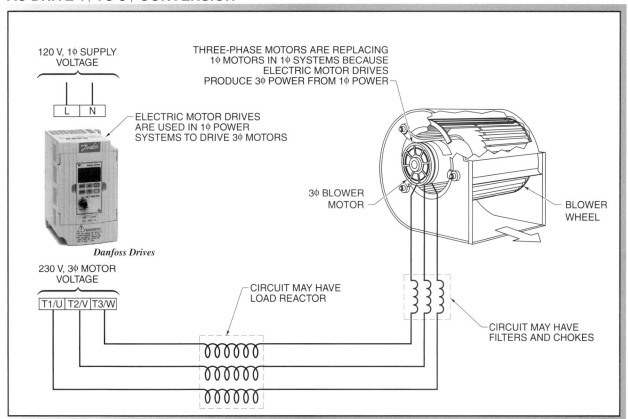

Figure 4-2. Electric motor drives can be supplied with 1ϕ power and create 3ϕ power to be used by 3ϕ motors.

that produces a rotating magnetic field. A *rotor* is the rotating part of an AC motor. The rotating magnetic field created by the stator rotates the rotor and shaft to produce work. Other parts of a motor are the frame, endbells, bearings, and a fan connected to the motor shaft. See Figure 4-3. Electric motors produce work by converting electrical energy into rotating mechanical energy.

ELECTRIC MOTOR PARTS

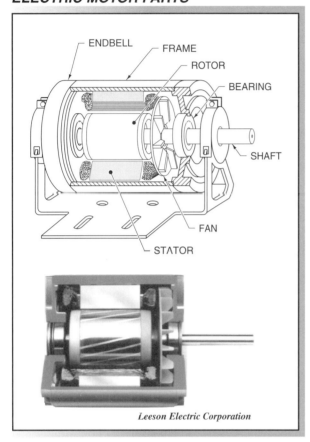

Leeson Electric Corporation

Figure 4-3. The stationary part of a 1φ motor is the stator, and the rotating part is the rotor.

Single-phase electric motors are used by industry in small horsepower and torque applications such as mixing ink in vats.

An electrical relationship exists between the stator and rotor of a 1φ motor. Electrical power applied to the stator windings creates a magnetic field. The magnetic field induces voltage in the rotor bars. The induced voltage causes current to flow in each of the rotor bars, producing a magnetic field in the rotor. The rotor magnetic field produces north (negative) and south (positive) poles, which are of opposite polarity to the stator poles. The unlike poles attract each other, producing a horizontal force between the stator and rotor. See Figure 4-4.

The alternating current alternates the magnetic field back and forth due to the change in the direction of current flow. Even though the magnetic field is alternating, no rotating magnetic field is produced. The motor cannot start, even with the horizontal force produced. No vertical force is produced without a rotating magnetic field. Even if the rotor started to rotate, it might not start in the right direction. A motor can start in either direction, depending on the position of the rotor when power is first applied to the stator.

The position of the rotor affects the direction of rotation when a motor starts rotating. When the north pole of the rotor is nearest the top or the north pole of the stator, the motor rotates in the clockwise direction. If the north pole of the rotor is nearest the bottom or the south pole of the stator, the motor rotates in the counterclockwise direction. See Figure 4-5.

Regardless of the position of the rotor when power is applied, the rotor cannot rotate. The stator field alternates back and forth and does so at such a rapid rate (60 times per second for 60 Hz AC) that the magnetic field of the rotor simply locks in step with the alternating magnetic field of the stator and no rotation occurs.

Locked in step is a motor condition that occurs when the field of the stator and the field of the rotor are parallel to one another, not allowing the shaft to rotate. Rotation is not possible because the magnetic attraction between the two fields is equal for both directions of rotation. The force trying to rotate the motor clockwise is equal to the force trying to rotate the motor counterclockwise. If the rotor is given a spin in one direction, the rotor continues to rotate in the direction of the spin. The rotor quickly accelerates until it reaches a speed slightly less than the rated synchronous speed of the motor.

Synchronous speed is the theoretical speed of a motor based on the number of poles of the motor and line frequency. *Line frequency* is the number of complete electrical cycles per second of a power source. The speed of a motor is measured in revolutions per minute (rpm). All motors except synchronous motors turn at a speed less than synchronous speed. The operating (actual) speed is listed on the nameplate of the motor. Nameplate speed (base speed) is the speed of a motor when the motor develops rated horsepower at rated voltage and frequency. A motor with a nameplate listed speed of 1725 rpm typically has a synchronous speed of 1800 rpm.

Slip is the difference between the synchronous speed and the actual speed of a motor. Slip is measured in percentage. To calculate slip of a motor, apply the formula:

$$Slip = \frac{rpm_{syn} - rpm_{act}}{rpm_{syn}} \times 100$$

where

$Slip$ = percentage (in %)

rpm_{syn} = synchronous speed (in rpm)

rpm_{act} = actual speed (in rpm)

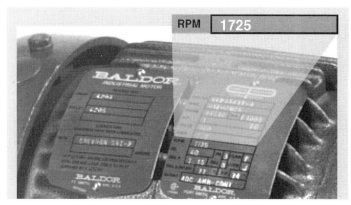

Baldor Electric Company
All electric motor nameplates are stamped with the listed rpm rating of the motor.

ELECTRIC MOTOR MAGNETIC FIELDS

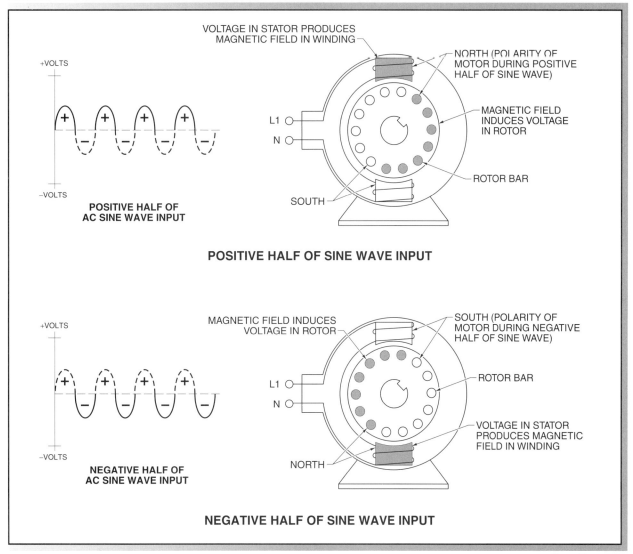

SEE APPENDIX FOR ⊕ TO ⊖ DIAGRAM

Figure 4-4. The stator is connected to the supply voltage and produces a magnetic field, which rotates the rotor.

STARTING MOTOR ROTATION

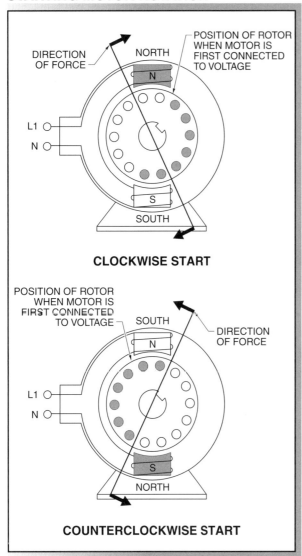

Figure 4-5. The position of the rotor and the polarity of the stator affect the starting rotation of the motor.

For example, what is the slip of a motor with a 1725 rpm nameplate rating and a synchronous speed of 1800 rpm?

$$Slip = \frac{rpm_{syn} - rpm_{act}}{rpm_{syn}} \times 100$$

$$Slip = \frac{1800 - 1725}{1800} \times 100$$

$$Slip = \frac{75}{1800} \times 100$$

$$Slip = \textbf{4.2\%}$$

TECH FACT

Motor frame numbers are listed on the motor nameplate. Motor frame numbers are used to indicate mounting and motor dimensions. Higher motor frame numbers indicate larger physical motor size and horsepower. NEMA® specifications are stated in inches and IEC specifications are stated in millimeters.

Calculating Motor Synchronous Speed

The synchronous speed of induction motors is based on the supply voltage frequency (Hz) and the number of poles in the motor winding. Motors designed for 60 Hz use have synchronous speeds of 3600, 1800, 1200, 900, 720, 600, 514, and 450 rpm. To calculate the synchronous speed of an induction motor, apply the formula.

$$rpm_{syn} = \frac{120 \times f}{N_p}$$

where

rpm_{syn} = synchronous speed (in rpm)

120 = constant

f = supply voltage frequency (in Hz)

N_p = number of motor poles

For example, what is the synchronous speed of a four-pole motor operating at 60 Hz?

$$rpm_{syn} = \frac{120 \times f}{N_p}$$

$$rpm_{syn} = \frac{120 \times 60}{4}$$

$$rpm_{syn} = \frac{7200}{4}$$

$$rpm_{syn} = \textbf{1800 rpm}$$

The formula proves that the only way to change the speed of an induction motor is to change the number of poles in the motor or change the line voltage frequency. The number of poles an induction motor can have is limited by space and is limited to being rated as four-pole two-speed, six-pole three-speed, or eight-pole four-speed. An electric motor drive used to control a motor changes the speed of the motor by changing the supply voltage frequency. See Figure 4-6.

Electric motor drives can lower the speed of a motor to below the nameplate rated speed, or increase the speed to synchronous speed and higher. Although electric motor drives can be used to increase the speed of a motor beyond base speed, the increased speed is not good for the motor. The increased speed allows lubricants to become unevenly distributed, and damage to the mechanical parts of the motor occurs. Motor manufacturers list the maximum speed at which a motor can safely be operated. Check with the motor manufacturer before operating an electrical motor at more than 5% above the motor base speed.

ELECTRIC MOTOR SPEED CONTROL

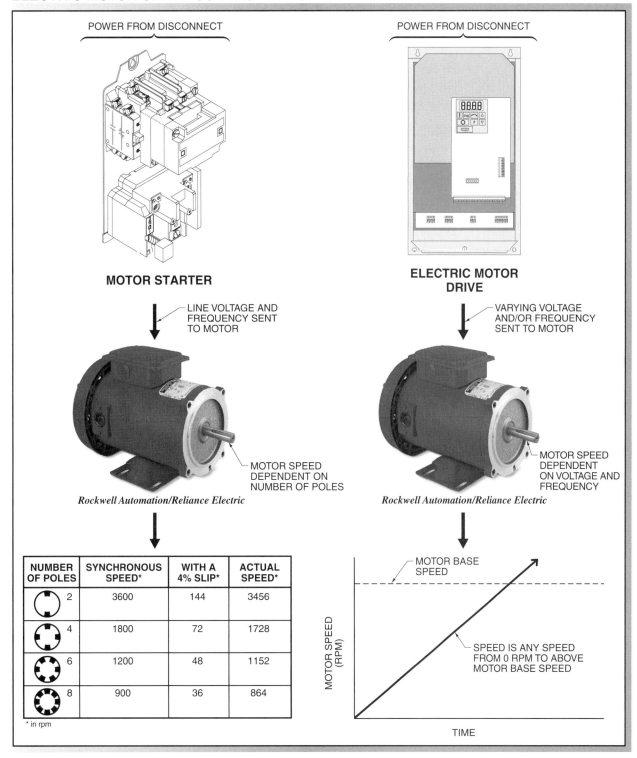

NUMBER OF POLES	SYNCHRONOUS SPEED*	WITH A 4% SLIP*	ACTUAL SPEED*
2	3600	144	3456
4	1800	72	1728
6	1200	48	1152
8	900	36	864

* in rpm

Figure 4-6. Without an electric motor drive, the speed of an AC motor is limited by the number of poles.

Single-phase motors cannot rotate without some way of first starting movement in the motor. Single-phase motor types are classified according to the method used to start them. Single-phase motor types are:

- shaded-pole motor
- split-phase motor
- capacitor-start motor
- capacitor-run motor
- capacitor start-and-run motor

Shaded-Pole Motors

A *shaded-pole motor* is an AC motor that uses a shaded pole for starting. Shading the stator poles is the simplest method used to start a 1φ motor rotating. The shaded poles produce a rotating effect on the rotor, but generate extremely low starting torque.

The low starting torque of shaded-pole motors limits their use to applications that require very small loads to be driven. See Figure 4-7. Torque is the force that produces rotation in a motor, and different designs of AC motors produce different amounts of starting torque. See Figure 4-8.

Shading the pole is accomplished by applying a short-circuited wire on one side of each stator pole. The shaded pole is usually a solid single turn or double turn of copper wire placed around a portion of the main pole laminations.

The function of the shaded pole is to delay the magnetic flux in the area of the pole that is shaded. Shading causes the magnetic flux at the shaded pole area to be approximately 90° away from the magnetic flux of the main stator pole. Since 360° equals the distance of the two poles, the shading pole is offset by one-fourth the distance of the main pole, causing a phase

SHADED-POLE MOTORS

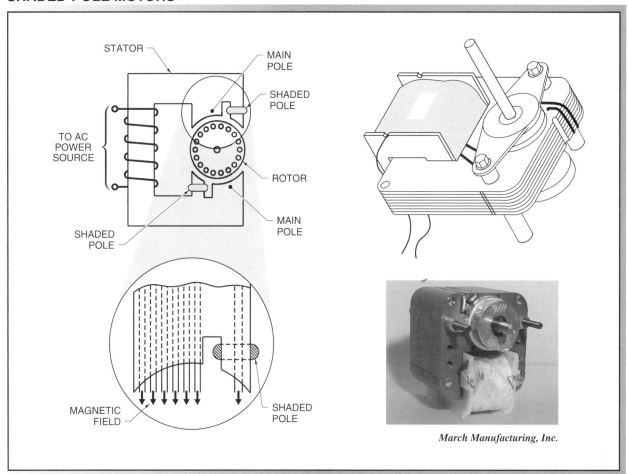

March Manufacturing, Inc.

Figure 4-7. Single-phase motors can be made to start in a specified direction by shading the stator poles.

displacement. The offset field moves the rotor from the main pole toward the shaded pole and determines the starting direction of the motor.

Shaded-pole motors are commonly ¹⁄₂₀ HP or less because of the low starting torque. The most common application for shaded-pole motors is for use as cooling fans in small appliances and computers. The only load the motor must turn is the fan blade. A shaded-pole motor cannot be reversed without moving the shaded poles to the opposite side of the poles. For this reason, shaded-pole motors are used in applications that never require the motor to be reversed.

Split-Phase Motors

A *split-phase motor* is an AC motor that has running and starting windings. A split-phase motor is made self-starting by adding a second stator winding. The main winding is the running winding and the auxiliary winding is the starting winding. The two windings are placed in the stator slots and spaced 90° apart. See Figure 4-9.

The running winding is made of larger gauge wire with a greater number of turns. When the motor is first connected to power, the inductive reactance (X_L) of the running winding is higher, and the resistance is lower than the starting winding. The difference in inductive reactance between the running and starting windings

forces the running winding current to lag the starting winding current (that is, there is a phase difference). The phase difference between the starting and running windings provides the motor with the necessary starting torque in one direction required to start the motor rotating.

A centrifugal switch is used to remove the starting winding when the motor reaches actual running speed to minimize energy loss and prevent heat buildup in the starting winding. A *centrifugal switch* is a switch that opens to disconnect the starting winding when the rotor reaches a preset speed and reconnects the starting winding when the speed falls below the preset value. After the starting winding is removed, the motor continues to operate on the running winding only.

The centrifugal switch located inside the motor housing is attached to the shaft and is connected in line (series) with the starting winding. The set speed is usually about 60% to 80% of the running speed.

Split-phase motors are generally available in sizes ranging from ¹⁄₁₆ to ¹⁄₃ HP. A split-phase motor is reversible if the starting winding leads are available. To reverse the rotation direction of a split-phase motor, either the starting or running winding connections are reversed. If both are reversed, the motor does not change direction of rotation. The industrial standard is to reverse the starting winding.

MOTOR STARTING TORQUE RATINGS

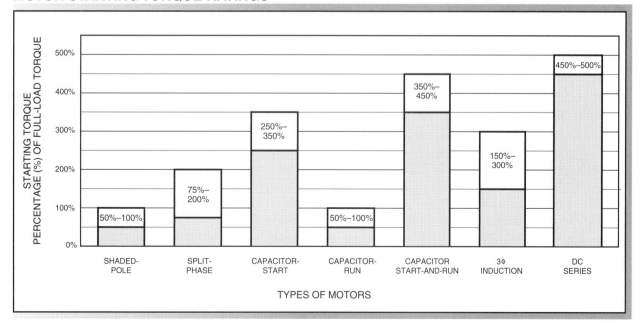

Figure 4-8. The motor starting torque rating varies with the type of motor.

SPLIT-PHASE MOTORS

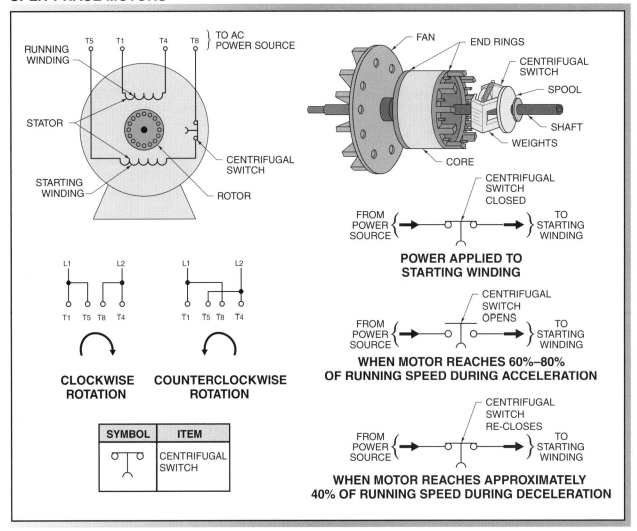

Figure 4-9. Split-phase motors are 1φ AC motors that include a running and starting winding.

Capacitor Motors

A *capacitor motor* is a 1φ motor with a capacitor connected in series with the stator windings to produce phase displacement in the starting or running winding and add higher starting and/or running torque. A capacitor motor is similar in design to a split-phase motor. Both motors have starting and running windings, but the capacitor motor also has a capacitor connected in series with the motor windings.

A capacitor-start motor is the most common type of capacitor motor. Capacitor-start motors are available in sizes from fractional to about 3 HP. Because the capacitor improves the power factor, the capacitor-start motor develops considerably more starting torque per ampere than the split-phase motor. When the motor is at speed, the capacitor and starting winding are removed from the circuit, making the two motor types equivalent when running. See Figure 4-10.

A capacitor-run motor leaves the starting winding and capacitor in the circuit at all times. The starting winding is not removed as the motor speed increases because there is no centrifugal switch. The advantage of leaving the capacitor in the circuit is that the motor has more running torque than the capacitor-start motor or split-phase motor. A capacitor-run motor has lower full-load speed than a capacitor-start motor and is used in sizes from fractional to 5 HP and for loads that require high running torque. See Figure 4-11.

CAPACITOR-START MOTORS

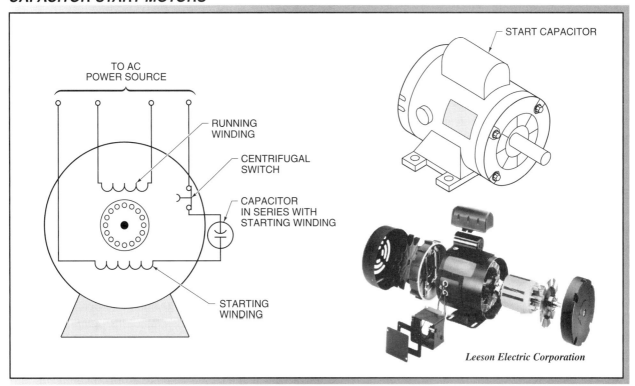

Figure 4-10. Capacitor-start motors remove the start capacitor from the circuit when the motor is running.

CAPACITOR-RUN MOTORS

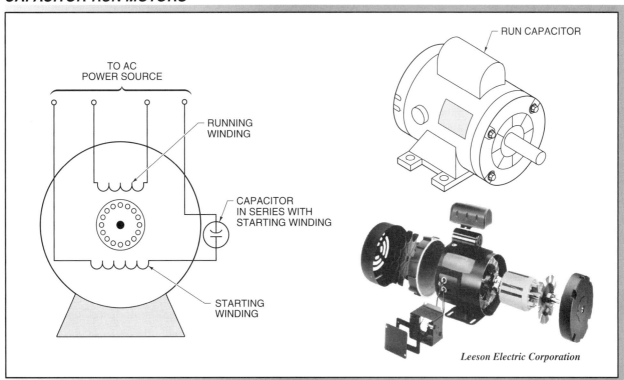

Figure 4-11. The capacitor is not removed from a capacitor-run motor when the motor is running.

A capacitor start-and-run motor uses two capacitors. The motor starts with one value capacitor in series with the starting winding and runs with a different value capacitor in series with the running winding. The capacitor start-and-run motor has the same starting torque as a capacitor-start motor. A capacitor start-and-run motor has more running torque than a capacitor-start motor or capacitor-run motor. See Figure 4-12.

THREE-PHASE MOTORS

Approximately 5% of the 3φ motors used in industry are repulsion motors. A *repulsion motor* is a motor with the rotor connected to the power supply through brushes that ride on a commutator.

The other 95% of 3φ motors used in industry are induction motors. An *induction motor* is a motor that has no physical electrical connection to the rotor because there are no brushes. Current in the rotor is induced by the magnetic field of the stator. Induction motors do not develop as much torque as repulsion motors.

Like the 1φ motor, a 3φ induction motor requires a rotating magnetic field but does not require any additional components to produce the rotating magnetic field. The rotating magnetic field is set up automatically in the stator when the motor is connected to 3φ power lines. The coils in the stator are connected to form three separate windings (phases). Each phase contains one-third of the total number of individual coils in the motor. These composite windings or phases are phase A, phase B, and phase C. See Figure 4-13.

Each phase is placed in the motor so that it is 120° from the other phases. Since each phase reaches its peak value 120° away from the other phases, a rotating magnetic field is produced in the stator.

Single-Voltage 3φ Motors

To develop a rotating magnetic field in the motor, the windings must be connected to the proper voltage. This voltage level is determined by the manufacturer and stamped on the motor nameplate.

TECH FACT

Electric motors typically can be mounted in any position or at any angle, but some motors such as dripproof motors are manufactured for a specific mounting position.

CAPACITOR START-AND-RUN MOTORS

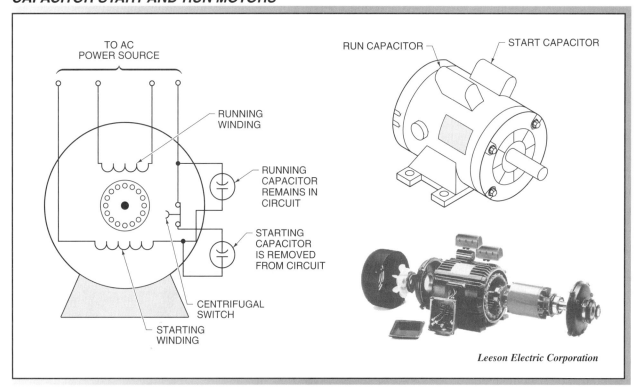

Leeson Electric Corporation

Figure 4-12. Capacitor start-and-run motors remove the starting capacitor from the circuit when the motor reaches speed.

THREE-PHASE MOTOR ROTATION

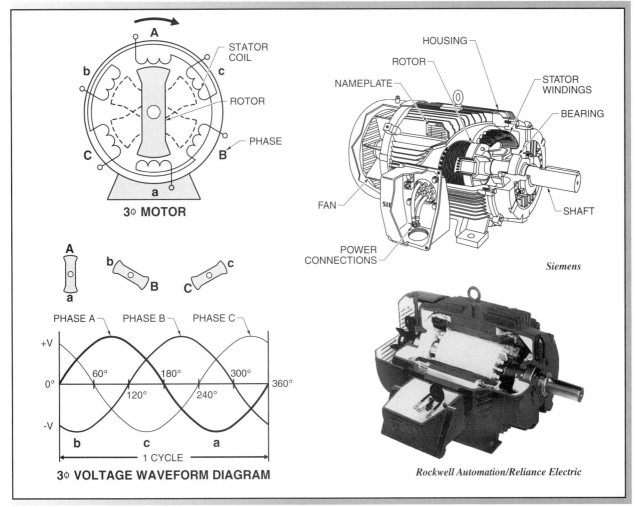

Figure 4-13. Three-phase motors have each phase 120° apart.

A *single-voltage motor* is a motor that operates at only one voltage level. Single-voltage 3φ motors are less expensive to manufacture than dual-voltage 3φ motors, but are limited to locations having the same voltage as the motor. Typical single-voltage, 3φ motor ratings are 230 V, 460 V, and 575 V. Other single-voltage 3φ motor ratings are 200 V, 208 V, 220 V, and 240 V.

Wye-Connected Motors. A wye-connected 3φ motor has one end of each of the three phases internally connected to the other phases. The remaining end of each phase is then brought out externally and connected to the incoming power source. The leads that are brought out externally are labeled terminals T1 (U), T2 (V), and T3 (W). Connecting a wye-connected 3φ motor to the 3φ power lines requires that the power lines and motor terminals be connected L1 to T1, L2 to T2, and L3 to T3 to achieve proper phase sequencing. See Figure 4-14.

ASI Robicon

Three-phase electric motors for industrial applications can be 500 HP and larger.

WYE-CONNECTED 3φ MOTORS

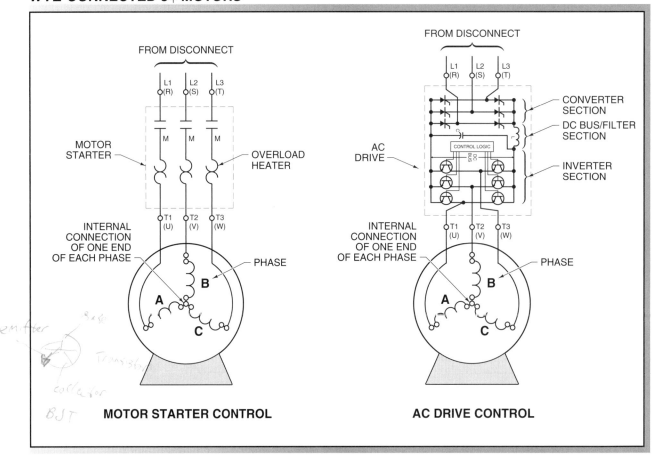

Figure 4-14. Wye-connected 3φ motors have one end of each of the three phases internally connected.

Delta-Connected Motors. A delta-connected 3φ motor has each phase wired end to end to form a completely closed loop circuit. Each point where the phases are connected has the leads brought out externally to form T1 (U), T2 (V), and T3 (W). T1, T2, and T3, are connected to the three power lines similar to a wye-connected motor, with L1 connected to T1, L2 to T2, and L3 to T3. The 3φ lines supplying power to the motor must have the same voltage and frequency rating as the motor. See Figure 4-15.

Dual-Voltage 3φ Motors

A *dual-voltage motor* is a motor that operates at more than one voltage level. Many 3φ motors are manufactured so that they can be connected to either of two voltages. The purpose in making motors for different voltages is to enable the same motor to be used with different power line voltages, requiring fewer motors to be kept in stock.

A typical dual-voltage, 3φ motor rating is 230/460 V. Other common dual-voltage, 3φ motor ratings are 240/480 V and 208-230/460 V. The dual-voltage rating of a motor is listed on the nameplate of the motor.

If both voltages are available, the higher voltage is usually preferred because the motor uses the same amount of power, given the same horsepower output, for either high or low voltage. As the voltage is doubled (e.g., 230 V to 460 V), the current drawn on the power lines is cut in half. With the reduced current, the wire size is reduced, and the material cost is decreased.

TECH FACT

As a motor gets hot, the horsepower is lower than the initial horsepower generated by a cool motor. Blowing enough air to keep the motor at approximately ambient temperature allows the horsepower rating to increase 10%–15%, with 20% possible with some motor designs.

DELTA-CONNECTED 3φ MOTORS

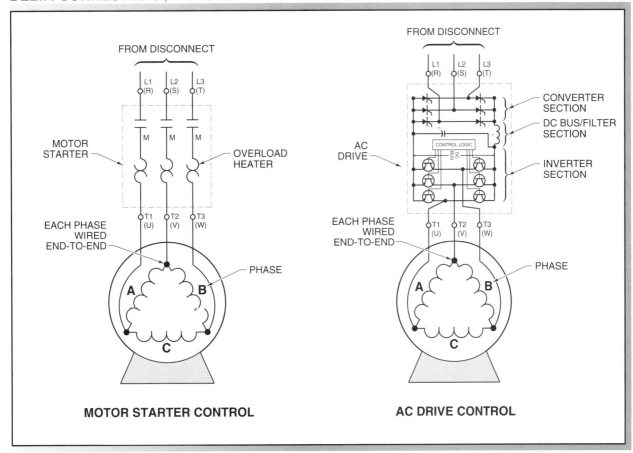

Figure 4-15. Delta-connected 3φ motors have each phase wired end to end to form a closed loop.

Dual-Voltage, Wye-Connected Motors. A dual-voltage, wye-connected, 3φ motor has each phase coil (A, B, and C) divided into two equal parts. By dividing the phase coils in two (six leads), plus three phase connection terminals, nine terminal leads are available. The motor leads are marked terminals one through nine (T1 – T9). The nine terminal leads are connected according to low or high voltage usage. See Figure 4-16.

To connect a dual-voltage, wye-connected, 3φ motor to low voltage, connect L1 to T1 and T7, L2 to T2 and T8, and L3 to T3 and T9 at the motor. Connect T4, T5, and T6 together. Connecting the terminals together for low voltage places the individual coils of each phase in parallel. With the coils connected in parallel, the applied voltage is present across each set of coils.

To connect a dual-voltage, wye-connected, 3φ motor to high voltage, connect L1 to T1, L2 to T2, and L3 to T3 at the motor. Connect T4 to T7, T5 to T8, and T6 to T9. Connecting the terminals together for high voltage

places the individual coils of each phase in series. With the coils connected in series, the applied voltage divides equally among the coils.

Dual-voltage motors allow companies to keep fewer motors in stock for replacement purposes because one motor can fit various voltage applications.

WYE-CONNECTED MOTOR WIRING DIAGRAM

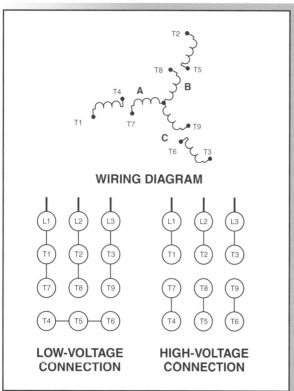

Figure 4-16. Dual-voltage, wye-connected 3ϕ motors have nine leads that must be connected.

Industry often connects electric motors to gearboxes to improve the torque rating of the motor for an application.

Dual-Voltage, Delta-Connected Motors. A dual-voltage, delta-connected, 3ϕ motor has each phase coil (A, B, and C) divided into two equal parts. By dividing the phase coils in two (six leads), plus three phase connection terminals, nine terminal leads are available. The motor leads are marked terminals one through nine (T1 – T9). The nine terminal leads are connected according to low or high voltage usage. See Figure 4-17.

To connect a dual-voltage, delta-connected, 3ϕ motor to low voltage, connect L1 to T1, T7, and T6; L2 to T2, T8, and T4; and L3 to T3, T9, and T5 at the motor. Connecting the terminals together for low voltage places the individual coils of each phase in parallel. With the coils connected in parallel, the applied voltage is present across each set of coils.

To connect a dual-voltage, delta-connected, 3ϕ motor to high voltage, connect L1 to T1, L2 to T2, and L3 to T3 at the motor. Connect T4 to T7, T5 to T8, and T6 to T9 together. Connecting the terminals together for high voltage places the individual coils of each phase in series. With the coils connected in series, the applied voltage divides equally among the coils.

DELTA-CONNECTED MOTOR WIRING DIAGRAM

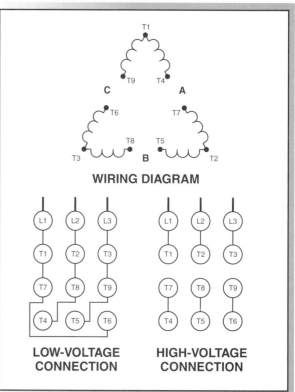

Figure 4-17. Dual-voltage, delta-connected 3ϕ motors have nine leads that must be connected.

Twelve-Lead, 3ϕ Motors

A 12-lead, wye-connected motor has the same nine motor leads (T1–T9) as a nine-lead, wye-connected motor, but there are three additional leads (T10 –T12). A 12-lead, delta-connected motor has the same nine motor leads (T1–T9) as a nine-lead, delta-connected motor, but there are three additional leads (T10 –T12). See Figure 4-18.

Typically, dual-voltage, 3ϕ motors have nine leads coming out of the motor box. A nine-lead, wye-connected motor and a nine-lead, delta-connected motor have internal connections made by the manufacturer. However, manufacturers of dual-voltage, 3ϕ, wye-connected and delta-connected motors sometimes do not make the internal connections. The internally unconnected motors have 12 leads coming out of the motor box.

Reversing 3ϕ Motors

Interchanging any two of the three power lines to a 3ϕ motor reverses the direction of rotation of the motor. Although any two lines can be interchanged, the industrial standard is to interchange T1 and T3. This standard came about because L2 and T2 were permanently connected and taped to provide an insulated barrier between the untaped L1 to T1 and L3 to T3 temporary connections that were made in order to test the motor for proper rotation.

> ⚠ **CAUTION**
>
> *Starting current for motors is approximately 10 times the full load amperage. Disconnects, fuses, wires, and contactors must be selected accordingly.*

TWELVE-LEAD MOTOR WIRING DIAGRAMS

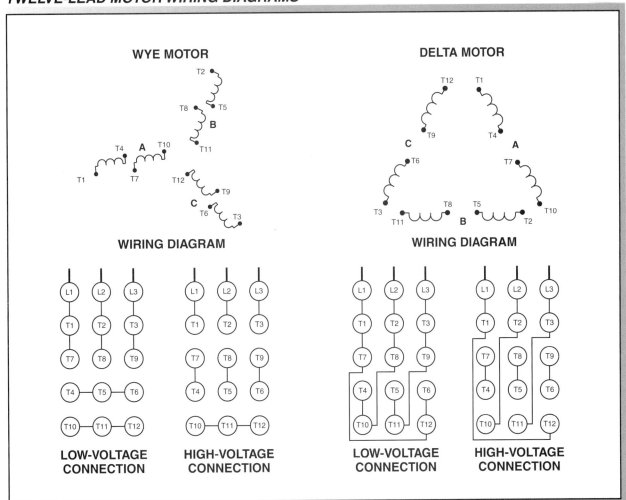

Figure 4-18. Some 3ϕ motors do not have three internal connections made, leaving 12 leads to be connected by the technician.

When motor starters are used to perform motor reversing, two starters are required. One starter (forward starter) is used for connecting the motor in the forward direction. The second starter (reversing starter) is used for connecting the motor in the reverse direction. The two starters must be installed in such a way (interlocked) as to prevent both starters from being energized at the same time. Energizing both starters at the same time would create a short circuit in the power lines. See Figure 4-19.

An electric motor drive used to control a motor reverses the motor without the need for any additional circuit components such as a reversing starter. Electric motor drives reverse two of the power lines to the motor internally in the drive unit, simplifying the installation of the motor control circuit.

TECH FACT

The ambient temperature of a motor is not to exceed 104°F (40°C) or –13°F (25°C) unless the nameplate lists other values.

AC DRIVE MOTORS

An *AC drive motor* is a motor specifically designed for use with an electric motor drive. It is recommended that motors controlled by electric motor drives be designed specifically for electric motor drive use. The high switching frequencies and fast voltage rise times of AC drives produce stress on standard AC motors. The stress is produced by the high-frequency electrical output of the electric motor drive and sometimes the low speed of the motor. High switching frequencies and fast voltage rise times produced by an electric motor drive destroy standard motor insulation even when the motor is properly cooled.

Most electric motors are self-cooled as air is forced over the motor windings by a fan attached to the motor shaft. Industry oversizes and operates motors at less than full load capability to allow for adequate motor cooling, even when the motor is operated at low speeds. Motors operating at maximum torque (or close to maximum torque) and/or at low speeds are often destroyed if the insulation of the motor becomes overheated. A vector

MOTOR REVERSING WIRING

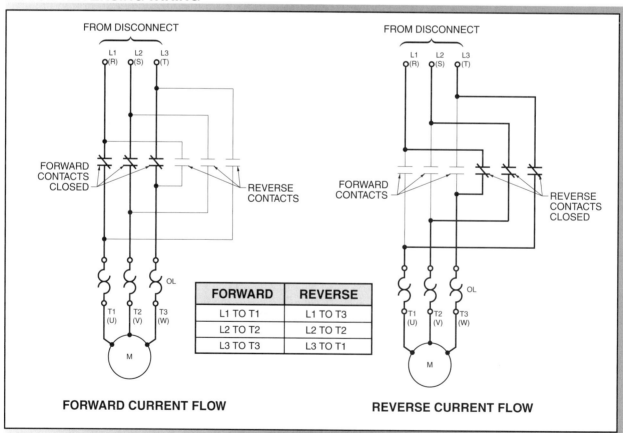

Figure 4-19. AC motors that are reversed during operation require a forward starter and a reversing starter.

or inverter duty-rated motor can be used to prevent overheating problems. A *vector* or *inverter duty-rated motor* is a motor made with wire and insulation that resist voltage spikes and high temperatures to extend the life expectancy of the motor.

TECH FACT

746 W equals 1 HP at 100% motor efficiency.

DIRECT CURRENT MOTORS

A *direct current (DC) motor* is a motor that uses direct current connected to the field and armature to produce shaft rotation. *Direct current* is current that flows in one direction only. Connecting voltage directly to the field and armature of a DC motor allows the motor to produce higher torque in a smaller frame than AC motors. DC motors do require more maintenance than AC motors because they have brushes that wear.

DC motors provide excellent speed control for acceleration and deceleration with effective and simple torque control. DC motors perform better than AC motors in most traction equipment applications. DC motors are used as the drive motor in mobile equipment such as golf carts, quarry and mining equipment, and locomotives. DC motors have the same external appearance as AC

motors, but differ in internal construction and output performance. The selection of the best type of DC motor to use is based on the mechanical requirements of the load to be driven.

DC Motor Construction

Any current-carrying conductor has a magnetic field around it. In a DC motor, a magnetic field is produced in the armature by current flowing through the armature coils. The armature magnetic field interacts with the magnetic field produced by the field. The magnetic field interaction rotates the armature. See Figure 4-20.

An *armature* is the rotating part of a DC motor. The *field* is the stationary windings, or magnets, of a DC motor. A *commutator* is the part of the armature that connects each armature coil to the brushes by using copper bars (segments) that are insulated from each other with pieces of mica. The commutator is mounted on the same shaft as the armature and rotates with the shaft. The brushes make contact with successive copper bars of the commutator, as the shaft, armature, and commutator rotate.

DC power is delivered to the armature coils through the brushes and commutator segments. The armature coils, commutator, and brushes are arranged so that the flow of current is in one direction in the loop on one side of the armature, and the flow of current is in the opposite direction in the loop on the other side of the armature.

DC MOTOR ROTATION

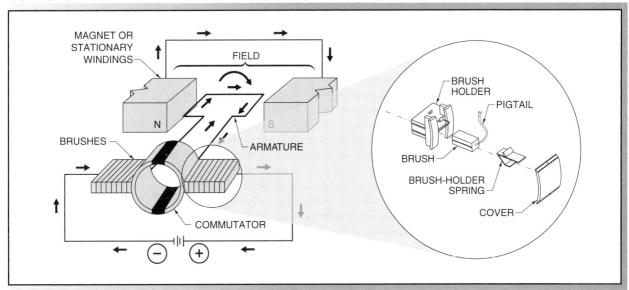

SEE APPENDIX FOR ⊕ TO ⊖ DIAGRAM

Figure 4-20. The stationary part of a DC motor is the field, and the rotating part is the armature.

A *brush* is the sliding contact that makes the connection between the rotating armature and the external circuit (power supply) to the DC motor. Brushes are made of carbon or graphite materials and are held in place by brush holders. A pigtail connects a brush to the power supply. A *pigtail* is an extended, flexible connection or a braided copper conductor. A brush is free to move up and down in a brush holder. The freedom allows the brush to follow irregularities in the surface of the commutator. A spring placed behind a brush forces the brush to make contact with the commutator. The spring pressure is usually adjustable, as is the entire brush-holder assembly, allowing shifting of the position of the brushes on the commutator.

The four types of DC motors are:
- series motor
- shunt motor
- compound motor
- permanent-magnet motor

DC Series Motors

A *DC series motor* is a motor with the field connected in series with the armature. The field must carry the load current passing through the armature. The field coil has comparatively few turns of heavy-gauge wire. The wires extending from the series coil are marked S1 and S2. The wires extending from the armature are marked A1 and A2. See Figure 4-21.

DC series motors are used as traction motors because DC series motors produce the highest torque of all DC motors. DC series motors can develop 500% of full-load torque upon starting. Typical applications include traction bridges, hoists, gates, and starting motors in automobiles.

The speed regulation of a DC series motor is poor. As the mechanical load on the motor is reduced, a simultaneous reduction of current occurs in the field and the armature. If the mechanical load is entirely removed, the speed of the motor increases without limit and may destroy the motor. For this reason, series motors are always permanently connected to the load the motor controls.

DC Shunt Motors

A *DC shunt motor* is a motor with the field connected in shunt (parallel) with the armature. The wires extending from the shunt field of a DC shunt motor are marked F1 and F2. The armature windings are marked A1 and A2. See Figure 4-22.

DC SERIES MOTORS

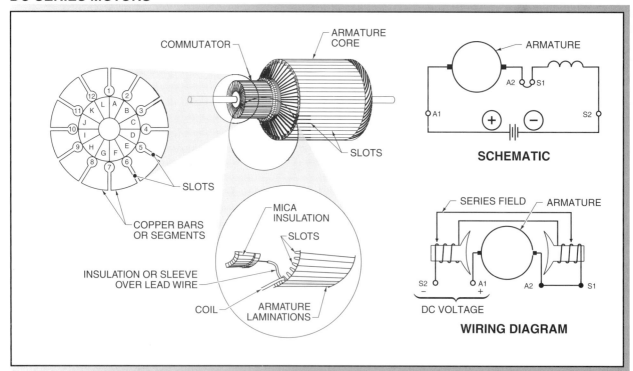

Figure 4-21. DC series motors have the field connected in series with the armature.

DC SHUNT MOTORS

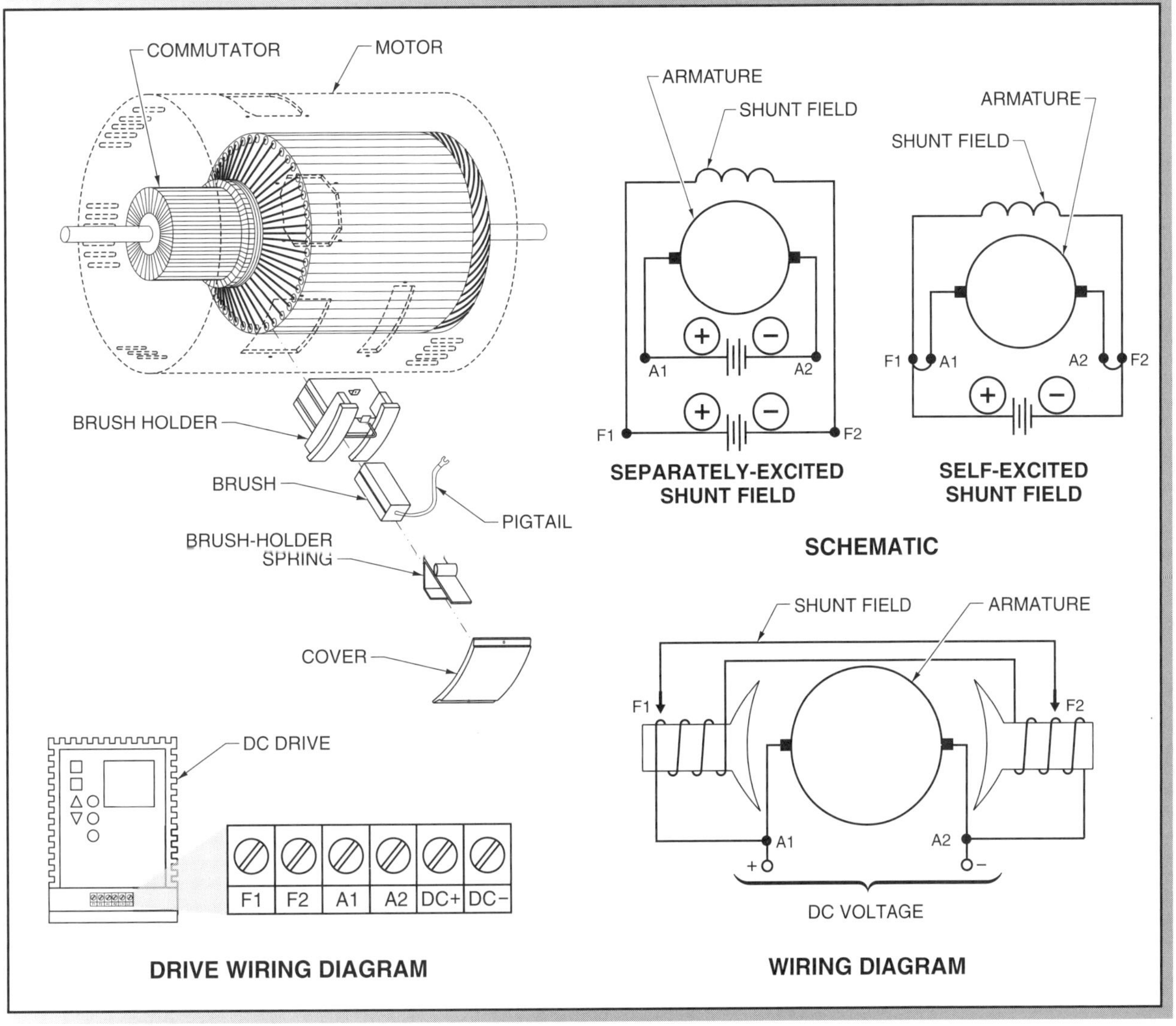

Figure 4-22. DC shunt motors have the field connected in parallel (shunt) with the armature.

The field has numerous turns of wire, and the current in the field is independent of the armature, providing the DC shunt motor with excellent speed control. The shunt field may be connected to the same power supply as the armature or may be connected to another power supply. A *self-excited shunt field* is a shunt field connected to the same power supply as the armature. A *separately excited shunt field* is a shunt field connected to a different power supply than the armature.

The DC shunt motor is used where constant or adjustable speed is required and starting conditions are moderate. Typical applications include fans, blowers, centrifugal pumps, conveyors, elevators, woodworking machinery, and metalworking machinery. In industrial applications, DC drives are used to control motor speed and provide motor protection.

In conventional DC speed control circuits, a variable external resistor is placed in series with the shunt field to control motor speed. When a DC drive is used to control a DC shunt motor, the armature leads and field leads are connected to separate terminals on the DC drive, making a separately-excited shunt field, so the drive can best control motor speed. A DC drive controls motor speed by varying armature voltage. DC drives provide far better control and protection of DC motors than conventional methods.

DC Compound Motors

A *DC compound motor* is a DC motor with the field connected in series and shunt with the armature. The field coil is a combination of the series field (S1 and S2) and shunt field (F1 and F2). The series field is connected in series with the armature. The shunt field is connected in parallel with the series field and armature combination. The arrangement gives the motor the advantages of the DC series motor (high torque) and the DC shunt motor (constant speed). The DC compound motor is used when high starting torque and constant speed are required. See Figure 4-23.

DC Permanent-Magnet Motors

A *DC permanent-magnet motor* is a motor that uses magnets, not a coil of wire, for the field winding. The DC permanent-magnet motor has molded magnets mounted into a steel shell. The permanent magnets are the field coils of the motor. DC power is supplied only to the armature. See Figure 4-24.

Permanent magnets become demagnetized if the armature is not rotating and the magnets are subjected to large amounts of current. The stalling of the motor shaft can occur during equipment jam-ups or breakdowns. Frequent starting of a permanent magnet motor demagnetizes the permanent magnets over time. Weakened magnets cause a reduction in speed and torque.

Following are some of the advantages of using DC drives over motor starters with permanent magnet motors:

• A DC drive can prevent magnet demagnetizing by detecting the high current (jam-up) and reducing or controlling the applied motor current.
• DC drive extends the life of the brushes by detecting and controlling the current applied to the motor through the brushes. The electronic circuits of the DC drive hold the applied current to safe levels, reducing arcing at the brushes.
• DC drive can detect problems instantly and take corrective action.

DC COMPOUND MOTORS

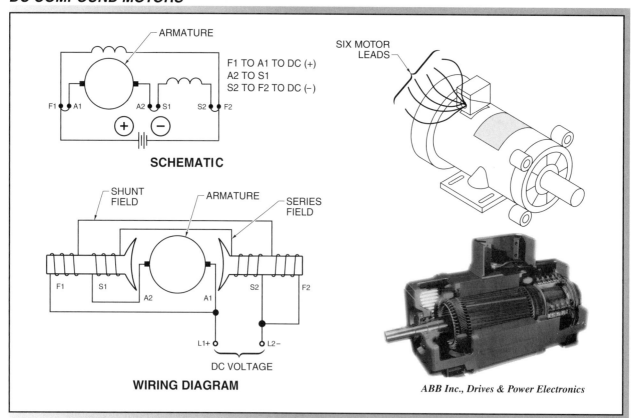

Figure 4-23. DC compound motors have the field connected in both series and shunt with the armature.

DC PERMANENT-MAGNET MOTORS

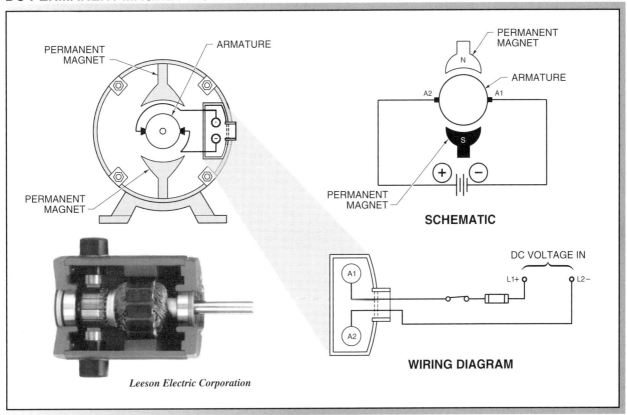

Figure 4-24. DC permanent-magnet motors use permanent magnets for the field.

Name _____ Date _____

True-False

(T)   F   1. The speed (rpm) of a DC motor is proportional to the voltage applied to the motor.

T   (F)   2. A rotor is the stationary part of an AC motor that produces a rotating magnetic field. *pg 90*

T   (F)   3. Shaded-pole motors produce high starting torque. *pg 94*

(T)   F   4. A capacitor motor is similar in design to a split-phase motor. *pg 96*

(T)   F   5. When connecting a dual-voltage motor, the higher voltage is usually preferred. *pg 100*

T   (F)   6. Typically, dual-voltage, 3ɸ motors have 12 leads coming out of the motor box. *pg 101*

T   (F)   7. DC motors require less maintenance than AC motors. *pg 105*

(T)   F   8. A separately excited shunt field is a shunt field connected to a different power supply than the armature. *pg 107*

T   (F)   9. The wires extended from the series coil of a DC motor are marked A1 and A2. *pg 106 coil = S1 & S2*

(T)   F   10. A DC shunt motor is a motor with the field connected in parallel with the armature. *pg 106*

(T)   F   11. Line frequency is the number of cycles per second of the supplied voltage. *pg 90*

T   (F)   12. The field is the part of the armature that connects each armature coil to the brushes by using copper bars that are insulated from each other with pieces of mica. *commutator pg 105*

(T)   F   13. A pigtail is an extended, flexible connection or braided copper conductor. *pg 106*

(T)   F   14. Locked in step is a motor condition that occurs when the field of the stator and the field of the rotor are parallel to one another, not allowing the shaft to rotate. *pg 90*

(T)   F   15. A shaded-pole motor is an AC motor that has running and starting windings. *pg 94*

Completion

_____ *slip* _____   1. ___ is the difference between the synchronous speed and the actual speed of a motor. *pg 91*

__ *centrifugal switch* __   2. A(n) ___ opens to disconnect the starting winding when the rotor reaches a preset speed. *pg 95*

__ *induction motor* __   3. A(n) ___ is a motor that has no physical electrical connection to the rotor because there are no brushes. *pg 98*

Delta-connected **4.** A(n) ___ 3φ motor has each phase wired end to end to form a completely closed loop circuit.

parallel **5.** When connecting a dual-voltage, delta-connected, 3φ motor to low voltage, the individual coils of each phase are in ___.

_2_ **6.** When motor starters are used to perform motor reversing, ___ starters are required.

armature **7.** A(n) ___ is the rotating part of a DC motor.

_50%_ **8.** DC series motors can develop ___% of full-load torque upon starting.

field winding **9.** A DC permanent-magnet motor uses magnets, not a coil of wire, for the ___.

Compound motor **10.** The DC ___ motor is used when high starting torque and constant speed are required.

Multiple Choice

_D_ **1.** ___ speed is the theoretical speed of a motor based on the number of poles of the motor and line frequency.
- A. Base
- B. Calculated
- C. Nameplate
- D. Synchronous

_C_ **2.** The industrial standard for reversing a 1φ motor is to reverse the ___ winding.
- A. commutator
- B. run
- C. starting
- D. synchronous

_C_ **3.** A capacitor start-run motor uses ___ capacitor(s) for successful operation.
- A. 0
- B. 1
- C. 2
- D. 3

_A_ **4.** When a dual-voltage, wye-connected, 3φ motor is connected to high voltage, the individual coils of each phase will be in ___.
- A. parallel
- B. series
- C. series-parallel
- D. shunt

_B_ **5.** Although any two lines can be interchanged to change the direction of rotation of a 3φ motor, the industrial standard is to interchange ___ and ___.
- A. T1; T2
- B. T1; T3
- C. T2; T3
- D. X2; X3

Name_____ Date _____

Activity 4-1. Motor Voltage Unbalance Derating

The voltage in the 3φ power lines must be balanced when delivering power to a motor. When the voltage in the power lines is not balanced, current is not balanced. As little as a 2% voltage unbalance causes a 20% current unbalance. A current unbalance creates extra heat in the stator of a motor. Manufacturers typically rate motors for an acceptable 1% voltage unbalance, with an assumed applied voltage of ±10% of the motor's rated voltage. When voltage unbalance is common, motors must be derated.

Determine the derated horsepower for each motor using data in the Motor Derating chart. Round derated horsepower to the lowest whole number.

MOTOR DERATING	
UNBALANCE*	MOTOR HP DERATING*
1	NONE
2	5
3	10
4	18
5	25
ABOVE 5	NOT ACCEPTABLE

* in percent

_____ **1.** A motor is derated to ___ HP when the voltage unbalance is 2%.

_____ **2.** A motor is derated to ___ HP when the voltage unbalance is 3%.

_____ **3.** A motor is derated to ___ HP when the voltage unbalance is 4%.

_____ **4.** A motor is derated to ___ HP when the voltage unbalance is 5%.

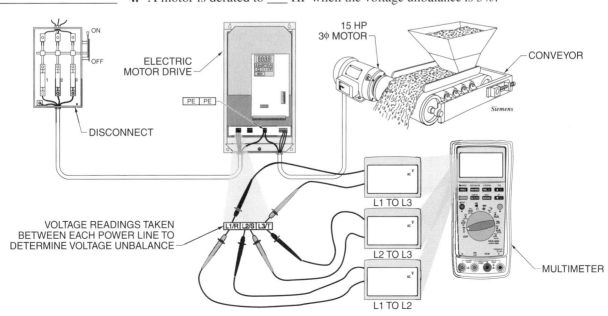

113

Activity 4-2. Operating Motor above Rated Speed

An AC motor can be operated at higher than 60 Hz when powered by an electric motor drive. The higher the frequency supplied to a motor, the faster the motor rotates. The upper limits are based on the voltage and mechanical balancing limits of the motor. In general, a motor can be operated at a speed that does not exceed 10% over the rated speed. To limit motor speed, an electric motor drive must be programmed to limit the upper frequency that a drive can output to a motor.

Determine the maximum frequency of an electric motor drive when limiting the operating speed to no more than 25% above rated speed. Round answer to lowest whole number.

_____ **1.** To safely operate Motor 1, the maximum electric motor drive frequency must be set for ___ Hz.

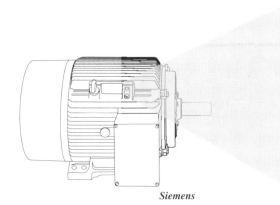

MANUFACTURER				
PE·21 PLUS™		PREMIUM EFFICIENCY		
ORD. NO.	1LA02864SE41	MAG.	21.0	
TYPE	RGZESDI	FRAME	286T	
HP	2	SERVICE FACTOR	1.0	3 PH
AMPS	9.6	VOLTS	230	
RPM	1725	HERTZ	60	
DUTY	CONT 40°C AMB.		DATE CODE	
CLASS INSUL	H	NEMA DESIGN B	KVA CODE G	NEMA NOM. EFF. 93.6
SH. END BRG.	50BC03JPP3	OPP. END BRG.	50BC03JPP3	

51-770-642

Inverter Duty AC Induction Motor   made in U.S.A.

Siemens

MOTOR 1

_____ **2.** To safely operate Motor 2, the maximum electric motor drive frequency must be set for ___ Hz.

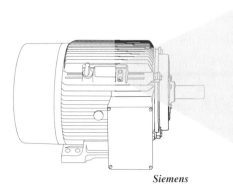

MANUFACTURER				
PE·21 PLUS™		PREMIUM EFFICIENCY		
ORD. NO.	1LA02864SE41	MAG.	21.8	
TYPE	RGZESDI	FRAME	286T	
HP	2	SERVICE FACTOR	1.0	3 PH
AMPS	9.6	VOLTS	230	
RPM	1150	HERTZ	50	
DUTY	CONT 40°C AMB.		DATE CODE	
CLASS INSUL	H	NEMA DESIGN B	KVA CODE G	NEMA NOM. EFF. 93.6
SH. END BRG.	50BC03JPP3	OPP. END BRG.	50BC03JPP3	

51-770-642

Inverter Duty AC Induction Motor   made in U.S.A.

Siemens

MOTOR 2

Activity 4-3. Connecting DC Permanent-Magnet Motor and Electric Motor Drive

Electric motor drive manufacturers provide wiring information with drives. The information shows required connections (supply power and motor power) and selected control connections that may be used (external forward, reverse, jog, and speed control).

1. Using the manufacturer information, connect the disconnect (including ground) to the required terminals to supply power to the electric motor drive and motor circuit. Connect the electric motor drive so that it operates using a three-position selector switch to control motor direction and an external potentiometer to control motor speed. Connect the DC permanent-magnet motor terminals for proper operation.

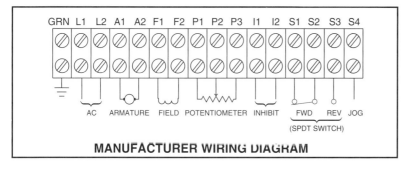

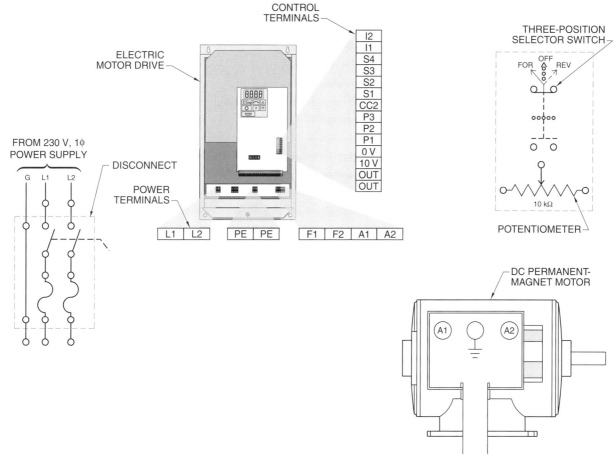

Activity 4-4. Connecting DC Shunt Motor and Electric Motor Drive

Electric motor drive manufacturers provide wiring information with drives. The information shows required connections (supply power and motor power) and selected control connections that may be used (external forward, reverse, jog, and speed control).

1. Using the manufacturer information, connect the disconnect (including ground) to the required terminals to supply power to the electric motor drive and motor circuit. Connect the electric motor drive so that it operates using a three-position selector switch to control motor direction and an external potentiometer to control motor speed. Connect the shunt motor terminals for proper operation.

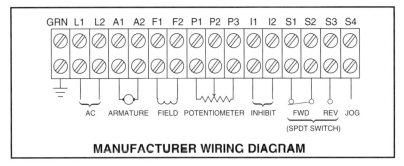

MANUFACTURER WIRING DIAGRAM

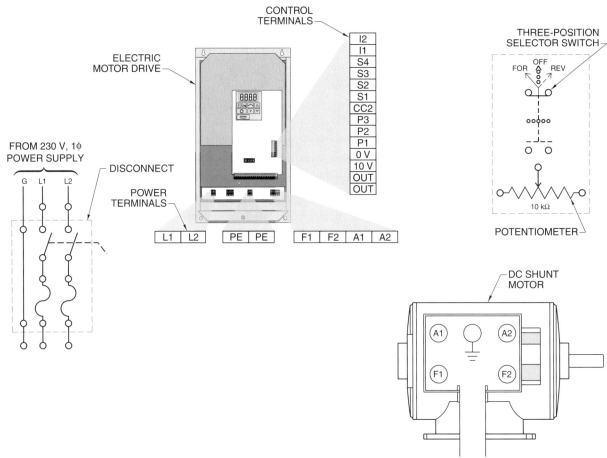

Activity 4-5. Connecting Wye Motor and Electric Motor Drive

Electric motor drive manufacturers provide wiring information with drives. The information shows required connections (supply power and motor power) and selected control connections that may be used (external forward, reverse, jog, and speed control).

1. Using the manufacturer information, connect the disconnect (including ground) to the required terminals to supply power to the electric motor drive and motor circuit. Connect the electric motor drive so that it operates using a liquid level switch to control forward direction and an external potentiometer to control motor speed. Connect the 3ϕ, dual-voltage wye motor terminals for high-voltage operation.

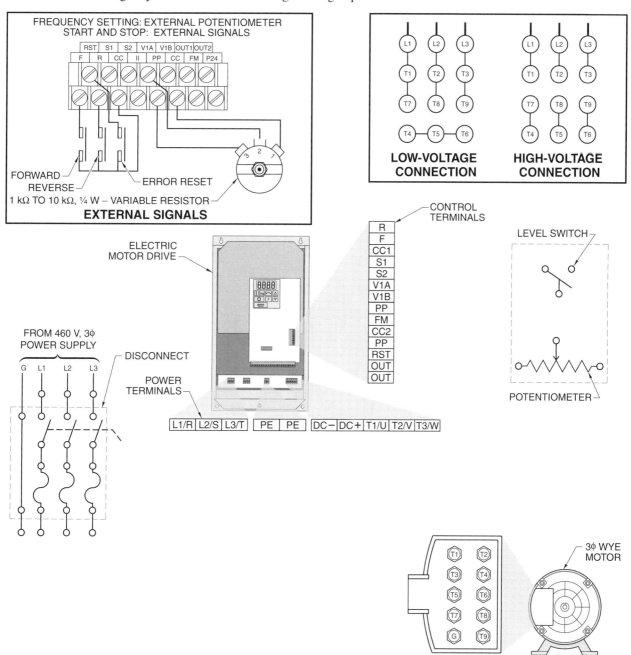

Activity 4-6. Connecting Delta Motor and Electric Motor Drive

Electric motor drive manufacturers provide wiring information with drives. The information shows required connections (supply power and motor power) and selected control connections that may be used (external forward, reverse, jog, and speed control).

1. Using the manufacturer information, connect the disconnect (including ground) to the required terminals to supply power to the electric motor drive and motor circuit. Connect the electric motor drive so that it operates using a temperature switch to control forward direction and a solar cell to control motor speed. The solar cell can also be placed where it will detect the strength of the sun. This allows the electric motor drive to automatically adjust motor speed as the sun increases in strength. Connect the 3ϕ, dual-voltage delta motor terminals for low-voltage operation.

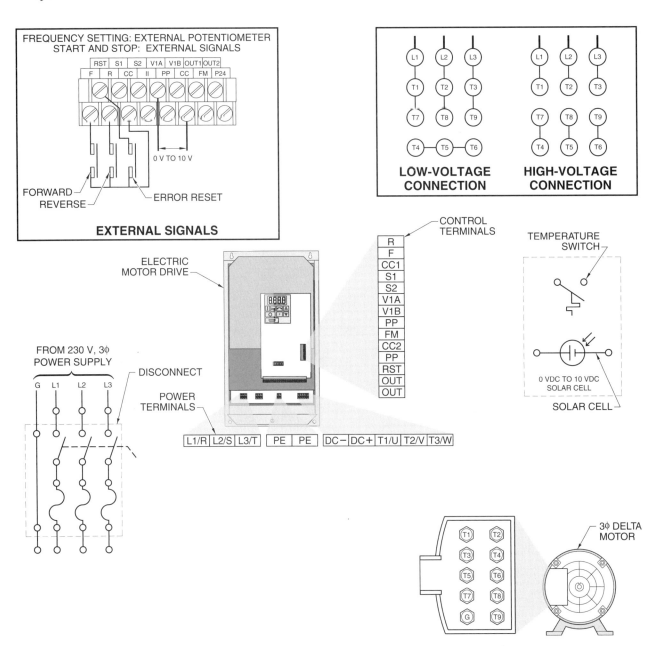

5 Electric Motor Control

The first motor controls used knife switches to connect a motor directly to power lines, but as motor sizes and safety concerns increased, better motor starters were developed. Today's high-tech world of faster and more complex but safer motor applications, and the desire to maintain a competitive edge, require that precise and reliable motor controls be used. Understanding the traditional methods of controlling motors and integrating new methods allows for a continuous smooth transition from old to new technologies.

METHODS OF MOTOR CONTROL

A motor must be correctly connected to an appropriate control circuit and power circuit to operate in a safe, efficient, and productive manner. A *motor control circuit* is a circuit that provides control functions such as starting, stopping, jogging, direction of rotation control, speed control, and motor protection. Motor control circuits vary from simple to complex. A basic on/off switch can be used to control small fractional horsepower motors that do not require any protection. Electric motor drives, programmable logic controllers (PLCs), personal computers, and integrated networks are used for complex motor control applications. Application parameters, motor size, cost limitations, and motor environment determine the type of motor control circuit used.

When a motor has been selected for an application, the type of motor control method must be determined. See Figure 5-1. Numerous specialized methods of controlling motors are available, but four control methods are commonly used for motor control. The control methods are used individually or in any combination. The four methods of motor control include the following:

- direct hard wiring
- terminal hard wiring
- using a programmable logic controller (PLC)
- using an electric motor drive

Each motor control method has advantages and disadvantages. Direct hard wiring is the most straightforward and oldest motor control method. Direct hard wiring to terminal strips allows for easier circuit changes and simplifies troubleshooting. Hard wiring to terminal strips was the standard method of wiring motor control circuits prior to the development of PLCs and electric motor drives. Using a PLC for motor control allows for great flexibility and circuit monitoring of a motor and control circuit. The main advantage of using PLCs is the flexibility of motor control, and not in the actual power circuit delivering power to a motor. A PLC can monitor and control all control functions that are connected to the PLC, but cannot directly monitor and display motor parameters such as voltage, current, frequency, and power. Electric motor drives allow for direct control of voltage, current, frequency, and power to a motor in addition to providing the basic control functions of starting, stopping, and jogging. Combining a PLC and an electric motor drive allows for maximum control and monitoring of a motor.

TECH FACT

Electric motors powered and controlled by AC drives should be "inverter rated."

MOTOR CONTROL METHODS

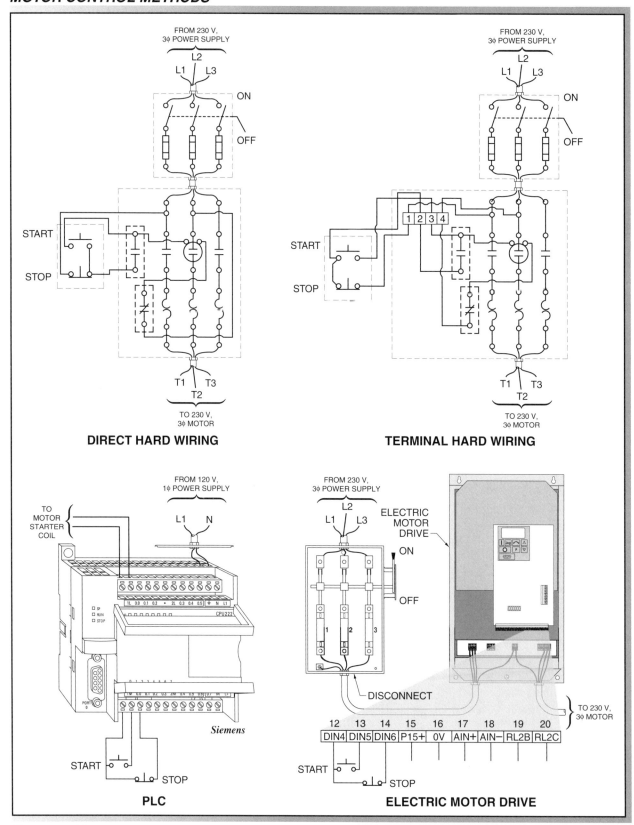

Figure 5-1. Motor control methods can be used individually, or in combination, to control a motor.

Manual Motor Starter Controls

A *manual control circuit* is a circuit that requires technicians or operators to initiate an action in order for the circuit and motor to operate. A *manual motor starter* is a control device used to control a motor by having technicians or operators control the motor directly at the location of the starter. Manual motor starters provide basic motor control and complex motor control such as soft start. See Figure 5-2.

A *manual motor starter* is a starter that has a manually operated switch (contactor) and includes motor overload protection. A *three-phase manual motor starter* is an electrical control device that manually energizes or de-energizes 3φ power to a load. Manual motor starters are used where simple motor control and overload protection (protection during operation) are required. To provide overload protection, a current-sensing element is added to the unit. Current-sensing elements are designed to detect the amount of current drawn by a motor. Current-sensing elements may be either thermal (heaters), magnetic, or solid state designs.

An overload protection device must be included in a motor starter because the National Electrical Code® (NEC®) requires that a control device not only turn a motor ON and OFF, but also protect the motor from destroying itself under an overloaded condition such as a locked rotor. A *locked rotor* is the condition in which a motor is loaded so heavily that the motor shaft cannot turn. The overload protection device senses the excessive current draw of the motor and opens the circuit to remove power from the motor.

Electric motor drives can be operated as manual motor starters because drives provide manual control of a motor directly from the keypad of the drive. Most electric motor drives provide manual control functions for starting, stopping, jogging, reversing, speeding up, and slowing down a motor directly from the keypad.

Manual motor controllers have an inherent problem. Manual motor controllers require technicians and operators to control a motor only from the motor starter and do not allow remote control or automatic control of a motor. Some motor control applications require that electrical control equipment such as pushbuttons, temperature switches, and limit switches be located in one area while the motor control devices such as starters or electric motor drives are in another location.

MANUAL MOTOR CONTROLS

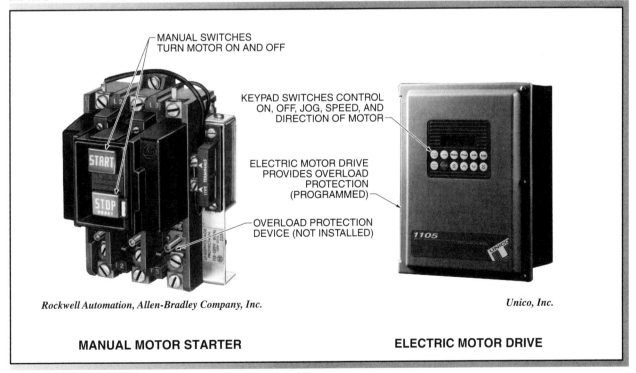

MANUAL SWITCHES TURN MOTOR ON AND OFF

KEYPAD SWITCHES CONTROL ON, OFF, JOG, SPEED, AND DIRECTION OF MOTOR

ELECTRIC MOTOR DRIVE PROVIDES OVERLOAD PROTECTION (PROGRAMMED)

OVERLOAD PROTECTION DEVICE (NOT INSTALLED)

Rockwell Automation, Allen-Bradley Company, Inc.

Unico, Inc.

MANUAL MOTOR STARTER

ELECTRIC MOTOR DRIVE

Figure 5-2. All electric motor drives provide manual motor starting through keypads.

Magnetic Motor Starter Controls

To allow a motor starter to be controlled from a remote location, magnetic motor starters are traditionally used. A *magnetic motor starter* is a starter that has an electrically operated switch (contactor) and includes motor overload protection. A magnetic motor starter is an electrical control device that uses a small control current to energize or de-energize 3ϕ power to a load. Magnetic motor starters include electrical contacts, a coil to magnetically open and close the contacts, and an overload protection device. See Figure 5-3.

The coil switches a motor ON or OFF when energized by a remotely located control switch or switches. Overload protection is provided by thermal, magnetic, or solid state overload elements that operate an overload contact that automatically removes power from a motor when an overload occurs. Magnetic motor starters include high-power contacts for switching a motor ON and OFF and typically include additional low-power auxiliary contacts that are used in the control circuit. The voltage of a coil in a magnetic starter is typically lower than the voltage used by the motor. When the coil and control circuit require lower voltage, a transformer is used to reduce the higher starter supply voltage to the lower voltage for coil and control circuit use. See Figure 5-4.

TECH FACT

Good motor starter coils measure 50 Ω–300 Ω, shorted coils measure 0 Ω–2 Ω, and open coils measure OL.

MAGNETIC MOTOR STARTER CONTROLS

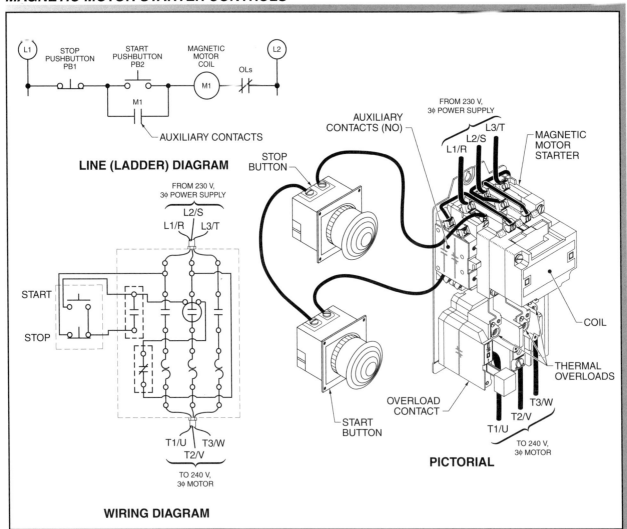

Figure 5-3. Magnetic motor starters allow for remote starting of a motor.

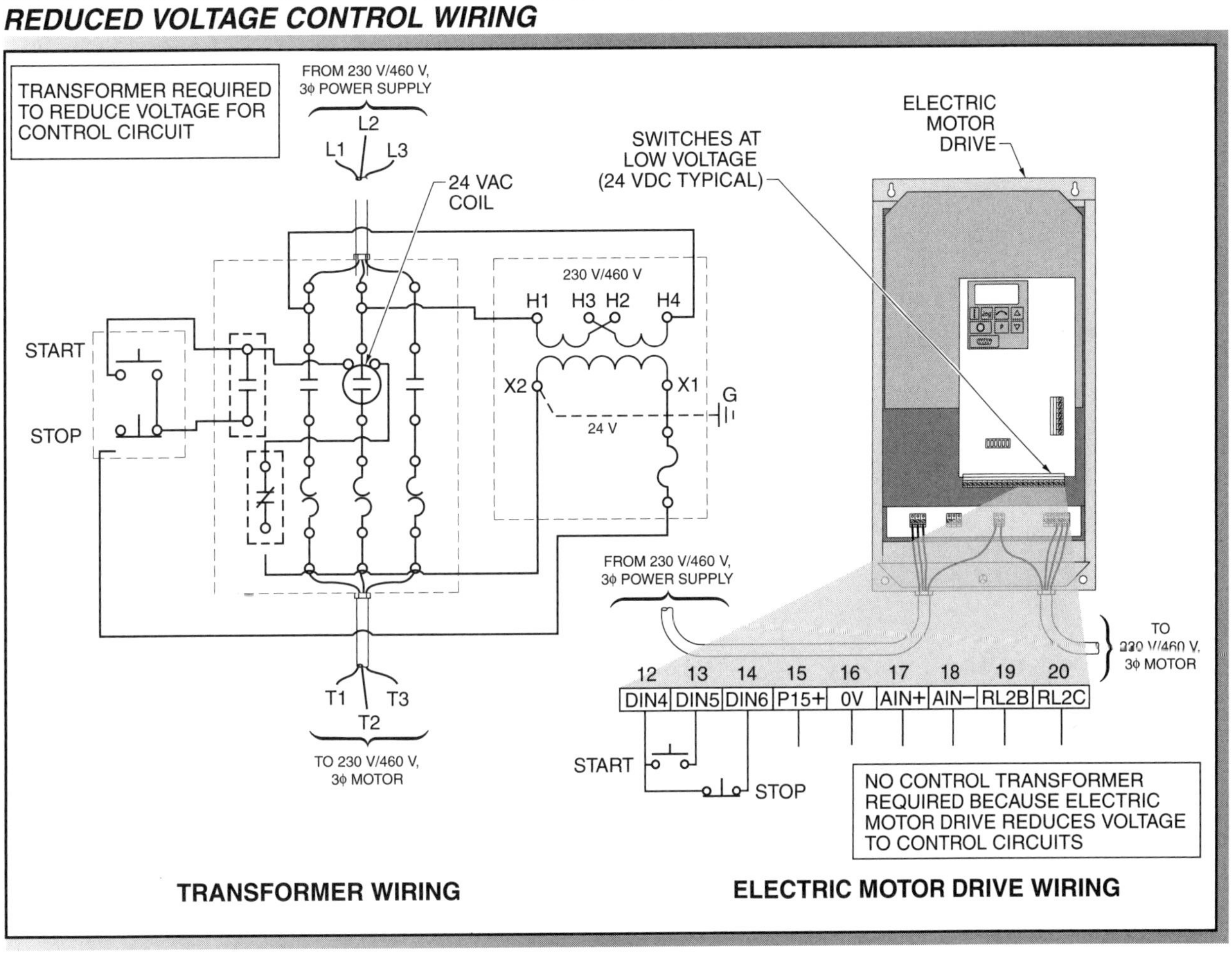

REDUCED VOLTAGE CONTROL WIRING

Figure 5-4. A transformer is used to reduce control circuit voltage for motor starters.

The voltage of a control circuit is lower than the rating of the motor in most motor control circuits. The standard is that 120 V control circuits are used to control 230 V or 460 V motors. Control voltages of 24 V or lower, whether AC or DC, are becoming standard because of the ease in wiring and greater safety of the lower voltage.

Electric motor drives are replacing magnetic motor starters and control circuits. Similar to magnetic motor starters, electric motor drives have remote control switches connected to the drive to control a motor from remote locations. Electric motor drives also use lower control voltages that allow for safer control circuits to control the drive and motor. The most common electric motor drive control voltage is 24 VDC.

Soft Start and Soft Stop Controls

A *soft starter* is a motor control device that provides a gradual voltage increase (ramp up) during AC motor starting and a gradual voltage decrease (ramp down) during stopping. Soft starters are used to control 1φ and 3φ motors. The capabilities and advantages of soft starters fall between a magnetic motor starter and an electric motor drive. A soft starter has advantages over a magnetic motor starter in that soft starters start and stop a motor gradually, causing less strain on the motor. A soft starter is the simplest solid state starter available, but provides fewer functions than basic electric motor drives.

Soft starting is achieved by increasing the voltage of a motor gradually in accordance with the setting of the ramp-up control. A potentiometer is used to set the ramp-up time (typically 1 sec to 20 sec). Soft stopping is achieved with a second potentiometer by decreasing the motor voltage gradually in accordance with the setting of the ramp-down time (typically 1 sec to 20 sec). A third potentiometer is used to adjust the starting level of the voltage to a motor with a value at which the motor starts to rotate immediately when soft starting is applied. See Figure 5-5.

SOFT STARTER CONTROLS

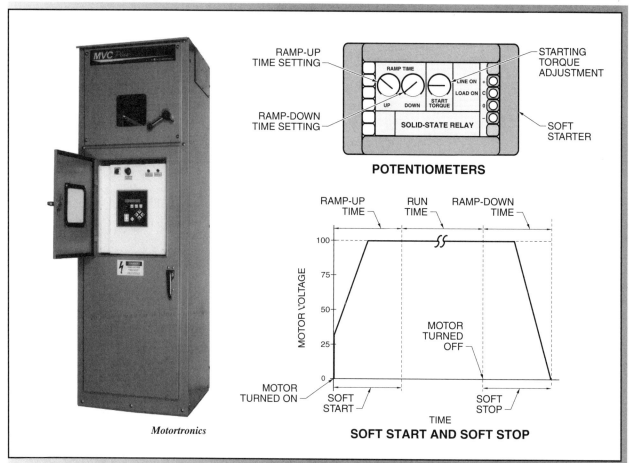

Figure 5-5. Soft starters provide a gradual voltage increase during motor starting.

Soft starters are used with newspaper webs to prevent damage to the web.

TECH FACT

Heat sinks of soft starters reach temperatures of 194°F (90°C). Allowances must be made for proper heat dissipation.

Similar to any electrical device or solid state switch, soft starters produce heat that must be dissipated for proper operation. Large heat sinks are required to dissipate the heat when high current loads (motors) are controlled. Typically, contactors are added in parallel with soft starters. The soft starter is used to control a motor when the motor is starting or stopping and the contactor is used to short out the soft starter when the motor is in operation. Placing contactors and soft starters in parallel allows for soft starting and soft stopping of motors without the need for large heat sinks during motor operation. Soft starters include an output signal which is used to control the time that a contactor is ON or OFF. See Figure 5-6.

SOFT STARTER WIRING

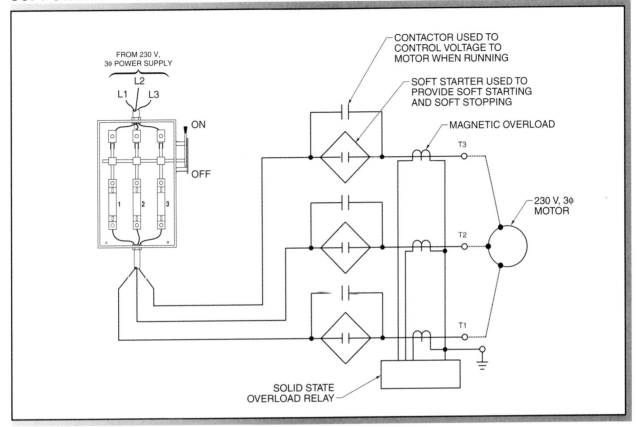

Figure 5-6. Contactors are used to bypass soft starters when a motor reaches speed.

Electric Motor Drives

Magnetic motor starters and soft starters are being replaced in many applications by electric motor drives. Electric motor drives are either AC drives or DC drives, with AC drives being the most common. AC drives are also referred to as adjustable-speed drives, variable-frequency drives, or inverters. See Figure 5-7.

Electric motor drives perform the same functions as motor starters, but also vary motor speed, reverse motors, provide additional protection features, and display operating information, and can be interfaced with other electrical equipment. Electric motor drives are used to control any size motor from fractional horsepower to hundreds of horsepower.

AC drives control and monitor motor speed by converting incoming AC voltage to DC voltage and then converting the DC voltage back to a variable-frequency AC voltage. To change the speed of a motor, electric motor drives vary the frequency (Hz) to the motor. A standard 60 Hz AC motor operates at full speed when connected to 60 Hz, at half speed when connected to 30 Hz, and at one-quarter speed when connected to 15 Hz.

DC drives control speed by controlling the DC voltage to a motor by varying the amount of voltage and current on the field and armature of the motor. AC power can also be connected to DC drives because the incoming AC voltage is converted to DC voltage.

The section of an electric motor drive that coverts AC voltage to DC voltage is the converter section, and the section that converts the DC voltage back to AC voltage is the inverter section. A *converter* is an electronic device that changes AC voltage into DC voltage. An *inverter* is an electronic device that changes DC voltage into AC voltage. See Figure 5-8. Silicon-controlled rectifiers (SCRs) or diodes are used in the converter section to convert the incoming AC voltage into DC voltage. In addition to converting AC voltage into DC voltage, SCRs can also control the level of the DC voltage.

TECH FACT

New electric motor drives are more energy-efficient than drives 5 yr old. Replacing old electric motor drives is cost-effective and reduces power plant emissions.

AC AND DC DRIVES

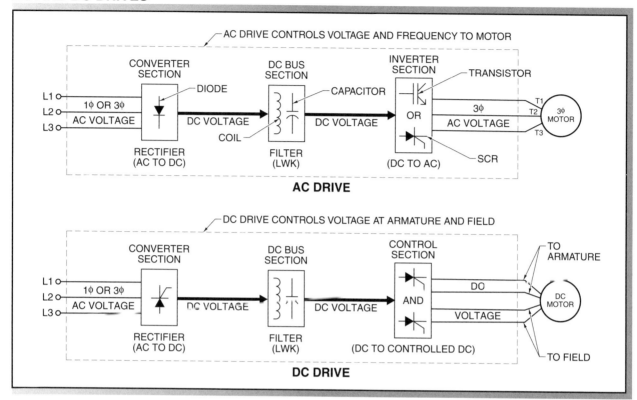

Figure 5-7. The two main types of electric motor drives are AC drives and DC drives.

AC drives use the inverter section switches to switch the DC voltage ON and OFF to reproduce simulated AC sine waves. The simulated AC sine waves are actually pulsating DC voltages at different time intervals. The pulsating DC voltage is used to operate AC motors and control the speed of the motors by varying the voltage and frequency applied to the motors. AC drives control the voltage and frequency sent to a motor by switching the DC voltage ON and OFF at the correct moments. Electronic switches (SCRs and transistors) are used because of the fast switching required to simulate AC sine waves. A microprocessor circuit located within an electric motor drive controls the switching.

Inverters are classified by the method used to change DC voltage to AC voltage. Inverter designs include variable voltage inverters (VVI), current source inverters (CSI), and pulse width modulated (PWM) inverters.

⚠ **CAUTION**

Verify all technicians and personnel are clear of equipment when testing electric motor drive control circuits.

ELECTRIC MOTOR DRIVE CONTROL SELECTION

Electric motor drives are typically controlled by any of three different methods. Electric motor drive control methods include local control, remote control, and PLC/PC/HIM control. A *programmable logic controller (PLC)* is a solid state control device that is programmed to automatically control an industrial process or machine. A *personal computer (PC)* is a desktop or laptop computer intended for personal use in a home, office, or factory. A *human interface module (HIM)* is a manually operated input control unit that includes programming keys, system operating keys, and normally a status display. When one of the three control methods is selected, the other two are not operable.

When using the local control method, motor speed (frequency) is controlled from the keypad of the electric motor drive. The UP and DOWN buttons on an electric motor drive keypad are used to control motor speed. The local control method is automatically set when power is first applied to an electric motor drive. To change motor speed and other control functions other than by using the keypad, an electric motor drive must be programmed for remote control.

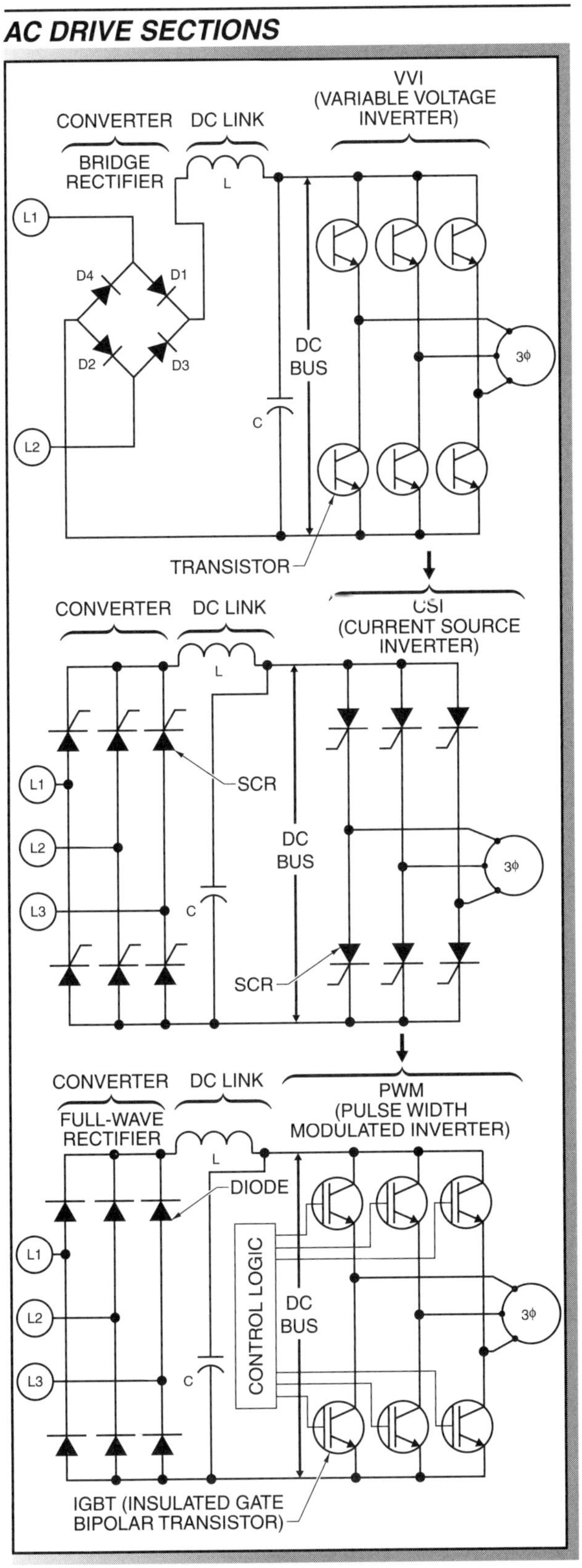

AC DRIVE SECTIONS

Figure 5-8. AC drives include a converter section, DC link (bus) section, and inverter section.

When using the remote control method, motor speed (frequency) may be controlled by an external potentiometer connected to the control terminals of an electric motor drive. Electric motor drives change motor speed based on several different variables such as variable resistance from thermistors, photoresistors, or other devices, with 0 V to 10 V signals, or 4 mA to 20 mA signals. Most electric motor drives allow a set number of external switches to be used to preselect motor speeds.

All electric motor drives allow for local control and at least some remote control functions. Some electric motor drives allow a PLC, PC, or HIM connected to the drive to control speed and other functions of a motor. The output signals of control devices are connected to the inputs of an electric motor drive where standard pushbuttons and other control devices are normally connected. Some electric motor drives provide separate ports that allow infinite speed control from an external control unit. Some electric motor drives allow speed control (frequency control) and other control functions to be controlled through standard communication ports such as RS-232C ports.

REDUCED-VOLTAGE STARTING CONTROLS

Full-voltage starting is the least expensive means of starting a motor. Full-voltage starting connects a motor directly to power supply lines through starters. Manual motor starters and magnetic motor starters provide full-voltage starting. Full-voltage starting of small horsepower motors usually does not cause any problems on electrical distribution systems that are properly sized. Full-voltage starting of large horsepower motors typically causes problems in power distribution systems and/or with the motors. To prevent or reduce problems, reduced-voltage starting is used. Reduced-voltage starting reduces interference in the power distribution system, motors, and the electrical environment near motors.

Reduced-voltage starting is used to reduce the large current drawn from the power company when full-voltage starting is applied to large horsepower motors. An AC motor acts like a short circuit on the secondary side of a transformer when the motor is started. The current drawn by a motor is several times higher than the rated nameplate current of the motor. The sudden demand for large current reflects back into the power distribution lines and creates problems. Reduced-voltage starting reduces the amount of current a motor draws and the current drawn from the power distribution system during starting. See Figure 5-9.

REDUCED-VOLTAGE STARTING

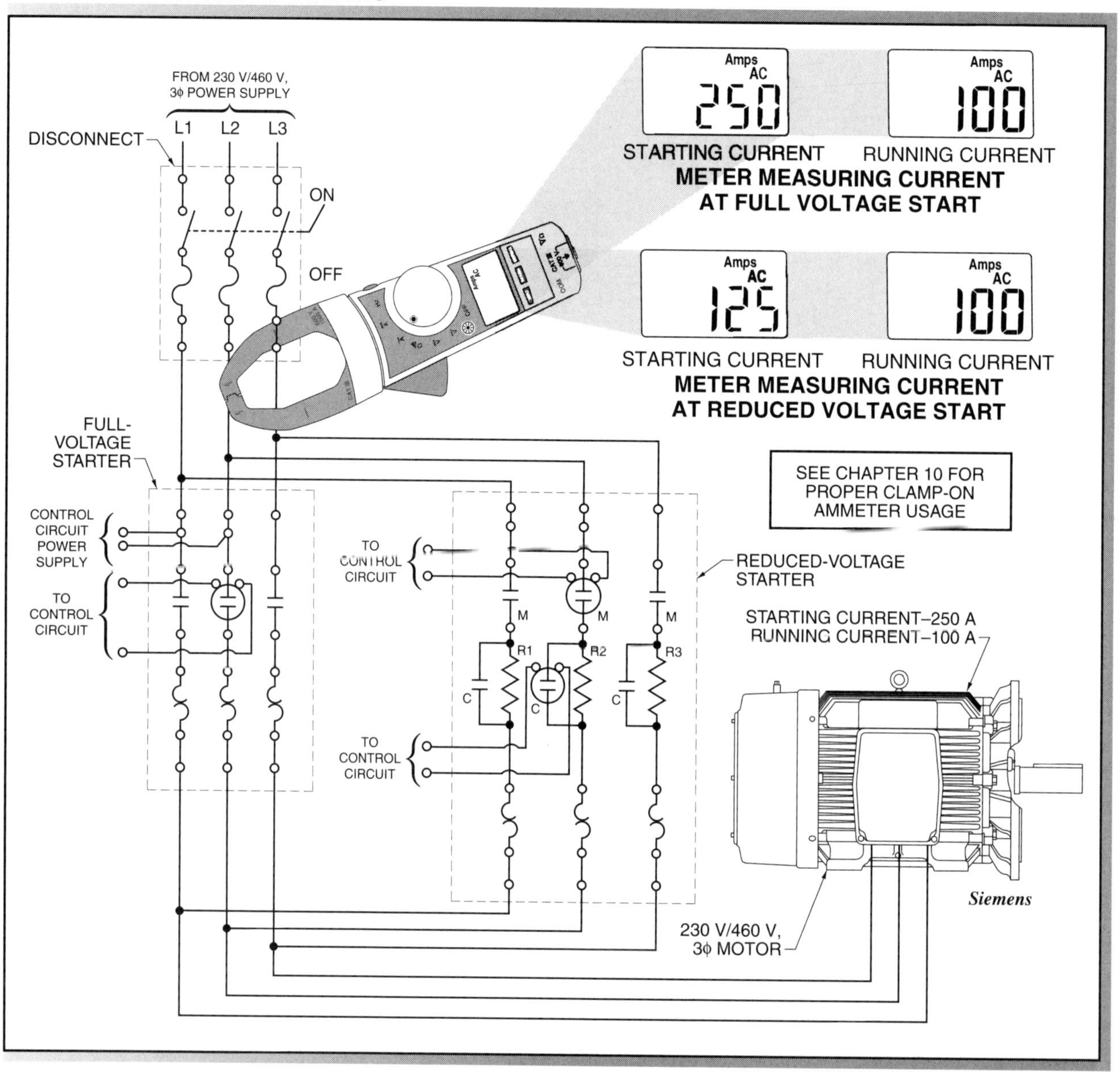

Figure 5-9. Reduced-voltage starting is used to reduce the amount of starting current and/or torque at a motor during starting.

Some applications such as controlling paper or fabric or other delicate materials require that care be taken to avoid sudden high starting torque (turning force). High torque stretches or tears delicate products. To prevent product damage or damage to gear drives, belt drives, and chain drives, technicians are required to limit starting torque surges. Reduced-voltage starting is used to overcome excessive starting torque by providing a gentle start and smooth acceleration of a motor. The traditional methods used to provide reduced-voltage starting of large horsepower motors are primary resistor starting, autotransformer starting, part-winding starting, wye-delta starting, and solid state starting.

Each of these reduced-voltage starting methods provide a way to reduce the amount of starting voltage and current (starting torque) drawn by a motor. Traditional methods only allow two or three incremental steps of current reduction. Soft starters provide a gradual voltage increase, but electric motor drives are typically used today to provide reduced-voltage starting of motors. Electric motor drives are not limited to a few incremental starting steps, but provide a smooth, controlled linear acceleration of the motor.

Primary Resistor Starting

Primary resistor starting is a motor starting method that uses resistors connected in the motor conductors to produce a voltage drop. The resistors reduce the starting current drawn by a motor. A timer is provided in the control circuit to short the resistors after the motor accelerates to a specified speed. The motor is started at reduced voltage but operates at full line voltage. See Figure 5-10.

> ⚠ **CAUTION**
>
> *Avoid contacting the resistors of a primary resistor starting unit, which reach temperatures of 700°F (371°C). Failure to observe the caution may result in a minor to moderate burn injury. Other than a laser, an electric arc is the hottest heat source in existence. Electric arcs are capable of producing temperatures up to 10,000°F. Temperatures of such intensity are capable of producing serious burns at distances up to 20'.*

PRIMARY RESISTOR STARTING

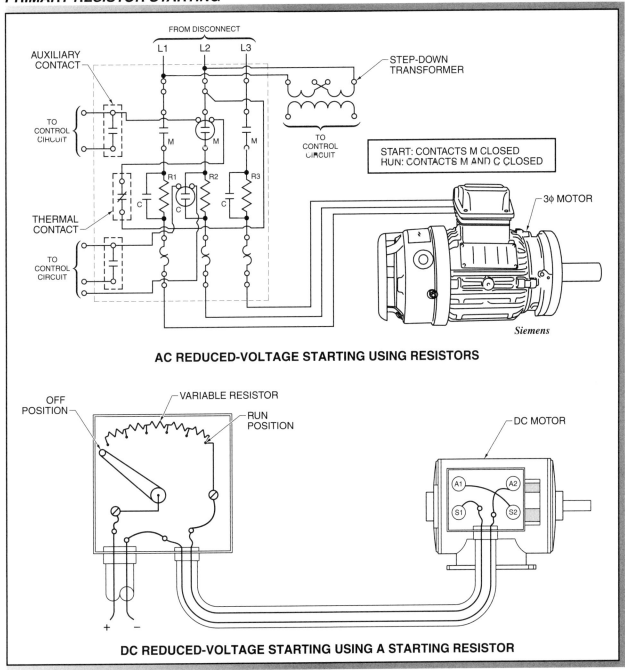

AC REDUCED-VOLTAGE STARTING USING RESISTORS

DC REDUCED-VOLTAGE STARTING USING A STARTING RESISTOR

Figure 5-10. External resistors are added to a starter circuit to reduce starting voltage and current.

TECH FACT

Do not use automatic restart circuits or parameters with reduced-voltage starting methods without consulting the drive manufacturer and applicable federal, state, and local codes.

Standard primary resistor starters provide two-point acceleration (one step of resistance), with approximately 70% of line voltage delivered to the terminals of the motor at the instant the motor starts. Multiple-step starting is possible by using additional contacts and resistors. As multiple steps are added, the cost and size of a primary

resistor starting circuit increases. Primary resistor starters are being replaced by electric motor drives that provide infinite steps without adding any additional size or cost to a system.

Autotransformer Starting

Autotransformer starting is a motor starting method that uses a tapped 3φ autotransformer to provide reduced-voltage starting. See Figure 5-11. Autotransformer starting is preferred over primary resistor starting when the starting current drawn from the supply lines must be held to a minimum value but the maximum starting torque per supply line ampere is required.

AUTOTRANSFORMER STARTING

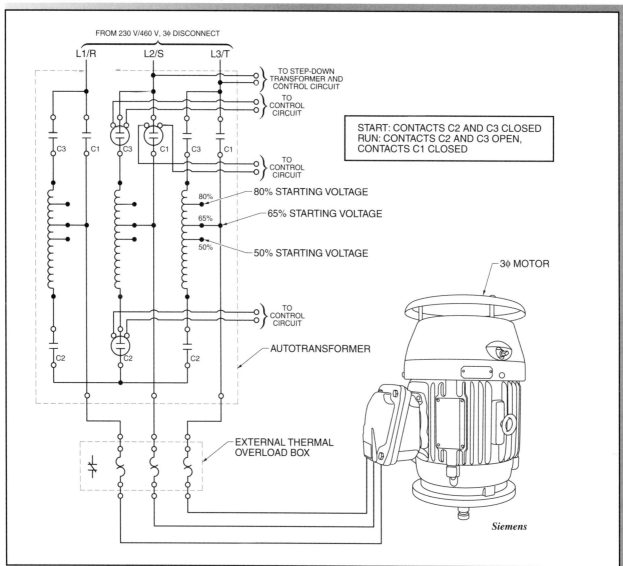

Figure 5-11. Autotransformers are used for reduced-voltage starting.

The two types of autotransformer control circuits are the open circuit transition and the closed circuit transition. A motor using an open circuit transition may be temporarily disconnected when moving from one incremental voltage to another. A motor using closed circuit transition is never removed from a source of voltage when moving from one incremental voltage to another. Closed circuit transition is preferred because it produces the least amount of interference to an electrical system compared to opening and reconnecting a motor in open circuit transition. All electric motor drives have the advantage of providing closed circuit transition through the output voltage ranges of the drive.

Part-Winding Starting

Part-winding starting is a motor starting method that first applies power to part of the coil windings of a motor for starting and then applies power to the remaining coil windings for normal operation. Motor stator windings must be divided into two or more equal parts for a motor to be started using part-winding starting. Each equal part must also have a terminal available for external connection to a power supply. In most applications, a dual-voltage, wye-connected motor is used, but a delta-connected motor can also be started using part-winding starting. See Figure 5-12.

Part-winding starting is not a true reduced-voltage starting method, because the actual voltage to the motor windings is never reduced. Part-winding starting is classified as reduced-voltage starting because of the resulting reduced current and torque of the motor. No additional resistors or autotransformers are required, allowing part-winding starting to be less expensive than other methods. Part-winding starting also has the advantage of closed circuit transition from part winding to full winding.

The disadvantage of part-winding starting is that part-winding starting produces poor starting torque because the motor is started with only half the motor windings connected to the power supply. Motor manufacturers do not recommend using part-winding starting with certain motors because of the internal winding design and characteristics of some motors. Electric motor drives are replacing part-winding starters because a drive connects all the windings of a motor for use during starting. Electric motor drives have all of the motor windings powered at all times.

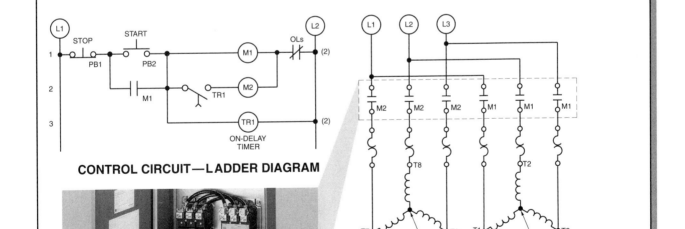

PART-WINDING STARTING

CONTROL CIRCUIT—LADDER DIAGRAM

Rockwell Automation – Allen-Bradley Company, Inc.

POWER CIRCUIT—SCHEMATIC

Figure 5-12. Part-winding starting starts a motor with part of the motor winding removed from the circuit until the motor reaches speed.

Wye-Delta Starting

Wye-delta starting accomplishes reduced-voltage starting by connecting a motor into a wye configuration for starting. When the motor reaches speed, the motor is reconnected to run as a delta motor. See Figure 5-13.

A wye-delta starting method does not require starting resistors or autotransformers, but does require a special wye-delta designed motor and control circuit. Wye-delta motors are specially wound with six leads extending from the motor to enable the windings to be connected in either a wye or delta configuration. Wye-delta motors are more expensive because wye-delta motors have a special design. Electric motor drives are replacing wye-delta starting because drives are used with most standard motors, which reduces cost.

Solid State Starting

A *solid state starter* is a motor starting device that uses solid state switches to control the voltage to a motor. A *solid state switch* is an electronic switching device that opens and closes circuits at a precise point in time to control current flow and voltage levels. The advantages of solid state switching include fast action, absence of moving parts, and long life. The exact type of solid state switching used depends upon the motor and control circuit design. All solid state switching are used as basic ON and OFF switches and some can be used to provide smooth, stepless acceleration as with the soft starting feature of electric motor drives. Transistors, silicon-controlled rectifiers (SCRs), triacs, and alternistors are used as solid state switches. See Figure 5-14.

WYE-DELTA STARTING

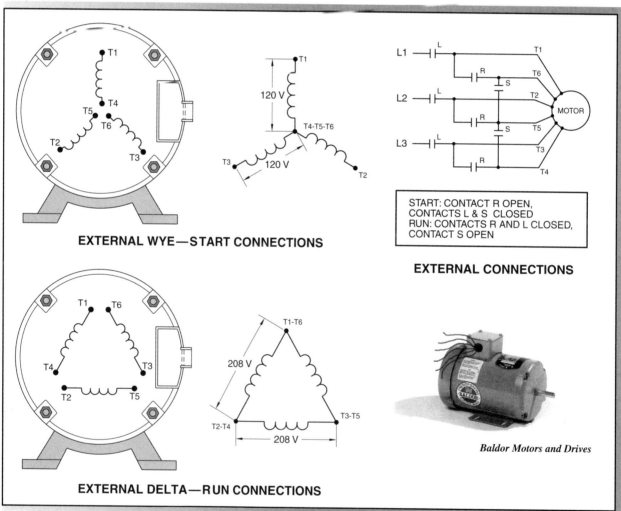

Baldor Motors and Drives

Figure 5-13. Windings of a special wye-delta motors are connected as a wye connection for starting and as a delta connection for normal operation.

SOLID STATE STARTING

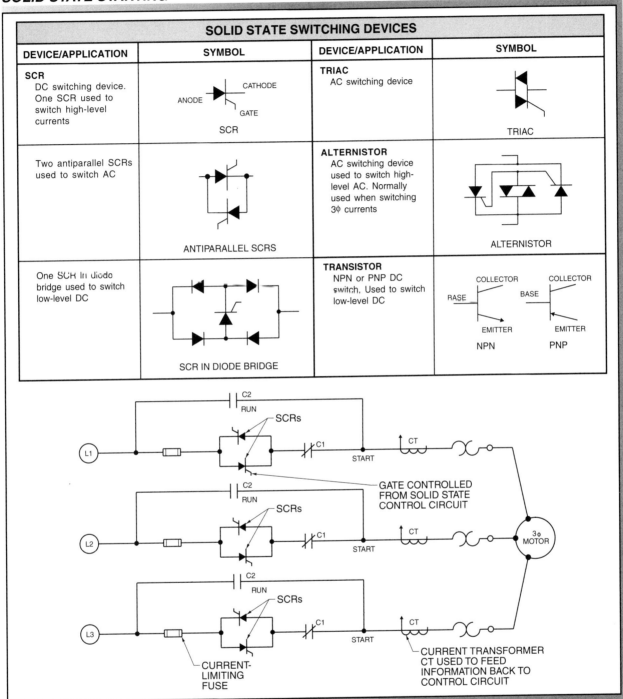

Figure 5-14. Solid state switches include silicon-controlled rectifiers, triacs, alternistors, and transistors.

Silicon-Controlled Rectifiers. Silicon-controlled rectifiers (SCRs) are solid state devices with the ability to rapidly switch heavy currents. SCRs are used as solid state switches for switching low and high DC currents. When controlling AC currents, two SCRs are mounted in an antiparallel configuration, with each SCR controlling one-half of the AC sine wave. The advantage of using two separate SCRs instead of one triac is the greater heat dissipation of two SCRs. One SCR in a diode bridge can be used when low-level current switching is required.

Triacs. Triacs are solid-state switching devices used to switch alternating current. Typically, triacs are used as solid-state AC switches. Triac switches are either fully ON or fully OFF, but are also used to control the level of output voltage over a range.

Alternistors. Alternistors are two antiparallel thyristors and a triac mounted as one unit. Alternistors were developed specifically for industrial AC high-current switching applications. Alternistors require less space than the antiparallel configuration of SCRs, and three alternistors are typically used when switching 3ϕ power.

Transistors. Transistors are three-terminal electronic switching devices that control current through the device by referencing the amount of applied voltage to the base. Transistors are NPN or PNP and can be switched ON and OFF at very fast speeds. Transistors have extremely high resistance when open and extremely low resistance when closed.

Cleveland Motion Controls
Solid state switches such as silicon-controlled rectifiers, triacs, alternistors, and transistors are typically used as ON and OFF switches.

SPEED CONTROL

Speed control is essential in most motor control applications. Industrial applications of speed control include mining machines, printing presses, cranes and hoists, elevators, assembly line conveyors, food processing equipment, and metalworking or woodworking machines. Traditionally, AC motors only provided up to five different speeds. Limited-speed AC motors are suitable for applications that only require a few different speeds, such as a three-speed washing machine motor. AC motors were not used in applications requiring variable speeds because to change the speed of an AC motor required changing the number of poles in the motor or changing the frequency of the voltage sent to the motor. Changing frequency was not practical or cost-effective before electric motor drives, and the number of poles a motor has is limited by space. The result was that DC motors were used when a range of speeds was required.

Changing the voltage applied to a DC motor changes the speed of the motor. Changing voltage is less complicated than changing frequency. The disadvantage of using DC motors is that DC motors require high maintenance due to brushes wearing.

Voltage and Frequency

The voltage applied to the stator of an AC motor must be decreased by the same amount as the frequency when controlling speed. An AC motor heats excessively and damage occurs to the windings of the motor when voltage is not reduced as frequency is reduced. An AC motor does not produce rated torque when the voltage is reduced more than required.

The ratio between the voltage applied to a stator and the frequency of the voltage applied to the stator must be constant. The ratio is referred to as the volts-per-hertz (V/Hz) ratio (constant volts-per-hertz characteristic). *Volts-per-hertz (V/Hz)* is a control mode that provides a linear voltage ratio to the frequency of a motor from 0 rpm to base speed. AC motors develop rated torque when the volts-per-hertz relationship is kept constant (linear).

The volts-per-hertz ratio for an induction motor is found by dividing the rated nameplate voltage by the rated nameplate frequency. To find the volts-per-hertz ratio for an AC induction motor, apply the formula:

$$V/Hz = \frac{V}{Hz}$$

where

V/Hz = volts-per-hertz ratio

V = rated nameplate voltage (in V)

Hz = rated nameplate frequency (in Hz)

For example, what is the volts-per-hertz ratio when a motor nameplate rates an AC motor for 230 VAC at 60 Hz?

$$V/Hz = \frac{V}{Hz}$$

$$V/Hz = \frac{230}{60}$$

$$V/Hz = \textbf{3.83}$$

> **⚠ CAUTION**
>
> *Applying voltage boost to motors from an electric motor drive for long periods of time causes damage and/or fires to motors because of motor insulation overheating. Electric motor drives do not provide direct thermal protection for motors.*

Above approximately 15 Hz, the amount of voltage needed to keep the volts-per-hertz ratio linear is a constant value. Below 15 Hz the voltage applied to a motor stator must be boosted to compensate for the large power loss AC motors experience at low speeds. The amount of voltage boost depends on the specific motor design. See Figure 5-15.

Electric motor drives are programmed to apply a voltage boost at low motor speed to compensate for the power loss of a motor. The voltage boost gives the motor additional rotor torque at very low speeds. The amount of torque boost depends on the voltage boost programmed into the electric motor drive. The higher the voltage boost, the greater the torque of the motor. See Figure 5-16.

Electric motor drives are also programmed to change the standard linear volts-per-hertz ratio to a nonlinear ratio. Nonlinear ratios produce customized motor torque patterns that are required by the operating characteristics of an application; for example, an electric motor drive may be programmed for two nonlinear ratios that are applied to fan or pump motors. Fans and pumps are typically classified as variable torque/variable horsepower loads. Variable torque and variable horsepower loads require varying torque and horsepower at different speeds.

SWITCH FUNCTIONS

All switches serve some function in an electrical circuit. The primary function of any switch is to start, stop, or redirect the flow of current in a circuit. Switches are typically designated by the function the switch performs such as on, off, start, stop, emergency stop, up, down, left, right, forward, reverse, fast, slow, or reset. To perform a function, switches are wired into an electrical circuit as single-function switches or dual-function switches. See Figure 5-17.

A *single-function switch* is a switch that performs only one switching function such as START or STOP. When performing one function, one position of the switch performs the designated function, such as starting the motor, and the other switch position performs no function at all.

VOLTAGE BOOST STARTING

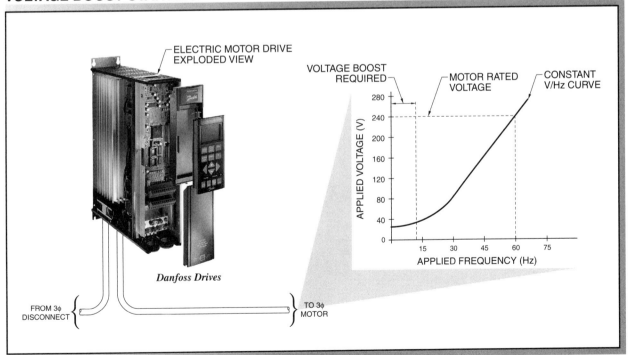

Danfoss Drives

Figure 5-15. The ratio of motor voltage and frequency must be constant to avoid motor overheating.

PROGRAMMED VOLTAGE BOOST STARTING

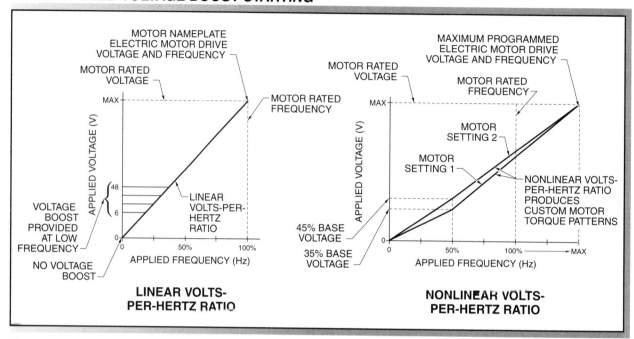

Figure 5-16. Electric motor drives provide voltage boost at low speeds to improve the starting torque of a motor.

SWITCH FUNCTIONS

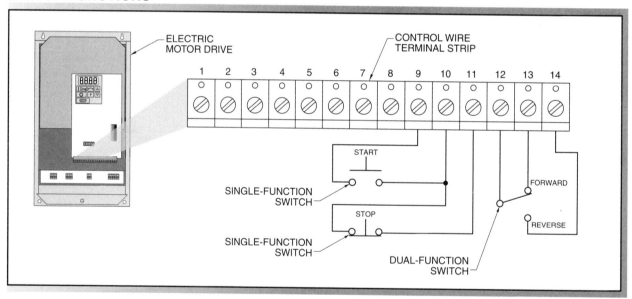

Figure 5-17. All switches are either dual-function or single-function.

A *dual-function switch* is a switch that performs two different switching functions such as forward or reverse. When performing two functions, one position of the switch is used for one function, such as turning the motor clockwise (forward), and the other position of the switch is used for the second function, such as turning the motor counterclockwise (reverse).

TWO-WIRE MOTOR CONTROL

Control circuits are often referred to by the number of conductors used in the control circuit, such as two-wire or three-wire control. *Two-wire control* is an input control for an electric motor drive requiring two conductors to complete a circuit. *Three-wire control* is an input control for an electric motor drive requiring three conductors to complete a circuit.

Two-wire control has two wires leading from a control device to a starter. See Figure 5-18. A control device can be a thermostat, float switch, or other device. Two-wire motor control circuits are used when motors are required to operate automatically from such control devices as liquid level switches, pressure switches, or temperature switches.

TECH FACT

Electric motor drives controlled by two-wire and three-wire controls, rapidly and precisely control machinery movements.

Two-wire control is also used when using an electric motor drive to control a motor. Electric motor drives automatically reduce the voltage to control switch circuits. Typically, the control voltage used by electric motor drives is low voltage DC (24 VDC). See Figure 5-19.

TWO-WIRE MOTOR STARTER CONTROL

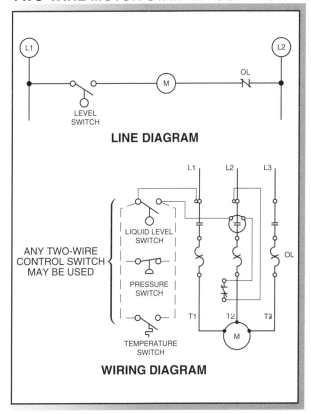

Figure 5-18. Two-wire starter control has two wires leading from the control switch to the starter.

ELECTRIC MOTOR DRIVE TWO-WIRE CONTROL

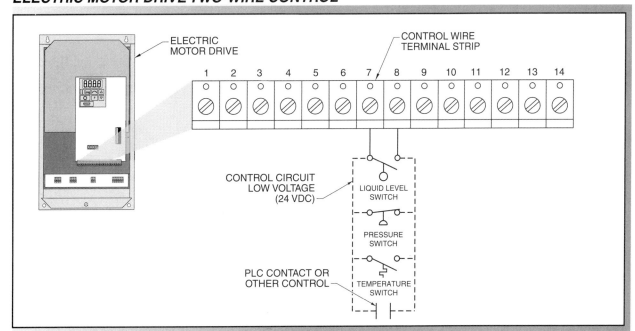

Figure 5-19. Electric motor drives use two-wire control from standard switches.

THREE-WIRE MOTOR CONTROL

Three-wire control has three wires leading from the control device to a starter to complete the circuit. Auxiliary contacts are added to starters to give memory to three-wire control circuits that use pushbuttons. When a motor starter coil (M) is energized, the coil causes the normally open contacts (NO) to close and remain closed (memory) until the coil is de-energized. *Memory is a control function that keeps a motor running after the start pushbutton is released.* Memory circuits are also known as holding or sealing circuits. When a memory circuit is ON, the circuit remains ON until turned OFF, and remains OFF until the circuit is turned back ON. See Figure 5-20.

ANALOG INPUTS

One major advantage of using electric motor drives to control motors is to provide speed control. Electric motor drives are connected to analog inputs to provide control over a full range of motor speeds. An *analog signal* is a type of input signal to an electric motor drive that can be either varying voltage or varying current. Analog input devices connected to electric motor drives include potentiometers, variable voltage inputs (commonly 0 VDC to 10 VDC), or variable current inputs (commonly 4 mA to 20 mA). See Figure 5-21.

THREE-WIRE MOTOR STARTER CONTROL

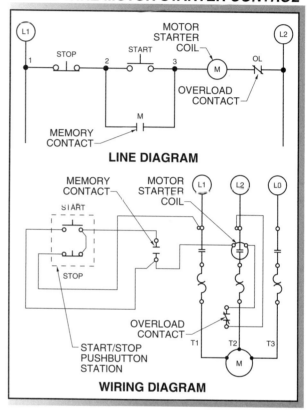

Figure 5-20. Three-wire starter control circuits add memory to the control circuit.

ELECTRIC MOTOR DRIVE-MOTOR SPEED CONTROL

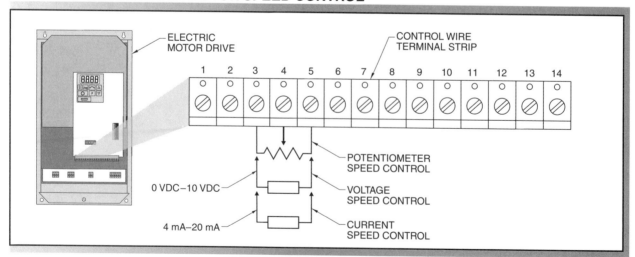

Figure 5-21. Electric motor drives can be connected to analog inputs for motor speed control.

Name_____ Date _____

True-False

T (F) 1. Direct hard wiring allows for easier circuit changes and simplifies troubleshooting.

(T) F 2. A locked rotor is the condition in which a motor is loaded so heavily that the motor shaft cannot turn. *Pg 121*

T F 3. Magnetic motor starters include high-power contacts for switching a motor ON and OFF and typically include auxiliary contacts that are used in the control circuit.

T F 4. The voltage of a control circuit is higher than the rating of the motor in most motor control circuits.

(T) F 5. DC drives control speed by varying the amount of voltage and current on the field and armature of the motor. *pg 125*

T F 6. Reduced-voltage starting of large horsepower motors typically causes problems in power distribution systems.

(T) F 7. Autotransformer starting is a motor starting method that uses a tapped 3ϕ autotransformer to provide reduced-voltage starting.

T (F) 8. Typically, triacs are used as solid state DC switches. *AC pg 134*

(T) F 9. Fans and pumps are typically classified as variable torque or variable horsepower loads.

T F 10. Electric motor drives are programmed to apply a voltage boost at low motor speed to compensate for the power loss of a motor.

(T) F 11. Placing contactors and soft starters in parallel allows for soft starting and soft stopping of motors without the need for large heat sinks during motor operation. *pg 124*

T (F) 12. Some electric motor drives allow control functions to be controlled through standard communication ports such as RS-202C ports. *pg 127*

T (F) 13. Wye-delta motors are specially wound with nine leads extending from the motor to enable the windings to be connected in either a wye or delta configuration.

T F 14. Typically, the control voltage used by electric motor drives is a low voltage, 24 VDC.

(T) F 15. Analog input devices/signals connected to electric motor drives include potentiometers and variable voltage inputs. *pg 138*

Completion

_____memory circuits_____ 1. ___ are also known as holding or sealing circuits. *pg 138*

_____solid state_____ 2. Transistors, silicon-controlled rectifiers, triacs, and alternistors are used as ___ switches. *pg 132*

_____motor control_____ 3. A(n) ___ circuit is a circuit that provides control functions such as starting, stopping, jogging, direction of rotation control, speed control, and motor protection. *pg 118* _____magnetic motor starter_____ 122

4. A(n) ___ is a starter that has an electrically operated switch (contactor) and that includes motor overload protection.

_____increasing_____

5. Soft starting is achieved by ___ a motor's voltage gradually in accordance with the setting of the ramp-up control. pg 123

_____AC_____

6. Electric motor drives are either AC drives or DC drives, with ___ drives being the most common. pg 125

_____local control_____

7. When using the ___ method, motor speed is controlled from the keypad of the electric motor drive. pg 126

_____Autotransformer starting_____

8. ___ starting is used to reduce the large current drawn from the power company when full-voltage starting is applied to large horsepower motors.

_____Primary resistor_____

9. ___ starting is a motor starting method that uses resistors connected in the motor conductors to produce a voltage drop.

_____Voltage ?_____

10. The ___ applied to the stator of an AC motor must be decreased by the same amount as the frequency when controlling speed.

Multiple Choice

_____C_____

1. Combining a ___ and an electric motor drive allows for maximum control and monitoring of a motor.
   - A. HIM
   - B. manual motor starter
   - C. PLC
   - D. soft starter

119

_____

2. Most ___ provide manual control functions for starting, stopping, jogging, reversing, speeding up, and slowing down directly from the keypad.
   - A. electric motor drives
   - B. magnetic motor starters
   - C. manual motor starters
   - D. PLCs

_____C_____

3. The section of an electric motor drive that converts AC voltage to DC voltage is the ___ section.
   - A. inverter
   - B. DC bus
   - C. converter
   - D. keypad

pg 125

_____B_____

4. The traditional methods used to provide reduced-voltage starting of large horsepower motors include all of the following except ___.
   - A. autotransformer starting
   - B. manual motor starting
   - C. primary resistor starting
   - D. solid state starting

_____B_____

5. Transistors are ___-terminal electronic switching devices that control current through the device by referencing the amount of applied voltage to the base.
   - A. two
   - B. three
   - C. four
   - D. five

c) 134

Name _____ Date _____

Activity 5-1. IEC Motor Markings

Power lines and motor terminals are marked using different letters and numbers. Power lines can be marked T1, T2, T3, or U, V, W, depending upon whether National Electrical Manufacturers Association (NEMA) or International Electrotechnical Commission (IEC) common practices are followed.

1. Connect Motor 1 to the electric motor drive using example 2.

2. Connect Motor 2 to the electric motor drive so that the drive is controlling the motor in low speed.

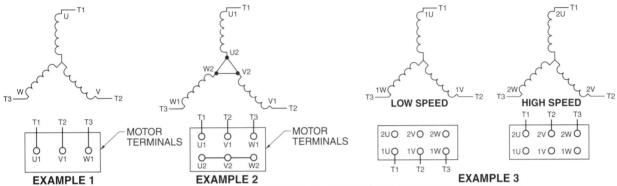

INTERNATIONAL ELECTROTECHNICAL COMMISSION MOTOR MARKINGS

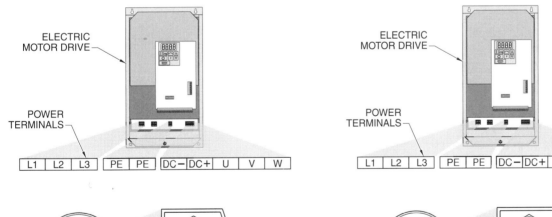

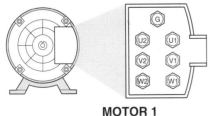

MOTOR 1

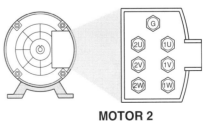

MOTOR 2

Activity 5-2. Wiring Three START/STOP Pushbutton Stations to a Motor Starter

In a hard-wired circuit, each component is interconnected using wires that run from component to component. Standard line diagrams provide the information required to hard wire motor control circuits.

1. Using the line diagram, connect the transformer secondary so that it supplies power to the motor starter coil and control circuit. Connect the magnetic motor starter coil so that it operates using any of the three START/STOP control stations. Connect the overload contact of the motor starter.

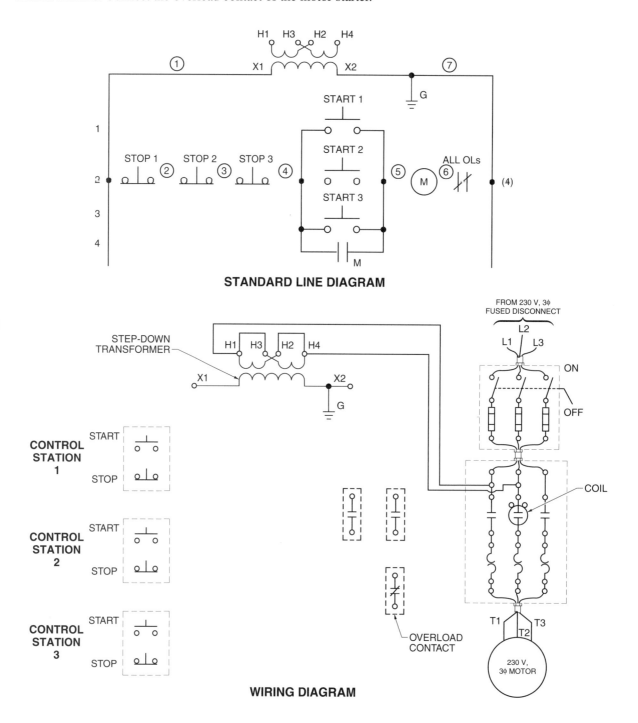

Activity 5-3. Wiring Three START/STOP Pushbutton Stations Using a Terminal Strip

To aid in troubleshooting, each component is connected to a set of terminals. Troubleshooting starts at the terminal strip in order to isolate the problem.

1. Using the line diagram, connect the transformer secondary to the required terminals to supply power to the motor starter and coil control circuit. Connect the magnetic motor starter coil to the terminals so that it operates using any of the three START/STOP control stations and connect the control stations to the terminals. Connect the overload contact of the motor starter to the terminals.

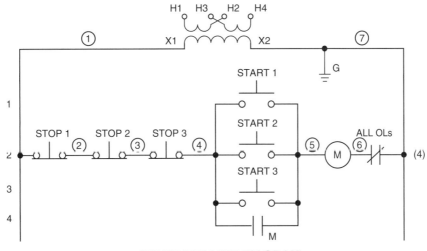

STANDARD LINE DIAGRAM

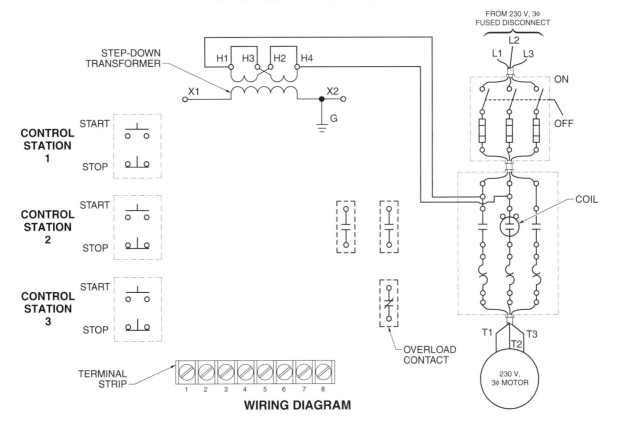

WIRING DIAGRAM

Activity 5-4. Wiring Three START/STOP Pushbutton Stations to an Electric Motor Drive

When using an electric motor drive to control a motor, external inputs (pushbuttons, selector switches, potentiometers, etc.) are connected to the control terminal strip of the drive.

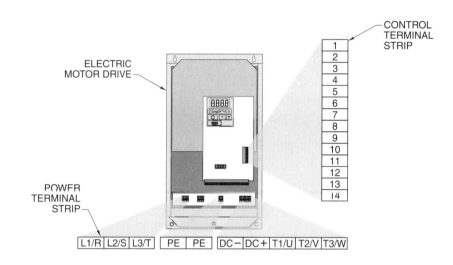

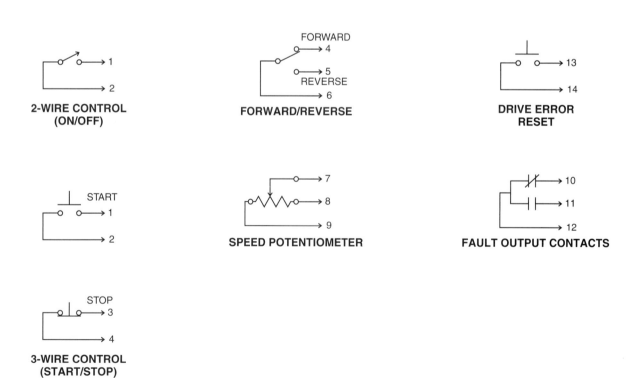

1. Connect the power supply lines to the required terminals to supply power to the electric motor drive and motor circuit. Connect the three START/STOP control stations to the control terminals of the electric motor drive. Connect the 3ϕ motor terminals to the terminals required for operation.

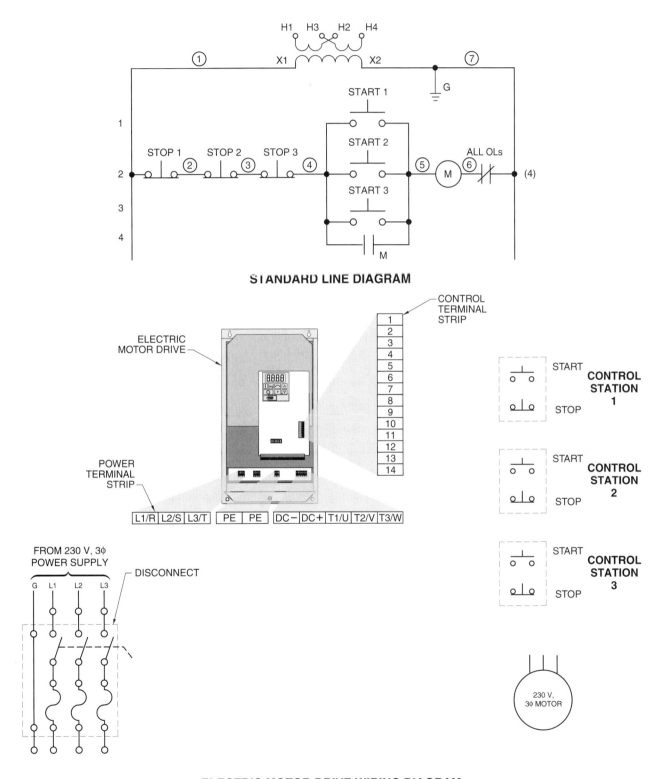

STANDARD LINE DIAGRAM

ELECTRIC MOTOR DRIVE WIRING DIAGRAM

Activity 5-5. Wiring Three START/STOP Pushbutton Stations to a PLC

To aid in programming or reprogramming, each component is connected to the input or output of a PLC. Each component is first identified (input 1, input 2, input 3, etc.) and then connected to the input numbers on the

1. Using the line diagram that identifies the input and output component numbers, connect the magnetic motor starter coil to the PLC terminals so that it operates using any of the three START/STOP control stations. Connect the three START/STOP control stations to the PLC terminals. Connect the overload contact of the motor starter to the PLC terminals.

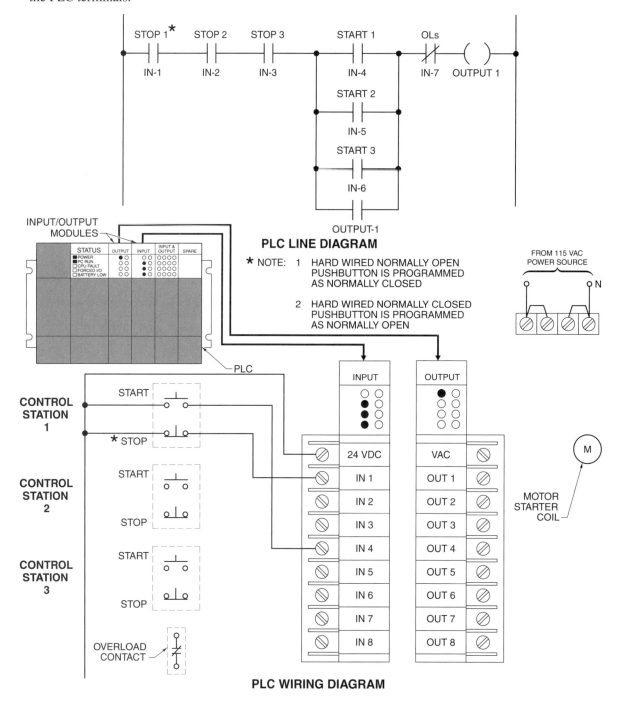

Electric Motor Drive
Installation and Troubleshooting

6 Electric Motor Installation

Connecting an electric motor drive and motor to control a load is straightforward when correct procedures are followed. Electric motor installations typically work at first, but some fail after a short time because of improper installation. Even the best electric motors are damaged and/or fail because of improper mounting, wrong coupling selection, bad alignment, or misapplication of the motor to the operating conditions. To guarantee safe and long-lasting operation, an understanding of the technical information provided on electric motor drive and motor nameplates is required.

MOTOR INSTALLATION

When an electric motor drive and motor are selected, the drive, motor, and load must be properly installed for long, safe, dependable operation. When installing or troubleshooting an electric motor drive and motor, motor nameplate information is used to provide electrical information, mechanical information, and a wiring diagram of the motor. Motor nameplate information such as voltage, current, speed, and frequency are used when selecting and programming an electric motor drive. Nameplate information is also used when taking electrical measurements to verify proper operation of a motor or to troubleshoot a problem. In addition to meeting electrical requirements, mechanical and application requirements of a motor must also be considered. Mechanical and application considerations such as type of motor enclosure, type of frame, mounting style, type of base, type of load, type of coupling, and alignment procedure are used to install motors to driven loads.

Motor Nameplates

A *motor nameplate* is a metal plate attached to a motor that lists the technical specifications of a motor. NEC® 430.7 specifies the minimum information that motor manufacturers must provide on the nameplate of a motor. Most manufacturers provide additional information beyond the minimum requirements.

The information listed on a motor nameplate contains all the performance data associated with a motor. Nameplate information must be verified when placing a motor in service, troubleshooting, or replacing a motor. Information commonly included on a motor nameplate is the name of the manufacturer, frame size, horsepower rating, duty cycle, phase rating, speed, line frequency, voltage rating, current rating, service factor, ambient temperature rating, and code letters.

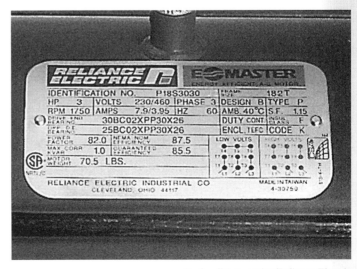

Rockwell Automation/Reliance Electric
Motor nameplates list the technical specifications of a motor per NEC® 430.7.

Motor Manufacturers. The nameplate of a motor includes the name of the manufacturer and company logo—for example, General Electric and the stylized GE logo. Catalog numbers, part numbers, and model numbers are included to quickly identify a motor for pricing or replacement. Numbers on the nameplate of a motor are coded by manufacturers, with each manufacturer using a separate, nonstandardized coding system.

Frame Sizes. All motors with frames provide some means of mounting the motor. Motor frames are classified by the National Electrical Manufacturers Association (NEMA) and the International Electrotechnical Commission (IEC). NEMA is primarily associated with equipment used in North America and the IEC is primarily associated with equipment used in Europe. Dimensionally, NEMA standards are expressed in English units and IEC standards are expressed in metric units.

NEMA and IEC frame sizes indicate physical size, construction, dimensions, and certain other physical characteristics of a motor. The frame size is abbreviated on a nameplate as FR. Common NEMA frame sizes for small motors include 42, 48, and 56 FR. Standardized frame sizes allow a motor from one manufacturer to be replaced with a motor from another manufacturer; for

example, a 56 FR from one manufacturer has features and dimensions similar to a 56 FR from another manufacturer. See Figure 6-1.

Frame numbers for most NEMA motors with base-to-shaft center distances of 3½″ and less have two frame numbers. The two digits of the frame number are determined by multiplying the base-to-shaft center distance by 16, such as with #48 and #56 frames. Motors with base-to-shaft center distances 3½″ or greater have a three- or four-digit frame number. The first two digits of a frame number are assigned by multiplying the base-to-shaft center distance by four. When the calculated number is not a whole number, round to the next higher whole number. The third digit of the frame number is assigned by doubling Dimension F and applying the Motor Frame Table; for example, a motor with a base-to-shaft dimension of 12½″ and an F dimension of 9″ requires a #505 frame. Letters immediately following a frame number indicate variations of the frame. See Appendix.

> **⚠ CAUTION**
> *DC and universal motors must never be used in hazardous locations that contain flammable materials or gases.*

MOTOR FRAMES

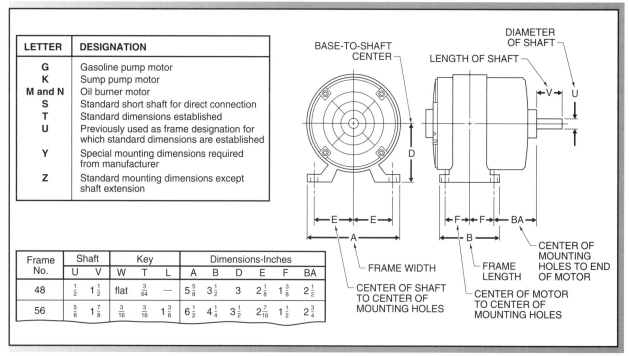

Figure 6-1. Frame size indicates the physical size, construction, dimensions, and certain other physical characteristics of a motor.

High NEMA or IEC frame numbers represent motors with large physical size and high horsepower ratings. Any letters or numbers that appear in front of a NEMA frame number are manufacturer numbers and do not apply to NEMA or IEC standards.

Horsepower Ratings. The horsepower rating of a motor is abbreviated on the nameplate as HP. Motors are rated for the amount of work or mechanical energy that can be delivered to a load. Motors are rated in horsepower (HP), with some small fractional-horsepower motors being rated in watts (W). The horsepower rating of a motor can be fractional, such as ¼ or ½ HP, or a whole number, such as 10 HP. Standard motor sizes are classified as milli (1–35 W), fractional (¹⁄₂₀–¾ HP), full (1–300 HP), and full-special order (350–50,000 HP). Motors rated by the IEC are rated in kilowatts (kW). See Figure 6-2.

TECH FACT

When sizing a motor, calculate the size needed for the load as closely as possible. Then, for best results, use the next largest available motor size. An oversized motor operating at partial load will operate cooler and last longer than a smaller motor operating at full load.

The horsepower rating of a motor is used when selecting or servicing an electric motor drive. Typical electric motor drive horsepower ratings start at 1 HP (possibly less) and go to 500 HP or more. When ordering an electric motor drive, the minimum horsepower rating of the drive must be equal to or greater than the horsepower rating of the motor. Electric motor drive manufacturers build drives to sizes that follow the standard horsepower ratings of motors.

Duty Cycles. Some motors deliver rated horsepower continuously while other motors only deliver rated horsepower for a short period of time. The duty cycle rating of a motor is determined by whether the motor delivers rated horsepower continuously or for short periods of time. The duty cycle is listed on the motor nameplate as DUTY, DUTY CYCLE, or TIME RATING.

Continuous duty cycle rated motors are marked CONT. Continuous duty cycle motors are operated constantly under full load for more than 1 hr. Intermittent duty cycle rated motors are marked INTER. Intermittent duty cycle motors are operated constantly under full load, but for less than 1 hr. When a duty cycle rating is not given on a nameplate, the motor is assumed to be continuous duty. See Figure 6-3.

MOTOR WATTAGE AND HORSEPOWER RATINGS

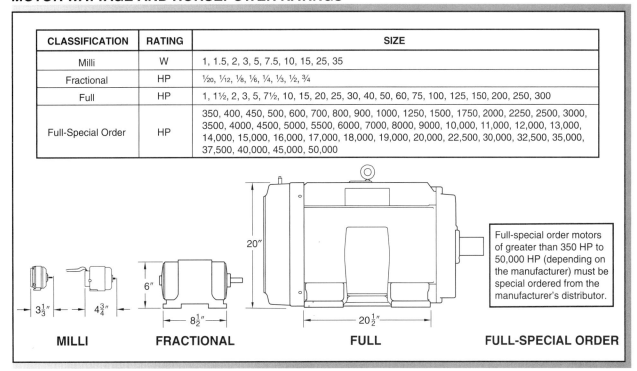

Figure 6-2. Standard motors are rated in horsepower (HP), with some small-fractional horsepower motors being rated in watts (W).

MOTOR DUTY CYCLES

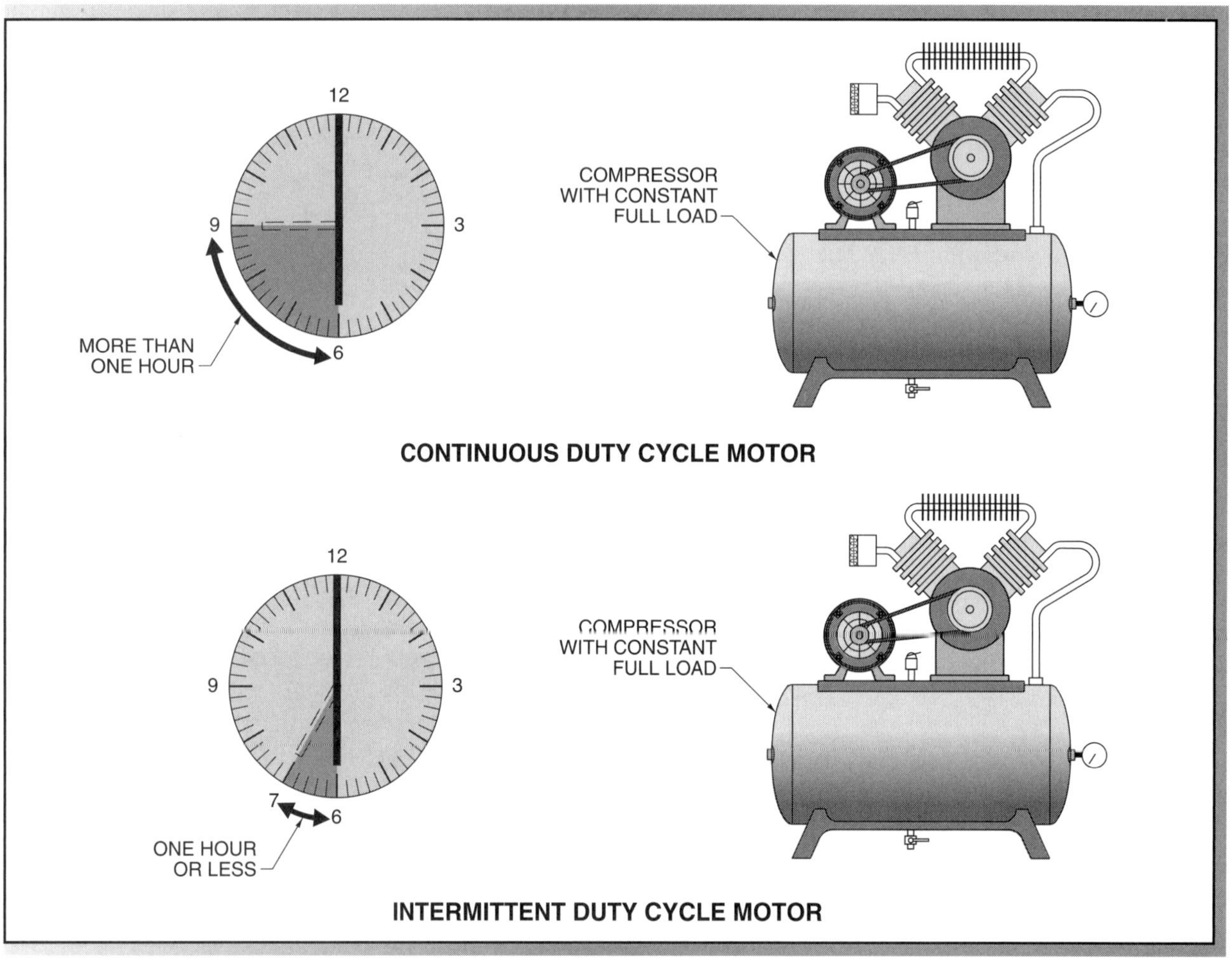

Figure 6-3. When a duty cycle is not given on a nameplate, the motor is assumed to be continuous duty.

Motors used with certain types of applications such as waste disposal, valve actuators, electric hoists, and other types of intermittent loads are rated for short-term duty such as 5 min, 15 min, 30 min, or 1 hr. A motor designed for intermittent duty is generally less expensive than an equivalent size, continuous duty motor because winding insulation and air cooling are not as critical. When an electric motor drive is used to control a motor, the drive must have a continuous duty rating regardless of the duty rating of the motor.

Phase Rating. The phase rating of a motor is abbre-viated on the nameplate as PH. All electric motors require direct current (DC), single-phase alternating current (1ɸ AC), or three-phase alternating current (3ɸ AC). The phase rating of a motor is listed as 1 (single-phase), 3 (three-phase), or DC (direct current).

Before electric motor drives, the type of current (DC, 1ɸ, or 3ɸ) and voltage level (120 V, 240 V, 480 V)

supplied to a motor starter had to match the nameplate rating of a motor; for example, a 3ɸ, 240 V rated motor controlled by a magnetic motor starter must be connected to a 3ɸ, 240 V power supply. When electric motor drives are used to control motors, the type and level of the input supply voltage does not have to match the rated type and voltage level of the motor. Some electric motor drives allow the input power to the drive to be of a different type and at a different voltage level than the power required by the motor; for example, an electric motor drive supplied with 1ɸ, 120 V power can be used to control a 3ɸ, 240 V motor. See Figure 6-4.

Motor Speed. The speed of an AC motor is determined by the number of poles in the motor and the frequency of the supply voltage. The speed of a DC motor is determined by the amount of supply voltage and/or the amount of field current.

ELECTRIC MOTOR DRIVE PHASE RATINGS

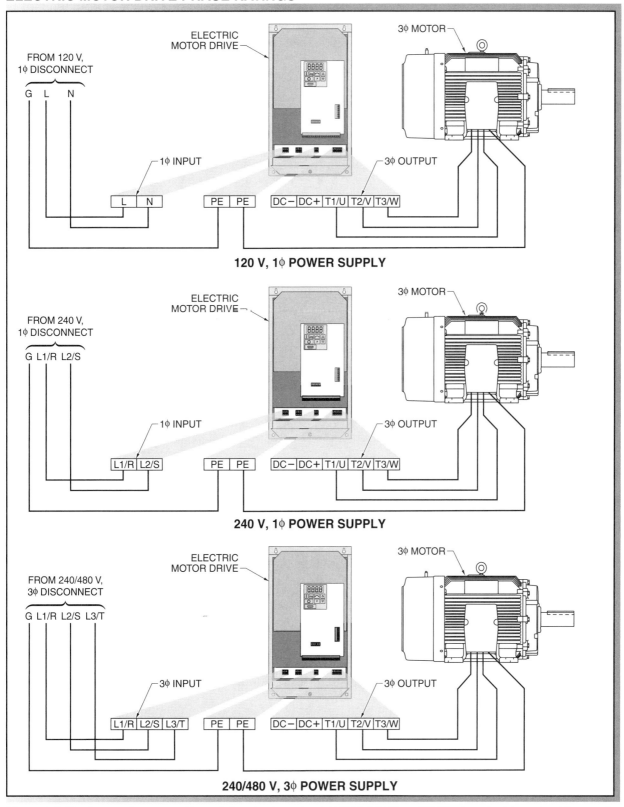

Figure 6-4. The input voltage type (1φ or 3φ) and voltage level (120 V, 240 V, or 480 V) of an electric motor drive can be different from the 3φ output of the drive.

The rated speed of a motor is given on the nameplate in revolutions per minute (rpm). The rated speed of a motor is not the exact operating speed, but the approximate speed at which a motor rotates when delivering rated horsepower to a load.

The rated speed of synchronous motors is the exact speed. The rated speed of all other motors is the synchronous speed minus the slip of the motor. *Slip* is the difference between the synchronous speed and rated speed of a motor. Non-synchronous motors typically have approximately 3% to 5% slip.

Line Frequency. *Line frequency* is the number of complete electrical cycles per second of a power source. See Figure 6-5. A *cycle* is one complete wave of alternating voltage or current. An *alternation* is one-half of a cycle. *Period (T)* is the time required to produce one complete cycle of a waveform. The line frequency rating of a motor is abbreviated on the nameplate as CY or CYC (cycle), or Hz (hertz). *Hertz* is the international unit of frequency and is equal to cycles per second.

In the United States, 60 Hz is the standard power source line frequency. Canada, Mexico, and most of the Caribbean (Bahamas and Cayman Islands, for example) also use 60 Hz as the standard power source line frequency. Frequency is 50 Hz for Europe, Asia, much of South America, and the rest of the world. Developing countries (Saudi Arabia, Colombia, Costa Rica, etc.) and countries using large numbers of USA-made electrical products are changing from 50 Hz to 60 Hz.

LINE FREQUENCY

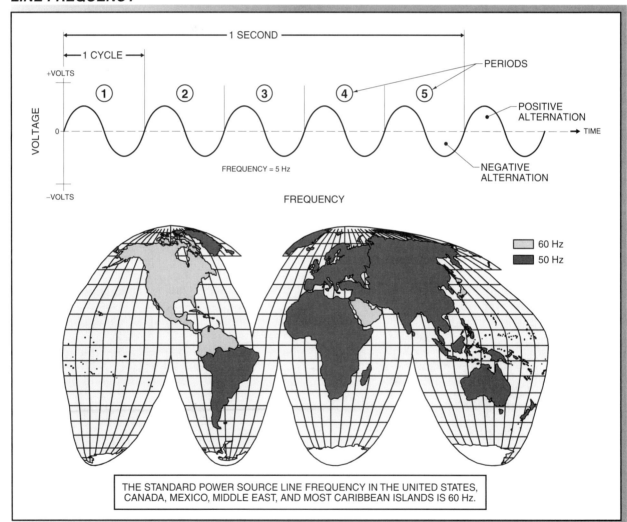

THE STANDARD POWER SOURCE LINE FREQUENCY IN THE UNITED STATES, CANADA, MEXICO, MIDDLE EAST, AND MOST CARIBBEAN ISLANDS IS 60 Hz.

Figure 6-5. Line frequency (in Hz) is the number of complete electrical cycles per second of a power source.

Electric motor drives typically have frequency range ratings that allow drives to be connected to standard 50 Hz or 60 Hz power supplies. Electric motor drives supplied with 50 Hz or 60 Hz power typically have frequency ratings of 48 Hz to 63 Hz. Regardless of what the supply power frequency into an electric motor drive is (50 Hz or 60 Hz), the frequency output of a drive (to the motor) is full range, such as 0 Hz to 250 Hz or more.

Voltage Ratings. Voltage rating is abbreviated V on the nameplate of a motor. The voltage rating may be a single rating such as 115 V, or a dual rating such as 230 V/460 V. NEMA design standard is for a motor to operate within +/- 10% of its nameplate voltage.

When a dual voltage rated motor is used, the higher voltage is preferred because a higher voltage rating has a lower current rating. The current rating of a motor determines the size of wires, size of conduit, and overcurrent protection to be used. When using an electric motor drive to control a motor, both the voltage and current into a drive (from the power supply) and out of a drive (to the motor) must be used when determining wire size and overcurrent protection.

Current Ratings. The nameplate current rating of a motor is abbreviated A or AMPS. Motors that have dual voltage ratings also have dual current ratings. Motors with high voltage ratings have low current ratings and motors with low voltage ratings have high current ratings. The current rating listed on the nameplate of a motor is the amount of current a motor draws when delivering full rated horsepower output. FLA or FLC is the current required by a motor to produce full-load torque at the motor's rated speed and voltage. Motors that are overloaded draw more current than rated nameplate current. See Figure 6-6.

that can be placed on a motor for short periods of time without damaging the motor. Service factor is a safety margin for motors. Some motor designs allow more than nameplate rated horsepower to be developed without encountering damage.

Motors with a service factor of 1.00 must not be overloaded beyond nameplate rated horsepower. Motors with service factors greater than 1.00 can be overloaded without the additional heat destroying insulation. For example, a 10 HP rated motor with a service factor of 1.25 safely develops 125% of rated power or 12.5 HP (10 × 125% = 12.5 HP).

Typical service factors are 1.00, 1.15, 1.25, and 1.35. When the nameplate of a motor does not list a service factor, a service factor of 1.00 is assumed and there is no built-in safety margin for the motor. Current and temperature rise when motors are operated at more than rated horsepower (above a service factor of 1.00). The amount of increased current is abbreviated on the nameplate as SFA (service factor amps).

Ambient Temperature Rating. The nameplate ambient temperature rating of a motor is abbreviated AMB or DEG. *Ambient temperature* is the temperature of the air surrounding an object. The temperature of a motor rises as work is performed. *Temperature rise* is the difference between the winding temperature of a running motor and ambient temperature. A motor's *permissible temperature rise* is the difference between the ambient temperature and the listed (nameplate) ambient rating of a motor. A typical ambient temperature rating for a motor is 104°F (40°C). See Figure 6-7.

MOTOR CURRENT DRAW		
VOLTAGE	**HORSEPOWER**	**CURRENT/HP**
115 V, 1φ	LESS THAN 1 HP	16 A TO 22 A/HP
115 V, 1φ	OVER 1 HP	10 A TO 13 A/HP
230 V, 1φ	LESS THAN 1 HP	8 A TO 11 A/HP
230 V, 1φ	OVER 1 HP	5 A TO 6.5 A/HP
230 V, 3φ	ANY HP	2.5 A TO 3.5 A/HP
460 V, 3φ	ANY HP	1.25 A TO 1.75 A/HP
120 VDC	ANY HP	6.5 A TO 9 A/HP

Figure 6-6. Motors that are overloaded draw more current than rated nameplate current.

Service Factors. Service factor is abbreviated SF on motor nameplates. Typically, motors can be overloaded for short periods of time without damage. A *service factor* is a multiplier that represents the percentage of extra load

ASI Robicon
Certain motor applications require a higher than normal ambient temperature rating.

AMBIENT TEMPERATURE RATING

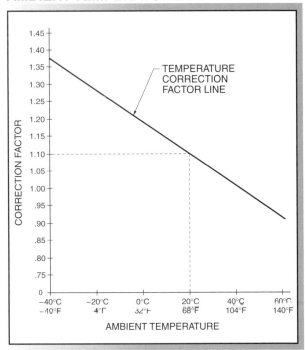

AMBIENT TEMPERATURE	LISTED MOTOR MAXIMUM AMBIENT TEMPERATURE	PERMISSIBLE TEMPERATURE RISE
MOTOR 1 = 77°F (25°C)	104°F (40°C)	27°F (15°C)*
MOTOR 2 = 86°F (30°C)	104°F (40°C)	18°F (10°C)
MOTOR 3 = 95°F (35°C)	104°F (40°C)	9°F (5°C)
MOTOR 4 = 104°F (40°C)	104°F (40°C)	0°F (0°C)†

* Least motor cooling required
† Most motor cooling required

Figure 6-7. The permissible temperature rise is the difference between the ambient temperature and the maximum rated ambient temperature of the motor.

Higher-than-rated temperatures destroy motor insulation and break down bearing lubricants. When motor insulation is destroyed, the windings become shorted and the motor is no longer functional. Motor installations with ambient temperatures above the rated ambient temperature of the motor require the use of a temperature correction factor. A correction factor derates motor specifications to prevent damage caused by environmental conditions other than what is stated by the manufacturer on the motor nameplate.

Ambient temperature correction charts provide temperature correction factors to derate motor specifications. Correction charts provide correction factors for above and below ambient temperature, but derating is typically applied when ambient temperature around the motor is above the listed ambient temperature rating of the motor. See Figure 6-8.

AMBIENT TEMPERATURE DERATING

Figure 6-8. An ambient temperature correction chart provides temperature correction factors to derate motors.

Electric motor drives have fixed or ranged ambient temperature ratings. An electric motor drive can have a fixed rating such as 104°F (40°C), or a ranged rating such as 32°F-104°F (0°C–40°C). Electric motor drives can also have a derating rating, such as 32°F–104°F (0°C–40°C) /derate up to 131°F (55°C). The derating rating allows an electric motor drive to be installed in an area that has a higher than 104°F (40°C) ambient temperature, up to 131°F (55°C) ambient temperature, but the drive must be derated. Derating charts are typically supplied by manufacturers.

Code Letters. Code letters on the nameplate of a motor are used to determine locked rotor current (LRC) in kVA/HP. Code letters from A through V are listed in NEC® Table 430.7(B). Code letters at the beginning of the alphabet indicate low starting current. Code letters at the end of the alphabet indicate high starting current. See Figure 6-9. To find the LRC of a motor, apply the following formula:

$$LRC = \frac{1000 \times HP \times kVA/HP}{V}$$

where
LRC = locked rotor current
HP = horsepower
V = voltage (in V)

MOTOR NAMEPLATE CODE LETTERS

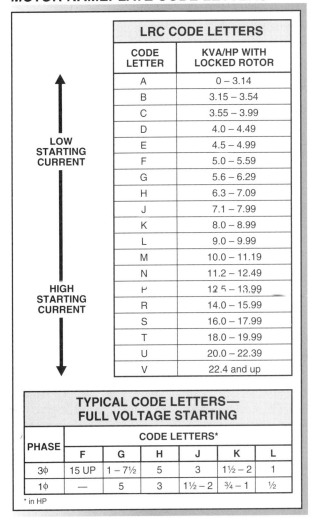

	LRC CODE LETTERS	
	CODE LETTER	KVA/HP WITH LOCKED ROTOR
	A	0 – 3.14
	B	3.15 – 3.54
	C	3.55 – 3.99
	D	4.0 – 4.49
LOW STARTING CURRENT	E	4.5 – 4.99
	F	5.0 – 5.59
	G	5.6 – 6.29
	H	6.3 – 7.09
	J	7.1 – 7.99
	K	8.0 – 8.99
	L	9.0 – 9.99
	M	10.0 – 11.19
	N	11.2 – 12.49
HIGH STARTING CURRENT	P	12.5 – 13.99
	R	14.0 – 15.99
	S	16.0 – 17.99
	T	18.0 – 19.99
	U	20.0 – 22.39
	V	22.4 and up

TYPICAL CODE LETTERS— FULL VOLTAGE STARTING						
PHASE	CODE LETTERS*					
	F	G	H	J	K	L
3φ	15 UP	1 – 7½	5	3	1½ – 2	1
1φ	—	5	3	1½ – 2	¾ – 1	½

* in HP

Figure 6-9. The code letter on a motor nameplate is used to determine locked rotor current (LRC).

For example, what is the maximum LRC of a 1 HP, 115 V motor with code letter K?

$$LRC = \frac{1000 \times HP \times kVA / HP}{V}$$

$$LRC = \frac{1000 \times 1 \times 8.99}{115}$$

$$LRC = \frac{8990}{115}$$

$$LRC = \textbf{78.17 A}$$

Insulation Class

Motor insulation prevents motor coils (windings) from shorting to each other or to ground (frame of motor).

All insulation deteriorates over time due to the effects of thermal stress (heat), high voltage spikes (transient voltages), contaminants, and mechanical stress. Heat, contaminants, and mechanical stress cause damage to insulation over time, but one high voltage transient can permanently damage insulation or destroy a motor. When motor insulation is damaged by heat, the windings eventually become shorted, causing motor failure. Heat buildup in a motor is caused by several factors such as the following:

- incorrect motor type or size for an application
- improper cooling, usually from dirt buildup
- excessive load, usually from improper use
- excessive friction, usually from misalignment or vibration
- electrical problems, typically voltage unbalance, phase loss, or surge voltages
- harmonics on the supply lines, especially negative sequence harmonics (5th, 11th, 17th, etc.)
- the nonsinusoidal waveforms produced by electric motor drives
- frequent starting and stopping

As the heat in a motor increases beyond the temperature rating of the insulation, the life of the insulation and of the motor is shortened. The higher the temperature, the sooner the insulation fails. Motor insulation temperature rating is listed on the nameplate as insulation class. Typically, for every 18°F (10°C) temperature rise above the NEMA rating, the usable service life of a motor is cut in half. Even a small rise in temperature above the insulation rating is damaging.

Insulation class is a measure of the resistance of insulation to breakdown due to heat. The four major classifications of motor insulation are A, B, F, and H. NEMA classifies insulation according to maximum allowable operating temperature. The temperature capabilities of individual classes are separated by 45°F (25°C) increments. Class F is the most common insulation type in use. See Figure 6-10.

MOTOR INSULATION OPERATING TEMPERATURES	
NEMA CLASSIFICATIONS	MAXIMUM OPERATING TEMPERATURES
A	221°F (105°C)
B	266°F (130°C)
F	311°F (155°C)
H	356°F (180°C)

Figure 6-10. NEMA classifies motor insulation according to maximum allowable operating temperature.

Motors must be replaced by motors that have the same insulation class or a higher temperature rating. Motors controlled by electric motor drives require Class F or higher insulation. Motor manufacturers produce motors, called inverter rated, for use with electric motor drives.

TECH FACT

An average standard-efficiency, totally-enclosed, fan cooled (TEFC) motor, rotating at 1750 rpm and operating 8750 hr per year at full load, with an energy cost of $.05 per kilowatt-hour, costs approximately $1.00 per horsepower per day to operate. The energy savings realized by using an energy efficient motor is approximately 6% for motors smaller than 10 HP and approximately 2% for motors greater than 200 HP. It is not typically recommended to replace standard motors that are working with energy efficient motors or rewind standard motors for higher efficiency.

Motor Design

Motors are suited for specific applications because each motor type has specific horsepower, torque, and speed characteristics. The basic characteristics of each motor are determined by the design of the motor. Motors are designed for high efficiency, high starting torque, or high power factor, but are not designed to have all three characteristics.

NEMA classifies motors using design letters. Motors are classified as Classes A through E. Each class of motor has different starting torque values. Classes B, C, and D are the most common, with Class B motors being the most widely used. Loads that require higher than normal starting torque use Class C or Class D motors. Class E

motors have the same basic electrical characteristics as Class A motors, but are more energy-efficient. See Figure 6-11.

Efficiency

Motor efficiency is the measure of the effectiveness with which a motor converts electrical energy to mechanical energy. Improvements in motor efficiency are achieved by reducing power losses in a motor. Power losses in a motor are a result of energy losses (heat loss) in the stator core, stator windings, bearings, and rotor. Power losses are considered part of converting electrical energy into mechanical energy. Power losses are always present to some degree.

Efficiency is included on the nameplate of most motors. The efficiency of a motor improves with speed because the electrical losses inside a motor are basically constant at any speed. Large motors are more efficient than small motors. See Figure 6-12. Efforts to improve energy efficiency and conservation of resources prompted the legislation of the Energy Policy Act (EPA), which was passed by Congress and signed into law October 24, 1992. The Energy Policy Act took effect on October 24, 1997, and mandates efficiency standards for 3φ, 1 HP to 200 HP general-purpose motors.

Motor Enclosures

Motor enclosures protect motors from environmental conditions, reduce maintenance, and increase safety. A wide variety of motor enclosures are available, with the type of enclosure affecting the cost of the motor. The two general classifications for motor enclosures are open motor enclosures and totally enclosed motor enclosures.

NEMA MOTORS TORQUE CHARACTERISTICS

NEMA DESIGN	LOCKED ROTOR TORQUE (%)		BREAKDOWN TORQUE (%)		LOCKED ROTOR CURRENT (%)		SLIP (%)		EFFICIENCY
	Small HP	Large HP	Small HP	Large HP	Small HP	Large HP	Small HP	Large HP	
A	275	70	300	175	800	700	5	0.5	MEDIUM OR HIGH
B	275	70	300	175	700	600	5	0.5	MEDIUM OR HIGH
C	250	200	225	190	700	600	5	1	MEDIUM
D	275	275	275	275	700	600	13	5	MEDIUM
E	190	75	200	160	600	500	3	0.5	HIGH

Figure 6-11. NEMA classifies motors according to starting torque characteristics.

TYPICAL MOTOR EFFICIENCIES

HP	STANDARD MOTOR (%)	ENERGY EFFICIENT MOTOR (%)	HP	STANDARD MOTOR (%)	ENERGY EFFICIENT MOTOR (%)
1	76.5	84.0	30	88.1	93.1
1.5	78.5	85.5	40	89.3	93.6
2	79.9	86.5	50	90.4	93.7
3	80.8	88.5	75	90.8	95.0
5	83.1	88.6	100	91.6	95.4
7.5	83.8	90.2	125	91.8	95.8
10	85.0	90.3	150	92.3	96.0
15	86.5	91.7	200	93.3	96.1
20	87.5	92.4	250	93.6	96.2
25	88.0	93.0	300	93.8	96.5

Figure 6-12. Motor efficiency charts are used to determine the efficiency characteristics of motors.

Open Motor Enclosures. An *open motor enclosure* is a motor enclosure with openings to allow passage of air to cool the windings. Open motor enclosures include general, dripproof, splashproof, guarded, semi-guarded, and dripproof fully guarded. See Figure 6-13.

Totally Enclosed Motor Enclosures. A *totally enclosed motor enclosure* is a motor enclosure that prevents air from entering the motor. Totally enclosed motor enclosures include fan-cooled, nonventilated, pipe-ventilated, water-cooled, explosionproof, dust-ignitionproof, and waterproof. See Figure 6-14.

Motor Mounting

Motors that are not mounted properly are likely to fail from mechanical problems. Motors must be mounted on a flat, stable base to reduce vibrations and misalignment problems. To ensure a long life span, motors must also be mounted in a manner that keeps the motor as clean as possible.

An adjustable motor base makes the installation, tensioning, maintenance, and replacement of belts easier. An adjustable motor base is a mounting base that allows a motor to be easily moved over a short distance. An adjustable motor base simplifies the installation of a motor and the tensioning of belts or chains. See Figure 6-15.

Motor Couplings

A motor coupling is a device that connects axially located motor shafts and equipment shafts. A flexible coupling allows a motor to operate the driven load while allowing for a slight misalignment between the motor and load shafts, and allowing for radial and axial movement of the shafts. Motor couplings are rated according to the amount of torque they can handle. Couplings are rated in pound-inches (lb-in) or pound-feet (lb-ft). The coupling torque rating must be correct for the application to prevent the coupling from bending or breaking. A bent coupling causes misalignment and vibration. A broken coupling prevents a motor from doing work.

Fluke Corporation
Motors must be securely mounted to flat, stable surfaces, using shims to correct height alignment where necessary.

OPEN MOTOR ENCLOSURES

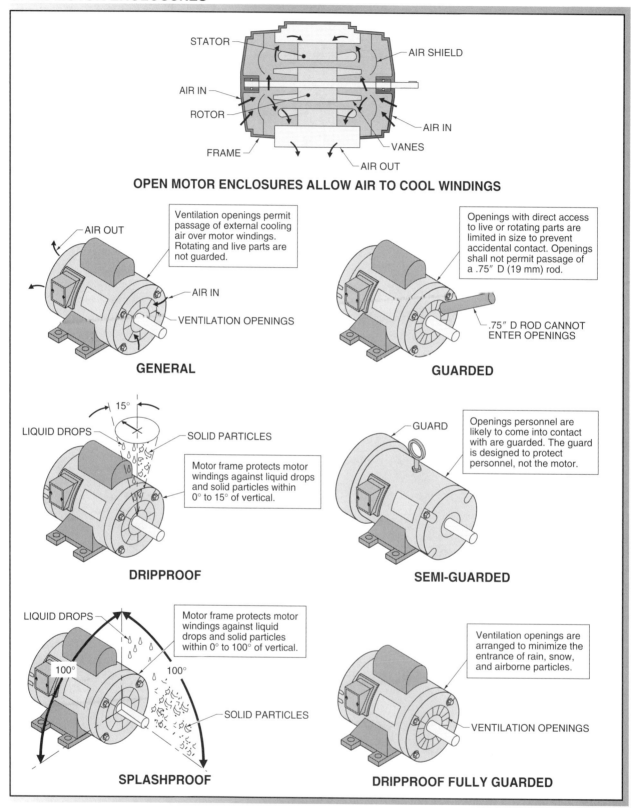

OPEN MOTOR ENCLOSURES ALLOW AIR TO COOL WINDINGS

Ventilation openings permit passage of external cooling air over motor windings. Rotating and live parts are not guarded.

GENERAL

Openings with direct access to live or rotating parts are limited in size to prevent accidental contact. Openings shall not permit passage of a .75″ D (19 mm) rod.

GUARDED

Motor frame protects motor windings against liquid drops and solid particles within 0° to 15° of vertical.

DRIPPROOF

Openings personnel are likely to come into contact with are guarded. The guard is designed to protect personnel, not the motor.

SEMI-GUARDED

Motor frame protects motor windings against liquid drops and solid particles within 0° to 100° of vertical.

SPLASHPROOF

Ventilation openings are arranged to minimize the entrance of rain, snow, and airborne particles.

DRIPPROOF FULLY GUARDED

Figure 6-13. An open motor enclosure allows air to flow through the motor to cool the windings.

TOTALLY ENCLOSED MOTOR ENCLOSURES

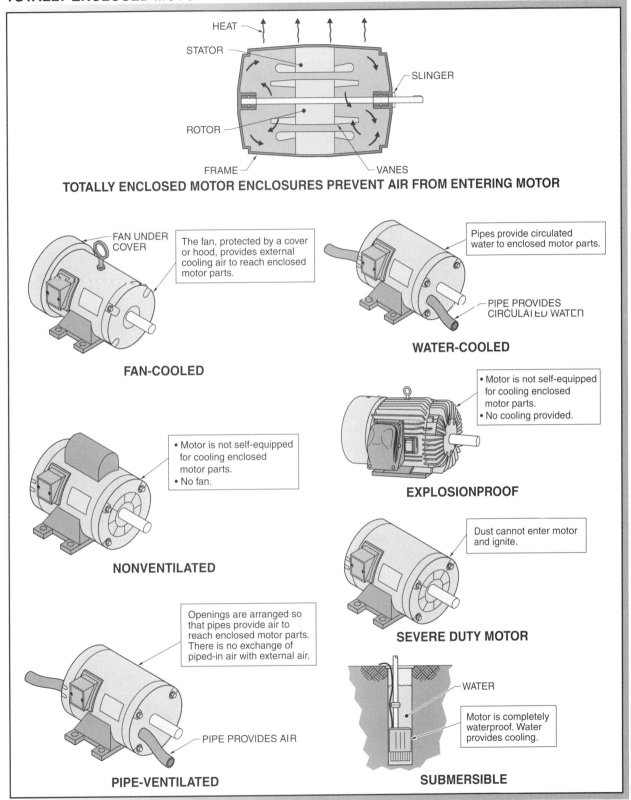

Figure 6-14. A totally enclosed motor enclosure prevents air from entering the motor.

ADJUSTABLE MOTOR BASES

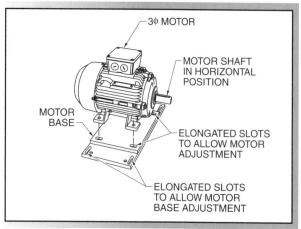

Figure 6-15. An adjustable motor base is a mounting base that allows a motor to be easily moved to tension belts and chains.

Several types of flexible couplings are used to connect rotating equipment shafts. The best coupling to use depends upon the application and size of the equipment. Different coupling designs allow different degrees of movements (radial and axial) and misalignment. Flexible couplings are designed to accommodate various amounts of misalignment but excessive misalignment still damages the motor, coupling, and load. Flexible couplings that are best used in motor applications include the SURE-FLEX®, gear, chain, universal, and grid. See Figure 6-16.

TECH FACT

Rigid couplings must not be used to connect motor shafts to load shafts. Rigid couplings are only used to connect shafts that have a common frame (machine housing).

FLEXIBLE COUPLINGS

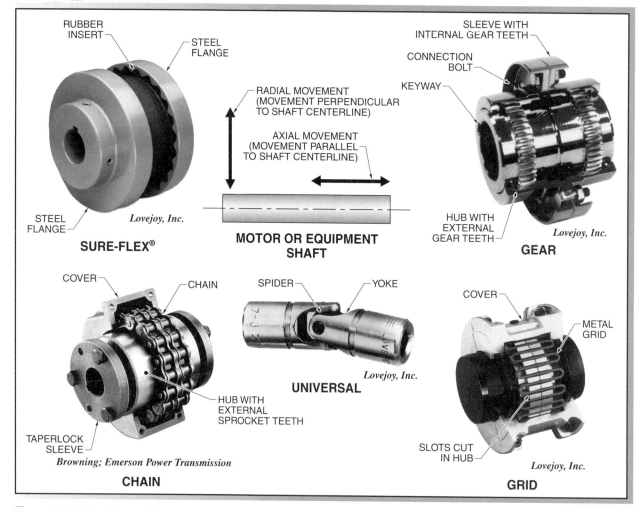

Figure 6-16. Flexible couplings accommodate various amounts of misalignment between the motor and driven load.

V Belts

Belt drive systems provide a quiet, compact, and durable form of power transmission. Belts are manufactured using a combination of fabric, cord, and/or metal reinforcement, vulcanized together with natural or synthetic rubber compounds. The type of belt used depends on the application. The most common belt used for power transmission is the V belt. V belts have a tapered cross-sectional area that allows V belts to wedge between the sides of the sheave (pulley) grooves while under load, providing frictional contact.

A V belt drive has the load side of the V belt before the load and the non-load side of the V belt after the load during rotation. Belt contact of 170° to 180° allows the belt to have maximum frictional contact with the drive pulley for maximum transfer of energy. Whenever possible, align pulley centerlines on the same horizontal plane. See Figure 6-17.

Vertical belt drive applications cause more problems than horizontal belt drive applications. Typically, the drive pulley is mounted on top, with the load side of the belt pulling up. Whenever possible, align pulley centers on the same vertical plane. When the pulley centers cannot be aligned on the same vertical plane, an angle of 45° or less between the pulley centers and the vertical plane is permissible.

V BELT TENSIONING CONDITIONS

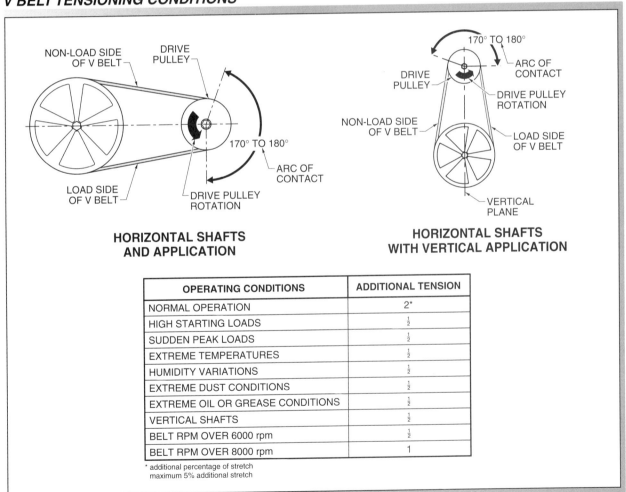

HORIZONTAL SHAFTS AND APPLICATION

HORIZONTAL SHAFTS WITH VERTICAL APPLICATION

OPERATING CONDITIONS	ADDITIONAL TENSION
NORMAL OPERATION	2*
HIGH STARTING LOADS	$\frac{1}{2}$
SUDDEN PEAK LOADS	$\frac{1}{2}$
EXTREME TEMPERATURES	$\frac{1}{2}$
HUMIDITY VARIATIONS	$\frac{1}{2}$
EXTREME DUST CONDITIONS	$\frac{1}{2}$
EXTREME OIL OR GREASE CONDITIONS	$\frac{1}{2}$
VERTICAL SHAFTS	$\frac{1}{2}$
BELT RPM OVER 6000 rpm	$\frac{1}{2}$
BELT RPM OVER 8000 rpm	1

* additional percentage of stretch
maximum 5% additional stretch

Figure 6-17. V belts must be properly installed to prevent excessive wear.

All belts wear over time, and worn or broken belts must be removed or replaced. When servicing a V belt-driven system, the belts must be checked for proper tension and alignment. V belts must be tight enough not to slip, but not so tight as to overload the bearings of the shafts.

V belt tension is checked by placing a straightedge across the sheaves and measuring the amount of deflection at the midpoint of the shafts' center-to-center distance, or by using a tension tester. Belt deflection must equal $\frac{1}{64}''$ per inch of shaft center-to-center distance; for example, two shafts with a center-to-center distance of 16″ must have a $\frac{1}{4}''$ deflection ($16 \times \frac{1}{64} = \frac{1}{4}''$). When V belt tension requires adjustment, the drive component is moved away from the driven component to increase belt tension (reducing deflection). See Figure 6-18.

Problems exist in a V belt drive system when the belts do not last as expected. Common problems in V belt systems include unmatched belts, unequally stretched belts, cracked or worn belts, belts with broken cords, worn pulley grooves, misalignment of pulleys, or overloading.

V BELT TENSIONING

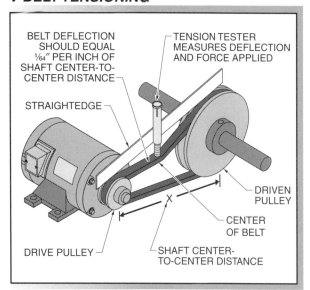

BELT DEFLECTION SHOULD EQUAL $\frac{1}{64}''$ PER INCH OF SHAFT CENTER-TO-CENTER DISTANCE

TENSION TESTER MEASURES DEFLECTION AND FORCE APPLIED

STRAIGHTEDGE

DRIVEN PULLEY

CENTER OF BELT

DRIVE PULLEY

SHAFT CENTER-TO-CENTER DISTANCE

Figure 6-18. A V belt must be tight enough not to slip, but not so tight as to overload the bearings of the shafts.

Motor Alignment

When installing equipment, angular misalignment and parallel misalignment occur. *Angular misalignment* is misalignment caused by two shafts that are not parallel. *Parallel misalignment* is misalignment caused by two shafts that are parallel but not on the same axis. See Figure 6-19.

SHAFT MISALIGNMENT

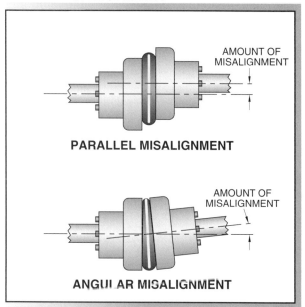

AMOUNT OF MISALIGNMENT

PARALLEL MISALIGNMENT

AMOUNT OF MISALIGNMENT

ANGULAR MISALIGNMENT

Figure 6-19. Misalignment of the motor and load shafts causes motor and electric motor drive failure.

Misalignment of a motor and driven load is a major cause of motor failure. When a motor and driven load are misaligned, premature failure of the motor bearings and/or load bearings occurs. Moving a motor or placing shims under the feet of a motor often corrects alignment problems. A *shim* is a thin piece of metal that is thousandths of an inch thick and made of steel or brass.

Environmental Conditions

Environmental conditions are the characteristics of the atmosphere and materials that surround a motor. Ambient temperature, altitude, humidity, chemicals in the atmosphere, mounting surface, and mechanical forces, such as shock and vibration, are environmental conditions that affect the operation of a motor. De-rating a motor compensates for abnormal environmental conditions such as high ambient temperatures and excessive altitude. Selecting the correct motor enclosure compensates for abnormal environmental conditions such as humidity and chemicals in the atmosphere. Abnormal environmental conditions such as uneven mounting surface and vibrations are compensated for during installation and maintenance.

Normal Service Conditions. Satisfying the electrical requirements of a motor ensure that a motor operates properly. Satisfying the mechanical and environmental requirements of a motor determines how long the motor operates. Standard motors are designed to operate under normal service conditions. See Figure 6-20.

Abnormal Service Conditions. Abnormal service conditions affect the operation of motors. Special-purpose motors are used when motors are required to operate in abnormal service conditions such as high ambient temperatures, extreme dampness, corrosive or combustible atmospheres, and/or high levels of dirt. See Figure 6-21.

NORMAL MOTOR SERVICE CONDITIONS

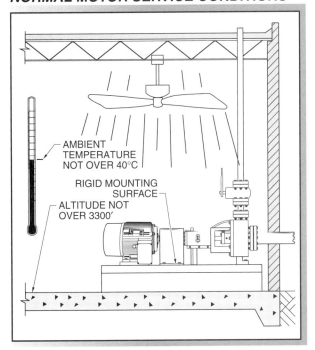

AMBIENT TEMPERATURE NOT OVER 40°C

RIGID MOUNTING SURFACE

ALTITUDE NOT OVER 3300′

Figure 6-20. Standard motors are designed to operate under normal service conditions.

TECH FACT

Bearing failure is a major cause of motor failure. Bearing failure is typically the result of misalignment.

ABNORMAL MOTOR SERVICE CONDITIONS

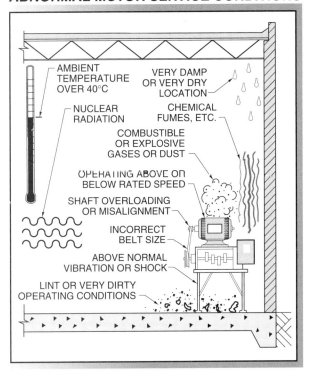

AMBIENT TEMPERATURE OVER 40°C

VERY DAMP OR VERY DRY LOCATION

NUCLEAR RADIATION

CHEMICAL FUMES, ETC.

COMBUSTIBLE OR EXPLOSIVE GASES OR DUST

OPERATING ABOVE OR BELOW RATED SPEED

SHAFT OVERLOADING OR MISALIGNMENT

INCORRECT BELT SIZE

ABOVE NORMAL VIBRATION OR SHOCK

LINT OR VERY DIRTY OPERATING CONDITIONS

Figure 6-21. Special service motors must be used for abnormal service conditions.

Name_____ Date _____

True-False

T   F   **1.** The nameplate of a motor includes the name of the manufacturer.

T   F   **2.** Electric motor drive manufacturers build drives to sizes that follow the standard horsepower ratings of motors.

T   F   **3.** Continuous duty cycle motors are marked INTER on the nameplate of a motor.

T   F   **4.** When electric motor drives are used to control motors, the type and level of the input supply voltage must match the rated type and voltage level of the motor.

T   F   **5.** When a dual voltage rated motor is used, the higher voltage is preferred.

T   F   **6.** Temperature rise is the temperature of the air surrounding a piece of equipment.

T   F   **7.** Each National Electrical Manufacturers Association (NEMA) design class of motor has different starting torque values.

T   F   **8.** An adjustable motor base complicates the installation of a motor and the tensioning of belts or chains.

T   F   **9.** Angular misalignment is misalignment caused by two shafts that are not parallel.

T   F   **10.** Moving a motor or placing shims under the feet of a motor often corrects alignment problems.

T   F   **11.** National Electrical Code® 400.7 specifies the minimum information that motor manufacturers must provide on the nameplate of a motor.

T   F   **12.** Line frequency is the number of complete electrical cycles per minute of a power source.

T   F   **13.** The most common belt used for power transmission is the V belt.

T   F   **14.** Belt deflection must equal 1/16″ per inch of shaft center-to-center distance.

T   F   **15.** Humidity is considered an environmental condition that affects the operation of a motor.

Completion

_____ **1.** ___ information is used when taking electrical measurements to verify proper operation of a motor or to troubleshoot a problem.

_____ **2.** NEMA is primarily associated with equipment used in North America and the ___ is primarily associated with equipment used in Europe.

Duty Cycle   **3.** The ___ rating of a motor is determined by whether the motor delivers rated horsepower continuously or for short periods of time. *pg149*

Poles   **4.** The speed of an AC motor is determined by the number of ___ in the motor and the frequency of the supply voltage.

Less   **5.** Motors that are not fully loaded draw ___ current than rated nameplate current.

motor efficiency   **6.** ___ is the measure of the effectiveness with which a motor converts electrical energy to mechanical energy. *pg156*

open motor enclosure   **7.** A(n) ___ is a motor enclosure with openings to allow passage of air to cool the windings. *pg157*

Flexible   **8.** A(n) ___ coupling allows a motor to operate the driven load while accommodating for a slight misalignment between the motor and load shafts. *pg157*

parallel   **9.** ___ misalignment is misalignment caused by two shafts that are parallel but not on the same axis. *pg162*

De-rating   **10.** ___ motors are used when motors are required to operate in abnormal service conditions such as high ambient temperature or extreme dampness. *pg162*

Multiple Choice

A   **1.** Motors with base-to-shaft center distances 3½″ or greater have a three- or ___-digit frame number.
       A. four   C. six
       B. five   *pg148*   D. seven

D   **2.** In (the) ___, 60 Hz is the standard power source line frequency.
       A. Asia   C. South America
       B. Europe   D. United States

A   **3.** Typical service factors include all of the following except ___.
       A. .95   C. 1.15
       B. 1.00   D. 1.25

B   **4.** Typically, for every ___°F temperature rise above the NEMA rating, the usable service life of a motor is cut in half.
       A. 16   C. 20   *pg155*
       B. 18   D. 22

C   **5.** Totally enclosed motor enclosures include all of the following except ___.
       A. dust-ignitionproof   C. splashproof
       B. explosionproof   D. water-cooled

pg157

Name_____ Date _____

Activity 6-1. Operating above Rated Speed

> AC motors can be operated at higher frequencies (speeds) than normal when powered by electric motor drives. The higher the frequency, the faster the speed of the motor. The upper speed limit of a motor is based on motor voltage and mechanical balancing limits. Motors can be operated at speeds up to 10% above the rated (nameplate) speed. The maximum frequency parameter setting of an electric motor drive limits the operating speed of motors.

_____ 1. When using a panel control on an electric motor drive, what is the factory default setting for the direction the motor will rotate?

_____ 2. What is the factory default setting for the acceleration time of a motor?

_____ 3. When a motor has a nameplate rating of 60 Hz/1730 rpm and is operated using the factory maximum frequency default setting (MFS), the motor is actually operating at ___ rpm at full speed.

_____ 4. What is the MFS of a 60 Hz/1730 rpm motor if the motor is not operated more than 10% above nameplate speed?

_____ 5. How fast would a 60 Hz/1730 rpm motor operate if set for the maximum allowable setting of the MFS parameter? *Note:* This is an unsafe condition that would destroy a motor.

TITLE	PARAMETERS	ADJUSTABLE RANGE	DEFAULT VALUE
AU1	Automatic acceleration and deceleration	0: NO 1: YES	0
AU2	Automatic torque boost*	0: NO 1: AUTOMATIC 50 Hz MOTOR 2: SENSORLESS VECTOR CONTROL 3: SENSORLESS VECTOR CONTROL AUTOMATIC TUNING	0
AU3	Automatic environment setting*	0: NO 1: AUTOMATIC 50 Hz MOTOR 2: AUTOMATIC 60 Hz MOTOR	0
FMC	FM terminal function selection	0: FREQUENCY METER 1: OUTPUT CURRENT METER	0
SMS	Standard mode selection*	0: NO ACTION 1: 50 Hz STANDARD 2: 60 Hz STANDARD 3: DEFAULT SETTING 4: CLEARING LOG ERRORS 5: CLEARING ACCUMULATED OPERATION TIME 6: INITIALIZE INVERTER TYPEFORM	3
FRS	Forward/reverse selection (panel)	0: FORWARD 1: REVERSE	0
ACC	Acceleration time #1 (sec)	0.1 ~ 3600	10.0
DAC	Deceleration time #1 (sec)	0.1 ~ 3600	10.0
MFS	Maximum frequency (Hz)*	30.0 ~ 320.0	80.0
BFS	Base frequency (Hz)	25.0 ~ 320.0	60.0

* These parameters cannot be changed while running

Activity 6-2. Motor Overload Protection

Electric motor drives must be programmed (using a parameter setting) to trip when the drive detects that a motor is overloaded. The time it takes for an electric motor drive to detect an overload and turn a motor OFF depends on the extent of the motor overload and motor operating frequency. Electric motor drive manufacturers use graphs to explain the approximate overload trip time based on the amount of overload and the output frequency (motor operating frequency) of an electric motor drive. The percent of motor overload is found by applying the following formula:

$$\%OL = \frac{I_{RUN}}{I_{FL}} \times 100$$

$\frac{72}{48} \cdot 100 = 150\%$

where

$\%OL$ = percent motor is overloaded (in %)

I_{RUN} = amount of current motor is drawing (in A)

I_{FL} = amount of current listed on the nameplate of the motor (in A)

100 = constant

_____ 1. Using the manufacturer-supplied graph, determine the approximate amount of trip time, in seconds, before an electric motor drive will automatically stop an overloaded motor when the motor is operating at nameplate speed.

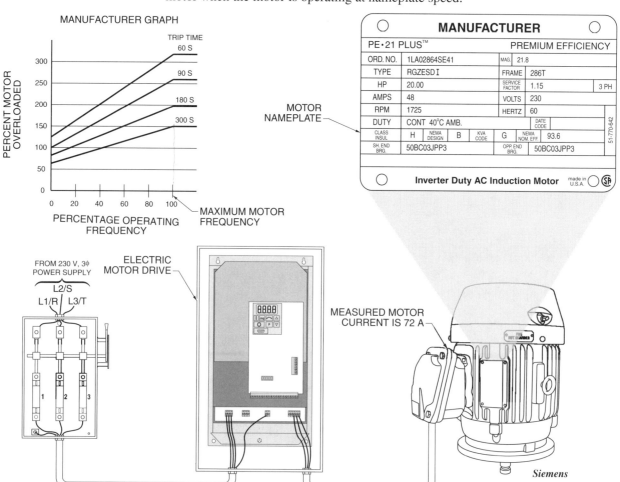

Siemens

Activity 6-3. Setting Electric Motor Drive DIP Switches

Some electric motor drive manufacturers include DIP switches with the drive, which are used for setting some of its functions. DIP switches must be set correctly for proper electric motor drive and motor performance.

1. Using the nameplate information of a motor, determine the position of each DIP switch for Setting A-2 and Setting B-2 so that the motor operates at a speed no greater than the nameplate rating and has an automatic boost at low speed and a standard coast to stop mode.

MANUFACTURER				
PE·21 PLUS™		PREMIUM EFFICIENCY		
ORD. NO.	1LA02864SE41	MAG. 21.8		
TYPE	RGZESDI	FRAME	286T	
HP	30.00	SERVICE FACTOR	1.0	3 PH
AMPS	77.5	VOLTS	230	
RPM	1765	HERTZ	60	
DUTY	CONT 40°C AMB.	DATE CODE		
CLASS INSUL H	NEMA DESIGN B	KVA CODE G	NEMA NOM. EFF. 93.6	
SH. END BRG.	50BC03JPP3	OPP. END BRG.	50BC03JPP3	

Inverter Duty AC Induction Motor made in U.S.A.

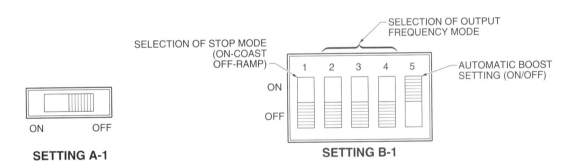

SELECTION OF STOP MODE (ON-COAST OFF-RAMP)

SELECTION OF OUTPUT FREQUENCY MODE

AUTOMATIC BOOST SETTING (ON/OFF)

SETTING A-1

SETTING B-1

Setting A Information		
Input Voltage	Switch	Drive Output
230 V	OFF	3.83 V/Hz (230 V at 60 Hz)
	ON	4.6 V/Hz (230 V at 50 Hz)
208 V	OFF	3.46 V/Hz (208 V at 60 Hz)
	ON	4.16 V/Hz (208 V at 50 Hz)
460 V	OFF	7.66 V/Hz (460 V at 60 Hz)
	ON	9.2 V/Hz (460 V at 50 Hz)
415 V	OFF	6.92 V/Hz (415 V at 60 Hz)
	ON	8.3 V/Hz (415 V at 50 Hz)
380 V	OFF	6.3 V/Hz (380 V at 60 Hz)
	ON	7.6 V/Hz (380 V at 50 Hz)

Setting B Information			
SW2	SW3	SW4	Description
OFF	OFF	OFF	0 Hz TO 60 Hz
ON	OFF	OFF	0 Hz TO 50 Hz
OFF	ON	OFF	0 Hz TO 120 Hz
OFF	OFF	ON	0 Hz TO 240 Hz

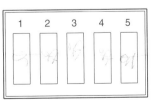

SETTING A-2

SETTING B-2

Activity 6-4. Motor Insulation Life

Motor insulation life is based on the class of insulation used (A, B, F, or H) and the maximum temperature the insulation can withstand. Motor insulation is typically rated for an average life of 20,000 hr to 60,000 hr. Insulation life is longer at low temperatures and is shorter at high temperatures. Temperature increases within motors are caused by motor overloading, higher or lower applied voltages, unbalanced voltage lines, high ambient temperatures, blocked ventilation passageways, and frequent starting.

The temperature rating of motor insulation is based on ambient temperature, maximum insulation rated temperature rise, and a hot spot allowance that takes into consideration the center of the motor windings where temperatures are higher. The temperature rating of Class A insulation has a 104°F (40°C) ambient temperature with an allowable temperature rise to 140°F (60°C) and a 41°F (5°C) hot spot allowance.

Manufacturers use charts to show the expected insulation life based on the type of insulation and the maximum temperature the insulation can withstand.

Using the Manufacturer Insulation Life Chart, determine the approximate insulation life for each motor application.

Motor Application 1—The motor must operate in an area of ambient temperature that will raise the motor insulation to an average temperature of about 347°F (175°C).

_____ **1.** What is the rated motor insulation life if a motor with Class F insulation is used?

_____ **2.** What is the rated motor insulation life if a motor with Class B insulation is used?

Motor Application 2—The motor must operate in an area of ambient temperature that will hold insulation to an average temperature of about 302°F (150°C).

_____ **3.** What is the rated motor insulation life if a motor with Class B insulation is used?

_____ **4.** What is the rated motor insulation life if a motor with Class F insulation is used?

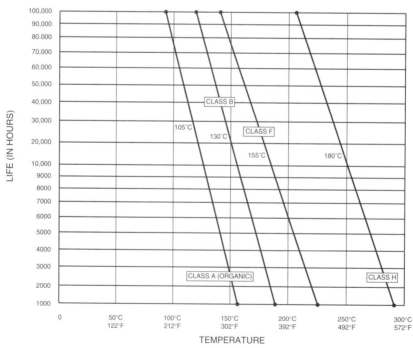

MANUFACTURER INSULATION LIFE CHART

Activity 6-5. Determining V Belt Tension

Determine what the belt tension tester should read for proper belt tension.

_____ **1.** Tension tester should measure___/64 for proper V belt tension.

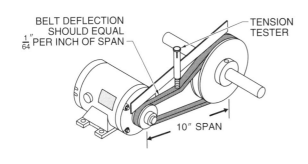

_____ **2.** Tension tester should measure ___/64 for proper V belt tension.

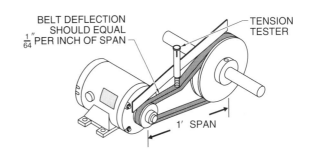

_____ **3.** Tension tester should measure ___/64 for proper V belt tension.

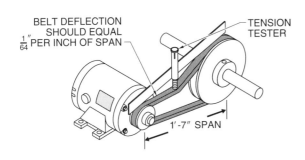

_____ **4.** Tension tester should measure ___/64 for proper V belt tension.

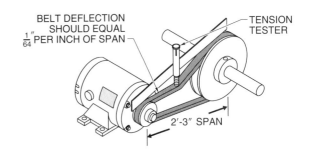

_____ **5.** The stretched measurement for proper V belt tension is ___.

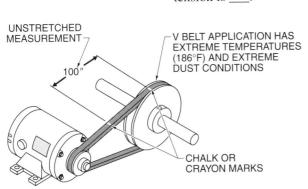

_____ **6.** The stretched measurement for proper V belt tension is ___.

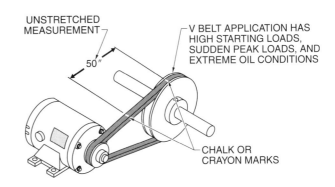

Activity 6-6. V Belt Mounting

Determine the proper rotation (CW or CCW) of the drive (small) pulleys.

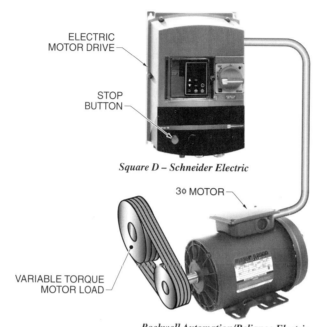

ELECTRIC
MOTOR DRIVE

STOP
BUTTON

Square D – Schneider Electric

3φ MOTOR

VARIABLE TORQUE
MOTOR LOAD

Rockwell Automation/Reliance Electric

_____ **1.** The direction of
rotation equals ___.

_____ **2.** The direction of
rotation equals ___.

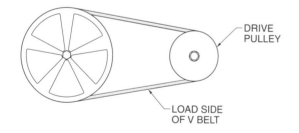

DRIVE
PULLEY

LOAD SIDE
OF V BELT

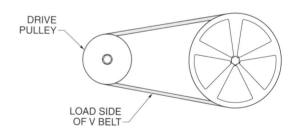

DRIVE
PULLEY

LOAD SIDE
OF V BELT

_____ **3.** The direction of
rotation equals ___.

_____ **4.** The direction of
rotation equals ___.

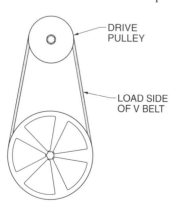

DRIVE
PULLEY

LOAD SIDE
OF V BELT

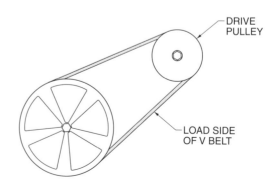

DRIVE
PULLEY

LOAD SIDE
OF V BELT

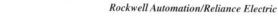

7 Solid State Electric Motor Drive Components

Electric motor drives operate by using solid state components to create a synthetic sine wave signal to control motor speed and torque. The type of solid state components used in an electric motor drive depends upon the drive type and cost, and the technology applied. Understanding the operation of the solid state components found in an electric motor drive aids in the troubleshooting of the drive system. The solid state components found in an electric motor drive include diodes, silicon-controlled rectifiers, gate turn-off thyristors, and transistors (BJTs, FETs, and IGBTs).

SOLID STATE SWITCHING

A *solid state switch* is an electronic switching device that opens and closes circuits at a precise point in time to control current flow and voltage levels. Solid state switches used in electric motor drives are high-speed switches that allow a DC voltage to be converted into usable AC voltage. An AC sine wave is electronically reproduced and controlled by solid state control devices and switches in the electric motor drive inverter section.

AC motors require an AC sine wave to operate properly. When contactors or magnetic motor starters switch an AC supply voltage ON, a pure AC sine wave is supplied to the motor directly from the power supply. An AC sine wave that is electronically reproduced and controlled from a DC voltage by an electric motor drive is altered and distorted. See Figure 7-1.

In order to vary motor speed, the reproduced sine wave must vary in voltage level and in frequency. At full speed, a motor needs full nameplate rated voltage (230 V, 460 V, etc.) and rated frequency (50 Hz or 60 Hz). At half speed, a motor needs half the voltage and half the frequency. The frequency must change as voltage changes to ensure motor torque rating. Other electric motor drive functions include monitoring voltage, current, power, drive temperature, and motor acceleration and deceleration times.

The advantages of using solid state switching include the following:

- long life because there are no moving parts to wear out
- high resistance to damage from shock and vibration because there are no moving parts
- high reliability when properly installed and used
- no contact arcing, which reduces electrical noise produced on the power lines
- fast switching compared to mechanical switches

Saftronics Inc.

Hundreds of solid state components, including solid state switching devices, can fit onto small PC boards for compact electric motor drive design.

AC SINE WAVE

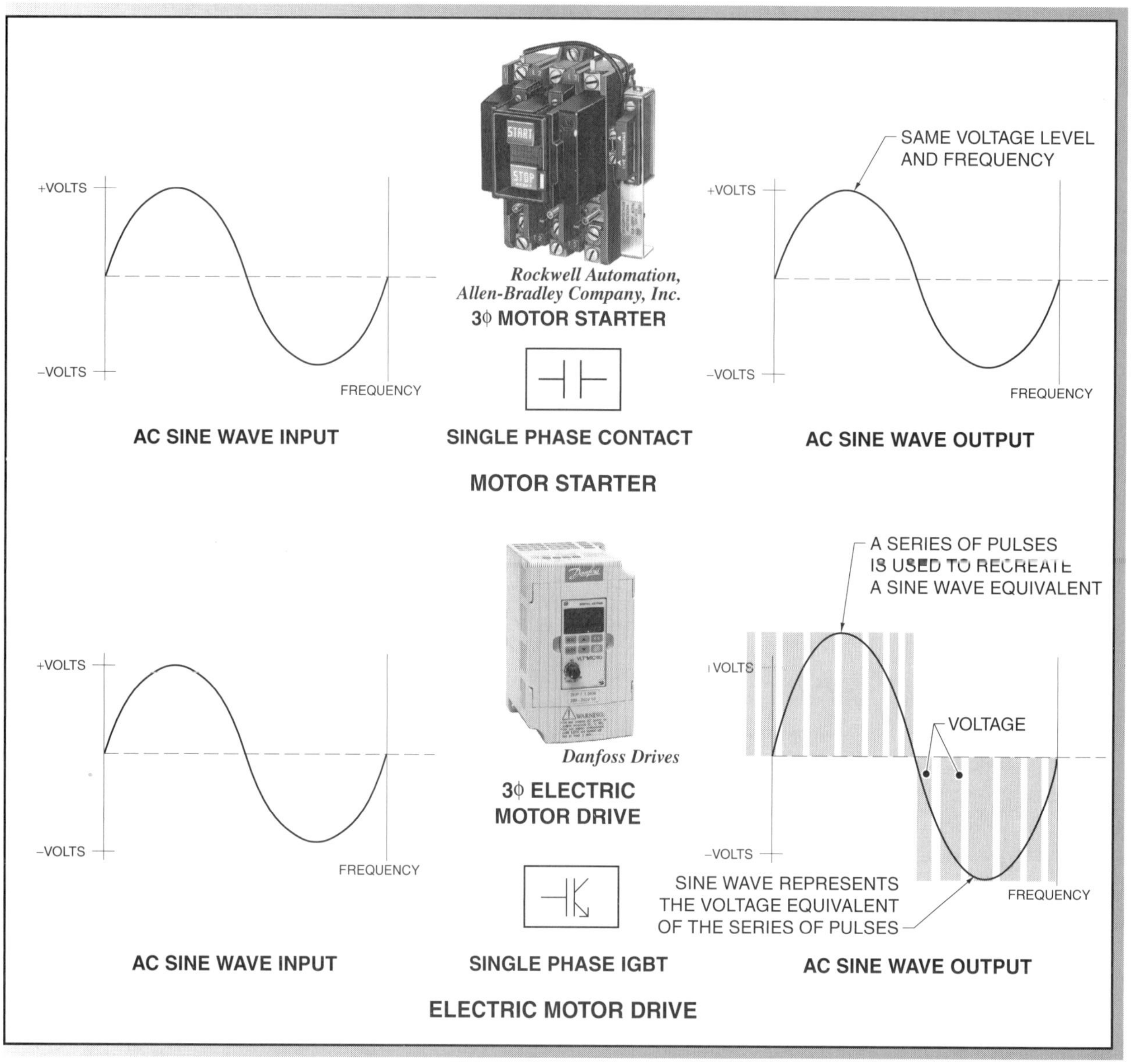

Figure 7-1. Mechanical switches or magnetic motor starters can turn ON or turn OFF the applied voltage, but a solid state switch of an electric motor drive can change the voltage level and frequency of the applied voltage.

Voltage Drop

Voltage drop is the amount of voltage consumed by a device or component as current passes through it. A solid state switching device has a voltage drop across it in the ON condition. A voltage drop of 3 mV or 4 mV across mechanical switch contacts typically does not affect circuit operation, but a voltage drop of 2 V or 4 V across a solid state switch can affect circuit operation.

Any voltage drop across a switch or component produces heat. The larger the current passing through the device or component, the greater the amount of heat produced. Heat affects the operation, life, and reliability of solid state switching devices. The heat produced must be removed to protect the solid state switch and other electric motor drive components.

A small voltage drop can produce high temperatures within a component. Heat is the result of power produced at a switch and is proportional to the power. See Figure 7-2. The power produced at a solid state switch is found by applying the formula:

$$P = E \times I$$

where

P = power
E = voltage drop across the switch
I = current flowing through the switch

For example what is the heat produced by a solid state switch with a 2 V voltage drop and a 10 A current flow?

$$P = E \times I$$
$$P = 2 \times 10$$
$$P = \mathbf{20\ W}$$

Heat is produced as current passes through a solid state switch by the friction of the electrons with the

VOLTAGE DROP

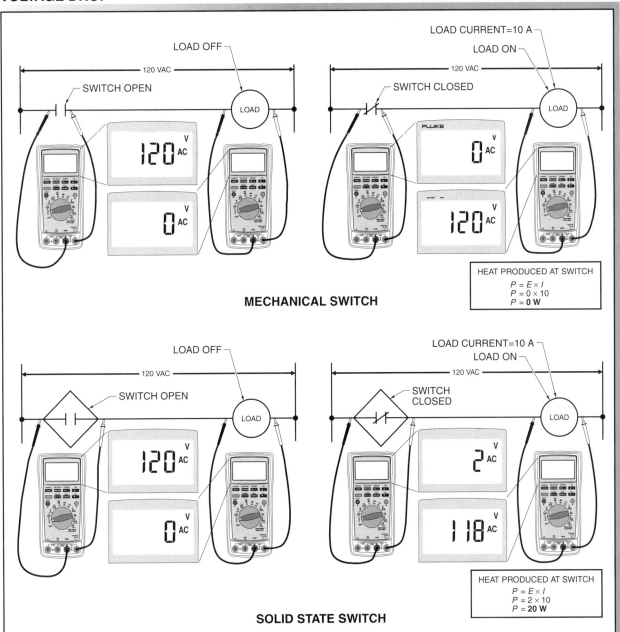

Figure 7-2. Even a small voltage drop across a closed solid state switch produces power loss in the form of heat.

device's solid state material. Heat in any solid state switch or electrical component must be kept to a safe level by limiting the amount of current passing through the solid state switch and/or removing the heat. The most effective way to control heat buildup in an electric motor drive is to remove (dissipate) the heat from the solid state switches and drive components.

Heat is dissipated from the electric motor drive components by heat sinks. A *heat sink* is a device that conducts and dissipates heat away from an electrical component. Solid state devices are mounted to heat sinks to ensure good heat conduction from the solid state device to the heat sink. Most electric motor drives also include one or more cooling fans to force air over the heat sinks for greater heat dissipation. See Figure 7-3.

All solid state component manufacturers specify the maximum temperature permitted before the component

fails. An electric motor drive must be properly cooled to keep the temperature below the specified safe maximum value. Electric motor drives monitor the heat at the heat sink(s), and the monitoring device shuts down the drive and motor if the temperature exceeds the preset limit. Electric motor drives also display the temperature of the heat sink(s) as one of the displayed values or read-only parameters. A *read-only parameter* is a parameter value that can be displayed, but not set or changed. When an electric motor drive shuts down a motor due to a high heat sink (drive) temperature, the electric motor drive is protecting the drive itself, not the motor. Before restarting the electric motor drive, the drive location, ambient temperature and ventilation, parameter settings such as acceleration time, and the electric motor drive ratings must be verified.

ELECTRIC MOTOR DRIVE HEAT DISSIPATION

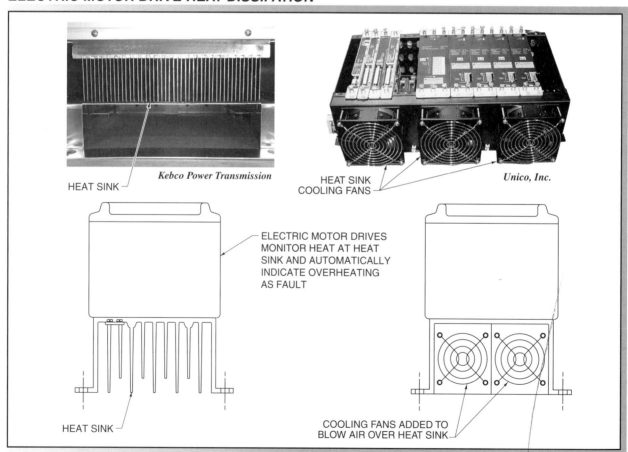

Figure 7-3. Electric motor drives use heat sinks and fans to dissipate the heat produced by the individual components of the drive.

Heat Effects

The heat produced at the solid state switch and the ambient temperature affect the performance of a solid state switch inside an electric motor drive. *Ambient temperature* is the temperature of the air surrounding an object. The type of enclosure used affects the temperature of the electric motor drive. The temperature inside an enclosure can be considerably higher than the ambient temperature of the air outside the enclosure. The electric motor drive shuts down if the enclosure does not allow the internal heat to be dissipated from the heat sinks. Proper airflow over the heat sinks is required to remove heat developed by electrical devices. Heat is removed from a heat sink by natural convection and/or forced-air cooling using a fan.

Following manufacturer electrical rating limits (voltage, current, and horsepower) and application requirements prevents the electric motor drive from producing excessive heat. In addition to the electric motor drive forced-air cooling, additional cooling may be required in some applications, such as enclosures exposed to sunlight or high ambient temperatures. Temperature problems can also develop when several electric motor drives are placed in the same cabinet. Heat shields (deflectors) are used to prevent heat from a lower mounted electric motor drive unit from affecting a drive unit mounted above. Following manufacturer mounting recommendations (spacing distances) allows for proper heat dissipation.

Electric motor drives have an ambient temperature rating. An electric motor drive operated at temperatures above its ambient rated temperature must be derated as specified by the drive manufacturer. An electric motor drive rated for proper operation in an ambient temperature range from 14°F (−10°C) to 104°F (40°C) can be operated in an ambient temperature above 104°F (40°C) if derated, such as a 20 HP drive opeating at an ambient temperature of 122° F (50°C). The electric motor drive is derated to 17.2 HP. An electric motor drive must not be operated in an ambient temperature above the manufacturer's highest derated temperature listing. High altitudes also reduce heat dissipation; typically, an electric motor drive must be derated at altitudes above 3300′. See Figure 7-4.

ELECTRIC MOTOR DRIVE DERATING

DRIVE TEMPERATURE DERATING		
AMBIENT TEMPERATURE		PERCENT DERATING
FAHRENHEIT	CELSIUS	
104°	40°	0
113°	45°	7
122°	50°	14
131°	55°	20

DRIVE TEMPERATURE DERATING		
AMBIENT TEMPERATURE		PERCENT DERATING
FEET	METERS	
3300	1005	0
4000	1220	2.3
4500	1370	4.0
5000	1525	5.6
7500	2285	13.9
8500	2590	19.0
10000	3050	22.0

Figure 7-4. Electric motor drives may need to be derated if temperature or altitude is above manufacturer application recommendations.

Kebco Power Transmission
Electric motor drive components, like all electronic components, can be damaged from overheating when installed without the proper cooling.

RESISTORS

A *resistor* is an electrical device that limits the current flowing in an electronic circuit. Resistors are classified by a resistance value in ohms (Ω) and power dissipation value in watts (W). Resistors are used for dividing voltage, reducing voltage, limiting current, and absorbing power in electric motor drives. Resistors may be fixed, variable, or tapped. A *fixed resistor* is a resistor with a set value. A *variable (adjustable) resistor* is a resistor with a set range of values. A *tapped resistor* is a resistor that contains fixed tap points of different resistances.

CAPACITORS

A *capacitor* is an electrical device designed to store a voltage charge of electrical energy. Capacitors are used as filters and to maintain voltage levels in electric motor drives and to improve torque in motors. Capacitors may be fixed or variable. A *fixed capacitor* is a capacitor with one capacitance value. A *variable capacitor* is a capacitor that can be varied in capacitance value.

A capacitor that has been discharged must be recharged before it can discharge again. The recharging of capacitors is why electric motor drives draw current from the power supply lines (L1, L2, and L3) in pulses and not as a pure sine wave. The current draw from the power supply lines takes place during peak voltage when the sine wave is at maximum voltage.

SEMICONDUCTOR DEVICES

A *semiconductor device* is an electronic device that has electrical conductivity between that of a conductor (high conductivity) and that of an insulator (low conductivity). Semiconductor devices are used in electric motor drives as electronic switches, rectifiers, filters, and inverters. Semiconductor devices used in electric motor drives include diodes, silicon controlled rectifiers (SCRs), gate turn-off thyristors (GTOs), bipolar junction transistors (BJTs), metal-oxide-semiconductor field-effect transistors (MOSFETs), and insulated-gate bipolar transistors (IGBTs). The specific device used depends on the electric motor drive type and size.

Diodes

A *diode* is an electronic device that allows current to pass through in only one direction. Diodes are also referred to as rectifiers because diodes change (rectify) AC voltage into pulsating DC voltage. A filter is normally added to smooth out the pulsating DC voltage from diodes. A diode has polarity as determined by the anode and cathode. The *anode* is the positive lead of a diode. The *cathode* is the negative lead of a diode. See Figure 7-5.

DIODE OPERATION

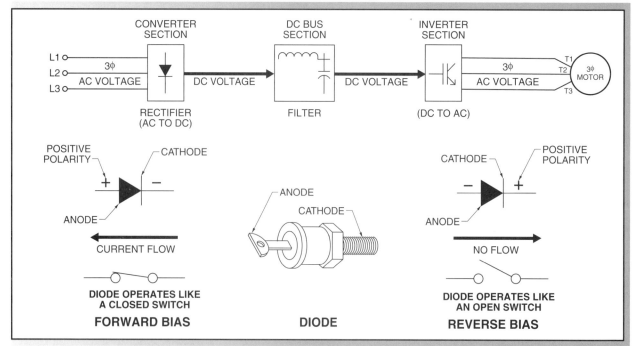

SEE APPENDIX FOR ⊕ TO ⊖ DIAGRAM

Figure 7-5. A diode allows current to flow in one direction only (forward bias).

Forward bias is the condition of a diode when the diode allows current flow. The anode in a forward-biased diode has a positive polarity and the cathode has a negative polarity. *Reverse bias* is the condition of a diode when it does not allow current flow and acts as an insulator. The cathode of a reverse-biased diode has a positive polarity and the anode has a negative polarity.

A diode allows current to flow only when it is forward-biased. A diode acts as a closed switch when it is forward-biased and as an open switch when it is reverse-biased. A diode is rated for the maximum forward-bias current it can safely conduct and the maximum reverse-bias voltage that can be applied. See Figure 7-6.

The maximum forward-bias current rating is the amount of current the diode can withstand while conducting electrons in the forward-bias direction. Exceeding the maximum forward-bias current rating causes the diode to overheat, resulting in diode failure. The maximum reverse-bias voltage rating is the amount of voltage that can be applied in the reverse direction without allowing current flow in the reverse direction (breakover voltage).

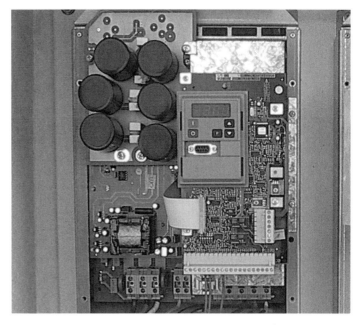

Electric motor drives consist of resistors, capacitors, coils, power terminal strips, control terminal strips, various integrated chips, SCRs, and gate turn-off thyristors or transistors.

DIODE RATING

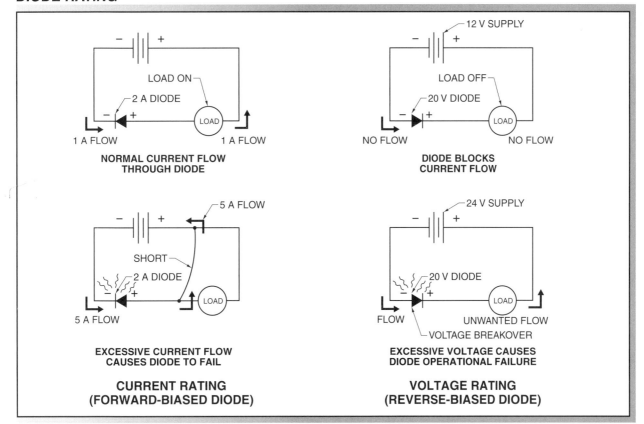

SEE APPENDIX FOR ⊕ TO ⊖ DIAGRAM

Figure 7-6. A diode can be damaged when current and/or voltage ratings are exceeded.

Three-phase bridge rectifiers are used in electric motor drives to convert AC power to DC bus power.

Rectifiers

A *rectifier* is a device that changes AC voltage into DC voltage. Alternating current (AC) power is used because it is more efficiently generated and transmitted over long distances than direct current (DC) power. AC voltage is rectified in the converter section of an electric motor drive. *Rectification* is the changing of AC voltage into DC voltage. Rectifiers can be half-wave, full-wave, or bridge rectifiers.

Half-Wave Rectifiers. A *half-wave rectifier* is a circuit containing one diode that allows only half of the input AC sine wave to pass. See Figure 7-7. Half-wave rectification is accomplished because current is allowed to flow only when the anode terminal has a positive polarity with respect to the cathode. Current is not allowed to flow through the rectifier when the cathode has a positive polarity with respect to the anode. Half-wave rectification is inefficient for most applications because one-half of the input sine wave is not used.

HALF-WAVE RECTIFIER

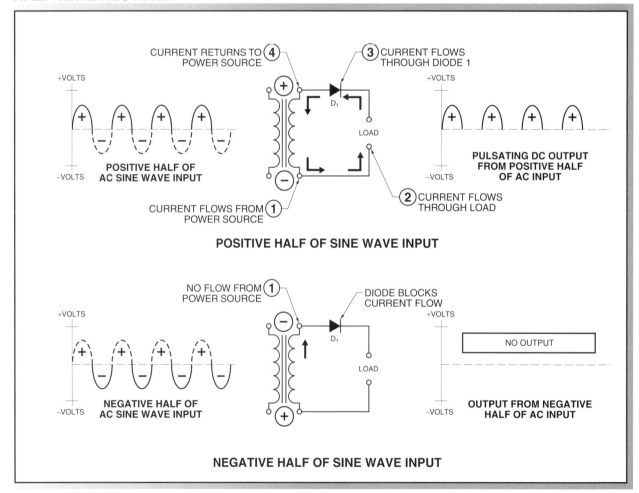

Figure 7-7. A half-wave rectifier converts AC voltage into pulsating DC voltage.   *SEE APPENDIX FOR ⊕ TO ⊖ DIAGRAM*

Full-Wave Rectifiers. A *full-wave rectifier* is a circuit containing two diodes and a center-tapped transformer that permits both halves of the input AC sine wave to pass. See Figure 7-8. Full-wave rectification is accomplished by one diode passing the positive half of the AC sine wave and the second diode passing the negative half of the AC sine wave. A full-wave rectifier is more efficient than a half-wave rectifier because both halves of the input sine wave are used.

Bridge Rectifiers. A *bridge rectifier* is a circuit containing four diodes that permits both halves of the input AC sine wave to pass. A bridge rectifier is more efficient than a half-wave or full-wave rectifier and is the most common rectifier circuit used in 1ϕ rectification circuits. See Figure 7-9.

The output of a bridge rectifier is a pulsating DC voltage that must be filtered (smoothed) before it can be used in most electronic equipment. A filter circuit connected to the output of the bridge rectifier filters the pulsating DC voltage. The filter circuit usually consists of one or more capacitors, inductors, and/or resistors connected in different combinations. Filtered DC eliminates pulsations and increases the output voltage.

TECH FACT

Use control signal wiring rated for 600 V even when the control signal is a low voltage (such as 12 VDC or 24 VDC) because electric motor drives contain voltages of 600 V or more.

FULL-WAVE RECTIFIER

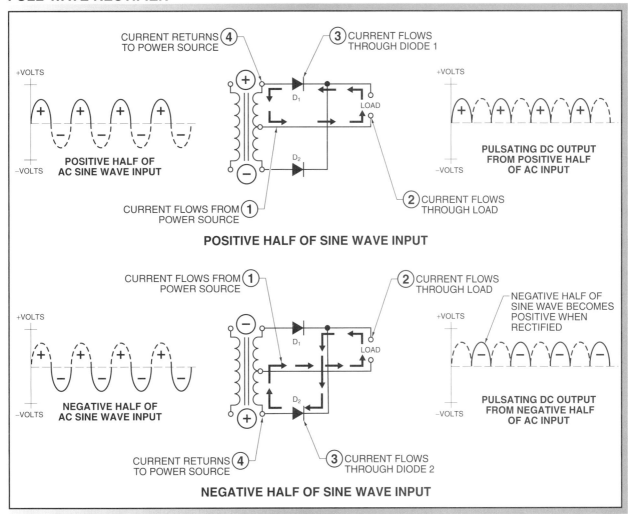

SEE APPENDIX FOR ⊕ TO ⊖ DIAGRAM

Figure 7-8. Full-wave rectification is obtained with two diodes and a center-tapped transformer.

BRIDGE RECTIFIER

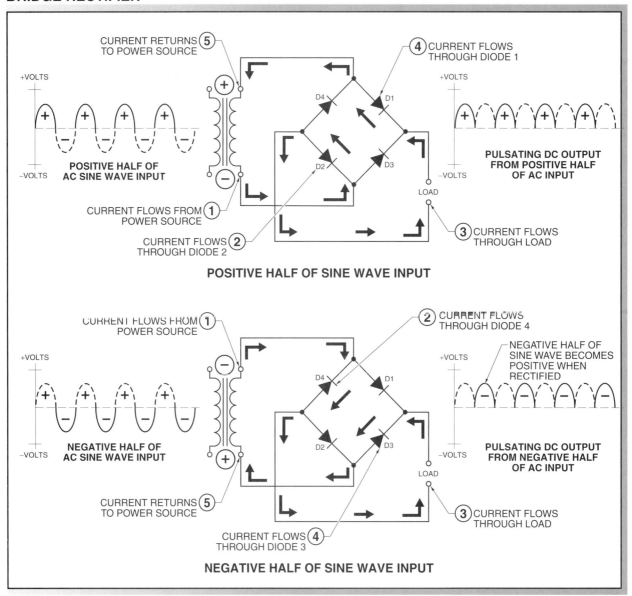

Figure 7-9. Full-wave rectification is obtained with four diodes.

SEE APPENDIX FOR ⊕ TO ⊖ DIAGRAM

Some small horsepower electric motor drives can be connected to a 1φ power supply. In residential and small commercial locations, only a 1φ power source is available. An electric motor drive can be used with 1φ power to control motor speed and conserve energy, but most electric motor drives are connected to 3φ power supplies. The advantage of using a 3φ power supply is that it is possible to obtain a smoother DC voltage output without the use of a filter circuit. In a 3φ circuit, when any one phase becomes negative, at least one of

the other phases is positive. When a filter is used, the result is a smooth output with little or no voltage variance. See Figure 7-10.

⚠ WARNING
DC bus capacitors do not discharge immediately. High voltage remains after power is removed.

ELECTRIC MOTOR DRIVE VOLTAGE CONVERSIONS

DC BUS
INDUCTOR

CAPACITOR

L1
L2
L3

T1 (U)
T2 (V)
T3 (W)

T1
T2
T3

3φ
MOTOR

RECTIFIER
(AC TO DC)
SECTION

DC FILTER
SECTION

INVERTER
(DC TO AC)
SECTION

+VOLTS
L1   L2   L3

0

60° 120° 180° 240° 300° 360° 60°
−VOLTS

3φ INPUT VOLTAGE

+VOLTS

**UNFILTERED DC
OUTPUT VOLTAGE**

+VOLTS

**FILTERED DC
OUTPUT VOLTAGE
(NO VOLTAGE
VARIANCE)**

+VOLTS

0

60° 120° 180° 240° 300° 360° 60°
−VOLTS

**3φ OUTPUT VOLTAGE
PER PHASE**

Figure 7-10. Electric motor drives use 3φ rectifier circuits to produce DC voltage.

Silicon-Controlled Rectifiers

A *silicon-controlled rectifier (SCR)* is a solid state device with the ability to rapidly switch high currents. SCRs belong to the thyristor family that also includes other solid state switching devices such as triacs and GTOs. See Figure 7-11. The three terminals of an SCR are the anode, cathode, and gate. The anode and cathode of the SCR are similar to the anode and cathode of a diode. The gate acts as the control for the SCR. When the gate is forward-biased and current begins to flow in the gate-cathode junction, the value of forward breakover voltage can be reduced. *Forward breakover voltage* is the forward bias voltage necessary for a semiconductor to go into conduction mode.

Low-current SCRs can operate with an anode current of less than 1 A. High-current SCRs can handle load currents of hundreds of amps. The size of an SCR increases with an increase in current rating.

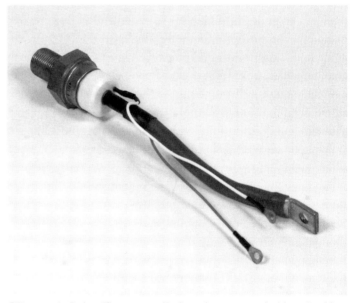

Silicon-controlled rectifiers are used by large-horsepower electric motor drives to convert AC voltage to DC bus voltage and for switching in DC drives.

THYRISTORS

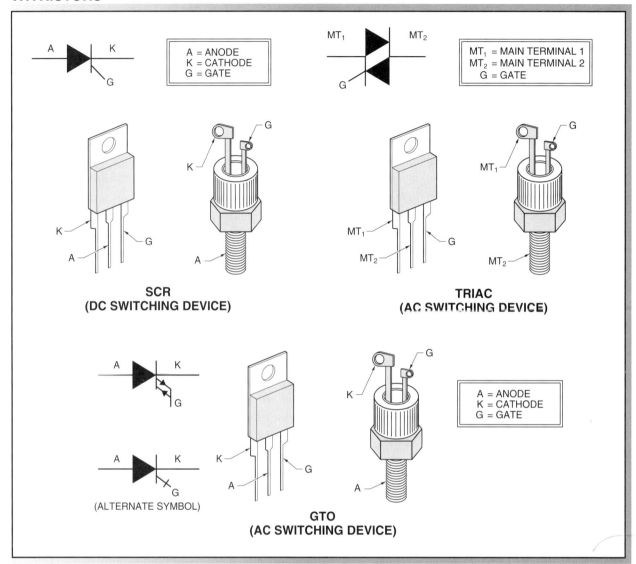

Figure 7-11. Thyristors operate as switches controlling the flow of DC voltage (SCRs) and AC voltage (triacs and GTOs).

An SCR is basically the same as a diode except the gate controls when the current will start flowing. An SCR operates much like a mechanical switch (ON or OFF). The SCR is switched ON when voltage is applied to the gate. The gate signal must be of a positive polarity with respect to the cathode for the SCR to turn ON. The SCR remains ON (latched on position), even when the applied voltage to the gate is removed, as long as the SCR remains forward-biased. Once the gate voltage has turned ON the flow of current through the SCR, the gate cannot be used to turn the flow of current OFF. The SCR will remain ON until the anode voltage has been significantly reduced to a level where the current is not large enough to maintain the proper level of latching current.

One characteristic of an SCR is that current flow can only stop when the anode current drops to zero. See Figure 7-12. When reverse-bias voltage is applied, there is no current flow through an SCR. Reverse bias exists when a positive polarity voltage is applied to the cathode and a negative polarity voltage is applied to the anode.

An SCR can vary voltage output by changing the firing angle (timing) of the gate. The later the gate pulse is applied in the sine wave cycle, the later the sine wave output voltage is turned ON, and the smaller the output voltage pulse becomes. The delay in the gate pulse lowers the average DC voltage output. See Figure 7-13.

SILICON-CONTROLLED RECTIFIER OPERATION

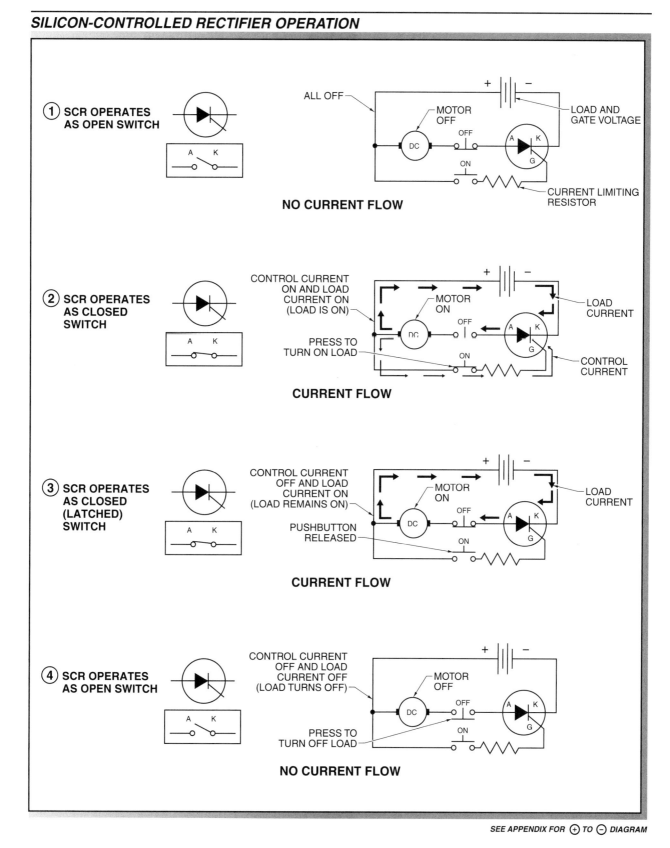

SEE APPENDIX FOR ⊕ *TO* ⊖ *DIAGRAM*

Figure 7-12. An SCR functions similar to a mechanical switch to control the flow of DC current to motors.

VARYING OUTPUT VOLTAGE OF SCR

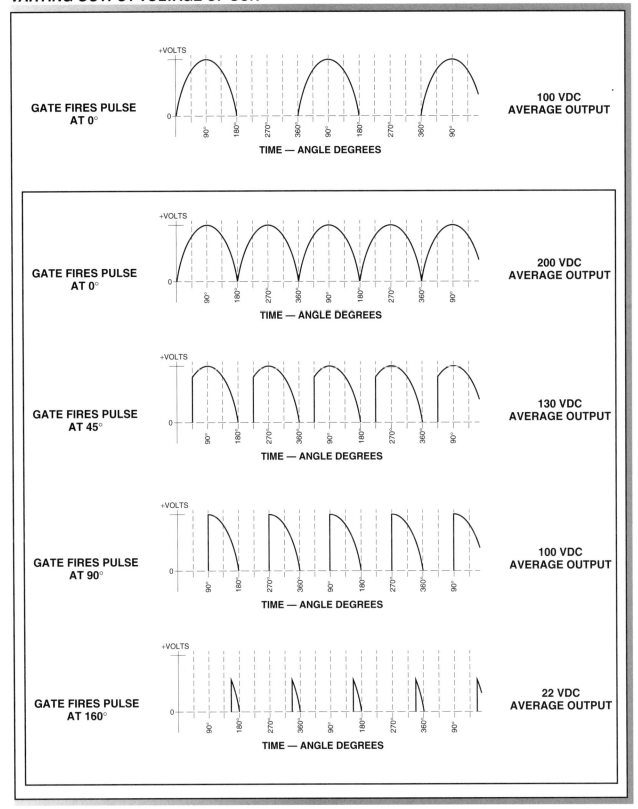

Figure 7-13. Average output voltage from an SCR is controlled by the firing angle of the gate.

SCRs are used to control DC loads or, when connected to 3φ AC power supplies, are used to produce DC voltages. Diodes are used in low-horsepower AC drives to rectify 3φ power, and SCRs rectify the 3φ power in large-horsepower AC drives to supply the DC bus voltage. See Figure 7-14.

SCRs start and stop the flow of current to DC motors, but also control the DC voltage level applied to the motor. Changing the applied voltage to a DC motor changes the motor's speed. The higher the applied voltage, the faster the motor rotates.

Two SCRs are used to control an AC load; one SCR is used to control the voltage level in the positive half of the sine wave alternation and the second is used to control the voltage level in the negative half of the sine wave alternation. SCRs controlling AC loads are connected in parallel and in opposite directions. Triacs can also control AC loads, and do so in the same way SCRs control DC loads. However, an advantage of using two SCRs instead of one triac to control an AC load is that there is less heat produced by the SCRs because each SCR operates only on every other half-cycle of the AC sine wave. See Figure 7-15.

TECH FACT

Oversizing an electric motor drive for an application will increase drive longevity. Undersizing an electric motor drive even for short-term operation shortens the life of the drive.

When solid state switching devices are used to control current, the devices require heat sink mounting because of the large amounts of heat generated.

Gate Turn-Off Thyristors

A *gate turn-off thyristor (GTO)* is a solid state device that allows for a controlled turn ON and controlled turn OFF of current using the gate. Applying a positive-polarity pulse to the gate of a GTO starts current flow and

SCRs AS RECTIFIERS

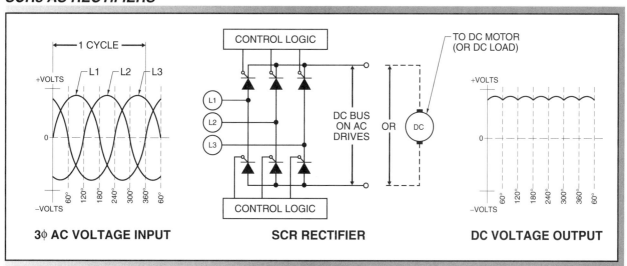

Figure 7-14. The DC voltage output of SCRs can be used to supply the DC bus voltage in AC drives.

a negative-polarity pulse to the gate stops current flow. The disadvantage of a GTO is the level of current required to switch OFF the device. A GTO requires one-third to one-fifth the normal anode current to switch OFF. Control of voltage and current is accomplished though the fast switching action of GTOs. See Figure 7-16.

Once the GTO is turned ON, the GTO continues conducting as long as the anode current remains above the minimum holding current, or until a negative-polarity voltage is applied to the gate. *Minimum holding current* is the minimum amount of current required to keep a device operating. GTOs can handle currents of several thousand amps at thousands of volts.

SCRs were the first solid state switches used in AC drive inverter sections (DC back to variable-frequency AC) but have been replaced by GTOs. Just as GTOs replaced SCRs in electric motor drives, transistors have replaced GTOs. Transistors switch faster and use less energy than SCRs or GTOs.

Transistors

A *transistor* is a solid state device that controls current according to the amount of voltage applied to the base. Transistors are used as amplifiers or as DC switches. *Amplification* is the process of taking a small signal and increasing the signal size. Transistors are used as switches in electric motor drives to control the voltage and current applied to motors. Transistors have the advantage of extremely fast switching with no moving parts.

The two types of transistors are NPN and PNP. See Figure 7-17. Each type includes an emitter (E), base (B), and collector (C). In the symbol for a transistor, the terminal with the arrow is the emitter. A transistor is an NPN transistor if the arrow points away from the base and is a PNP transistor if the arrow points towards the base. The collector/emitter circuit of a transistor is the section that operates like a switch and is equivalent to the mechanical contacts of a relay.

AC SWITCHING USING SILICON-CONTROLLED RECTIFIERS

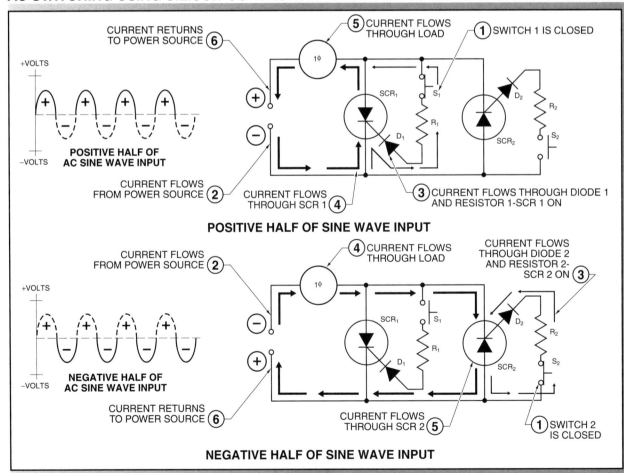

SEE APPENDIX FOR ⊕ TO ⊖ DIAGRAM

Figure 7-15. Two SCRs are required for AC voltage switching.

GATE TURN-OFF THYRISTOR OPERATION

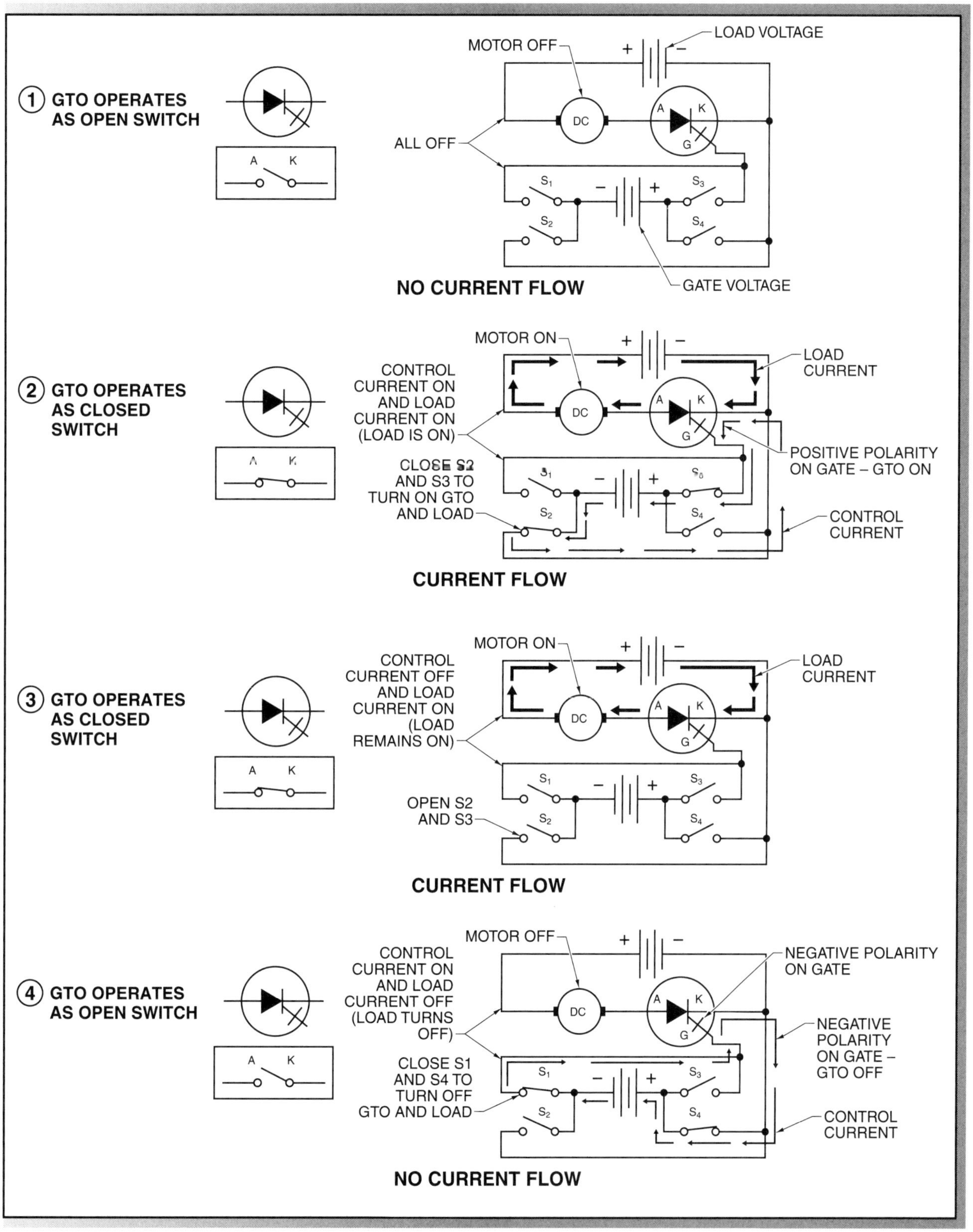

SEE APPENDIX FOR ⊕ TO ⊖ DIAGRAM

Figure 7-16. GTOs turn ON a load with a positive polarity pulse at the gate, and turn OFF a load with a negative polarity pulse at the gate.

TYPES OF TRANSISTORS

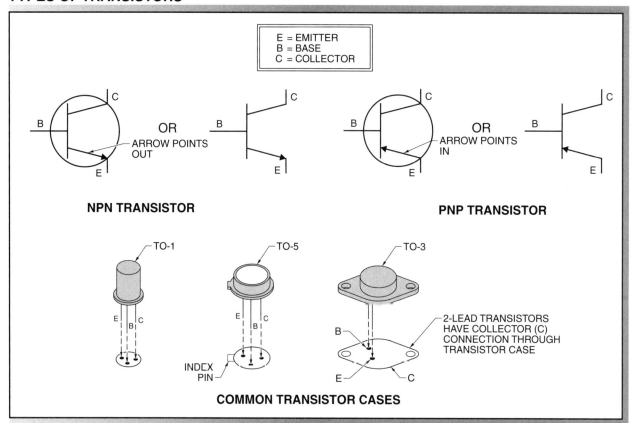

Figure 7-17. The two types of transistors are NPN and PNP.

Transistor Operation. Transistors were developed to replace mechanical switches. Transistors have no moving parts and can switch ON and OFF quickly. Transistors, like switches, have a very low resistance when ON and a very high resistance when OFF. See Figure 7-18.

When a transistor is operated as a switch, the resistance between the collector (C) and the emitter (E) is determined by the current flow between the base (B) and emitter (E). When the base/emitter resistance is high, like that of an open switch, no current flows between C and E. The transistor is OFF due to no flow between B and E, allowing no current flow in the load circuit (C and E).

When current flows between B and E, the collector/emitter resistance is reduced to a very low value, like that of a closed switch. A transistor switched ON is operating in the saturation region, which turns on the load circuit (C and E). The *saturation region* is the maximum current that can flow in the transistor load circuit.

The resistance (impedance) of the load is the only current-limiting device in the circuit when the circuit reaches saturation. The transistor is operating in the cutoff region when the transistor is switched OFF. The *cutoff region* is the point at which the transistor is turned OFF and no current flows.

Transistor Motor Control. Transistors are used in DC and AC drives. Four transistors are required to control the direction of a DC motor. See Figure 7-19. Two transistors are switched ON for one direction of rotation and the other two transistors are switched ON for the opposite direction of rotation.

AC drives use transistors in the inverter section (DC voltage to AC voltage) to control the voltage applied to an AC motor. To control AC voltages, two transistors are used per phase. One transistor controls the positive alternation of the sine wave and the second transistor controls the negative alternation of the sine wave. Electric motor drives that control 3ϕ motors require six transistors (two per phase).

TRANSISTOR CIRCUIT CURRENT FLOW

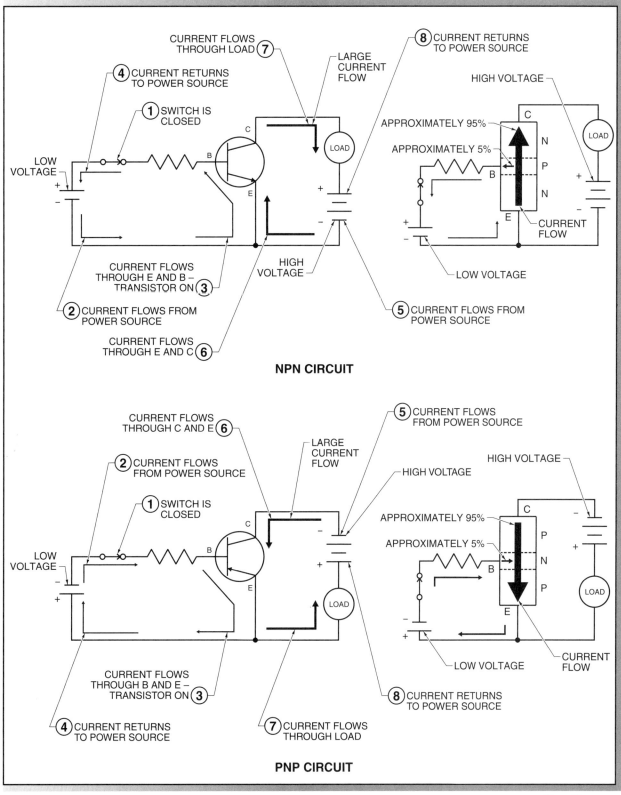

Figure 7-18. Transistors are used like mechanical switches to turn loads ON and OFF.

SEE APPENDIX FOR ⊕ TO ⊖ DIAGRAM

DC MOTOR CONTROL USING TRANSISTORS

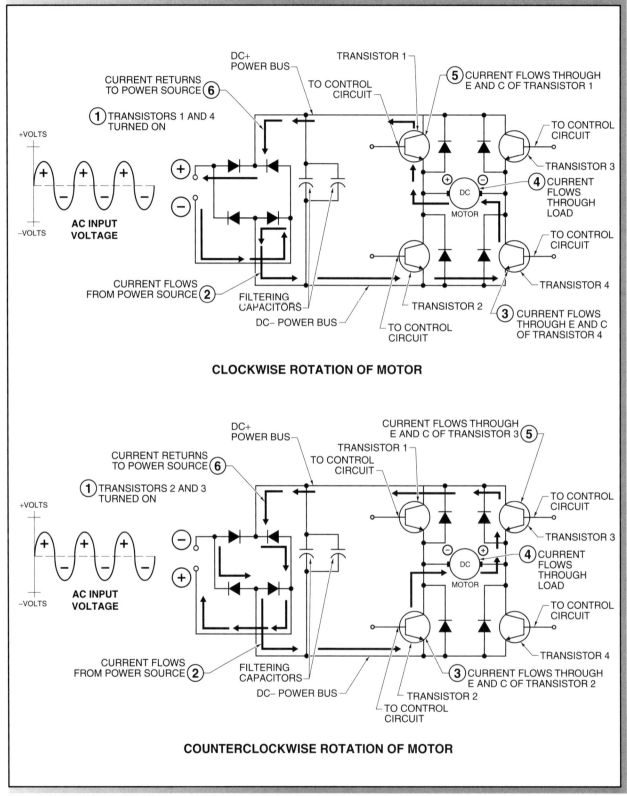

SEE APPENDIX FOR ⊕ TO ⊖ DIAGRAM

Figure 7-19. Four transistors are used to control the direction of current to a DC motor.

A *fly-back (freewheeling) diode* is a diode that handles regenerated voltage when an AC motor decelerates. See Figure 7-20. When the transistors are switched ON and the motor is operating, no current flows through any diode because each diode is reversed-biased. When the motor is pulsed OFF, the diodes act like closed switches, allowing the voltage spike to flow back into the DC power bus and be safely dissipated.

Transistors are engineered to different designs and cases to achieve better heat dissipation and faster switching characteristics in a smaller space. See Figure 7-21. The three basic designs of transistors used in electric motor drives are:

- Bipolar junction transistor (BJT)
- Metal-oxide semiconductor field-effect transistor (MOSFET)
- Insulated-gate bipolar transistor (IGBT)

TECH FACT

An electric motor drive designed for 3ϕ and 1ϕ power supplies must be derated when connected to a 1ϕ power supply. Consult the drive instructions for derating values.

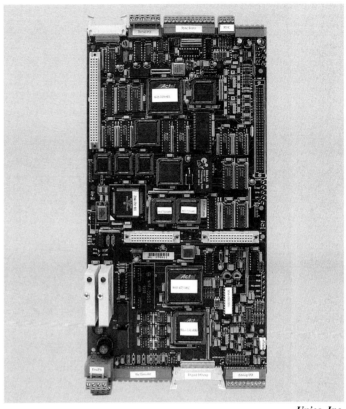

Unico, Inc.

An electric motor drive motherboard can have multiple processors and registers, depending on the drive design.

FLY-BACK DIODE PROTECTION

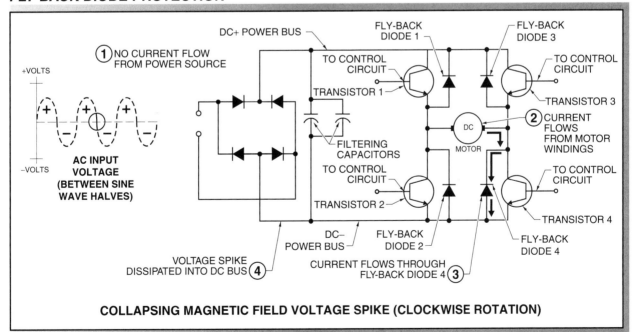

COLLAPSING MAGNETIC FIELD VOLTAGE SPIKE (CLOCKWISE ROTATION)

SEE APPENDIX FOR ⊕ TO ⊖ DIAGRAM

Figure 7-20. Fly-back diodes act as closed switches to dissipate motor voltage spikes into the DC bus.

TRANSISTOR DESIGNS

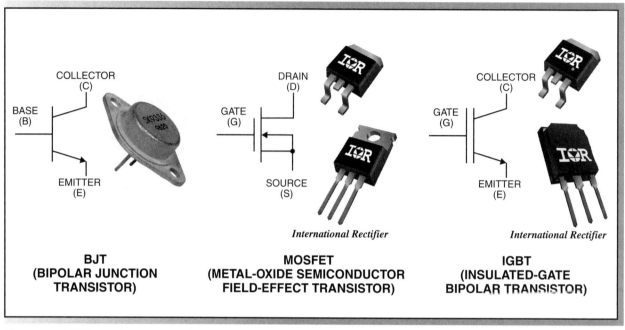

Figure 7-21. Different types of transistors are used to control current flow in electric motor drives.

Bipolar Junction Transistor. A *bipolar junction transistor (BJT)* is a transistor that controls the flow of current through the emitter (E) and collector (C) with a properly biased base (B). The bipolar junction transistor (BJT) is also called a bipolar transistor, or transistor, since it is the oldest and most common type. The NPN type BJT was the first transistor used for switching power in variable voltage inverters (VVI), but have been replaced by insulated gate bipolar transistors (IGBTs) in pulse-width-modulated inverters (PWM).

BJTs are available for switching several hundred amps at 1000 V or more. In AC and DC drives, BJTs were used as fast-acting, high-current switches, but have been replaced by faster switching types of transistors, such as FETs and IGBTs.

Field-Effect Transistor. A *field-effect transistor (FET)* is a transistor that controls the flow of current through the drain (D) and source (S) with a properly biased gate (G). The two subtypes of FETs are junction FET (JFET) and the metal-oxide-semiconductor FET (MOSFET). MOSFETs replaced BJTs in electric motor drives.

Applying voltage to the gate controls current flow through the drain and source. Gate current is small compared to the current flowing through the drain and source. MOSFETs can switch higher frequencies and turn ON and OFF faster than BJTs. The electric motor drive's carrier (switching) frequency can now operate at higher frequencies, reducing objectionable noise in the motor, and producing a waveform that closely represents a sine wave.

Insulated-Gate Bipolar Transistor. An *insulated-gate bipolar transistor (IGBT)* is a high power switching device that can switch high currents and high voltages. The insulated-gate bipolar transistor (IGBT) controls the flow of current through the collector (C) and emitter (E) with a properly biased gate (G). IGBTs have replaced BJTs and MOSFETs in electric motor drives because IGBTs require less operating power and have lower thermal resistance. See Figure 7-22. Less operating power and low thermal resistance produce less heat, allowing smaller size electric motor drives.

IGBTs can turn ON and OFF faster than MOSFETs, allowing the electric motor drive to use much higher carrier frequencies and control more power than MOSFETs. The higher carrier frequencies mean less noise, which is important in HVAC applications. IGBTs produce the cleanest sine wave waveform of any switching devices.

AC MOTOR DRIVES

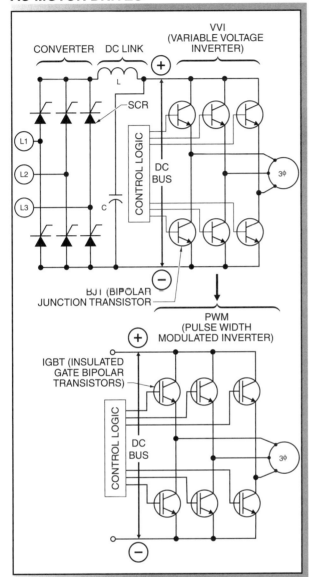

Figure 7-22. Transistors are used in the inverter section of an AC drive to control the voltage level and frequency applied to the motor.

INTEGRATED POWER MODULES

An electric motor drive, like any electronic device, can be built using separate individual components. Electric motor drives are built using both individual components and integrated modules. An *integrated power module* is grouping of separate parts, such as IGBTs, that are designed to perform the function of switching the power supplied to the load (motor). See Figure 7-23. The power section of an electric motor drive can be designed and assembled using individual IGBTs, diodes, etc. The

power switching section of an electric motor drive is also available as one complete integrated power module. An integrated power module increases reliability and allows for faster assembly and repair of the power section. Integrated power modules are commonly used in smaller horsepower drives that are typically replaced, not repaired.

> ⚠ **WARNING**
>
> *Solid state components leak small amounts of current when in the OFF state. Do not touch the output terminals T1 (U), T2 (V), and T3 (W) of an electric motor drive because the component leakage current is high enough to cause electrocution.*

INTEGRATED POWER MODULES

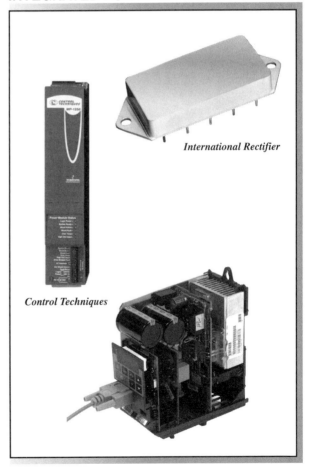

International Rectifier

Control Techniques

Figure 7-23. An integrated power module is a grouping of separate parts designed to perform a set task.

ELECTROSTATIC DISCHARGE PRECAUTIONS

Electrostatic discharge (ESD) is the movement of static electricity (electricity at rest) from the surface of one object to another object. Static electricity can exist on plastic, fabric, paper, or other objects and can be discharged by human skin contact.

Electronic controls (SCRs, GTOs, BJTs, FETs, and IGBTs) that are part of an integrated electronic control circuit can be damaged by currents as little as 1 mA (.001 A) and voltages of 10 V or more. On a dry day a person can develop a static charge of several thousand volts, and touching an electronic circuit discharges the static charge into the circuit components. ESD can damage or destroy semiconductors and other sensitive electronic components; this potential for damage must be understood when working with or around electric motor drives.

The two general categories of logic integrated circuits are the transistor-transistor logic (TTL) family and the metal-oxide-semiconductor (MOS) family. Digital logic probes used to test for digital signals include a switch to select the type of integrated circuit under test, TTL or MOS. See Figure 7-24.

MOS-based integrated circuits are extremely susceptible to damage from ESD. Sensitive electronic components are shipped in static-shielded wrapping that helps guard against ESD damage. Electronic components are susceptible to damage from improper handling before and after installation into a circuit. It is uncommon that an electronic component located on a printed circuit (PC) board of an electric motor drive will ever be replaced (the entire board is changed out), but care must be taken when working near or replacing PC boards.

TECH FACT

Store electric motor drives in a static-shielded wrapping and in a well-ventilated location, avoiding high humidity, dust, metal particles, and high temperatures that damage drive components.

TESTING INTEGRATED CIRCUITS WITH LOGIC PROBE

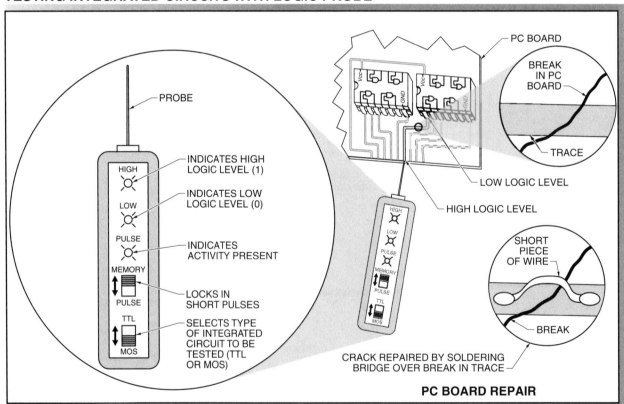

Figure 7-24. Digital logic probes used to test for digital signals include a switch to select the type of integrated circuit under test, TTL or MOS.

To prevent an electrostatic discharge from damaging electronic components inside an electric motor drive, take the following precautions:

- Discharge any static charge buildup by first touching a conductive surface, such as grounded conduit, before touching electronic components inside or outside an electric motor drive. Manufacturers recommend wearing a wrist strap when working with sensitive electronic components. See Figure 7-25.
- Touch the insulated edge of PC boards and not the components when replacing PC boards inside an electric motor drive.
- Ground electric motor drives to ensure the electronic circuits in the drive are grounded.

ELECTROSTATIC DISCHARGE PREVENTION

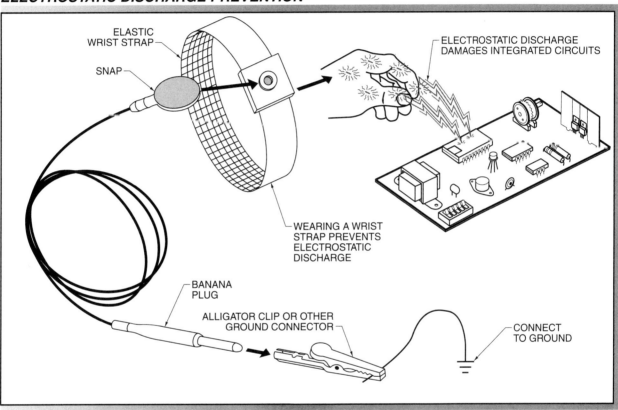

Figure 7-25. A wrist strap grounds a technician when working on sensitive electronic components.

Name _____ Date _____

True-False

T F **1.** A heat sink is a device that conducts heat away from an electrical component. *pg 176*

T F **2.** The type of enclosure used does not affect the temperature of an electric motor drive. *pg 177*

T F **3.** A semiconductor device is an electronic device that has electrical conductivity between that of a conductor and that of an insulator. *pg 178*

T F **4.** A rectifier changes DC voltage into AC voltage. *AC to DC* *pg 180*

T F **5.** The three terminals of a silicon-controlled rectifier are the anode, cathode, and emitter. *pg 183* *anode, cathode, gate*

T F **6.** Transistors have no moving parts and can switch ON and OFF quickly. *pg 180*

T F **7.** The two general categories of logic integrated circuits are the transistor-transistor logic family and metal-oxide-semiconductor family.

T F **8.** Static electricity cannot be discharged by human skin contact. *pg 196*

T F **9.** To control AC voltages, four transistors are used per phase.

T F **10.** An SCR can vary voltage output by changing the firing angle of the gate. *pg 184*

T F **11.** A fixed resistor is a resistor that contains fixed tap points of different resistances.

T F **12.** Reverse bias is the condition of a diode when it does not allow current flow and acts as an insulator. *pg 179*

T F **13.** An SCR is a solid state device that controls current according to the amount of voltage applied to the base.

T F **14.** The saturation region is the maximum current that can flow in the transistor load circuit. *pg 190*

T F **15.** A fly-back diode is a diode placed across a transistor to prevent high voltage spikes produced by the collapsing magnetic field of the motor windings. *pg 193*

Completion

Solid state switch **1.** A(n) ___ switch is a device that uses electronic components to open and close circuits at a precise point in time to control current flow and voltage levels. *pg 173*

Voltage Drop **2.** ___ is the amount of voltage consumed by a device or component as current passes through it. *pg 174* _ohms & watts_ **3.** Resistors are classified by a resistance value in ___ and power dissipation value in ___. *pg 177*

Diode **4.** A(n) ___ is an electronic device that allows current to pass through in only one direction. *pg 178*

DC voltage **5.** The output of a bridge rectifier is a pulsating ___ that must be filtered before it can be used in most electronic equipment. *pg 181*

SCR **6.** A(n) ___ is a solid state device with the ability to rapidly switch high currents. *pg 183*

Four **7.** ___ transistors are required to control the direction of a DC motor. *pg 190*

IGBT **8.** A(n) ___ is a high power switching device that can switch high currents and high voltages. *pg 184*

Integrated power module **9.** A(n) ___ is a grouping of separate parts designed to perform a set function or task, such as providing a time delay, power gain, or filtering a signal. *pg 195*

saturation Region **10.** The ___ region is the maximum current that can flow in the transistor load circuit. *pg 190*

Multiple Choice

_A_ **1.** A ___ parameter is a parameter value that can be displayed, but not set or changed.
  A. read-only   C. write
  B. read-write   D. universal   *pg 176*

_C_ **2.** Typically, electric motor drives must be derated at altitudes above ___′.
  A. 1000   C. 3300
  B. 2200   D. 4000   *pg 177*

_B_ **3.** A(n) ___ is an electrical device designed to store electrical energy.
  A. resistor   C. diode
  B. capacitor   D. inductor   *pg 178*

_A_ **4.** A diode allows current to flow only when it is ___.
  A. forward-biased   C. open
  B. reverse-biased   D. unbiased   *pg 178*

_A_ **5.** In residential and small commercial locations, only a(n) ___-phase power source is available.
  A. single   C. three
  B. two   D. open

pg 182

Name _____ Date _____

Activity 7-1. Connecting Electric Motor Drive to Electromechanical Outputs

Electric motor drives include outputs that can be used to turn loads ON and OFF within a system. Setting the electric motor drive parameter that controls the output determines when that output will switch ON and OFF. Electric motor drive outputs can be programmed to activate at a set frequency such as 20 Hz, 60 Hz, or 85 Hz or an electrical condition such as 2 A, 6 A, or 10 A. When an electric motor drive has electromechanical outputs (contacts), the outputs can be used to control AC or DC loads. The AC or DC loads must be within the voltage and current rating of the outputs. Listed voltage and current ratings for outputs specify the maximum voltage and current that can be switched. Before working on an electric motor drive and motor circuit, technicians and qualified persons must understand the electrical system and test equipment used.

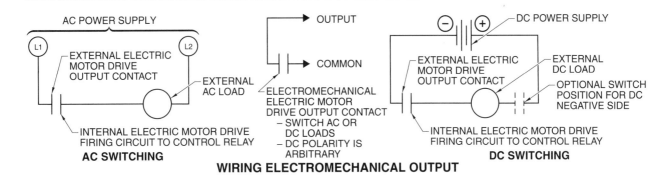

1. Connect electric motor drive Output 1 so that the drive controls the solenoid. Connect Output 2 so that the drive controls the alarm.

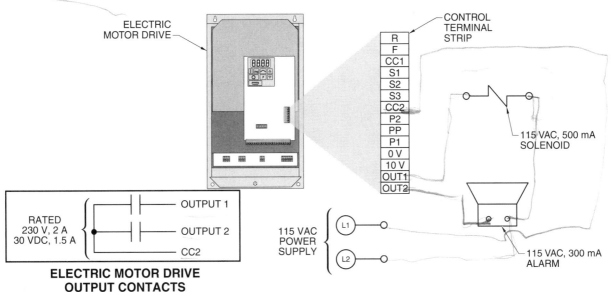

Activity 7-2. Connecting Electric Motor Drive Electromechanical Outputs to High Current Loads

Electric motor drive outputs have limited voltage and current ratings. When the ratings are exceeded, the contacts can be destroyed. Some electric motor drive outputs cannot be repaired or replaced. The drive must be replaced. To avoid damaging electric motor drive outputs, relays are used as interfaces between the drive output and high-voltage and/or current loads. Applications with relay interfaces use electric motor drives to control the coils of a relay, and the contacts of the relay are used to switch the high-voltage and/or current loads. Relays include normally open and normally closed contacts that allow a circuit greater versatility.

1. Connect electric motor drive Output 1 so that the drive controls the solenoid using the control relay as an interface. Connect the solenoid to turn ON anytime Output 1 turns ON.

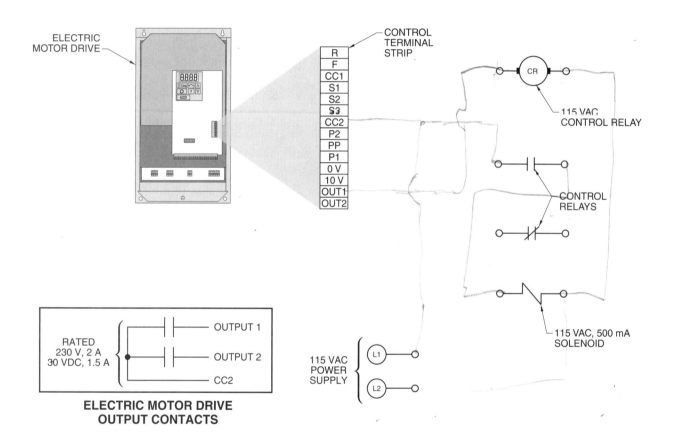

Activity 7-3. Connecting Electric Motor Drive to NPN Transistor Outputs

Transistors are used as DC switches. Transistors have no moving parts and can switch ON and OFF at very fast speeds. Transistors are available as NPN or PNP types. NPN transistors are the most common transistors used for input and output switches on electric motor drives. The emitter (E) and collector (C) are the switching parts of the transistor. Electric motor drive manufacturers use the term "open collector" to describe a transistor output and a symbol to indicate whether an NPN or PNP transistor is used. Because transistors are used to switch DC loads, polarity must be maintained. The voltage and current of a load switched by a transistor must be less than the maximum voltage and current rating of the transistor.

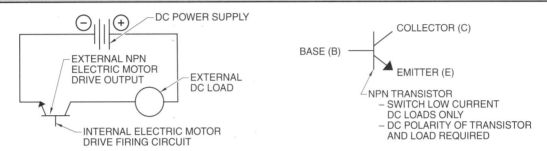

WIRING NPN TRANSISTOR OUTPUT

1. Connect electric motor drive Output 1 so that the drive controls Load 1. Connect Output 2 so that it controls Load 2.

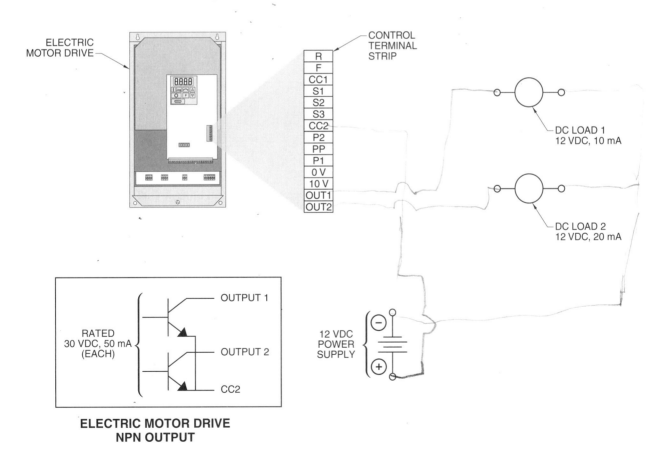

ELECTRIC MOTOR DRIVE NPN OUTPUT

Activity 7-4. Connecting Electric Motor Drive NPN Transistor Outputs to High Current Loads

Electric motor drive transistor outputs have limited voltage and current ratings. When the ratings are exceeded, the transistor can be destroyed. To avoid damaging electric motor drive transistor outputs, solid state relays are used as interfaces between the drive output and the higher voltage and/or current load. An electric motor drive is used to control the solid state relay, and the contacts of the solid state relay are used to switch the high-voltage and/or current load(s).

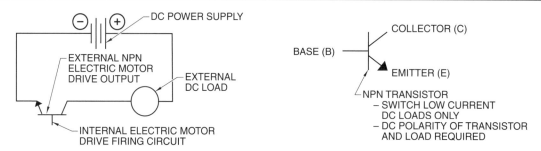

WIRING NPN TRANSISTOR OUTPUT

1. Connect electric motor drive Output 1 so that the drive controls the solenoid using the solid state relay as an interface. Connect the solenoid to turn ON anytime Output 1 turns ON.

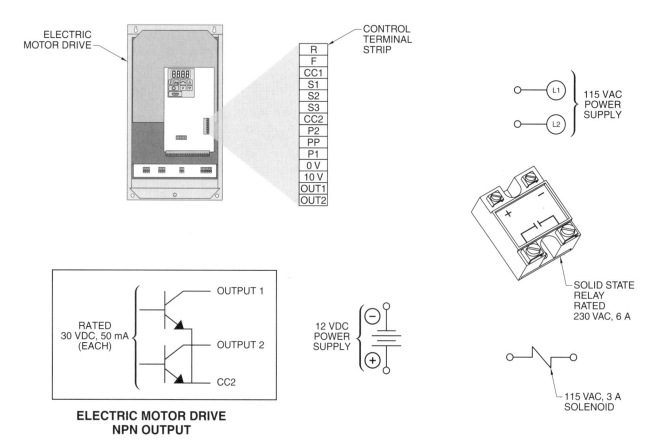

Activity 7-5. Connecting Electric Motor Drive to PNP Transistor Outputs

Transistors are used as DC switches. Transistors have no moving parts and can switch ON and OFF at very fast speeds. Transistors are available as NPN or PNP transistor types. Electricians must understand how to wire PNP outputs.

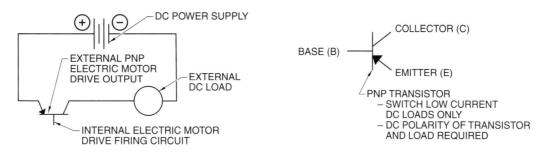

WIRING PNP TRANSISTOR OUTPUT

1. Connect electric motor drive Output 1 so that the drive controls Load 1. Connect Output 2 so that it controls Load 2.

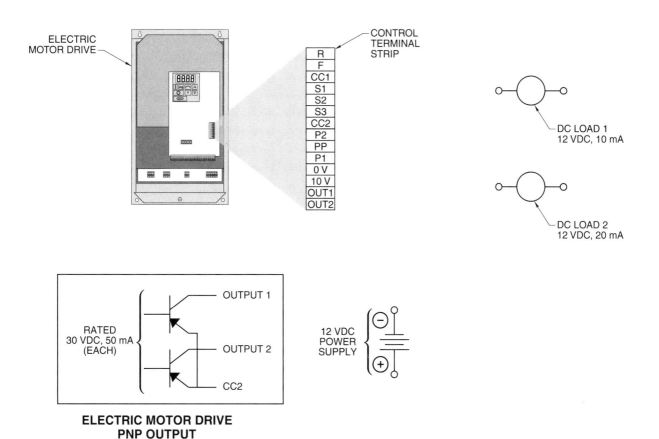

ELECTRIC MOTOR DRIVE PNP OUTPUT

Activity 7-6. Connecting Electric Motor Drive PNP Transistor Outputs to High Current Loads

Electric motor drive transistor outputs have limited voltage and current ratings. When the ratings are exceeded, the transistor can be destroyed. To avoid damaging electric motor drive transistor outputs, solid state relays are used as interfaces between the drive output and the higher voltage and/or current load. An electric motor drive is used to control the solid state relay, and the contacts of the solid state relay are used to switch the high voltage and/or current load(s).

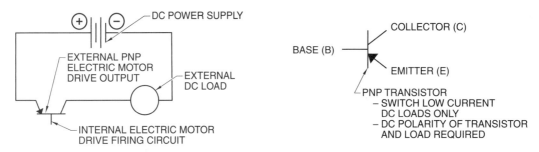

WIRING PNP TRANSISTOR OUTPUT

1. Connect the electric motor drive Output 1 so that the drive controls the solenoid using the solid state relay as an interface. Connect the solenoid to turn ON anytime Output 1 turns ON.

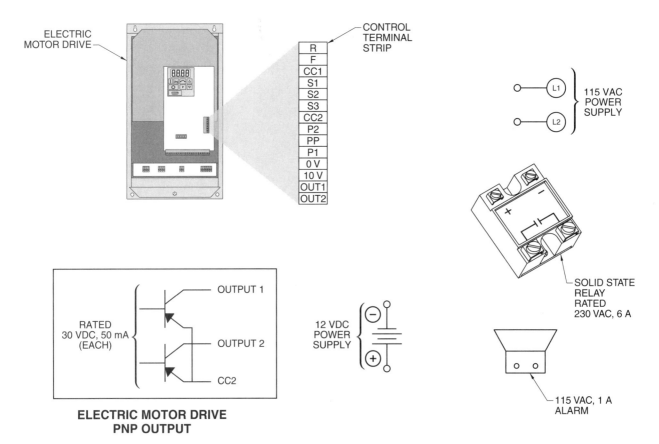

8 Electric Motor Drive Operation Fundamentals

Although all electric motor drives are designed with some common features and functions, each individual type has advantages. Advantages are low cost, reduced operating noise, improved torque at low speeds, precise speed control through a feedback system, advanced motor and electric motor drive monitoring and protection, and the ability to interface with other systems. Understanding the different types of electric motor drives that are available, drive operating principles, and advantages allows for proper drive selection and troubleshooting.

ELECTRIC MOTOR DRIVE NAMES

Often a piece of electrical equipment is called by several different names. The name might refer to the general name of the equipment or describe the type of equipment, the equipment's function, or the application (system) of the piece of electrical equipment. A 3ϕ totally enclosed fan cooled induction motor designed for a pump application may be called a motor, 3ϕ motor, TEFC 3ϕ motor, or 3ϕ pump motor. Regardless of what the motor is called, the primary function of the motor is to produce a mechanical force (torque) to safely drive the driven load connected to the motor.

An electric motor drive is one piece of electrical equipment that is often called by many names depending upon the type of drive, type of motor to be controlled, or the type of incoming supply voltage to the drive. Electric motor drives are manufactured as variable frequency drives, adjustable frequency drives, inverter drives, vector drives, direct torque

control drives, closed loop drives, and regenerative drives. The general types of motors are AC motors and DC motors. Supply voltage to an electric motor drive is either AC voltage or DC voltage. Regardless of what an electric motor drive is called, the primary function of a drive is to convert the incoming supply power to an altered voltage level and frequency that can safely control the motor connected to the drive and the load connected to the motor. See Figure 8-1.

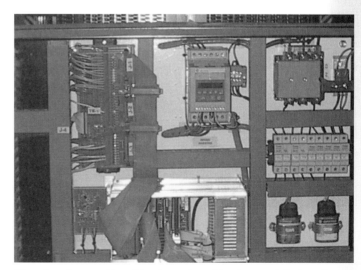

The strapping machine drive is a micro drive in an original equipment manufacturer (OEM) cabinet.

TECH FACT

The output of electric motor drives contains high frequencies. To prevent inaccurate voltage and current measurements, verify that the meter used is designed to be used with electric motor drives.

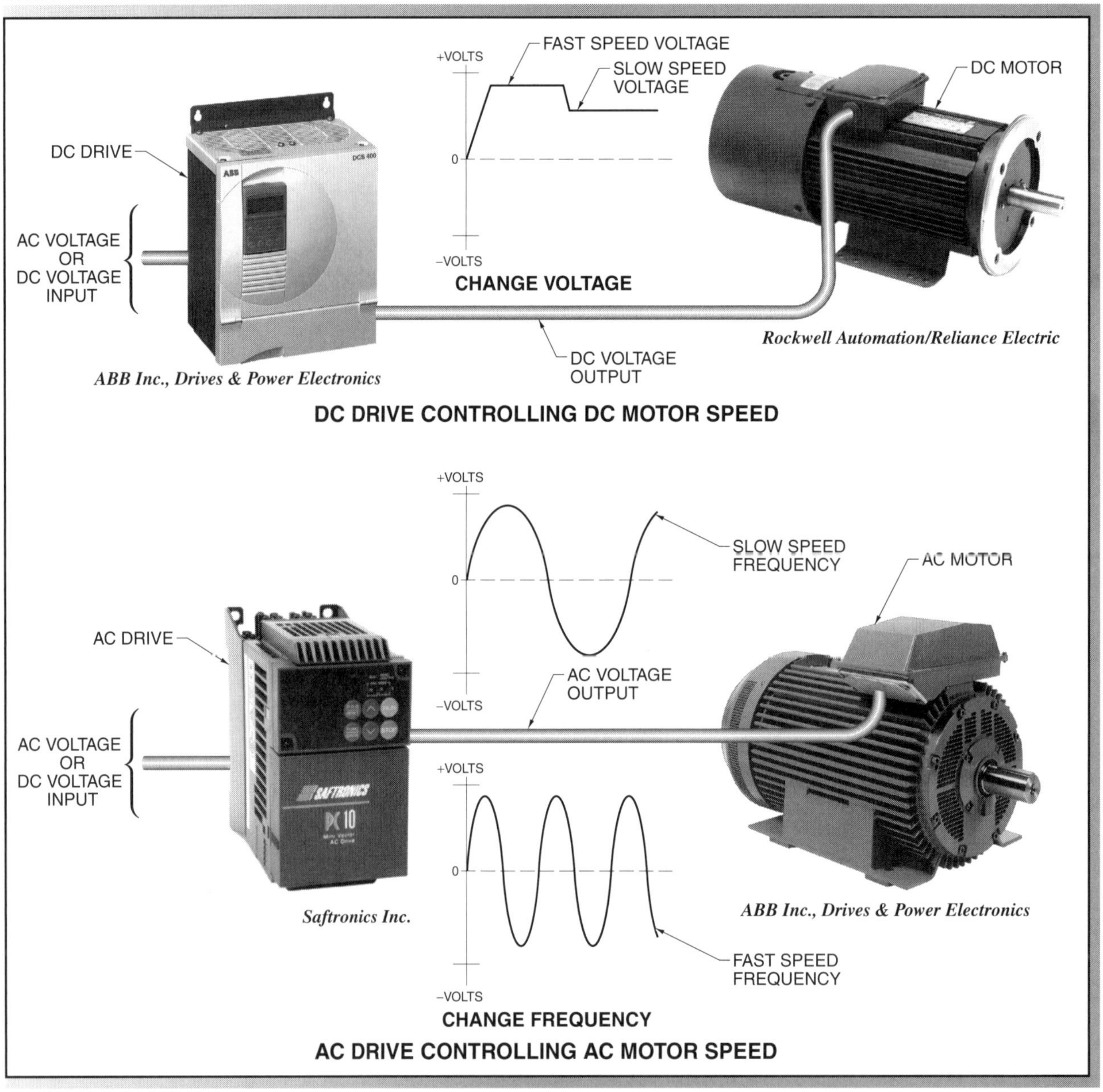

CONTROLLING MOTOR SPEED

Figure 8-1. Electric motor drives control motors by controlling the voltage and/or frequency to the motor.

ELECTRIC MOTOR DRIVE COMPONENTS

The three main sections of electric motor drives are the converter section, DC bus section, and inverter section. The converter section (rectifier) receives the incoming AC voltage and changes the voltage to DC. AC input voltage that is different than the AC output voltage sent to a motor requires that the converter section first step up or step down the AC voltage to the proper level. For example, an electric motor drive supplied with 115 VAC that must deliver 230 VAC to a motor requires a transformer to step up the input voltage. See Figure 8-2.

The DC bus section filters the voltage and maintains the proper DC voltage level. The DC bus section delivers the DC voltage to the inverter section for conversion back to AC voltage. The inverter section determines the speed of a motor by controlling frequency and controls motor torque by controlling the voltage sent to a motor.

CHANGING VOLTAGES IN AN ELECTRIC MOTOR DRIVE

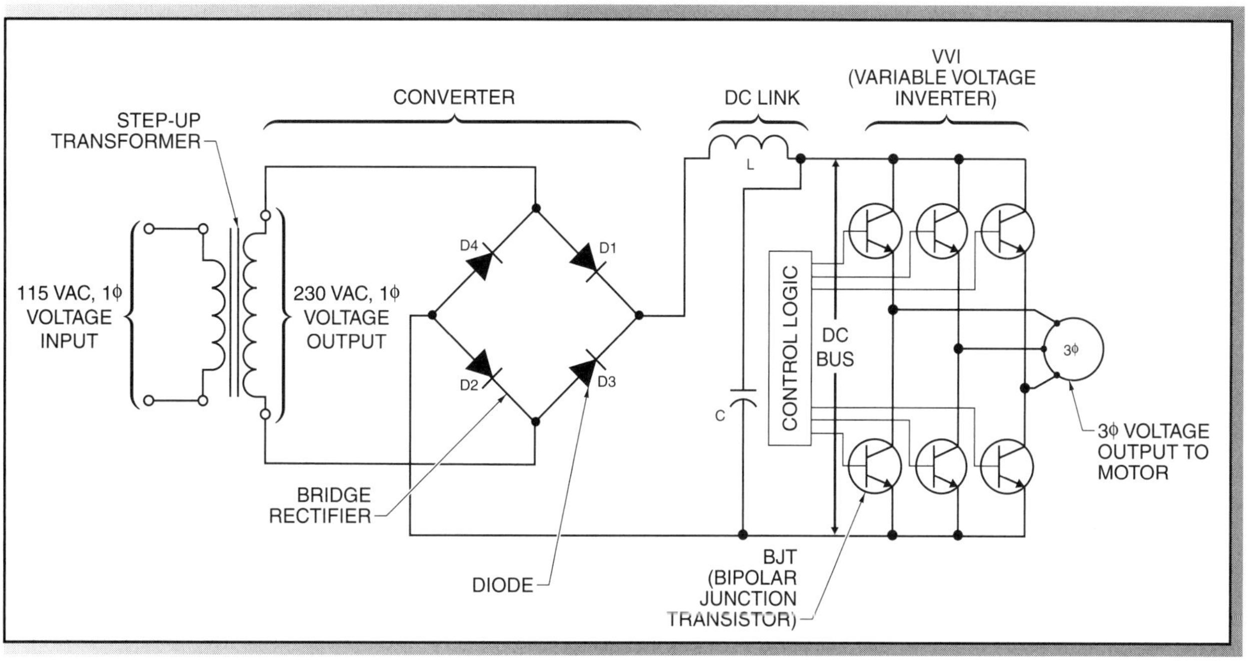

Figure 8-2. An electric motor drive converter section may have step-up or step-down transformers.

Converter Section

Converter sections of electric motor drives are 1φ full-wave rectifiers, 1φ bridge rectifiers, or 3φ full-wave rectifiers. Small electric motor drives supplied with 1φ power use 1φ full-wave or bridge rectifiers. Most electric motor drives are supplied with 3φ power requiring 3φ full-wave rectifiers. See Figure 8-3.

Electric Motor Drive Power Supply Requirements.

In order for a converter section to deliver the proper DC voltage to the DC bus section of an electric motor drive, the rectifier section must be connected to the proper power supply. Electric motor drives operate satisfactorily only when connected to the proper power supply. The proper power supply must not only be at the correct voltage level and frequency, but also provide enough current to operate an electric motor drive at full power. When a power supply cannot deliver enough current, the available voltage to an electric motor drive drops when the drive is required to deliver full power. Current to an electric motor drive is limited by the size of the conductors feeding the drive, fuse and circuit breaker sizes, and the transformer(s) delivering power to the system.

Supply voltage at an electric motor drive must be checked when installing a drive, servicing a drive, or adding additional loads or drives to a system. To determine if an electric motor drive is underpowered, the voltage into the drive is measured under no-load and then full-load operating conditions. See Figure 8-4. A voltage difference greater than 3% between no-load and full-load conditions indicates that the electric motor drive is underpowered and/or overloaded. Voltage difference is found by applying the formula:

$$V_D = V_{NL} - V_{FL}$$
where
V_D = volts dropped (in V)
V_{NL} = no load voltage (in V)
V_{FL} = full load voltage (in V)

For example, what is the voltage dropped when an electric motor drive is measured to have 230 V with no load and 226 V under full load?

$$V_D = V_{NL} - V_{FL}$$
$$V_D = 230 - 226$$
$$V_D = \textbf{4 V}$$

⚠ **EXPLOSION WARNING**

Do not connect a power supply to an electric motor drive with a voltage that exceeds the specified voltage rating of the drive. Overvoltage damages electric motor drives and may cause converter parts to explode, causing death or serious injury.

THREE-PHASE FULL-WAVE RECTIFIER

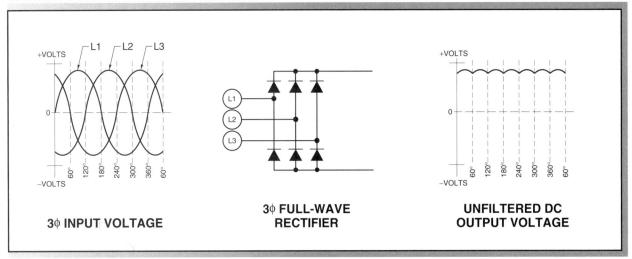

Figure 8-3. Industrial electric motor drives use 3φ rectifiers to convert AC voltage to DC voltage to supply the DC bus.

FULL-LOAD VOLTAGE DROP

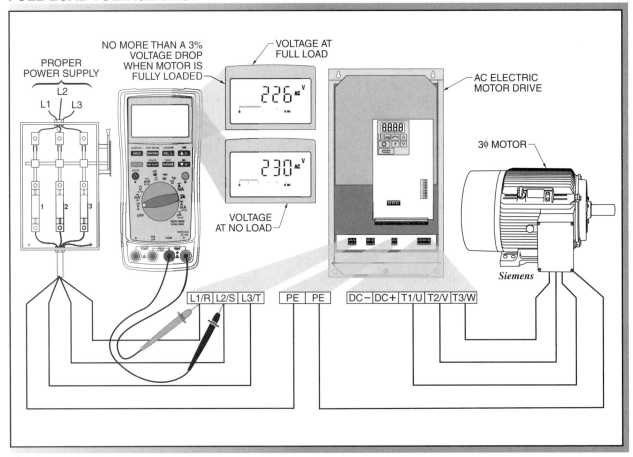

Figure 8-4. An electric motor drive is underpowered or overloaded if there is more than a 3% voltage drop between full-load and no-load conditions.

Voltage difference percentage is found by applying the formula:

$$V_\% = \frac{V_D}{V_{NL}} \times 100$$

where

$V_\%$ = Percentage of voltage drop (in %)
V_D = volts dropped (in V)
V_{NL} = no load voltage (in V)
100 = constant

For example, what is the voltage difference percentage when an electric motor drive has a 4 V voltage drop with a 230 V no-load measurement?

$$V_\% = \frac{V_D}{V_{NL}} \times 100$$

$$V_\% = \frac{4}{230} \times 100$$

$$V_\% = .01739 \times 100$$

$$V_\% = \mathbf{1.739\%}$$

When measuring the supply voltage to an electric motor drive, it is recommended to check the measured voltage against the drive's rated input voltage. Large horsepower electric motor drives are connected to high voltages to reduce the amount of current required. See Figure 8-5.

ELECTRIC MOTOR DRIVE SUPPLY VOLTAGES

PHASE & FREQUENCY	VOLTAGES*
1φ, 60 Hz	115, 208, OR 230
1φ, 60 Hz	110, 220, OR 240
3φ, 60 Hz	208, 230, 460, 575, 2300, 4160, OR 4600
3φ, 60 Hz	190, 220, 380, 415, 440, OR 4000

* in VAC

Figure 8-5. Large horsepower electric motor drives are connected to higher supply voltages to reduce the amount of required current.

AC voltage sources vary because of fluctuations within the power distribution system. AC loads, including electric motor drives and motors, are designed to operate within a specified voltage range. Operating outside the specified voltage range can cause an electric motor drive to not operate correctly and/or incur damage over time.

Electrical loads operating at lower rated voltages are less likely to be damaged than loads operating at higher rated voltages. Operating at a voltage less than

the rated voltage causes lamps to dim, heating elements to produce less heat, computers to lose memory or reboot, and motors to produce less torque. Operating at less than rated voltage is not desirable, but typically does not cause damage. Operating at a higher than rated voltage causes lamps to fail, heating elements to burn out (open), computer circuits to be permanently damaged, and motor insulation to be destroyed.

AC loads are rated for proper operation at a voltage that is ±10% of the device's rated voltage. Because higher voltages are more damaging, some higher voltage rated devices have a +5% to –10% voltage rating. See Figure 8-6.

ELECTRIC MOTOR DRIVE OPERATING VOLTAGES

DRIVE VOLTAGE*	TOLERANCE
ALL 1φ AC DRIVES	± 10%
200 TO 240, 3φ DRIVES	± 10%
400 TO 480, 3φ DRIVES	± 10%
500 TO 600, 3φ DRIVES	+ 5% TO – 10%

* in VAC

Figure 8-6. Electric motor drives typically have ±10% voltage tolerances, but high-voltage drives have + 5% to – 10% voltage tolerances.

Circuit Protection. A bridge rectifier receives incoming AC supply power and converts the AC voltage to a fixed DC voltage. The fixed DC voltage powers the DC bus of the electric motor drive. To prevent damage to the diodes in the converter section and to the electric motor drive's electronic circuits, protection against transient voltages must be included in the drive.

A *transient voltage* is a high-energy, high-voltage, short-duration spike in an electrical system. All electrical systems experience some type of transient voltage. Lightning strikes and utility switching cause high-energy-level transient voltages. High-energy-level transients seldom occur but are quite damaging to equipment if allowed to travel through a power distribution system and into electrical equipment. Low-energy-level transients are transient voltages commonly caused when switching motors and equipment ON and OFF. Low-energy-level transients occur often but do not cause immediate equipment damage. Low-energy-level transients cause malfunctions such as processing errors and damage to equipment over time.

The electronic circuits of an electric motor drive require protection against transient voltages. Protection methods include proper electric motor drive wiring, grounding, shielding of power lines, and using surge suppressors. A *surge suppressor* is an electrical device that provides protection from transient voltages by limiting the level of voltage allowed downstream from the surge suppressor.

Surge suppressors are installed at service entrance panels, at distribution panels feeding electric motor drives, and/or at the incoming power lines to a drive. Electric motor drive manufacturers cannot assume that surge suppressors are installed and that the incoming power lines are free of damaging transients. Electric motor drive manufacturers must include surge suppression inside electric motor drives. Typically, a surge suppressor consists of metal oxide varistors (MOVs) connected to the converter section of an electric motor drive. See Figure 8-7.

MOVs are designed for surge suppression of damaging transient voltages. When high-voltage transients enter an electric motor drive system, the MOVs change electrical state from high resistance (open switch) to low resistance (closed switch). In a low resistant state, MOVs absorb and/or divert transient voltage spikes. MOVs limit the level of transient voltages so voltages do not exceed the maximum voltage rating of rectifier diodes.

DC Bus Section

The DC bus (link) section includes DC voltage supplied by the converter section and the DC filter components. Capacitors and inductors in the DC bus section filter and maintain the proper DC bus voltage level. The DC bus voltage is typically about 1.4 times the AC supply voltage to an electric motor drive. See Figure 8-8.

ELECTRIC MOTOR DRIVE DC BUS OPERATING VOLTAGES	
AC SUPPLY VOLTAGE*	DC BUS VOLTAGE†
208	291
220	308
230	322
460	644
480	672

* in VAC
† in VDC

Figure 8-8. DC bus voltage is typically about 1.4 times the AC supply voltage to an electric motor drive.

METAL OXIDE VARISTOR SURGE SUPPRESSION

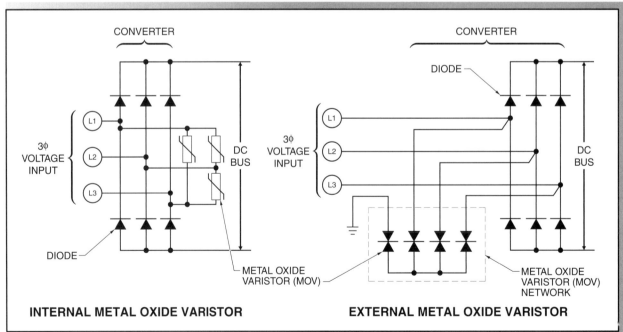

Figure 8-7. Metal oxide varistors (MOVs) are added to the rectifier section of an electric motor drive to reduce the amount of transient voltage entering an electric motor drive.

Inductors. Capacitors and inductors (coils) are used together in DC filter circuits. Working together, capacitors and coils maintain a smooth wave form because capacitors oppose a change in voltage and coils oppose a change in current. Capacitors oppose a change in voltage by holding a voltage charge that is discharged back into the circuit any time the circuit voltage decreases. As current flows through a coil, a magnetic field is produced. The magnetic field remains at maximum potential until the current in the circuit is reduced. As the circuit current is reduced, the collapsing magnetic field around the coil induces current back into the circuit. The coil smooths or filters the power by building a magnetic field as current is applied and adding current back into the circuit as the magnetic field collapses.

Capacitors. A *capacitor* is an electrical device designed to store a voltage charge of electrical energy. *Capacitance (C)* is the ability of a component or circuit to store energy in the form of an electrical charge. Capacitors in a DC bus section are charged from the rectified DC voltage produced by the converter section. When DC bus voltage starts to drop, capacitors discharge a voltage back into the system to stop the drop in voltage. The main function of capacitors is to maintain proper DC bus voltage levels even when voltage attempts to fluctuate. See Figure 8-9.

Coils used in the DC filter section of an electric motor drive are used to stabilize the current of the DC bus.

Resistors. A *resistor* is an electrical device used in the DC bus section to limit the charging current to capacitors, to discharge capacitors, and to absorb unwanted voltages. Current-limiting resistors prevent capacitors from drawing too much current during charging. Resistors are also used to discharge capacitors when power is removed from an electric motor drive. Braking resistors absorb voltage when a motor becomes a generator after a stop button is pushed.

Inverter Section

The inverter section of an electric motor drive is the most important part of the drive because the inverter section determines the voltage level, voltage frequency, and amount of current that a motor receives. Inverter sections have undergone changes in recent years while the rectifier section and DC bus have not changed. Electric motor drive manufacturers are continuously developing inverter sections that can control motor speed and torque with the fewest problems. The main problem for manufacturers is to find a high-current, fast-acting solid state switch that has the least amount of power loss (voltage drop).

Controlling Motor Speed and Torque. Both DC and AC motors produce work (deliver force) to drive a load by a rotating shaft. The amount of work produced is a function of the amount of torque produced by the motor shaft and the speed of the shaft. The primary function of all electric motor drives is to control the speed and torque of a motor.

DC BUS FILTER CAPACITORS

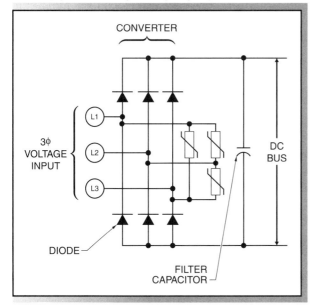

Figure 8-9. Capacitors are added to filter or smooth the DC bus voltage.

ABB Inc., Drives & Power Electronics
Electric motor drives can include specialty functions that are sent to the drive through on-board communication with a PC.

To safely control a motor, an electric motor drive must monitor electrical characteristics such as motor current, motor voltage, drive temperature, and other operating conditions. All electric motor drives are designed to remove power when there is a problem, and some drives allow conditions and faults to be monitored and displayed. In addition to controlling motor speed and torque, an electric motor drive can include additional specialty functions that are built-in, programmed into the drive, or sent to the drive through on-board communication with a PC or PLC.

Controlling the voltage to a DC motor controls the speed of the motor. The higher the applied voltage, the faster a DC motor rotates. DC drives normally control the voltage applied to a motor over the range 0 VDC to the maximum nameplate voltage rating of the motor. If an electric motor drive can deliver more voltage than the rating of the motor, the drive should be set to limit the output voltage to prevent motor damage.

Controlling the amount of current in the armature of a DC motor controls motor torque. Motor torque is proportional to the current in the armature. DC drives are designed to control the amount of voltage and current applied to the armature of DC motors to produce wanted torque and prevent motor damage. The ideal operating con-dition is to deliver current to a motor to produce enough torque to operate the load without overloading the motor, electric motor drive, or electrical distribution system.

Controlling the frequency (Hz) to an AC motor controls the speed of the motor. AC drives control the frequency applied to a motor over the range 0 Hz to several hundred hertz. AC drives are programmed for a minimum operating speed and a maximum operating speed to prevent damage to the motor or driven load. Damage occurs when a motor is driven faster than its rated nameplate speed. AC motors should not be driven at a speed higher than 10% above the motor's rated nameplate speed, unless a licensed engineer is consulted.

Controlling the volts-per-hertz ratio (V/Hz) applied to an AC motor controls motor torque. An AC motor develops rated torque when the V/Hz ratio is maintained. During acceleration (any speed between 1 Hz and 60 Hz), the motor shaft delivers constant torque because the voltage is increased at the same rate as the increase in frequency. Once an electric motor drive reaches the point of delivering full motor voltage, increasing the frequency does not increase torque on the motor shaft because voltage cannot be increased any more to maintain the volts-per-hertz ratio. See Figure 8-10.

TORQUE VOLTS-PER-HERTZ RATIO

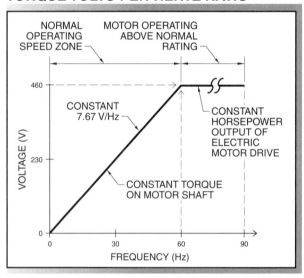

Figure 8-10. Controlling the volts-per-hertz ratio (V/Hz) applied to an AC motor controls motor torque.

PULSE WIDTH MODULATION

Pulse width modulation (PWM) is a method of controlling the amount of voltage sent to a motor. DC drives and AC drives must control the amount of voltage produced in order to control the speed and torque of a motor. Over the years, different methods of controlling the amount of voltage produced have been used in electric motor drives. Some methods have been replaced by newer technologies, but several older methods are still in use.

Silicon-controlled rectifiers (SCRs) were first used in DC drives to control the amount of voltage applied to a motor. An SCR controls the amount of DC voltage output by controlling the amount of AC voltage that is rectified into DC voltage. The amount of DC voltage output is determined by when the gate of an SCR allows current to flow.

Pulse width modulation control offers better performance than SCR control and is used in newer DC drives. PWM controls the amount of voltage output by converting the DC voltage into fixed values of individual DC pulses. The fixed-value pulses are produced by the high-speed switching of transistors (typically IGBT transistors) turning ON and OFF. By varying the width of each pulse (time ON) and/or frequency, the amount of voltage can be increased or decreased. The wider the individual pulses, the higher the DC voltage output. The higher the DC output, the faster a DC motor operates.

PWM of a DC voltage is also used to reproduce AC sine waves. See Figure 8-11. When PWM is used with AC voltage, two IGBTs are used for each phase. One IGBT is used to produce the positive pulses and another IGBT is used to produce the negative pulses of the sine wave. Because AC drives are typically used to control 3ϕ motors, six IGBTs—two per phase—are used. See Figure 8-12. The higher the switching frequency of the IGBTs, the closer the simulated AC sine wave is to a real sine wave. The closer the simulated sine wave is to a pure sine wave, the lower the amount of heat produced by the motor.

A single-pole single-throw (SPST) mechanical switch can be used to convert pure DC voltage into pulsating DC at varying voltage levels. If the switch remains closed, the DC voltage output would equal the applied DC voltage input. As the switch is opened and closed, the DC voltage output is equal to a voltage level less than the applied input voltage and greater than 0 V. The longer a switch is left open, the lower the average DC voltage, and the longer a switch is left closed, the higher the average DC voltage. Filters are added into circuits to smooth out waveforms and improve efficiency. When capacitors are added into a circuit, the waveform is smoothed as the capacitor discharges back into the circuit every time the voltage tries to return to zero (switch opened).

PULSE WIDTH MODULATION

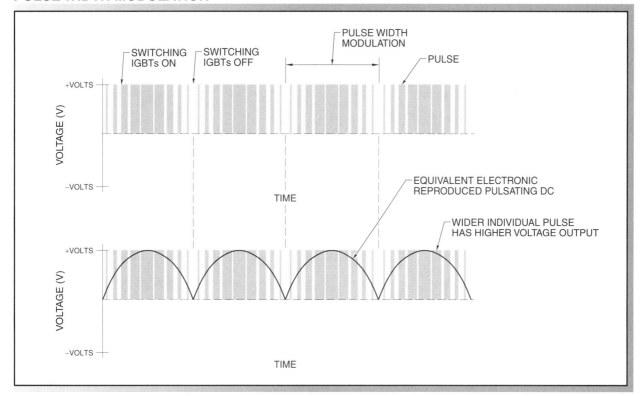

Figure 8-11. Pulse width modulation is used to produce a pulsating DC output.

IGBT PRODUCED SINE WAVE

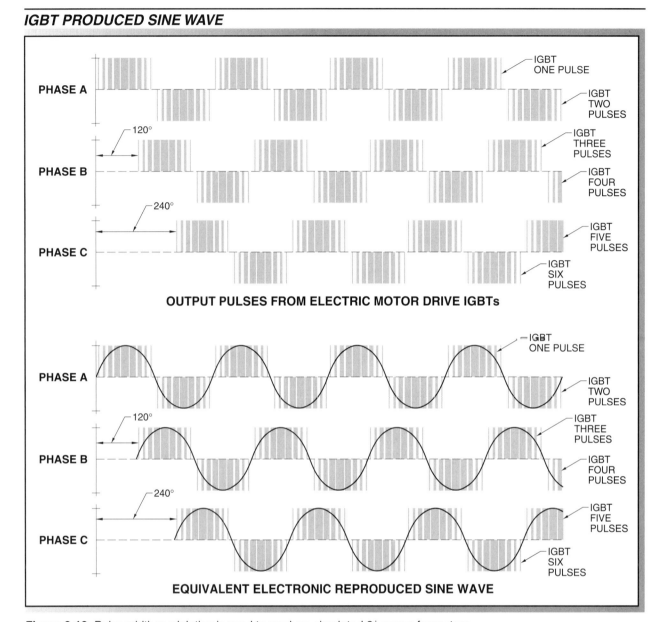

Figure 8-12. Pulse width modulation is used to produce simulated 3ϕ power for motors.

A SPST switch could be used to convert a pure DC voltage into pulsating DC voltage, but SPST switches could never be switched fast enough to produce a usable voltage. Electric motor drives replace SPDT switches with fast-acting electronic switches (IGBTs). When two IGBTs are used, an electronically reproduced AC voltage is produced. The fast-acting transistors are performing the same function as mechanical switches. See Figure 8-13.

Switches (two IGBTs or two SCRs) are used to place either a positive (+) or negative (–) voltage on the motor leads. Three switches (one for each phase) are always closed to apply either a positive or negative voltage on

each of the three motor leads (T1, T2, and T3). The positive and negative polarities produce current flow in a fixed direction through the stator windings of a motor. The switches keep changing the direction of current flow to produce a rotating magnetic field around the motor. The rotating magnetic field forces the rotor to rotate.

TECH FACT

The Joule rating of an MOV defines the amount of energy an MOV can absorb. The higher the number, the greater the energy absorbed.

IGBTs CONTROLLING MOTOR ROTATION

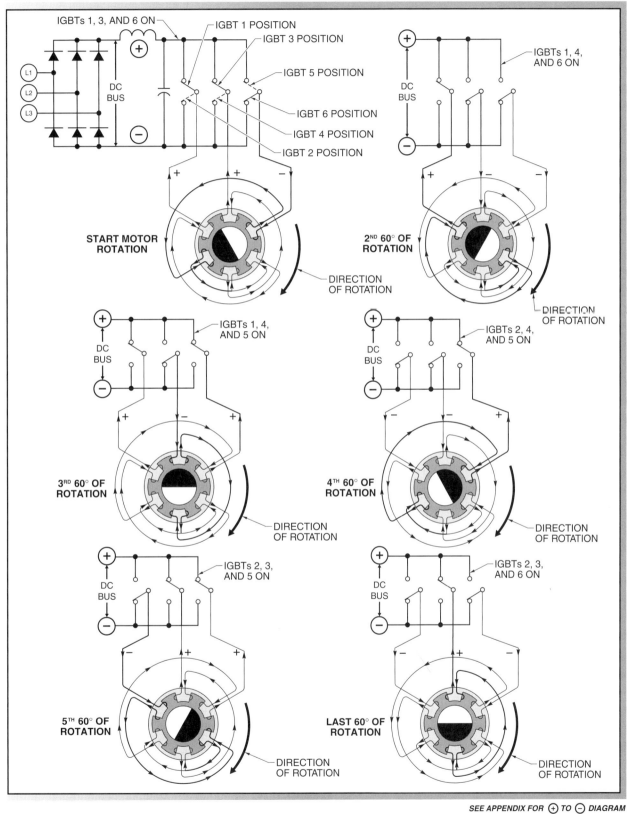

SEE APPENDIX FOR ⊕ TO ⊖ DIAGRAM

Figure 8-13. The rotating magnetic field in the stator windings causes the rotor to follow the rotating field.

Carrier Frequencies

Carrier frequency is the frequency that controls the number of times the solid state switches in the inverter section of a PWM electric motor drive turn ON and turn OFF. The higher the carrier frequency, the more individual pulses there are to reproduce the fundamental frequency. *Fundamental frequency* is the frequency of the voltage used to control motor speed. Carrier frequency pulses per fundamental frequency are found by applying the formula:

$$P = \frac{F_{CARR}}{F_{FUND}}$$

where

P = pulses

F_{CARR} = carrier frequency

F_{FUND} = fundamental frequency

For example, what is the number of pulses per fundamental frequency when a carrier frequency of 1 kHz is used to produce a 60 Hz fundamental frequency?

$$P = \frac{F_{CARR}}{F_{FUND}}$$

$$P = \frac{1000}{60}$$

$P = \textbf{16.66 pulses}$

A carrier frequency of 6 kHz used to produce a 60 Hz fundamental frequency would have 100 individual pulses per fundamental cycle. See Figure 8-14.

TECH FACT

To help overcome heat and poor power problems, motors with higher service factor ratings are used.

ELECTRIC MOTOR DRIVE FREQUENCIES

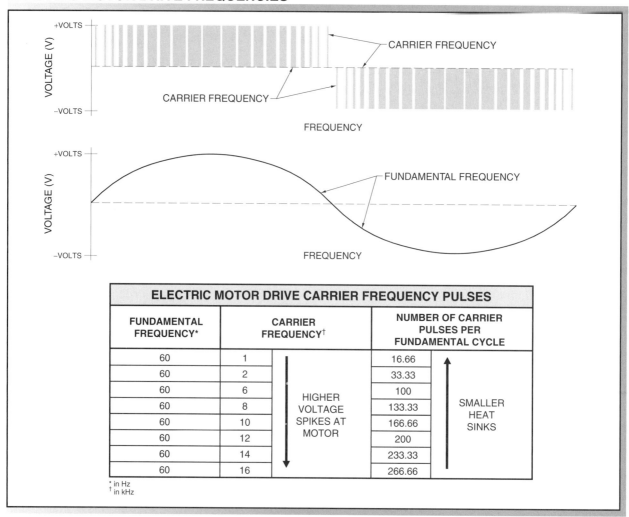

ELECTRIC MOTOR DRIVE CARRIER FREQUENCY PULSES		
FUNDAMENTAL FREQUENCY*	CARRIER FREQUENCY†	NUMBER OF CARRIER PULSES PER FUNDAMENTAL CYCLE
60	1	16.66
60	2	33.33
60	6	100
60	8	133.33
60	10	166.66
60	12	200
60	14	233.33
60	16	266.66

HIGHER VOLTAGE SPIKES AT MOTOR

SMALLER HEAT SINKS

* in Hz
† in kHz

Figure 8-14. Carrier frequencies of electric motor drives range from 1 kHz to 16 kHz.

Fundamental frequency is the frequency of the voltage a motor uses, but the carrier frequency actually delivers the fundamental frequency voltage to the motor. The carrier frequency of most electric motor drives can range from 1 kHz to about 16 kHz. The higher the carrier frequency, the closer the output sine wave is to a pure fundamental frequency sine wave.

Increasing the frequency to a motor above the standard 60 Hz also increases the noise produced by the motor. Noise is noticeable in the 1 kHz to 2 kHz range because it is within the range of human hearing and is amplified by the motor. A motor connected to an electric motor drive delivering a 60 Hz fundamental frequency with a carrier frequency of 2 kHz is about three times louder than the same motor connected directly to a pure 60 Hz sine wave with a magnetic motor starter. Motor noise is a problem in electric motor drive applications such as HVAC systems in which the noise can carry throughout an entire building.

Most people can hear a tone from a motor when it is in the 1 kHz to 3 kHz range. The sound is heard as a high-pitched whine. A person can hear frequencies above 3 kHz, but the higher frequencies are not amplified as loudly as the lower frequencies by the motor.

Manufacturers have raised the carrier frequency beyond the range of human hearing to solve the noise problem. High carrier frequencies cause greater power losses (thermal losses) in an electric motor drive because of the solid state switches in the inverter section. Electric motor drives must be slightly derated or the size of heat sinks increased because of the increased thermal losses. Derating an electric motor drive decreases the power rating of a drive and increasing heat sink size adds additional cost to a drive. See Figure 8-15.

Higher carrier frequencies are better, but only up to a point. A 6 kHz to 8 kHz carrier frequency simulates a pure sine wave better than a 1 kHz to 3 kHz carrier frequency and reduces heating in a motor. The more closely the voltage delivered to a motor simulates a pure sine wave, the cooler a motor operates. Even slightly reducing the temperature in a motor increases insulation life.

Carrier frequency can be changed at an electric motor drive to meet particular requirements. The factory default value is usually the highest frequency, and changing to a lower is done through a parameter change, such as changing 12 kHz to 2.2 kHz. One effect of high carrier frequencies is that the fast switching of the inverter produces larger voltage spikes that damage motor insulation. The voltage spikes become more of a problem as cable length between an electric motor drive and motor increases.

CARRIER FREQUENCY POWER DERATING CURVE

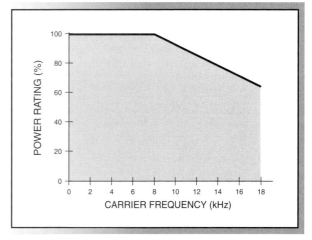

Figure 8-15. An electric motor drive may have to be derated at higher carrier frequencies due to power losses in the drive.

Motor Lead Length

In any electrical system, the distance between components affects operation. The primary limit to the distance between a magnetic motor starter and the motor is the voltage drop of the conductors. The voltage drop of conductors should not exceed 3% for any type of motor circuit.

When an electric motor drive is used to control a motor, the distance between the drive and motor may be limited by other factors besides the voltage drop of the conductors. Conductors between an electric motor drive and motor have line-to-line (phase-to-phase) capacitance and line-to-ground (phase-to-ground) capacitance. Longer conductors produce higher capacitance. The capacitance produced by conductors causes high voltage spikes in the voltage to a motor. Since voltage spikes are reflected into the system, the voltage spikes are often called reflective wave spikes, or reflective waves (reflected waves). As the length of conductors increases and/or an electric motor drive's output carrier frequency increases, the voltage spikes become larger. See Figure 8-16.

Determining the cable length at which voltage spikes become a problem is difficult. When cable lengths between an electric motor drive and motor are kept less than 100′, problems typically do not occur. Smaller horsepower motors and multiple motors connected to one electric motor drive can be more susceptable to voltage spikes. Voltage spikes are problems because spikes stress motor insulation. The insulation of a motor is only as good as its weakest point. When voltage spikes become a problem, the lead length should be reduced and/or filters that suppress voltage spikes should be added. Reducing the carrier frequency also reduces reflected wave voltage spikes. Inverter rated motors that have spike-resistant insulation reduce the damage caused by voltage spikes.

TECH FACT

Air used to cool an electric motor drive must be clean, dry, and free of dirt and corrosive materials. When heat sinks become dirty, power is removed and the heat sinks are cleaned with a vacuum and brush.

CONDUCTOR LENGTH VOLTAGE SPIKES

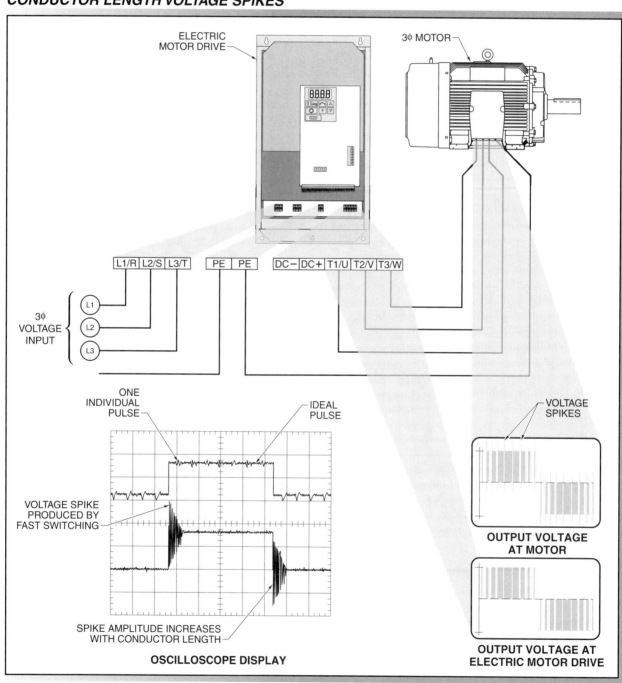

Figure 8-16. Longer lead lengths between an electric motor drive and a motor and/or higher carrier frequencies produce higher voltage spikes.

AC DRIVE TYPES

AC drives control motor speed and torque by converting incoming AC voltage to DC voltage and then converting the DC voltage to a variable-frequency AC voltage. The inverter section of an electric motor drive converts the DC voltage of the DC bus to AC voltage. AC drives control the voltage and frequency at the motor by switching the DC voltage ON and OFF at the proper time. Electronic switches such as SCRs, GTOs, or IGBTs are used to switch the DC voltage ON and OFF. A microprocessor circuit located within an electric motor drive controls the electronic switching.

Inverter drives are referred to according to the method used to change the frequency of the voltage. Inverter drives include variable voltage inverters (VVI), current source inverters (CSI), and pulse width modulated inverters (PWM). Some electric motor drives are referred to as VVI, CSI, or PWM drives. See Figure 8-17.

Variable voltage inverter drives control an AC motor but produce a square wave instead of a sine wave. Square waves produce high torque pulsations at the motor that cause a motor to have jerky movements at low speeds and operate at high temperatures. To overcome the square wave problem, electric motor drives use IGBTs to produce a more accurate sine wave.

TECH FACT

The operating expectancy of electric motor drives varies according to environment and operating time. Most electric motor drives that operate continuously and within the specified limits of the drive have the DC bus capacitors as the weakest link, with a life expectancy of about five years. Electric motor drives operating for seven years should have the DC bus capacitors replaced to prevent drive failure and system downtime.

In addition to naming electric motor drives by the type of inverter used (VVI, CSI, or PWM), drives are also referred to by the technology used by the drive to control motor torque. Vector electric motor drive (closed loop vector drive or open loop vector drive) is a name given to certain drives to better describe the operation of the drive.

A *closed loop drive* is an electric motor drive that operates using a feedback sensor such as an encoder or tachometer connected to the shaft of the motor to send information about motor speed back to the drive. An *open loop drive* is an electric motor drive that operates

without any feedback to the drive about motor speed. To simplify electric motor drive terminology, AC drives are referred to as either inverter drives or vector drives.

Both AC inverter and vector drives control motor speed and frequency. Vector drives apply newer technologies to improve motor torque performance over the full motor speed range and for different load types. As new technologies emerge, manufacturers advertise electric motor drives using different names to better describe the new technology used, such as a type of improved open loop vector drive called direct torque control (DTC) that is designed to provide better torque control. DTC drives perform the same function of controlling motor speed and torque as inverter drives and open loop vector drives, but apply technologies that improve motor torque control without requiring any feedback from motor sensors.

Inverter Drives

Inverter drives are the oldest type of AC drive. Older six-step inverter drives used SCRs in the inverter section of the AC drive to produce the AC sine wave. PWM inverters use transistors in the inverter section to produce the AC sine wave. Transistors operate at much faster speeds than SCRs, allowing higher switching frequencies to produce electronically reproduced sine waves that closely resemble a pure sine wave. The improved sine wave produces less heating in a motor than six-step inverters. See Figure 8-18.

Motortronics
Electric motor drive manufacturers produce drives in various horsepower, as AC or DC, and as PWM, VVI, CSI, or vector types.

TYPES OF OUTPUT VOLTAGES AND OUTPUT CURRENTS

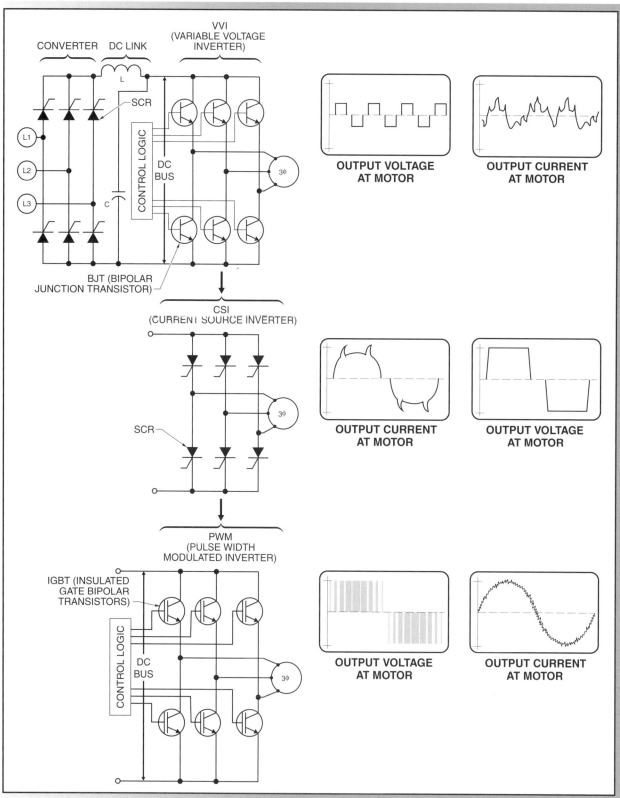

Figure 8-17. Variable voltage inverters, current source inverters, and pulse width modulated inverters are used to control the voltage, frequency, and current to a motor.

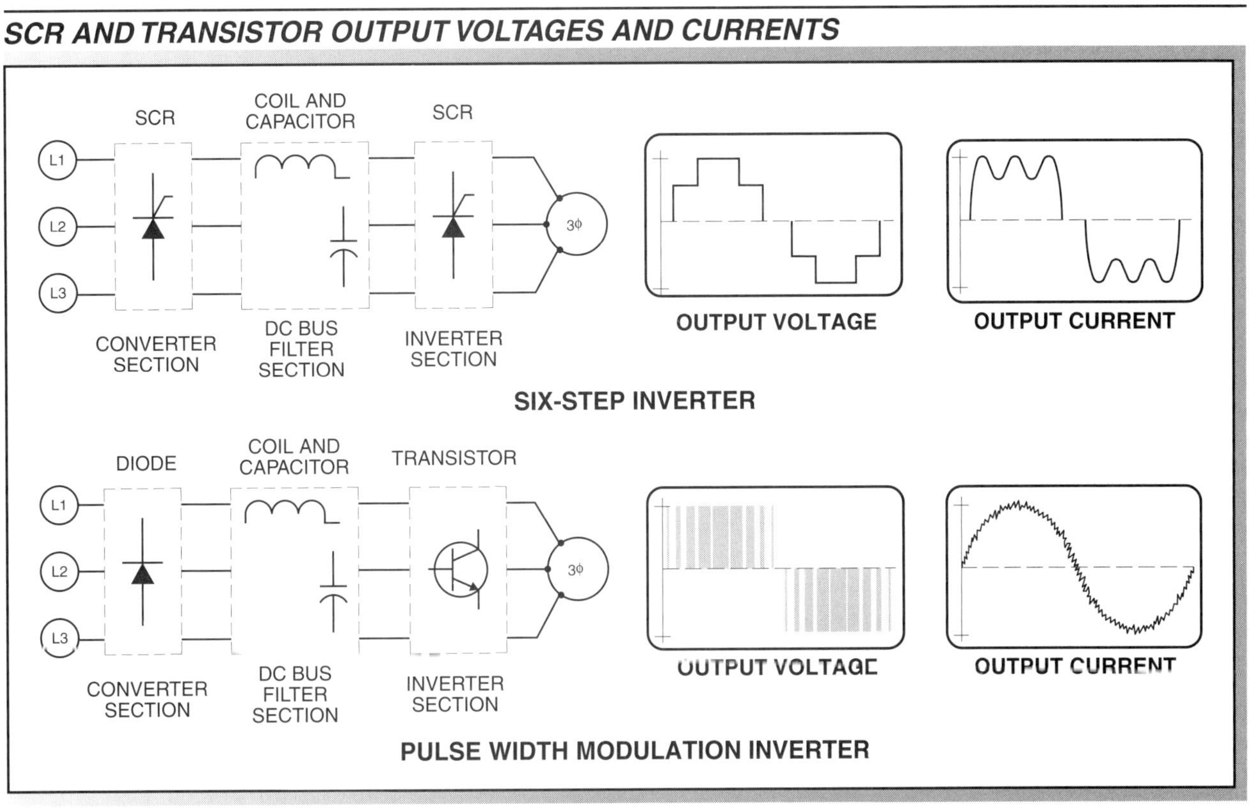

SCR AND TRANSISTOR OUTPUT VOLTAGES AND CURRENTS

Figure 8-18. Transistors used in the inverter section of electric motor drives produce sine waves that generate less heat in motors than SCRs.

Vector Drives

Inverter drives that control motor speed by setting motor voltage and frequency work well for most applications. Applications that required better motor torque control at different speeds use DC motors. To overcome the problem of having to use DC motors with higher maintenance costs, vector drives were developed.

Closed Loop Vector Drives. Closed loop vector drives use shaft-mounted encoders or tachometers to determine the rotor position and speed of a motor and send the information back to the electric motor drive. A *closed loop system* is a system with feedback from the motor sensors to the electric motor drive. Monitoring information from sensors allows an electric motor drive to automatically make adjustments to better meet the needs of the motor.

An *encoder* is a sensor (transducer) that produces discrete electrical pulses during each increment of shaft rotation. A *tachometer* is a sensor that monitors the speed of a rotating shaft. Encoders provide feedback to electric motor drives so drives can better control motor speed and torque. This is especially true when low-speed

control of torque is required in applications such as machine tools and presses. Encoders and tachometers are used in applications such as winders and rolling mills because encoders and tachometers allow electric motor drives to deliver maximum torque up to motor base speed. When required, closed loop vector drives can hold a motor shaft stationary against an applied torque.

Vector control (also called flux vector or field orientation) drives are designed to operate AC motors at the same performance as DC motors. Vector control drives achieve the performance by measuring the current drawn by a motor at a known speed (encoder feedback) and comparing the current draw to the applied voltage. Induction motors produce torque at the rotor and shaft when the rotor rotates at a slower speed than the rotating magnetic field of the stator. The difference in speed is known as motor slip. Induction motor torque is directly proportional to the amount of slip. When motor speed is known from encoder feedback, slip can be regulated and thus torque on the motor shaft can be controlled. AC motor performance using a vector drive is comparable with that of a DC motor. See Figure 8-19.

MOTOR TORQUE CHARACTERISTICS

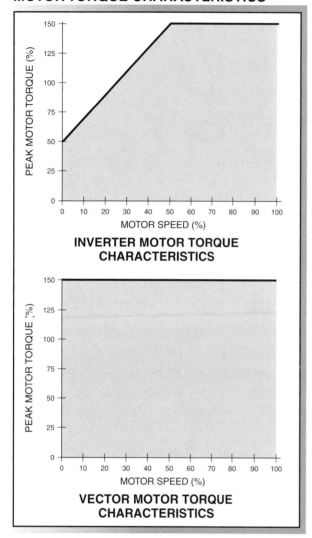

Figure 8-19. AC motors controlled by vector drives have performance characteristics similar to DC motors.

Vector drives are capable of performing the same functions as inverter drives and more. Vector drives are a good choice for many applications, but inverter drives are more economical and are better for applications such as centrifugal pumps, fans, and blowers or any applications that do not require full torque control.

Advantages of vector drives include the following:
- Full rated torque can be delivered at zero speed. Inver-ter drives cannot maintain full rated torque at zero speed.
- Motor speed can be accurately controlled to within ±.01% of the set speed of the drive when operating in a closed loop condition. Inverter drives can typically control motor speed within ±1% of the set speed of the drive.
- Motors run cooler.

Disadvantages of vector drives include the following:
- They are more expensive than inverter drives.
- They cannot be used to operate multiple motors from the same drive. Inverter drives can operate multiple motors from the same drive when individual motor protection is used with each motor.

The main disadvantage of a vector drive is that an encoder or tachometer is required. Mounting and maintaining the encoder adds cost to the system. To achieve the improved performance of vector drives when compared to inverter drives without the cost penalty, a vector drive that does not require an encoder was developed. Electric motor drives without encoders are called open loop vector drives, or sensorless vector drives.

Open Loop Vector Drives. An *open loop vector drive* is an electric motor drive that has no feedback method. Programming the conditions believed necessary to achieve the desired results and accepting the results of the motor control is a must in the open loop system. Open loop vector drives can be programmed using motor and application data to anticipate how best to control the motor and load.

Open loop vector drives are used where speed regulation is important, but not important enough to require encoder feedback. Open loop vector drives typically can control speed within .1% of the drive setting and a closed loop vector drive typically can control speed within .01% of the drive setting.

Future Electric Motor Drives

DC motors have always been the ideal motors for applications that require accurate torque and speed control, especially at lower speeds. Ever since AC drives were first used, it has been the goal of electric motor drive manufacturers to make a drive that allows an AC motor to have the same (or better) performance characteristics as a DC motor. Inverter drives perform well, but have problems when applied in applications that are better served by DC motors. Closed loop vector drives solved most problems, but required an encoder or tachometer mounted to the motor shaft. Open loop vector drives tried to solve the problems by reducing the need for an encoder or tachometer, but problems still existed in some applications.

With the increased capabilities of microcomputers, electric motor drive performance is continuously improving. Electric motor drive programs include known motor performance data that is used along with measured and nameplate electrical quantities to continuously calculate motor torque and adjust electric motor drive settings automatically. Electric motor drive manufacturers keep applying new technologies in an effort to improve drives and develop drives that work in all applications.

Electric motor drives called smart drives (or by some name brand) use the latest technology to make the drive as advanced as possible, but are still easy to apply and use. Some electric motor drives lead users from beginning to end by asking a few simple questions about the motor application, and the drive basically programs itself.

Smart drives still perform the same basic functions of any electric motor drive in that motor speed and motor torque are controlled. Better motor control is achieved because smart drives use the latest technologies and are building on past electric motor drive experience. Cost (both direct and indirect) and knowledge of the labor force (maintenance upkeep, etc.), motor application, and past performance (product line knowledge) factor into the final selection of the best type of electric motor drive to use.

Name_____ Date _____

True-False

T   F   **1.** The converter section of a drive receives the incoming AC voltage and changes the voltage to DC. *pg 209*

T   F   **2.** Electrical loads operating at lower rated voltages are more likely to be damaged than loads operating at higher rated voltages. *pg 211* *loss*

T   F   **3.** Lightning strikes and utility switching cause high-energy-level transient voltages. *pg 211*

T   F   **4.** Capacitors oppose a change in voltage and coils oppose a change in current.

T   F   **5.** The lower the carrier frequency, the closer the output sine wave is to a pure fundamental frequency sine wave.

T   F   **6.** Higher carrier frequencies cause smaller power losses in an electric motor drive.

T   F   **7.** As the length of conductors increases and/or an electric motor drive's output carrier frequency increases, the voltage spikes become larger.

T   F   **8.** The DC bus voltage of an electric motor drive is typically 1.4 times the AC supply voltage to a drive.

T   F   **9.** Transistors operate at much faster speeds than silicon-controlled rectifiers (SCRs), allowing higher switching frequencies.

T   F   **10.** Electric motor drives operate satisfactorily only when connected to the proper power supply.

T   F   **11.** A voltage sag is a high-energy, high-voltage, short-duration spike in an electrical system.

T   F   **12.** A capacitor is an electronic device specifically designed to store a charge of energy.

T   F   **13.** Most people can hear a tone from a motor when it is in the 1 kHz to 3 kHz range.

T   F   **14.** Inverter drives can typically control motor speed within ±1% of the set speed of the drive.

T   F   **15.** The frequency applied to a DC motor controls the motor speed.

Completion

_____ **1.** The primary function of a motor is to produce ___ to safely drive the driven load connected to the motor.

_____ **2.** The DC bus section of a drive delivers the DC voltage to the ___ section for conversion back to AC voltage.

Surge Suppressor **3.** A(n) ___ is an electrical device that provides protection from transient voltages by limiting the level of voltage allowed downstream. *pg 212*

Carrier Frequency **4.** ___ is the frequency that controls the number of times the solid state switches in the inverter section of a PWM electric motor drive turn ON and turn OFF. *218*

Voltage Drop **5.** The primary limit to the distance between a magnetic motor starter and the motor is the ___ of the conductors.

variable voltage **6.** ___ inverter drives control an AC motor but produce a square wave instead of a sine wave. *pg 221*

Open loop **7.** A(n) ___ drive is an electric motor drive that operates without any feedback to the drive about motor speed. *pg 221*

multiple **8.** Vector drives cannot be used to operate ___ motors from the same drive. *pg 223*

encoder **9.** A(n) ___ is a sensor that produces discrete electrical pulses during each increment of shaft rotation. *pg 223*

Current Source **10.** Inverter drives include variable voltage inverters, ___ inverters, and pulse width modulated inverters. *pg 222*

Multiple Choice

___D___ **1.** A voltage difference greater than ___ % between no-load and full-load conditions indicates that the electric motor drive is underpowered and/or overloaded.
 A. 1 C. 2
 B. 1.5 D. 3   *pg 209*

___B___ **2.** ___-energy-level transients cause malfunctions such as processing errors and damage to equipment over time.
 A. High C. Medium
 B. Low D. Moderate   *pg 211*

___C___ **3.** Controlling the frequency to an AC motor controls the ___ of the motor.
 A. armature voltage C. speed
 B. horsepower D. watts

___B___ **4.** When using PWM with AC voltage, ___ IGBT(s) is/are used for each phase.
 A. one C. three
 B. two D. four   *pg 216*

___B___ **5.** ___ rated motors that have spike-resistant insulation reduce the damage caused by voltage spikes.
 A. Explosion C. Moisture
 B. Inverter D. Thermal   *pg 220*

Name _____ Date _____

Activity 8-1. Full Load and Voltage Drop

To determine whether an electric motor drive is underpowered, the voltage into the drive under no-load conditions must be measured and the voltage into the drive under full-load conditions must be measured. A voltage difference greater than 3% between no-load and full-load conditions indicates that the electric motor drive is underpowered and/or overloaded.

Determine the following using the provided multimeter measurements.

_____ 9   **1.** The voltage drop between no load and full load is ___ V.

_____   **2.** What is the percentage of voltage drop?

_____   **3.** Is the electric motor drive underpowered?

_____   **4.** Is the DC bus voltage within normal range?

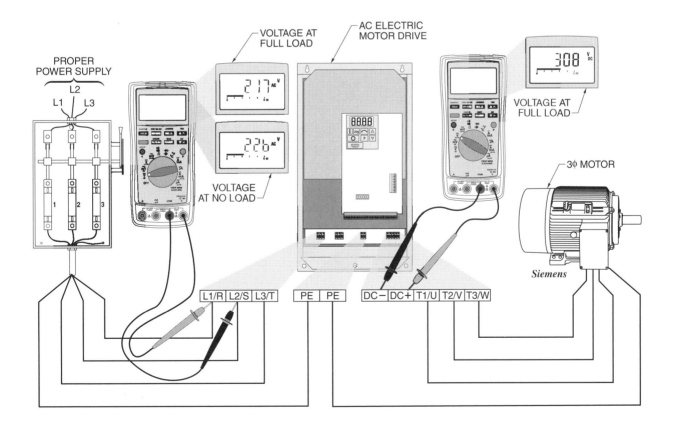

Activity 8-2. Circuit Protection

Surge suppressors are installed to protect circuits and/or components from damage due to transient voltages. Surge suppressors are typically installed at service entrance panels or at the incoming power supply of a machine or system. Snubber circuits may also be connected in parallel with individual loads that cause transients, such as motor starter coils and contactors, to protect the load from transient voltages.

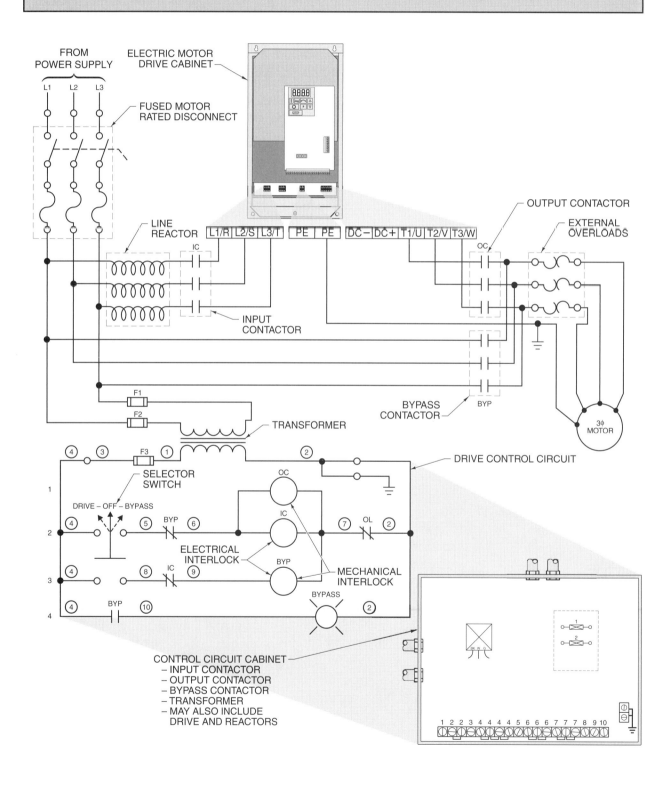

1. Connect the surge suppressor to the proper terminals so that the surge suppressor protects the entire control circuit from incoming transient voltages. Connect Snubber Circuit 1 so that Snubber Circuit 1 prevents the input and output contactors from producing damaging transient voltages. Connect Snubber Circuit 2 so that Snubber Circuit 2 prevents the bypass starter from producing damaging transient voltages.

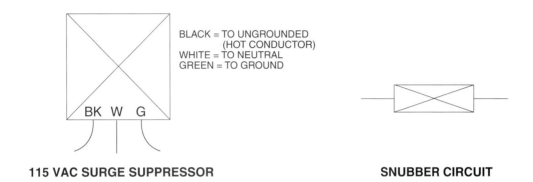

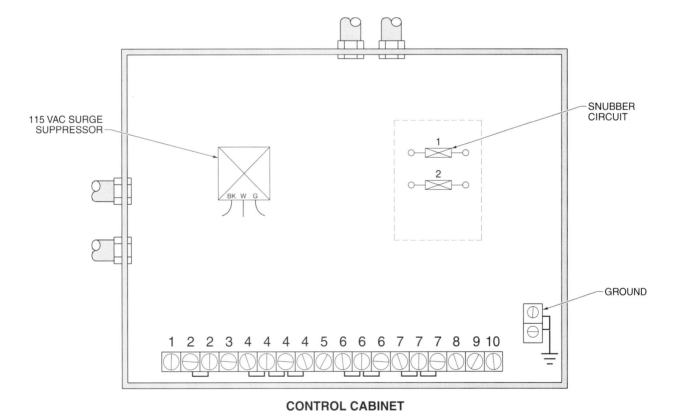

Activity 8-3. Cable Length Conversion

Electric motor drive manufacturers list the recommended maximum cable length between the drive and motor. Depending upon the electric motor drive manufacturer, cable lengths may be listed in feet or meters.

1 m = 3.3′ (meters × 3.2808 = feet)
1′ = 0.3 m (feet × 0.3048 = meters)

MAXIMUM MOTOR CABLE LENGTHS

DRIVE PART NUMBER	MAXIMUM MOTOR CABLE LENGTHS*		
	4 kHz	8 kHz	16 kHz
01	30	20	10
02	40	25	15
03	30	20	15
04	40	30	20

* in m

Determine the following using the electric motor drive operating information and the Maximum Motor Cable Lengths chart.

_____10_____ **1.** Circuit 1 maximum cable length is ___ ′.

_____40_____ **2.** Circuit 2 maximum cable length is ___ ′.

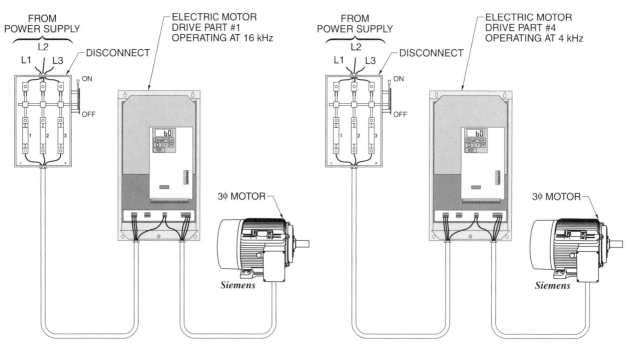

CIRCUIT 1   CIRCUIT 2

9 Electric Motor Drive Installation Procedures

Several items need to be considered in order to properly install and wire an electric motor drive and motor circuit. The motor controlled by the electric motor drive, system power quality, safety, installation factors, and both power and control wiring all need to be checked for compatibility with the motor and drive. A properly installed electric motor drive can function correctly and safely for a long time without introducing any problems into an existing electrical distribution system.

INSTALLATION CONSIDERATIONS

Special considerations are necessary when installing an electric motor drive and motor circuit. These considerations involve motors, power quality, safety, and engineered applications. The items that need to be addressed to ensure that the motor and electric motor drive function properly are the electrical distribution system, the electric motor drive, the motor connected to the drive, and the load connected to the motor.

ELECTRIC MOTOR CONSIDERATIONS

Electric motors connected to AC drives receive a simulated AC sine wave known as pulse width modulation (PWM). The PWM waveform is created by the AC drive rapidly pulsing the DC bus ON and OFF using insulated gate bipolar transistors (IGBTs). The height of each pulse is equal to the DC bus voltage. The width of the pulses is controlled to provide an effective rms voltage, similar to that of a pure sine wave. See Figure 9-1. The number of pulses in the PWM waveform is controlled to vary the output frequency of the voltage to the motor. The carrier frequency or pulse frequency (PWM frequency) of an electric motor drive determines how quickly the IGBTs must turn ON and OFF. The carrier frequency is constant and independent of the voltage frequency applied to the motor. The carrier frequency is selectable and can vary from 2 kHz to 20 kHz. The higher the carrier frequency, the quieter a motor runs.

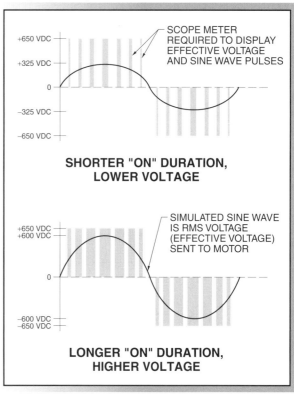

EFFECTIVE PULSE WIDTH MODULATION RMS VOLTAGE

SHORTER "ON" DURATION, LOWER VOLTAGE

SCOPE METER REQUIRED TO DISPLAY EFFECTIVE VOLTAGE AND SINE WAVE PULSES

LONGER "ON" DURATION, HIGHER VOLTAGE

SIMULATED SINE WAVE IS RMS VOLTAGE (EFFECTIVE VOLTAGE) SENT TO MOTOR

Figure 9-1. The width of the pulses determines the voltage to the motor; the wider the pulse, the higher the voltage.

Inverter Duty Motors

An *inverter duty motor* is an electric motor specifically designed to work with electric motor drives. A PWM waveform creates insulation problems for standard electric motors because the leading edge of each PWM pulse begins with a voltage spike. A voltage spike is due to the rapid rise time of the IGBTs. A voltage spike is typically twice the DC bus voltage of the AC drive. See Figure 9-2. Voltage spikes create damage insulation and shorten the life expectancy of standard electric motors. Inverter duty motors are designed to withstand voltage spikes.

PWM VOLTAGE SPIKES

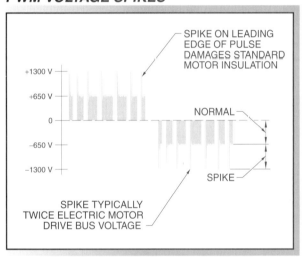

Figure 9-2. The spike on the leading edge of each pulse is caused by the rapid rise time of the IGBTs.

The National Electrical Manufacturers Association (NEMA) has developed a set of specifications for electric motors used with electric motor drives. The NEMA standard MG-1, Section IV, Part 31 provides an enhanced insulation requirement for electric motors, allowing compatibility with electric motor drives. Most motor manufacturers produce motors to comply with NEMA MG-1, *Motors and Generators,* Section IV, Part 31; these motors are referred to as "inverter rated" or "inverter duty" motors.

Standard motors that are not compliant with the NEMA MG-1, Section IV, Part 31 standard, and are used with electric motor drives may fail prematurely. The speed at which a standard motor fails depends upon the specific motor application. Before connecting

power from an electric motor drive to a standard motor, contact the manufacturer to verify compatibility of the motor in a drive circuit.

Service Factor. *Service factor (SF),* is a multiplier that represents the percentage of extra load that can be placed on a motor for short periods of time without damaging the motor. A motor with a service factor of 1.25 can be overloaded by 25% for a short time. Motors that are NEMA MG-1, Section IV, Part 31 compliant have a service factor of 1.0 per NEMA requirements. A motor that is used with an electric motor drive that is not NEMA MG-1, Section IV, Part 31 compliant must not be operated beyond a service factor of 1.0 regardless of the nameplate rating.

Lead Length. *Lead length* is the length of the conductors (motor leads) between the electric motor drive and the motor. The motor leads T1, T2, and T3 have a line-to-line capacitance and a line-to-ground capacitance. Capacitance increases the magnitude of the voltage spikes at the motor terminals. The longer the motor leads, the greater the capacitance and the greater the magnitude of the voltage spikes.

Long lead lengths also cause reflected waves, known as standing waves or voltage ring-up. Reflected waves are dependent on lead length and carrier frequency. The longer the lead length and the higher the carrier frequency, the more pronounced the problem. Reflected waves are the result of a portion of the voltage waveform being reflected back from the motor terminal due to an impedance mismatch. The reflected portion combines with the voltage from the AC drive, increasing the voltage at the motor terminals. The voltage can be as high as 1500 V at the motor terminals when fed from a 480 V electric motor drive.

The easiest way to avoid problems caused by lead length is to keep the lead lengths as short as possible. Typically, lengths less than 100 ft should not pose a problem. It is not always possible to avoid long lead lengths. Electric motors may be located in harsh environments with the electric motor drive located a long distance away in a clean, air-conditioned control room. In this case an output reactor, also referred to as a load reactor, can be installed at the electric motor drive. A load reactor helps eliminate voltage spikes by slowing down the rate of change in the output voltage.

Load reactors are installed as close as possible to the output of the electric motor drive. Load reactors are sized based on the electric motor drive input voltage and frequency, the drive horsepower, and the impedance of the reactor. See Figure 9-3.

LOAD REACTORS

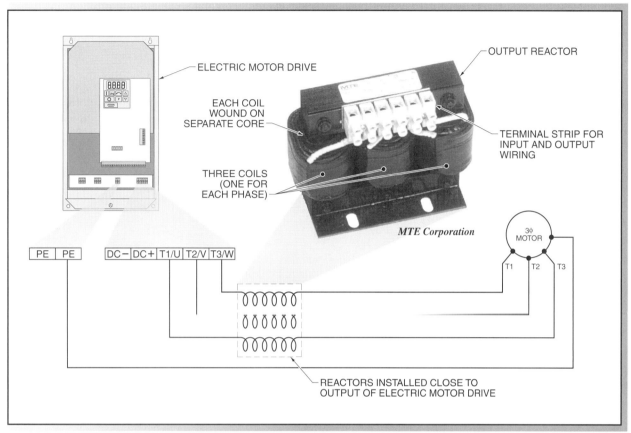

Figure 9-3. Output reactors are installed as close as possible to the output of an electric motor drive to reduce the level of voltage spikes at the motor.

Electric motor drive manufacturers may recommend output filters instead of load reactors. Output filters are also referred to as motor terminators. Output filters consist of inductance, capacitance, and resistance. An output filter may be installed at the output of the electric motor drive or at the motor. See Figure 9-4.

Another method to diminish the effects of lead length is to lower the carrier frequency. The result is fewer voltage pulses per unit of time and less stress on the motor insulation. At lower carrier frequencies, motors make more audible noise. Noise can be a problem for people working near the motor or if an HVAC fan motor transmits the noise through the ductwork.

TECH FACT

Follow manufacturer installation guidelines on motor lead length to avoid premature failure of motors.

AC drives are often used in industrial and commercial HVAC systems, and are increasingly used in residential HVAC systems.

Electric motor drive manufacturers provide specifications for maximum lead lengths for various carrier frequencies and for shielded or unshielded cable with output reactors or output filters.

Bearing Currents. A *bearing current* is the result of induced voltage in the motor rotor created by the electric motor drive. Bearing current causes premature failure of motor bearings. The spikes on the leading edge of the PWM waveform are induced into the rotor, and a voltage potential is developed between the rotor and the stator. The bearing race and balls or rollers serve as the current path for the voltage potential, and a continuous flow of current through the bearing occurs. The current flow causes atoms of metal to be removed. Over time, roughness, known as fluting, develops in the race, which eventually destroys the bearing.

OUTPUT FILTERS

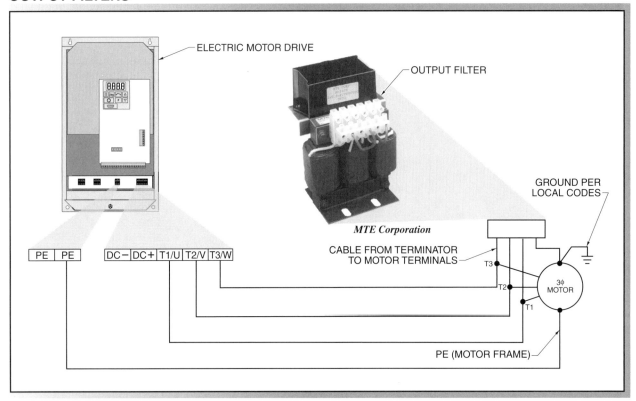

Figure 9-4. An output filter/terminator installed at the motor termination box reduces reflected waves.

Electric motor manufacturers have developed shaft grounding systems to eliminate the problem. A *shaft grounding system* is a system that connects (shorts) the rotor voltage to ground via a brush or other device in contact with the shaft to discharge unwanted voltage. See Figure 9-5. Another method to reduce or eliminate bearing currents is to lower the carrier frequency of the electric motor drive. Bearing currents are common in HVAC fan applications. Bearings are the only path for the potential to discharge to ground in fan applications.

Low-Speed Operation. The cooling effect of a fan attached to the motor shaft, and/or fins that are part of the rotor end ring, is reduced when a motor is operated at low speeds. Electric motors that are continuously operated at low speeds must be derated or provided with auxiliary cooling. Auxiliary cooling is accomplished by using a separate blower and motor mounted to the motor to provide cooling. Auxiliary blower motors operate on fixed voltages, and are either ON or OFF. The blower motor must be connected to a power source independent of the variable voltage output of an electric motor drive. Temperature sensors are installed in the stator windings of a motor and are connected to an

electric motor drive through a relay. The leads of a temperature sensor are labeled J1 (P1) and J2 (P2). When internal temperatures of a motor exceed safe limits, the temperature sensor opens, de-energizing the relay, shutting down the electric motor drive and motor. See Figure 9-6.

Operating Temperature. NEMA has standardized the maximum ambient temperature for motors as 104°F (40°C). Ambient temperature is the temperature of the air surrounding an object. Electric motors operated at a higher than ambient temperature must be derated. NEMA has also standardized altitude ratings for motor operation. Electric motors are designed to operate up to 3300′ (1000 m). Electric motors operated above 3300′ must be derated. The derating is requried because at higher altitudes, air is thinner and does not dissipate heat as quickly. Consult the charts provided by motor manufacturers for specific derating tables.

TECH FACT

Failure to derate or provide auxiliary cooling to motors operating at low speeds results in premature failure of motors.

SHAFT GROUNDING SYSTEMS

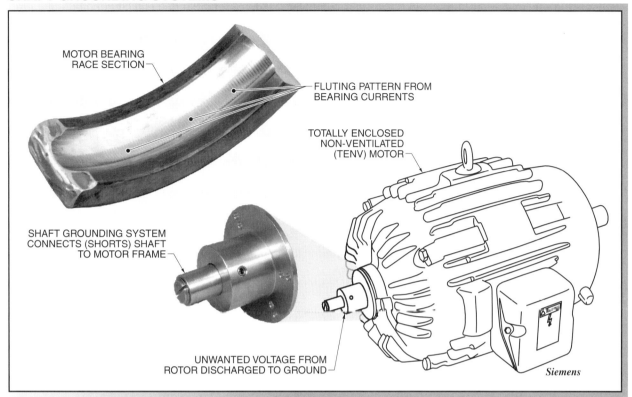

Figure 9-5. A shaft grounding system eliminates bearing currents that lead to bearing failures.

TEMPERATURE SENSOR

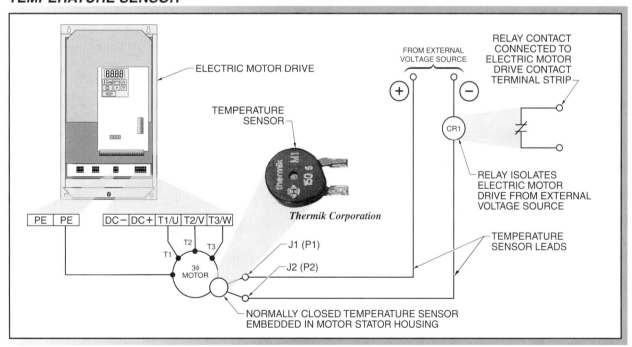

Figure 9-6. When excessive motor heat is detected, the temperature sensor opens, de-energizing the relay and sending a signal to the electric motor drive to stop the motor.

ASI Robicon

Power quality cabinets contain transformers, reactors, power factor correction capacitors, and filters to prepare electrical distribution power for electric motor drive use.

Power Quality

Electric motor drives are adversely affected by poor power quality. Electric motor drives can also introduce power quality problems into a power distribution system. Power quality is affected by voltage and frequency variations, and by whether a system is grounded or ungrounded.

Voltage and Frequency. Electric motor drives are designed to operate over a wide range of input voltages. Most electric motor drives are designed to operate with input frequencies of either 50 Hz or 60 Hz. Operating outside the specified frequency range can result in damage to an electric motor drive or cause various drive faults.

Grounded and Ungrounded Systems. Electric motor drives are intended to be powered from grounded systems with 3φ power lines that are electrically symmetrical with respect to ground. The NEC® permits ungrounded distribution systems in a limited number of instances. Electric motor drives contain metal oxide varistors (MOVs) to provide phase-to-phase and phase-to-ground protection from voltage surges. In an ungrounded system, there is no path back to the power source to activate an overcurrent device. The phase-to-ground MOV can become a continuous path for current, resulting in damage to the electric motor drive. See Figure 9-7.

Current. An electric motor drive is a nonlinear load. A *nonlinear load* is any load where the instantaneous load current is not proportional to the instantaneous voltage.

Voltage and current are not proportional because nonlinear loads draw current in short pulses, even when the source voltage is a sine wave. It is important that the power source for an electric motor drive has sufficient current capacity. If the power supply cannot deliver enough current, voltage flat-topping occurs. *Flat-topping* is the lowering of the peaks of the voltage sine wave. Flat-topping results in lower supply voltage to an electric motor drive. The lower supply voltage causes undervoltage faults and undependable motor operation. See Figure 9-8.

Input Reactors. Input reactors or isolation transformers are recommended for AC drives in a variety of situations. Input reactors are sized based on the electric motor drive input voltage and frequency, the drive horsepower, and the impedance of the input reactor. Input reactors are used in the following situations:

• When line impedance is too low. A reactor raises the line impedance and protects an electric motor drive from voltage spikes in the electrical distribution system caused by large loads or lighting being switched ON and OFF.
• When power factor correction capacitors are installed.
• When the electric motor drive causes harmonics in the electrical distribution system. Harmonics produce power quality problems on the electrical distribution system. The effects of harmonics are cumulative; the greater the number of electric motor drives installed in a facility, the greater the harmful effects. Harmonics can be either voltage harmonics or current harmonics.

Input reactors may also be used as output reactors. Input reactors should be part of every drive installation because they provide multiple benefits and add little to the overall motor system installation costs.

Power Factor Correction Capacitors. A *power factor correction capacitor* is a capacitor used to improve a facility's power factor by improving voltage levels, increasing system capacity, and reducing line losses. *Power factor* is the ratio of true power to apparent power. *True power* is the power that actually performs work. *Apparent power* is the total power delivered. Utility companies penalize customers with low power factors. Facilities with many inductive loads have a poor power factor because voltage is leading the current. The common practice is to install power factor correction capacitors to improve the power factor of a facility because capacitors cause the current to lead voltage.

Power factor correction capacitors can be placed ahead of an electric motor drive in the AC supply lines but not between the drive and motor. Power factor correction capacitor units with automatic switching must not be used unless specifically recommended by the manufacturer. See Figure 9-9.

METAL OXIDE VARISTOR (MOV) VOLTAGE SURGE PROTECTION

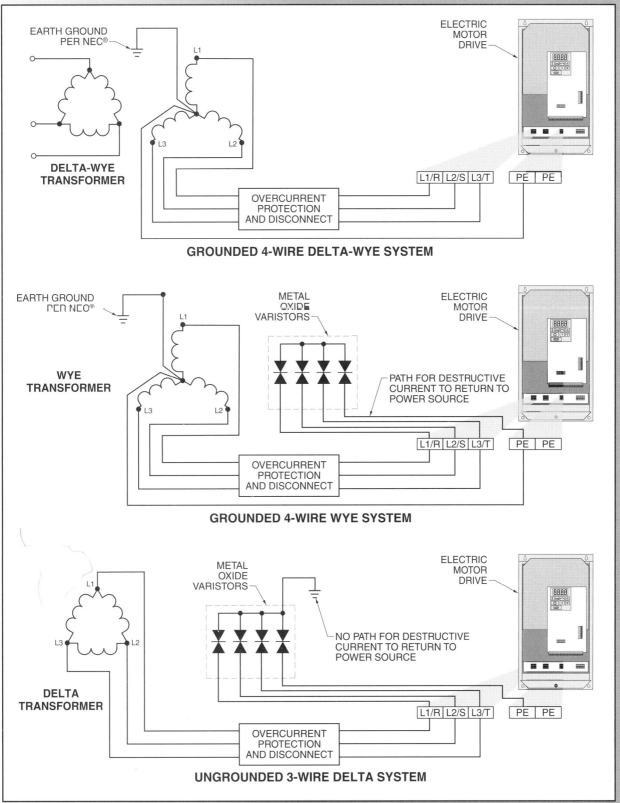

Figure 9-7. Electric motor drives are intended to be powered from 3φ lines with a common ground.

VOLTAGE AND CURRENT POWER SUPPLY

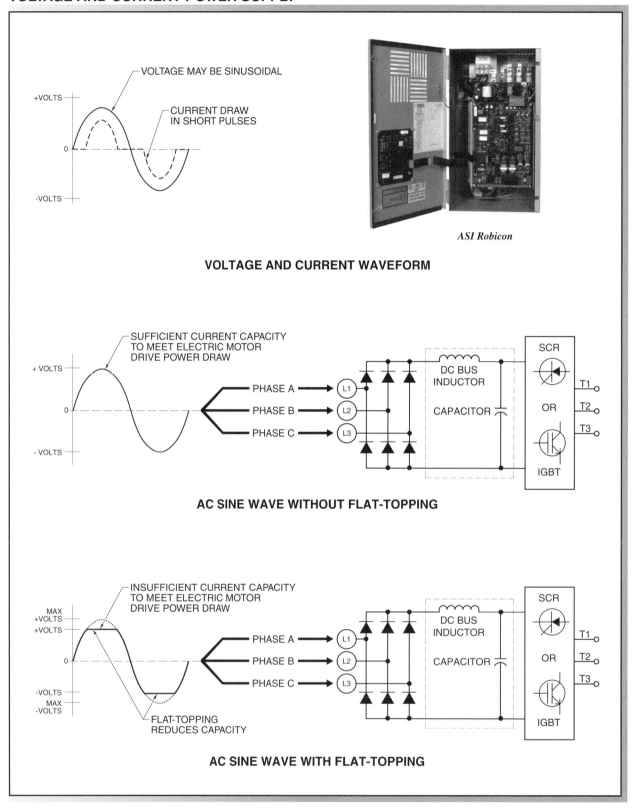

Figure 9-8. In nonlinear loads, instantaneous load current is not proportional to the instantaneous voltage, allowing flat-topping to occur and causing electric motor drive faults.

POWER FACTOR CORRECTION CAPACITORS

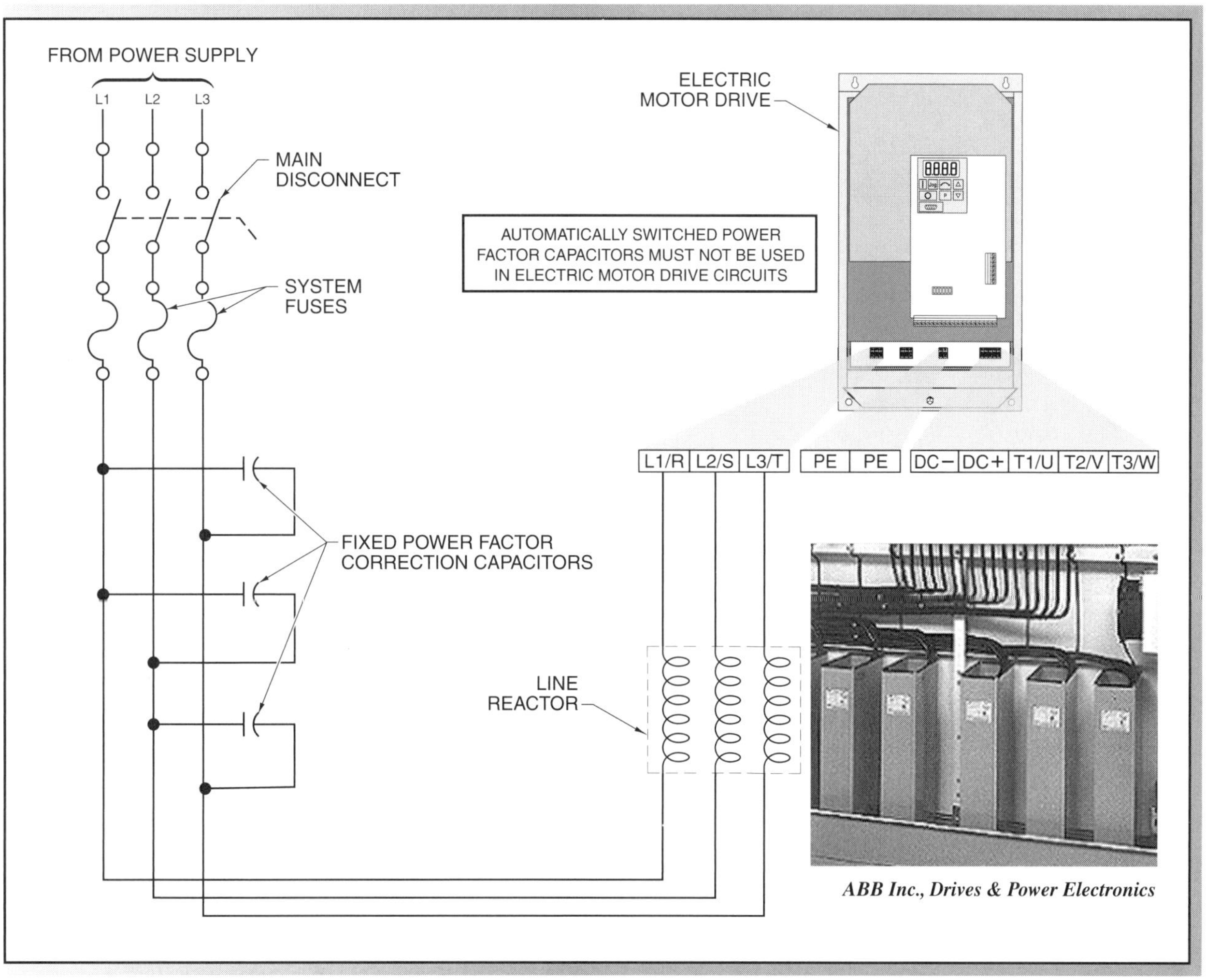

Figure 9-9. The effects of inductive and nonlinear loads can be compensated using properly sized power factor capacitors.

SAFETY

Electric motor drives are part of a system that includes a motor, a driven load, and possibly external controls. Care must be taken to ensure that all items are compatible, and that any item or combination of items does not introduce a safety hazard. Electric motor drive control circuits are not to be used as emergency stops. An emergency stop is used where accidental contact with moving equipment, or the unintended flow of product, can result in physical injury or property damage. If an emergency stop is required, an additional hard-wired stop circuit that removes AC line power to the electric motor drive is necessary. When AC line power is removed, the braking effect of the electric motor drive is lost and the motor coasts to a stop. The motor may require an auxiliary electromechanical brake to stop quickly or to stop and hold a load. See Figure 9-10.

The brake is mounted on the end of the motor opposite the shaft. When power is applied (solenoid energized), the brake disengages and allows the motor and load to rotate. When power is removed (solenoid de-energized), springs cause the brake to engage and hold the motor and load. The brake operates on an ON or OFF fixed voltage. The brake must be connected to a power source independent of the variable voltage output of the electric motor drive.

> **⚠ WARNING**
> *DC injection braking and dynamic braking are auxiliary braking functions and are not intended to take the place of electromechanical brakes.*

AUXILIARY ELECTROMECHANICAL BRAKES

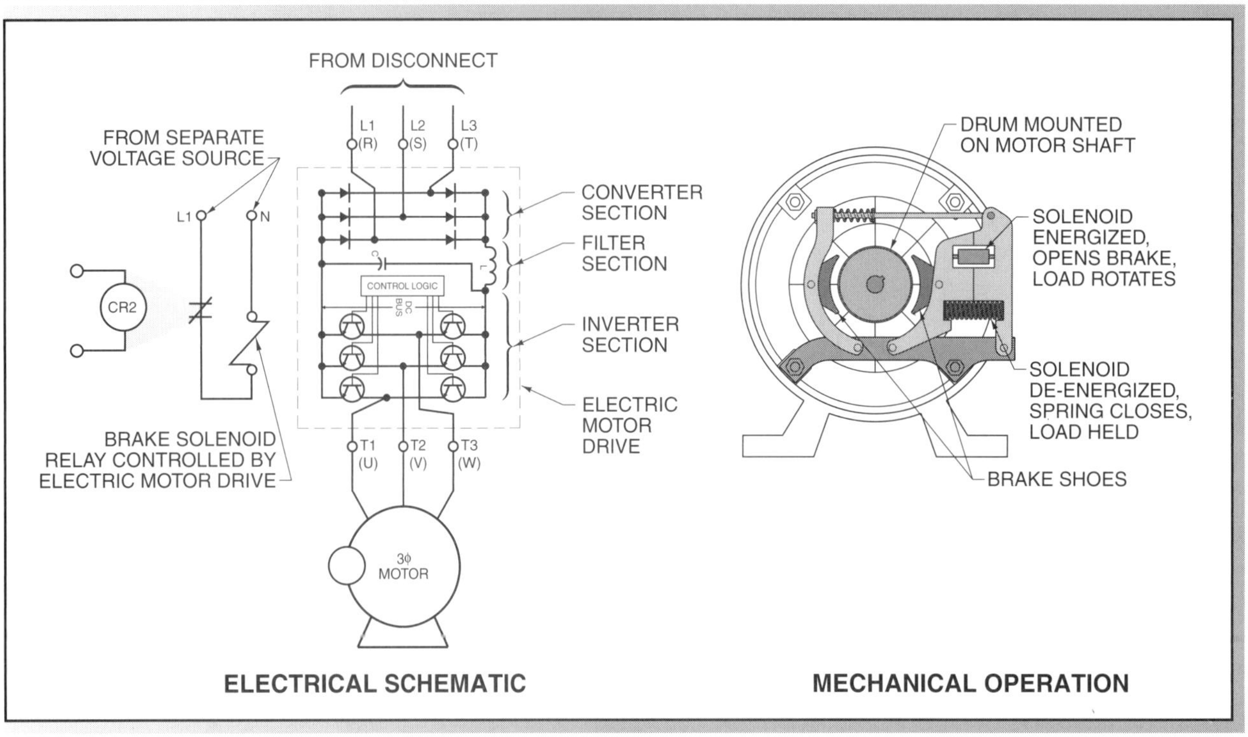

ELECTRICAL SCHEMATIC   **MECHANICAL OPERATION**

Figure 9-10. An electromechanical brake is used to hold a motor shaft stationary after voltage is removed from the motor.

ENGINEERED APPLICATIONS

An *engineered application* is an electric motor drive application that requires a licensed engineer to be safely implemented. A licensed engineer provides drawings, answers questions, and oversees the installation to ensure that it conforms to all applicable codes. Engineered applications include those where a motor is operated above its rated base speed, where an emergency stop button is required, or where a drive is controlling people-moving equipment. *Base speed* is the nameplate speed (rpm) at which a motor develops rated horsepower at rated load and voltage. When using an AC motor, base speed is typically the point where nameplate voltage and nameplate frequency are applied. When using a DC motor, it is typically the point where full armature voltage is applied with full rated field excitation. Operation of a motor above base speed is governed by the mechanical limits of the rotor, armature, and bearings.

ELECTRIC MOTOR DRIVE INSTALLATION

An electric motor drive installation process requires receiving shipment of a drive from a manufacturer, selecting a suitable location and enclosure, mounting the drive (taking into account derating factors), and minimizing electromagnetic interference. Paying close attention to all installation factors results in a smooth start-up and years of trouble-free service.

Receiving Drive and Hardware

Upon receipt of an electric motor drive and any associated hardware, there are several steps to follow:

- Thoroughly inspect the items for any damaged or missing parts. Immediately report any problems to the freight company.
- Verify the size, rating, configuration, etc., of the items ordered.
- Remove the instructions and other documentation, and store in a safe location for future reference.

⚠ **WARNING**

Always consult the electric motor drive manufacturer or licensed engineer regarding drive application questions. Incorrect application of an electric motor drive can result in serious personal injury or damage to the drive and/or driven load.

- Store the electric motor drive and related hardware in a clean, dry, and secure location that conforms to the operating manufacturer specifications.
- Do not store an electric motor drive for longer than a year. Electric motor drives contain electrolytic capacitors, which deteriorate if not powered for long periods of time.

Location

Electric motor drives should be mounted in a clean and dry location. High temperatures, high humidity, dust, particles or fibers in the air, corrosive or explosive vapors, constant vibration, and direct sunlight should be avoided. The location should have sufficient lighting and sufficient working space to facilitate the installation, start-up, and maintenance of the electric motor drive.

Enclosures

NEMA rates enclosures based on use and service. Most electric motor drives are delivered from a manufacturer in NEMA Type 1 enclosures. An electric motor drive in a NEMA Type 1 enclosure may be placed in a suitable environment or in another encosure. Other NEMA rated enclosures are available for installing electric motor drives in hostile environments. See Figure 9-11.

An electric motor drive in a NEMA Type 1 enclosure may require a NEMA Type 3R enclosure if the enclosure is subjected to falling rain, ice, or other damaging environments. The enclosure that is to house the electric motor drive should be mounted, and any metal debris must be discarded before the drive is installed in the enclosure. All precautions must be taken to prevent debris from falling into an electric motor drive and causing a problem. The person responsible for installing the electric motor drive must decide whether the drive can be installed as is or whether it requires installation in another enclosure.

When an electric motor drive is installed in an enclosure, heat buildup can be a problem. The electric motor drive radiates heat and the enclosure may not be able to dissipate the heat. The heat may cause electric motor drive faults, or premature component failure. Extra cooling may be required to supplement the fan(s) that are integral to the electric motor drive. The additional cooling can be in the form of a fan or a cooling unit.

Mounting

An electric motor drive should be mounted on a smooth, nonflammable vertical surface, with the name of the manufacturer facing out and right side up. Small electric motor drives are mounted in rack slots or on a DIN rail. Medium-size electric motor drives may be mounted directly to a motor or in a cabinet using mounting holes. Larger electric motor drives have separate mounting holes for individual fasteners. See Figure 9-12. The fastening method should be adequate to support the weight of the electric motor drive.

Heat is generated during the normal operation of an electric motor drive. An electric motor drive is mounted to allow the free flow of air across the heat sink(s), possibly aided by integral cooling fans. Adequate clearances must be maintained around the electric motor drive for the free flow of air. Follow manufacturer specifications for mounting an electric motor drive.

Derating

Electric motor drives are designed for specific operating ranges of temperature, voltage, altitude, and humidity. Derating is required if an electric motor drive is operated outside the normal operating ranges specified by the manufacturer. Electric motor drives operated above the normal temperature range, from a 1ϕ power source or from a power source with reduced voltage, or at a high altitude, or higher carrier frequencies, must be derated. Manufacturers supply charts and derating multipliers for each particular condition.

Omron IDM Controls
Cabinets with multiple electric motor drives require extra attention to mounting positions, airflow, and derating to avoid heat buildup and drive failures.

ENCLOSURE SELECTION

ENCLOSURES					
Type	Use	Service Conditions	Tests	Comments	Type
1	Indoor	No unusual	Rod entry, rust resistance		
3	Outdoor	Windblown dust, rain, sleet, and ice on enclosure	Rain, external icing, dust, and rust resistance	Do not provide protection against internal condensation or internal icing	
3R	Outdoor	Falling rain and ice on enclosure	Rod entry, rain, external icing, and rust resistance	Do not provide protection against dust, internal condensation, or internal icing	
4	Indoor/outdoor	Windblown dust and rain, splashing water, hose-directed water, and ice on enclosure	Hosedown, external icing, and rust resistance	Do not provide protection against internal condensation or internal icing	
4X	Indoor/outdoor	Corrosion, windblown dust and rain, splashing water, hose-directed water, and ice on enclosure	Hosedown, external icing, and corrosion resistance	Do not provide protection against internal condensation or internal icing	
6	Indoor/outdoor	Occasional temporary submersion at a limited depth			
6P	Indoor/outdoor	Prolonged submersion at a limited depth			
7	Indoor locations classified as Class I, Groups A, B, C, or D, as defined in the NEC®	Withstand and contain an internal explosion of specified gases, contain an explosion sufficiently so an explosive gas-air mixture in the atmosphere is not ignited	Explosion, hydrostatic, and temperature	Enclosed heat-generating devices shall not cause external surfaces to reach temperatures capable of igniting explosive gas-air mixtures in the atmosphere	
9	Indoor locations classified as Class II, Groups E or G, as defined in the NEC®	Dust	Dust penetration, temperature, and gasket aging	Enclosed heat generating devices shall not cause external surfaces to reach temperatures capable of igniting explosive gas-air mixtures in the atmosphere	
12	Indoor	Dust, falling dirt, and dripping noncorrosive liquids	Drip, dust, and rust resistance	Do not provide protection against internal condensation	
13	Indoor	Dust, spraying water, oil, and noncorrosive coolant	Oil explosion and rust resistance	Do not provide protection against internal condensation	

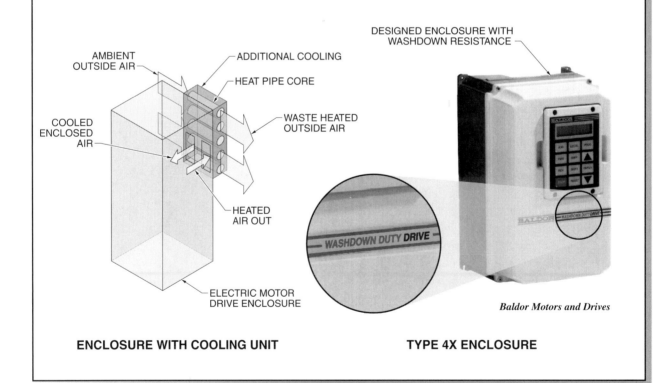

DESIGNED ENCLOSURE WITH
WASHDOWN RESISTANCE

AMBIENT OUTSIDE AIR

ADDITIONAL COOLING

HEAT PIPE CORE

COOLED ENCLOSED AIR

WASTE HEATED OUTSIDE AIR

HEATED AIR OUT

ELECTRIC MOTOR DRIVE ENCLOSURE

WASHDOWN DUTY DRIVE

Baldor Motors and Drives

ENCLOSURE WITH COOLING UNIT

TYPE 4X ENCLOSURE

Figure 9-11. Depending on the application, an enclosure with a cooling unit may be required.

ELECTRIC MOTOR DRIVE MOUNTS

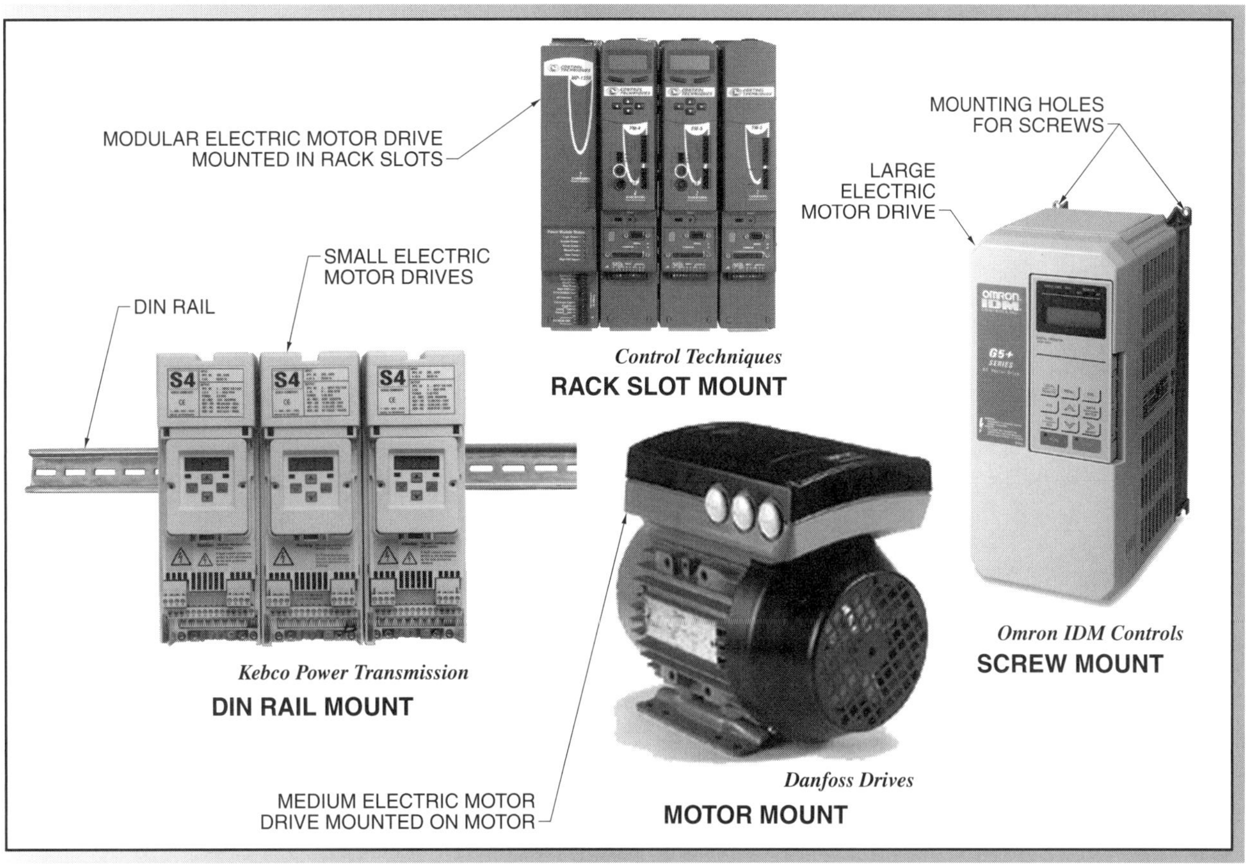

Figure 9-12. Mounting methods vary with the size of electric motor drive.

An electric motor drive model may have multiple output ratings, depending on the type of load and the carrier frequency used. Constant torque loads require the same current from zero to base speed. Variable torque loads require less current at lower speeds. Consequently, an electric motor drive model typically has higher output ratings for a variable torque load than for a constant torque load.

Power losses occur when the IGBTs of an electric motor drive are between states (neither ON nor OFF). The higher the carrier frequency, the more pulses per unit of time, and the more loss of power the IGBTs create. An electric motor drive has lower output ratings at higher carrier frequencies because of IGBT power losses.

Electromagnetic Interference

Electromagnetic interference (EMI), also known as electrical noise, is the unwanted signals generated by electrical and electronic equipment. All electrical and electronic equipment generate EMI. Low frequency magnetic-field EMI can be emitted from the cables connected to equipment, or from inductive loads in proximity to equipment.

High frequency electric-field EMI can emanate from equipment as electromagnetic radiation, also known as radio frequency interference (RFI). EMI can be received by other equipment and interfere with its proper operation. EMI problems range from corrupted data transmissions to electric motor drive damage.

All electrical and electronic equipment should have some level of immunity to EMI. *Electromagnetic compatibility (EMC)* is a comparison of how different pieces of equipment work together with varying levels of interference. As use of electronic equipment in industrial environments has increased, manufacturers have taken steps to both minimize emissions and maximize EMI immunity in the design of equipment.

Regulations regarding EMC/EMI are very complicated and vary from country to country. A properly installed electric motor drive minimizes potential problems from EMI. Installation techniques and supplemental devices that minimize emissions and maximize immunity include wiring techniques, grounding, shielding, and isolation of power and signal or control cable.

Power Wiring. The drive output power wiring (load conductors) contains high-voltage switching waveforms which can affect the line conductors. The line and load conductors of an electric motor drive need to be installed in separate metal conduits. Metal conduit serves as a shield, preventing conductors from radiating EMI and protecting conductors from any radiated EMI. Nonmetallic conduit does not provide the shielding effect. A separate grounding conductor must be installed in each conduit and connected at each end. In the event that multiple electric motor drives feed multiple motors, the line and load conductors of each drive need to be installed in separate metal conduits.

When cable trays are used, armored metal cable with a separate grounding conductor, or shielded power cable with a separate grounding conductor, must be used. The

grounding conductors contained in the cable need to be connected at each end.

Control Wiring. *Control wiring* is all external wiring connected to an electric motor drive, excluding the line and load conductors. Electric motor drives have control wiring that communicate signals to start and stop the drive, communicate sensor conditions to the drive, and carry output signals from drive auxiliary contacts. See Figure 9-13. Control wiring can communicate four types of signals: AC digital signals, DC digital signals, DC analog signals, and serial communication signals. *Digital signals* are signals that have only two states, ON or OFF like a momentary pushbutton. *Analog signals* are signals that vary over a range of values such as 0 VDC to 10 VDC for speed reference. *Serial communication signals* are digital data signals from an external source such as a PLC or PC.

ELECTRIC MOTOR DRIVE CONTROL WIRING

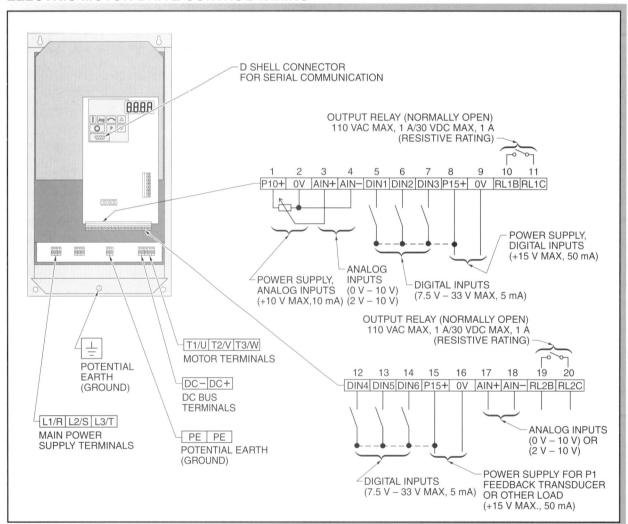

Figure 9-13. Power wiring and control wiring are terminated at separate locations on an electric motor drive.

The harmful effects of EMI from control wiring can be minimized by the following procedure:

• Control and power wiring should be separated as much as possible inside an electric motor drive enclosure. If control and power wiring must cross, they should cross at a 90° angle.

• Control wiring should not be installed in the same conduit as power wiring. Control wiring should be installed in a separate metal conduit.

• Install each category of control wiring in its own metal conduit.

• DC analog signals and serial communication signals should run in twisted shielded pair cable.

Grounding. Proper grounding is required for the safe and reliable operation of electric motor drives, motors, and related equipment. Grounding requirements can be divided into equipment grounding and signal wire grounding.

Equipment grounding is an equal potential between all metal components of an installation and a low-impedance path for fault currents to operate overcurrent protective devices (OCPDs). Equipment grounding also aids in containing EMI. All equipment grounds should be brought back to a single point at the electric motor drive. The connections should be tight and mechanically sound to guarantee a good electrical connection. When making a grounding connection to the interior of a metal enclosure, scrape away any protective coating to ensure bare metal-to-metal contact. See Figure 9-14.

Control wires carry analog signals and serial communication signals using twisted shielded pair cable. A shield is constructed of metal foil or mesh and a bare wire wrapped around the twisted conductors. The bare wire is also referred to as the drain wire. The shield/drain wire and the twisting of the conductors provide enhanced noise protection. The shield/drain wire is only grounded at the electric motor drive end. A shield/drain wire that is grounded at both ends introduces EMI into the signal circuit because of the difference of potential between the two points of grounding.

ELECTRIC MOTOR DRIVE AND MOTOR CIRCUIT GROUNDING

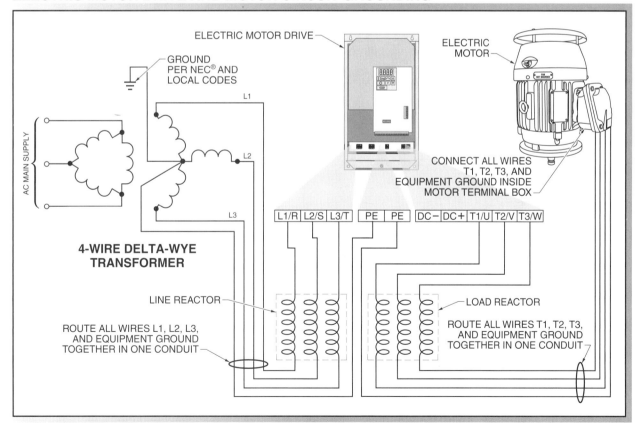

Figure 9-14. Equipment grounds for line and load conductors are terminated at the electric motor drive.

Inductive Loads. Inductive loads such as solenoids or the coils of relays and contactors are a source of EMI. When inductive loads are de-energized, a large counter electromotive force (CEMF) is generated. *CEMF* is a voltage spike created by the inductive loads when de-energized that can cause electric motor drive faults. Coils in the same enclosure as an electric motor drive or coils controlled by relay contacts of a drive need to suppress voltage spikes.

Devices are installed across the coils to dissipate the CEMF and suppress EMI. Resistor-capacitor networks (RC snubber networks) or MOVs are connected across the AC coils. Diodes are connected across DC coils to suppress EMI. The diodes are connected in reverse bias across the DC coils in the same way fly-back diodes are used with transistors. See Figure 9-15.

Other EMI/RFI Reducing Devices. Other devices and techniques are used for reducing EMI/RFI. An output reactor or an output choke may be used to reduce interference with sensitive equipment. An RFI filter installed at the input of an electric motor drive reduces RFI interference. Lowering the PWM frequency also reduces EMI. See Figure 9-16.

WIRING

The National Electrical Code (NEC®) must be followed when connecting an electric motor drive. The wiring process of an electric motor drive is divided into NEC® and manufacturer requirements.

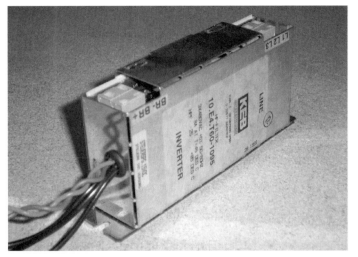

Kebco Power Transmission
EMI filters installed across EMI-producing devices suppress EMI and prevent it from disrupting control signals.

INDUCTIVE LOAD EMI SUPPRESSION

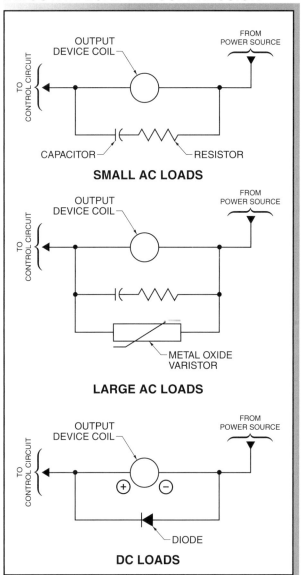

Figure 9-15. A snubber circuit can be used to suppress EMI.

NEC® and Electric Motor Drives

The NEC® refers to drives as adjustable speed drive systems or conversion equipment. NEC® Article 430, Motors, Motor Circuits and Controllers, is the principal code source for drives. Part X of Article 430, Adjustable-Speed Drive Systems, contains specific requirements for the installation of electric motor drives. Other sections of Article 430 apply as well, as an electric motor drive is a controller that powers a motor. Other portions of the NEC®, even when not specifically named, also apply to drives. See Figure 8-17.

OTHER EMI/RFI REDUCING DEVICES

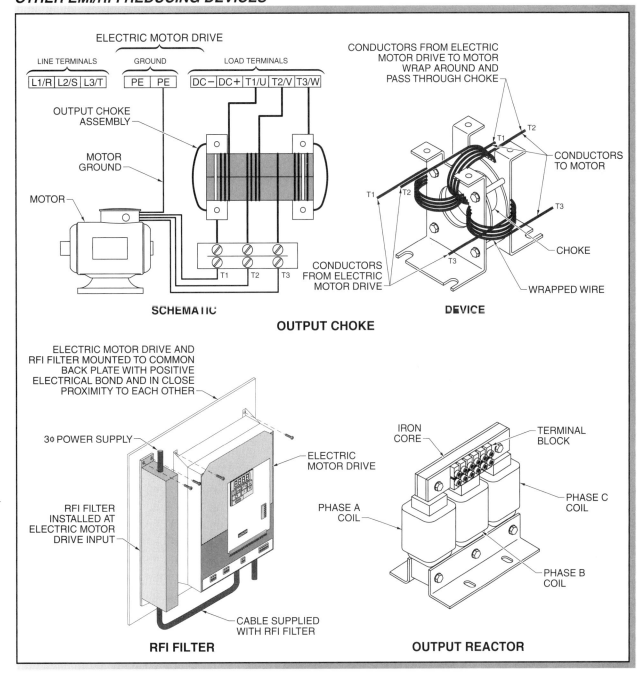

Figure 9-16. Output chokes, RFI filters, and output reactors are used to reduce EMI.

Manufacturer Instructions. NEC® Section 110.3(B) states, "Listed or labeled equipment shall be installed and used in accordance with any instructions included in the listing or labeling." NEC® Section 110.3(B) makes the instructions of the manufacturer the primary reference for installation. Manufacturer instructions provide information on wire size, recommended mounting location, and the size and type of overcurrent devices. Electric motor drives may use fuses or circuit breakers as circuit overcurrent protection. Electric motor drive manufacturers specify the type of overcurrent protection to be used with particular drives.

ELECTRIC MOTOR DRIVE SYSTEMS – NEC® REFERENCES

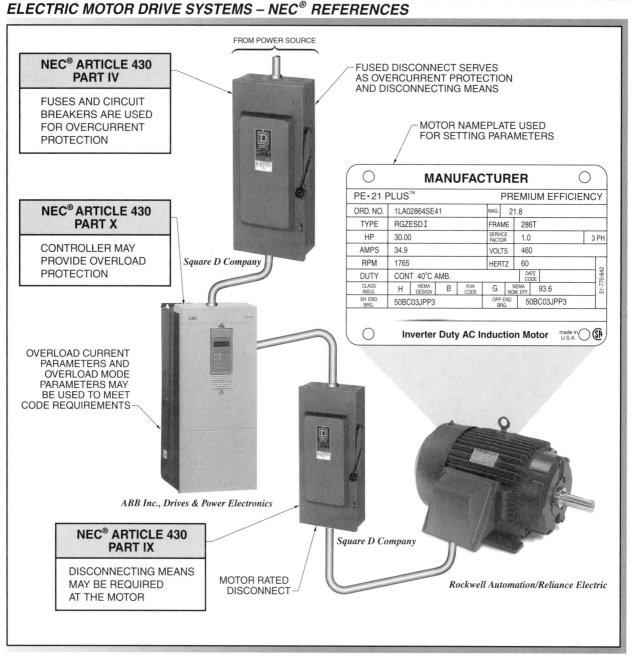

Figure 9-17. NEC® Article 430 contains several parts that apply to electric motor drives.

Wire Size. In the event the instructions do not specify wire size, NEC® Section 430.122(A) states, "Circuit conductors supplying power conversion equipment included as part of an adjustable-speed drive system shall have an ampacity not less than 125% of the rated input to the power conversion unit." The wire size can be calculated using the electric motor drive (power conversion equipment) nameplate current and NEC® Table 310.16.

Disconnects. A *disconnect* is a device that isolates an electric motor drive and/or motor from the voltage source to allow safe access for maintenance or repair. NEC® Article 430 Part IX contains general requirements for disconnects. Article 430 Part X contains specific requirements for the disconnect located in the incoming line to the electric motor drive. NEC® Section 430.128 states "The disconnecting means shall be permitted to be in the incoming line to the conversion equipment and shall have a rating of not less than 115 percent of the rated input current of the conversion unit." Circuit breakers and motor-circuit switches rated in horsepower are devices used as disconnecting means. A circuit breaker or fused motor-rated disconnect can serve as both a disconnecting means and overcurrent device. A motor-circuit switch is commonly referred to as a motor-rated disconnect.

NEC® Article 430 Part IX has specific rules regarding disconnects for various applications. One of the more commonly applied sections, NEC® 430.102(B) exception, allows a controller disconnect that is capable of being locked open to also serve as the motor disconnecting means. Articles and exceptions in the NEC® may be superseded by other building codes such as a mechanical code requiring a disconnect be located at all HVAC equipment.

Overload Protection. In a majority of electric motor drive applications, the drive provides overload protection for the motor. See Figure 9-18. NEC® Section 430.124(A) states, "Where the power conversion equipment is marked to indicate that overload protection is included, additional overload protection shall not be required." In order for the electric motor drive to provide overload protection, the nameplate information of the motor must be programmed into the drive. The electric motor drive control board provides the overload protection based on the information entered.

If an electric motor drive is not approved for use as an overload, or if multiple motors are fed from the drive, an external overload relay(s) must be provided. See Figure 9-19. NEC® Article 430 contains information on selecting and sizing overcurrent devices and overload protection. The normally closed contacts of the overload relays are connected in series and terminate at the control terminal strip of the electric motor drive. An overload in any motor opens the circuit and causes the electric motor drive to stop outputting voltage to the motors.

ELECTRIC MOTOR DRIVE AS OVERLOAD PROTECTION

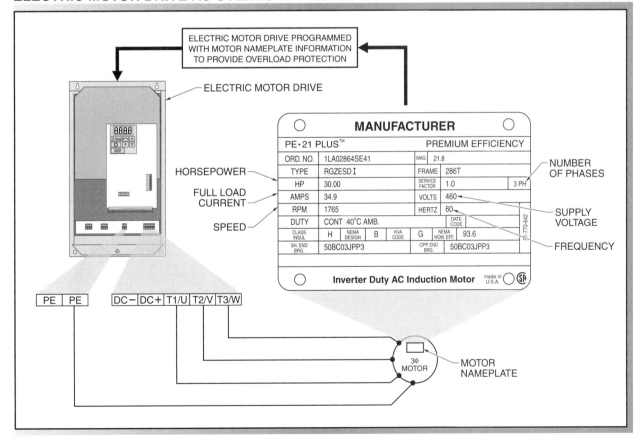

Figure 9-18. Motor nameplate information is entered into an electric motor drive and the information is processed by the CPU to calculate overload protection levels.

OVERLOAD PROTECTION PROVIDED BY OVERLOAD RELAY

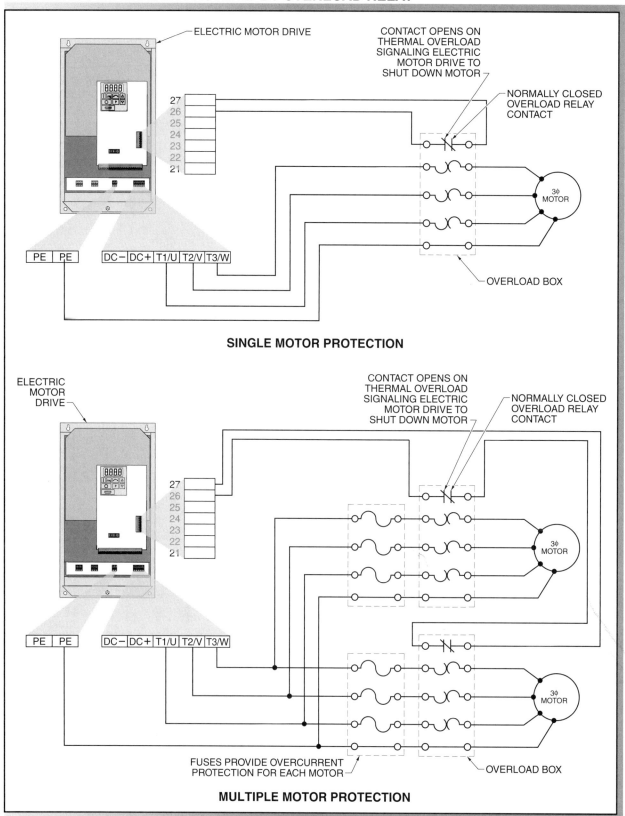

Figure 9-19. External fuses and overload relays may be required along with the electric motor drive in certain applications.

Clearances. NEC® Article 110 Part II lists requirements for dedicated working space and dedicated electrical space. The working space requirements are intended to protect the technician who must adjust or service the equipment by providing minimum working spaces, and include illumination requirements. The equipment space requirements are intended to protect the electrical equipment by limiting nonelectrical piping systems above electrical equipment. Electric motor drive installations must comply with NEC® Article 110 Part II requirements. See Figure 9-20.

Grounding. Proper grounding of an electric motor drive provides equal potential between all metal surfaces, a low-impedance path to activate overcurrent devices, and EMI reduction. NEC® Article 100 defines an equipment grounding conductor as "The conductor used to connect the non-current-carrying metal parts of equipment, raceways, and other enclosures to the system grounded conductor, the grounding electrode conductor, or both, at the service equipment or at the source of a separately derived system." The NEC® permits certain metal raceways to be used as equipment grounding conductors. For added safety and enhanced EMI reduction, electric motor drive manufacturers require an equipment grounding conductor in addition to the metal raceway containing the power and control conductors of the drive. Grounding requirements and the sizing of equipment grounding conductors are covered in NEC® Article 250.

> **⚠ CAUTION**
> *Consult the electric motor drive manufacturer before powering a drive from an ungrounded system or a 3ϕ supply whose line voltages are not symmetrical to ground to avoid injuries to personnel and avoid damaging the drive.*

Power Wiring

Power wiring for an electric motor drive consists of the conductors supplying power to the drive (line conductors) and the conductors supplying power to the motor (load conductors). A variety of wiring methods are available, depending upon installation considerations. The load conductors may power a single motor or multiple motors. Certain applications use a line bypass contactor to provide standard 60 Hz power to the motor in the event the electric motor drive fails.

NEC® EQUIPMENT SPACE REQUIREMENTS

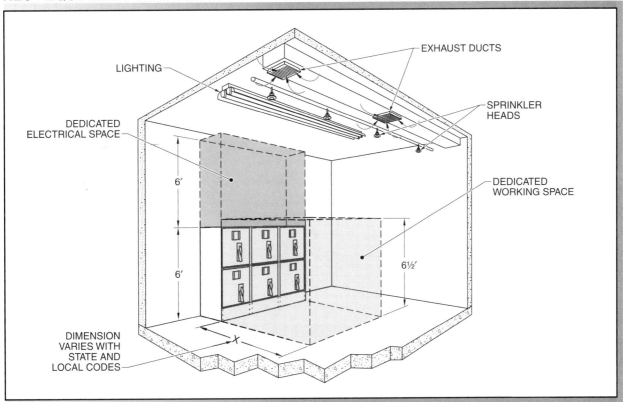

Figure 9-20. The NEC® requires a dedicated working space and a dedicated electrical space for indoor installations of electrical equipment.

Power Terminal Strip. The power terminal strip and the control wiring terminal strip are separated from one another to reduce EMI. The line conductors and load conductors are terminated at the power terminal strips. Many electric motor drives have a location on the power terminal strip to connect an auxiliary braking unit directly to the DC bus. The North American designations for conductors are L1, L2, and L3 for line; and T1, T2, T3 for load. The European designations for conductors are R, S, and T for line; and U, V, W for load. Care must be taken to ensure the line conductors are connected at the line terminals, and the load conductors are connected at the load terminals. The DC bus connection is designated by DC+, DC–. Some drive manufacturers denote the DC bus with B+ and B–. This is done to indicate the location where a dynamic braking resistor can be installed. The ground terminal is designated by potential earth (PE), or the ground symbol ($\perp$). See Figure 9-21.

ELECTRIC MOTOR DRIVE TERMINAL STRIPS

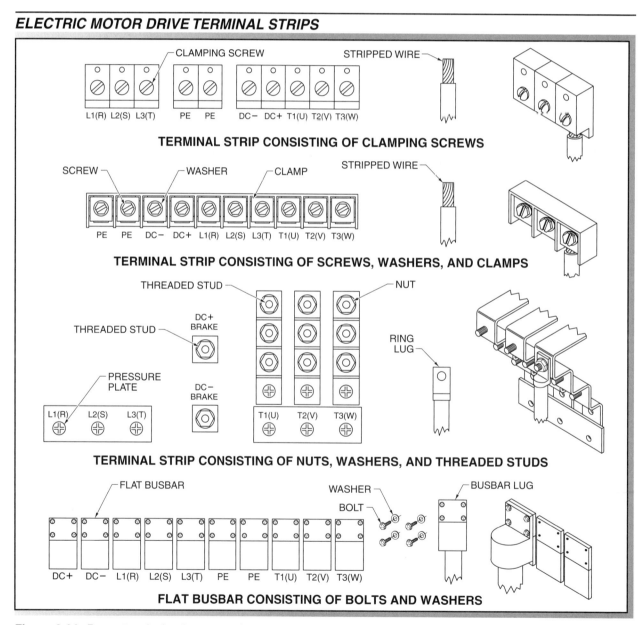

Figure 9-21. Power terminal strips use various termination methods according to horsepower requirements and have multiple designations for the line and load terminations.

The type of terminal strip required varies depending on the horsepower rating of the electric motor drive. Small electric motor drives have screws with pressure plates. Large electric motor drives have threaded studs or busbars with nuts and washers. Ring lugs are also used to terminate conductors onto threaded studs. Ring lugs are used instead of fork lugs because they do not fall off the threaded stud if the nut and washer become loose. Always use the lugs recommended by the electric motor drive manufacturer. The terminals need to be tightened for a secure electrical connection.

TECH FACT

> *Nonmetallic conduits and raceways must not be used for power wiring and control wiring when nonshielded cable (wiring) is used.*

Wiring Methods. The NEC® permits conductors 1/0 AWG and larger to be paralleled. Conductors may be paralleled for ease of installation, or where wire-bending space for large conductors is limited. Where conductors are paralleled, conductors must be kept together in sets with T1(1), T2(1), T3(1) in one conduit, and T1(2), T2(2), T3(2) in another conduit. NEC® Section 310.4 contains the specific requirements for paralleling conductors.

Commonly used wiring conduit for power conductors include armored cable (Type AC), metal-clad cable (Type MC), rigid metal conduit, electrical metallic tubing (EMT), intermediate metal conduit (IMC), and shielded power cable. The specific location and application determines the wiring conduit to be used, such as in commercial HVAC installation, industrial installation, or an original equipment manufacturer (OEM) piece of machinery. See Figure 9-22. Plastic raceways are not recommended because plastic raceways do not provide shielding to limit EMI.

The final connection to the motor should be made with a short length of flexible conduit (6′ or less) and stranded wire. Either flexible metal conduit (flex) or liquidtight flexible metal conduit (Sealtite™) can be used. Each method allows the motor to be moved for minor adjustments and prevents vibration from the motor and driven load from damaging a solid raceway or solid conductor. See Figure 9-23.

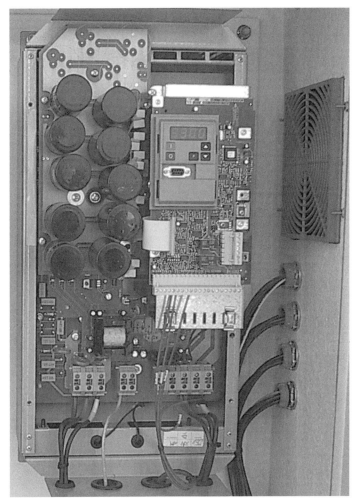

Control wiring and power wiring must be in separate metal conduits. Control wiring should also be separated by voltage into separate metal conduits.

Single-Motor Installation. The most common electric motor drive application involves a single motor powered by a drive. The overcurrent device that protects the electric motor drive also protects the load conductors and motor. The electric motor drive provides the overload protection to the motor if it is approved for this purpose. An electric motor drive controlling an HVAC fan motor is a common single-motor installation. See Figure 9-24.

> **⚠ WARNING**
>
> *Always follow the instructions and recommendations of electric motor drive manufacturers and applicable federal, state, and local codes. Failure to do so can result in serious physical injury and/or equipment damage. Only qualified personnel are allowed to install, start up, and troubleshoot electric motor drives.*

ELECTRIC MOTOR DRIVE TO MOTOR WIRING METHODS

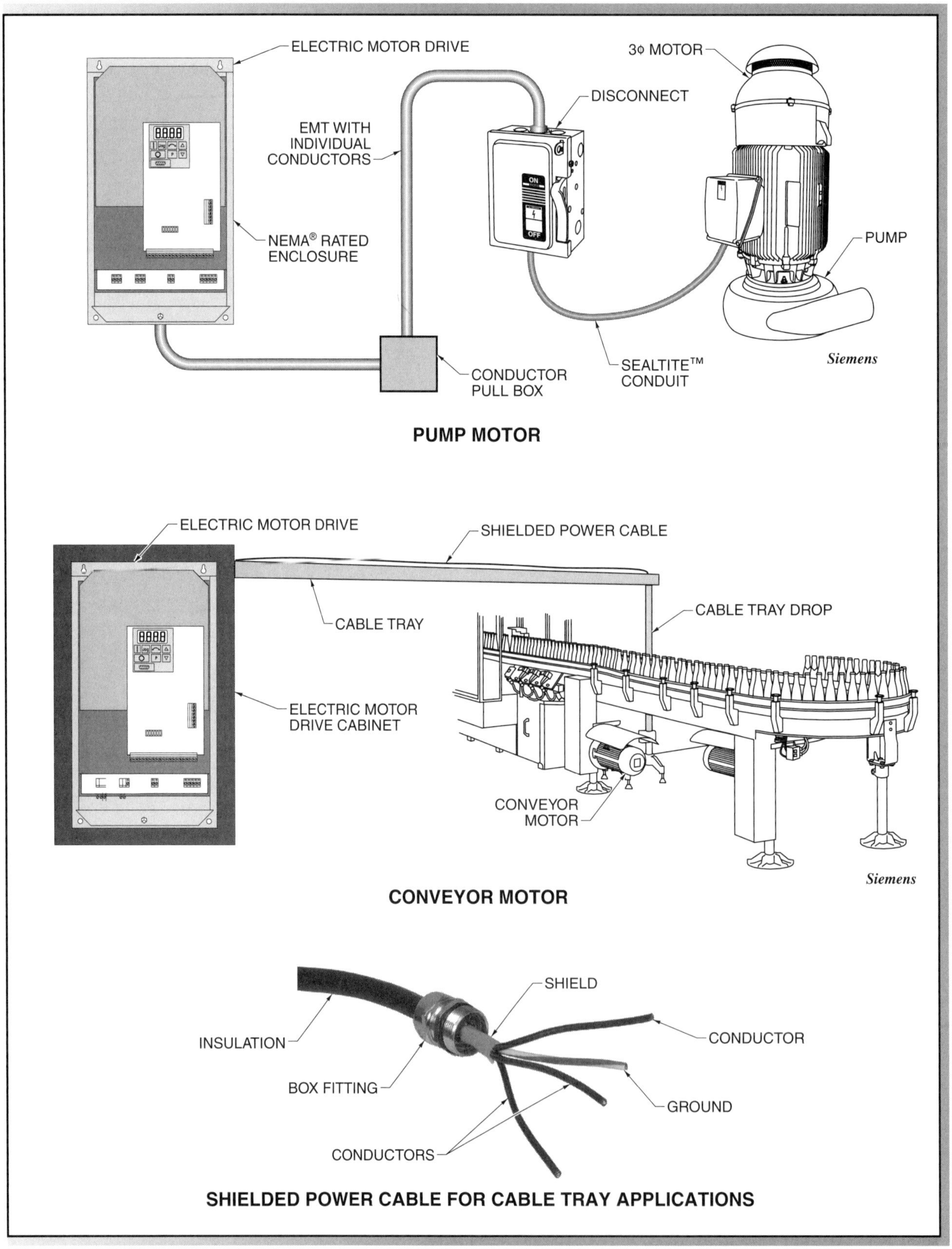

PUMP MOTOR

Siemens

CONVEYOR MOTOR

Siemens

SHIELDED POWER CABLE FOR CABLE TRAY APPLICATIONS

Figure 9-22. A wide variety of wiring methods are available for installing an electric motor drive and motor.

FLEXIBLE MOTOR CONDUIT

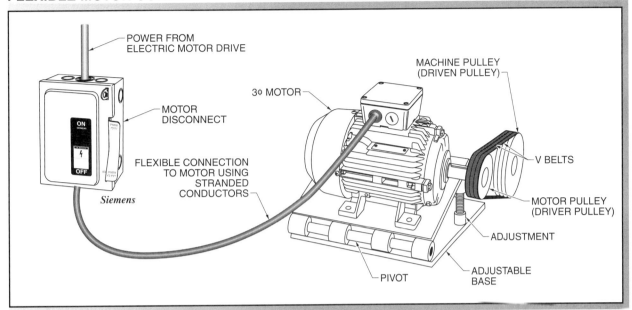

Figure 9-23. A flexible connection to a motor allows motor adjustment and negates the effects of vibration from the motor or building.

ELECTRIC MOTOR DRIVE AND SINGLE-MOTOR SYSTEM INSTALLATION

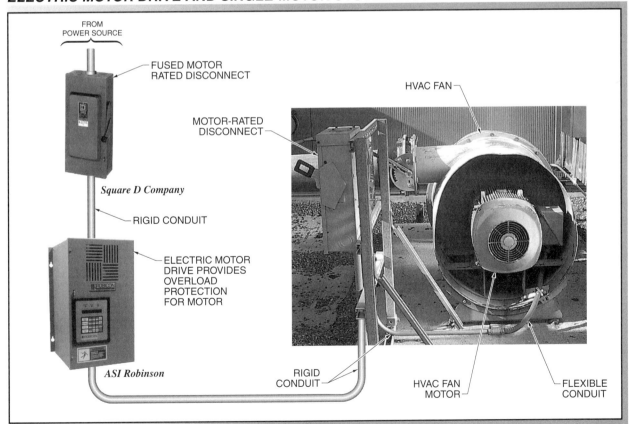

Figure 9-24. A basic electric motor drive application involves a single motor powered by a drive.

Multiple-Motor Installation. A multiple-motor installation involves multiple motors being fed from a single electric motor drive. An electric motor drive controlling a conveyor with multiple motors that operate at the same speed is a common multiple-motor installation. Each motor has its own set of conductors. Supplemental overcurrent devices and overload protection must be installed, as the rating of the individual motors is less than the electric motor drive rating. A multiple-motor installation requires that all the motors have the same NEMA design letter. Open-loop or closed-loop vector drives cannot be used in multiple motor installations. See Figure 9-25. NEC® Article 430 contains information on sizing overcurrent devices and overload protection for multiple-motor installations.

Bypass Contactor. A *bypass contactor* is a contactor that allows line power to a motor that is normally controlled by an electric motor drive. A bypass contactor allows a critical load to operate during electric motor drive maintenance, or when a drive fails. An electric motor drive controlling an exhaust fan that removes fumes from a parking garage is a common application requiring a bypass contactor. Speed control is not possible when operating in bypass mode.

The bypass contactor works in conjunction with an input contactor, output contactor, overload relay, and other control components. See Figure 9-26. When in the electric motor drive mode, the bypass contactor is open, the input contactor and output contactor are closed, and the motor is powered by the drive. In the bypass mode, the bypass contactor is closed, the input and output contactors are open, and the motor is powered by the line voltage, not the electric motor drive.

The bypass contactor is mechanically interlocked to the output contactor and electrically interlocked to the input contactor. The interlocking ensures that both the output contactor and input contactor are de-energized (open) when the bypass contactor is closed. Isolating the electric motor drive prevents line power from reaching

MULTIPLE-MOTOR INSTALLATION

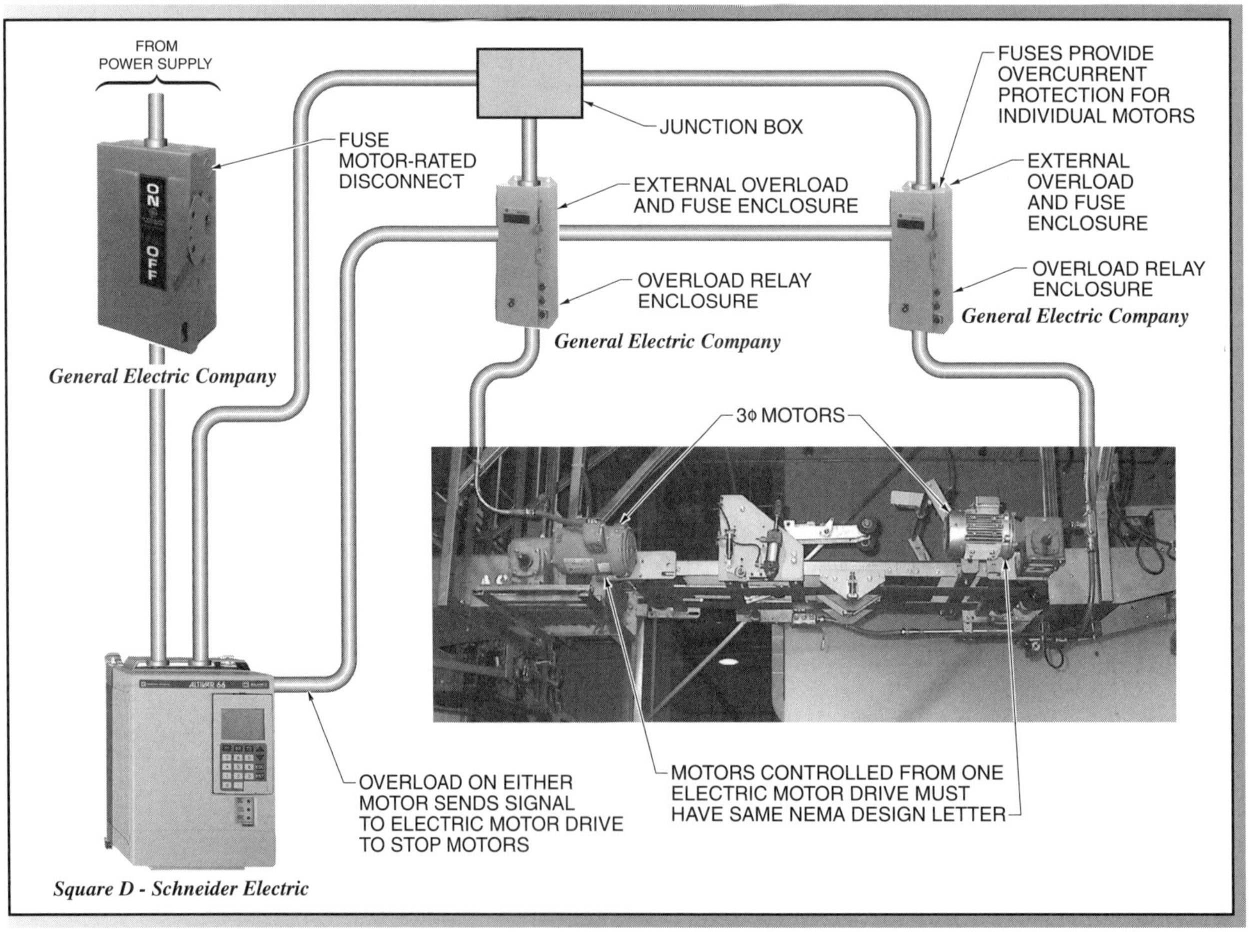

Figure 9-25. Multiple motors fed from a single electric motor drive require supplemental overcurrent and overload protection.

the drive output. The overload relay is provided to protect the motor when the electric motor drive with its integral overload protection is bypassed. The controls to initiate the bypass function may be manual or automatic. Other bypass control systems are found in industry for specific applications.

The input contactor and output contactor do not control stopping and starting of an electric motor drive. The starting and stopping of an electric motor drive is controlled by inputs to the drive. The output contactor must be closed prior to starting an electric motor drive, and the drive must be OFF before opening the output contactor.

BYPASS CONTACTOR CIRCUIT

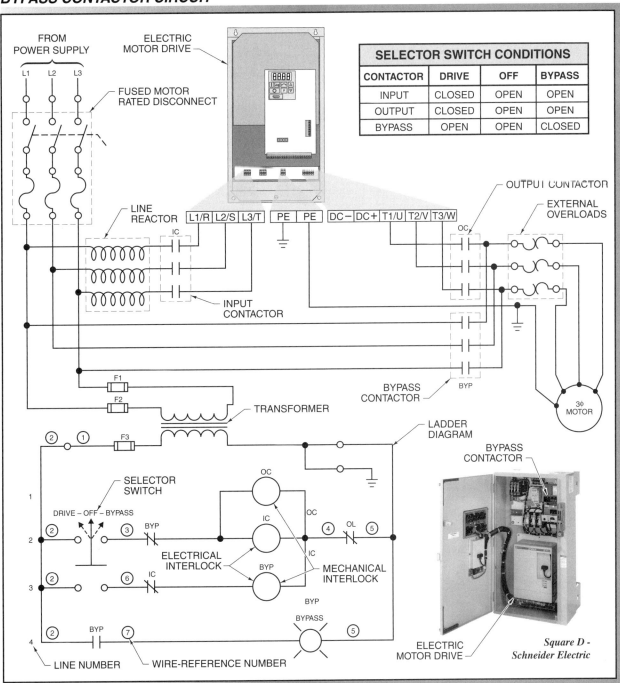

Figure 9-26. A bypass contactor provides line power to a motor if the electric motor drive fails.

Control Wiring

Control wiring consists of inputs and outputs connected to the electric motor drive, excluding the power wiring. The inputs and outputs are connected to the electric motor drive at the control terminal strips. The type of inputs and outputs determine the wiring method. Various wiring schemes are used, depending upon the specific application.

Control Terminal Strips. Digital inputs, analog inputs, digital outputs, and analog outputs are connected to the control terminal strips. The number and type of inputs and outputs varies with the complexity of the electric motor drive. The functionality of an input or output may be controlled by the programming parameters of an electric motor drive. The functionality of a relay output (digital output) may be programmed to indicate motor over-current, motor overload, or an electric motor drive fault.

Electric motor drive inputs control the starting, stopping, and speed of a motor. Digital inputs consist of control devices that provide an on/off signal such as a selector switch or pushbutton. Analog inputs consist of signals that vary over a range of values, such as a potentiometer with a 0 VDC–10 VDC signal. A combination of digital and analog inputs can control an electric motor drive by using a selector switch to start a motor and a 4 mA–20 mA signal to control the speed of the motor.

The outputs can control devices related to the electric motor drive or provide metering and annunciation functions. A digital output such as a relay can be used to start and stop an external electromechanical motor brake. An analog output can be used with an external meter that displays frequency or motor current. See Figure 9-27.

Wiring Methods. Digital inputs and outputs are connected to an electric motor drive with individual conductors. Analog inputs and outputs are connected to an electric motor drive with twisted shielded pair cables. See Figure 9-28. The following items must be considered when installing control wiring:

• Control wiring and power wiring should be installed in separate metal conduits.

• Wiring for digital signals, analog signals, and serial communication signals should be installed in separate metal conduits.

• Do not mix wiring for AC and DC control signals in the same conduit.

• Inside an electric motor drive enclosure, control and power wiring should be separated as much as possible. If control and power wiring must cross, they should cross at a 90° angle.

• Care must be taken when terminating the twisted shielded pair cables to ensure the shield is terminated only at the electric motor drive end, and the pairs should be twisted as close as possible to the terminals.

TERMINAL STRIP FUNCTIONALITY

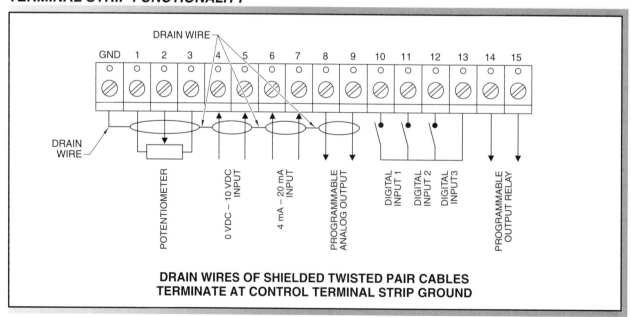

DRAIN WIRES OF SHIELDED TWISTED PAIR CABLES TERMINATE AT CONTROL TERMINAL STRIP GROUND

Figure 9-27. Programming the parameters of an electric motor drive allows the functions of the inputs and outputs to be modified.

CABLE DRAIN AND SHIELD CONNECTIONS

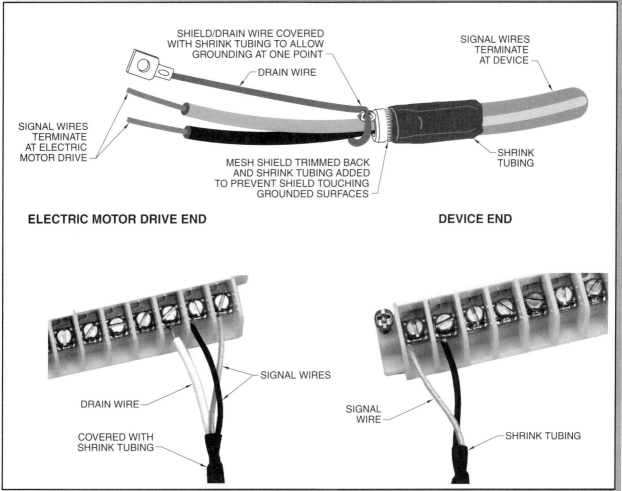

Figure 9-28. The termination of twisted shielded pair cable requires that shrink tubing be added correctly.

Control Wiring Schemes. Wiring schemes involving digital inputs fall into two categories, 2-wire control and 3-wire control. Two-wire control is the simplest. Two-wire control consists of two wires to an electric motor drive from a single digital input such as a selector switch. Opening and closing the digital input stops and starts the electric motor drive operation. Three-wire control consists of three wires from two digital inputs. One of the inputs is a normally open start and the other is a normally closed stop. The scheme is similar to the 3-wire control used with magnetic motor starters. See Figure 9-29.

TECH FACT

Electric motor drives on HVAC fans include a bypass contactor that is used to power the fan with utility power when drive failure occurs.

Wiring schemes involving analog signals fall into two categories, 2-wire control and 3-wire control. Certain analog signals only require a two-conductor twisted shielded pair cable with a 4 mA–20 mA or a 0 VDC–10 VDC signal. Other analog signals require a three-conductor twisted shielded pair cable as with a potentiometer.

Many electric motor drives have the capability to communicate with PLCs, other drives, building automation systems, HMIs, and PCs via serial communication. The serial communication may be RS-232 or RS-485. The serial communication may be integral to the electric motor drive or require an interface board. An electric motor drive is connected to other devices with twisted shielded pair cables. The advantages of serial communication include less wiring between control equipment, and the electric motor drive can be monitored and controlled from a central location.

CONTROL WIRING SCHEMES

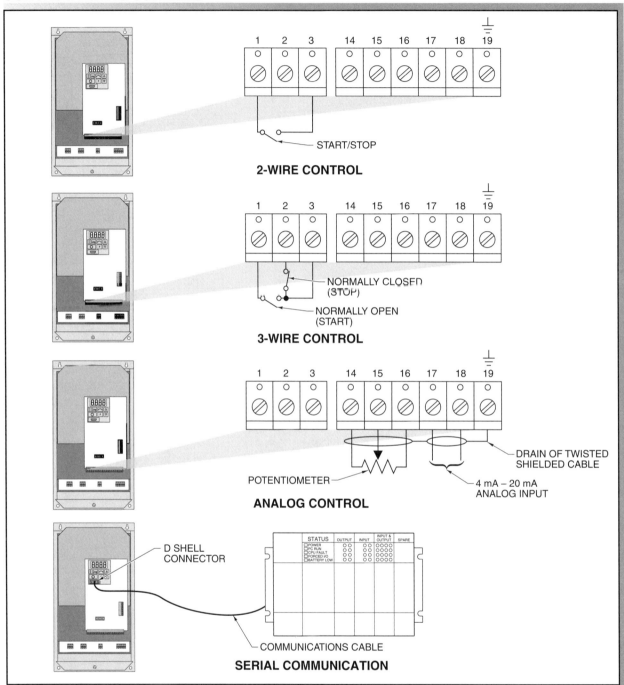

Figure 9-29. Electric motor drive control wiring schemes include 2-wire control, 3-wire control, analog control, and serial communication.

Name_____ Date _____

True-False

(T)   F   **1.** Service factor is a multiplier that represents the percentage of extra load that can be placed on a motor for short periods of time without damaging the motor.

T   F   **2.** Electric motor drives cannot introduce power quality problems into a power distribution system.

(T)   F   **3.** An input reactor is installed as close as possible to the input of an electric motor drive.

T   F   **4.** Most electric motor drives are delivered from a manufacturer in NEMA type 4X enclosures.

T   (F)   **5.** Control wiring should be installed in the same conduit as power wiring. *pg 260*

T   (F)   **6.** Grounding of an electric motor drive provides equal potential between all metal surfaces and a high-impedance path to activate overcurrent devices.

T   F   **7.** A multiple-motor installation requires that all the motors have the same NEMA design letter.

T   F   **8.** The final connection to the motor should be made with a short length of flexible conduit and stranded wire.

T   F   **9.** When terminating twisted shielded pair cables, the shield is terminated only at the device end.

T   (F)   **10.** If control and power wiring must cross, they should cross at a 45° angle. *pg 247* *90°*

(T)   F   **11.** A bearing current is the result of induced voltage in the motor rotor created by the electric motor drive. *pg 235*

T   (F)   **12.** Power wiring is all external wiring connected to an electric motor drive, excluding the line and load conductors. *pg 253*

T   F   **13.** A disconnect is a device that isolates an electric motor drive and/or motor from the voltage source to allow safe access for maintenance or repair.

(T)   F   **14.** A bypass contactor is a contactor that allows line power to a motor that is normally controlled by an electric motor drive. *pg 258*

(T)   F   **15.** Inductive loads such as solenoids or the coils of relays and contactors are a common source of electromagnetic interference (EMI) in electric motor drives.

Completion

_____ *lower, reduce* **1.** One method to diminish the harmful effects of long lead lengths between an electric motor drive and a motor is to ___ the carrier frequency.

_____ *ambient temperature* **2.** ___ is the temperature of the air surrounding a motor.

_____ *Emergency stop* **3.** A(n) ___ is used where accidental contact with moving equipment, or the unintended flow of product, can result in physical injury or property damage.

_____ *Metal conduit* **4.** ___ serves as a shield, preventing conductors from radiating EMI and protecting conductors from any radiated EMI.

_____ *Drive* **5.** In a majority of electric motor drive applications, the ___ provides the overload protection.

_____ *Plastic* **6.** ___ raceways are not recommended for electric motor drive installations because they do not provide shielding.

_____ *Speed control* **7.** ___ control is not possible when operating in bypass mode.

_____ *Metal* **8.** Wiring for digital signals, analog signals, and serial communication signals should be installed in separate ___ conduits.

_____ *Digital* **9.** ___ inputs consist of control devices that provide an on/off signal such as a selector switch.

_____ *Serial* **10.** Many electric motor drives have the capability to communicate with PLCs, other drives, and PCs via ___ communication.

Multiple Choice

_____ *D* **1.** NEMA standard MG-1, Section IV, ___ provides an enhanced insulation requirement for electric motors, allowing compatibility with electric motor drives.
  A. Part 10 C. Part 25
  B. Part 16 D. Part 31

_____ *B* **2.** Electric motors that are continuously operated at ___ speeds must be derated or provided with auxiliary cooling.
  A. high C. maximum
  B. low D. nameplate

_____ *C* **3.** A(n) ___ load is any load where the instantaneous load current is not proportional to the instantaneous voltage.
  A. inductive C. nonlinear
  B. linear D. resistive

_____ *A* **4.** ___ speed is the speed at which a motor develops rated horsepower at rated load and voltage.
  A. Base C. Constant
  B. Braking D. Maximum

_____ *C* **5.** NEC® Article ___ is the main code source for electric motor drives.
  A. 230 C. 430
  B. 240 D. 440

Name_____ Date _____

Activity 9-1. Required Enclosure Types

Many applications allow one of several different NEMA enclosures to be used for an electric motor drive. Better (sealing method and design) enclosures can be used for more demanding applications. A NEMA 12 indoor enclosure can be placed anywhere a NEMA 1 indoor enclosure can be placed. The difference is that the NEMA 12 enclosure costs more than the NEMA 1 enclosure. Selecting the correct enclosure for an application involves considering the price and service conditions of the enclosure.

Answer the questions using enclosure chart in appendix.

Sample Enclosure Cost

1. NEMA 1 (least costly)
2. NEMA 12
3. NEMA 4 and 4X
4. NEMA 3 and 3R
5. NEMA 7 and 9 (most costly)

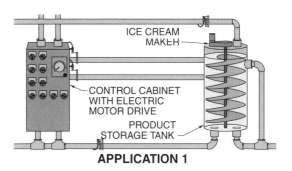

APPLICATION 1

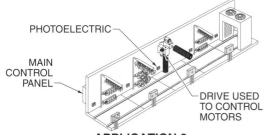

APPLICATION 2

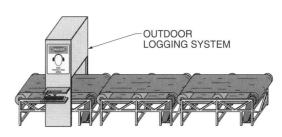

APPLICATION 3

_____*NO*_____   **1.** In application 1, an electric motor drive is used in a food production area that makes ice cream. Can a NEMA 1 enclosure be used?

_____*NO*_____   **2.** In application 1, an electric motor drive is used in a food production area that makes ice cream. Which is the best choice, a NEMA 4 or a NEMA 7?

_____*NO*_____   **3.** In application 2, an electric motor drive is used in a car wash application. Can a NEMA 3 enclosure be used?

_____*NEMA 4X*_____   **4.** In application 2, an electric motor drive is used in a car wash application. Would a NEMA 4 or a NEMA 4X enclosure work better for this application?

_____*NO*_____   **5.** In application 3, an electric motor drive is used in an outdoor logging area to move logs to a chipping area. Can the more expensive NEMA 7 enclosure be used over a NEMA 4 enclosure?

Activity 9-2. Electric Motor Drive Power Source

The best power source for most electric motor drive installations is the grounded 4-wire wye system. A transformer bank can be used to change (increase or decrease) voltage level and change a delta system into a wye system.

1. Connect the power line terminals to the transformer primary terminals in a delta configuration.

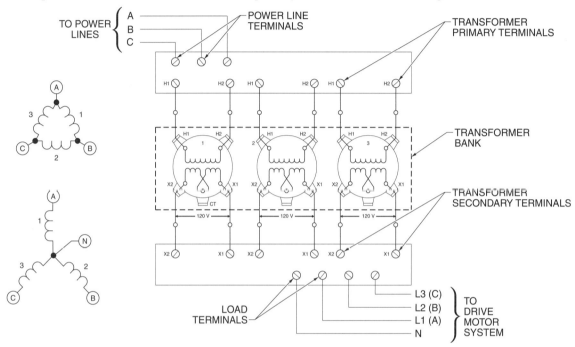

2. Connect the secondary side terminals of the transformer in a wye configuration.

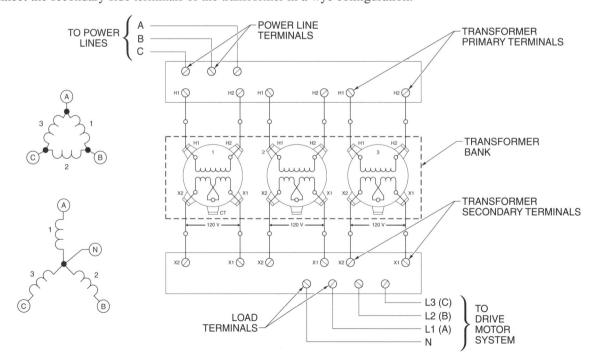

Activity 9-3. Connecting Control Input Switches

Numerous different control devices control electric motors, including pushbuttons, pressure switches, and temperature switches. Understanding electrical symbols is important when designing, installing, and trouble-shooting electric motor drive control circuits.

Draw the correct symbols using industrial electrical symbols in the appendix.

1. Draw the symbols for a variable resistor and a thermocouple and connect them to the correct terminals.

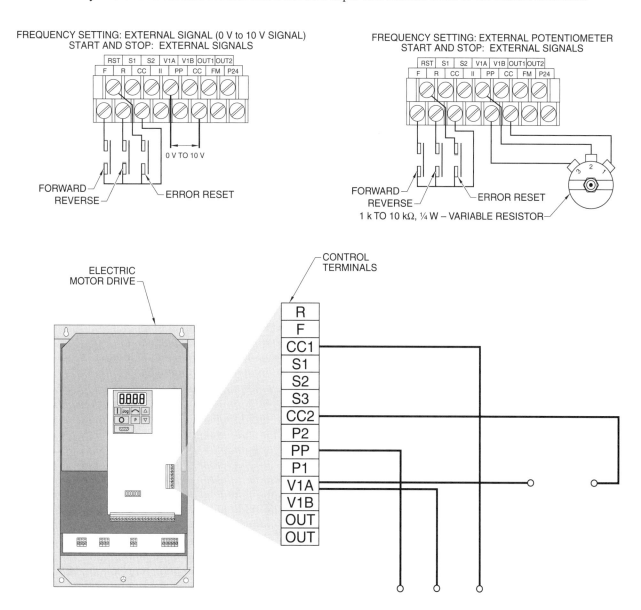

Activity 9-4. Connecting Analog Inputs

Numerous different control devices control electric motors, including analog inputs. Understanding correct electrical symbols is important when designing, installing, and troubleshooting electric motor drive control circuits.

Draw the correct symbols using industrial electrical symbols in the appendix.

1. N.O. foot switch

2. N.O. pressure switch

3. N.O. temperature switch

4. N.C. pressure switch

5. N.O. liquid level switch

6. N.C. temperature switch

10 Electric Motor Drive Programming

Electric motor drive programming incorporates a number of programming devices, menu formats, and parameters that are common among drive models and drive manufacturers. An electric motor drive application may require that a few basic parameters be programmed or that advanced parameters be programmed for complex applications. The software used is the only limit to the functionality of an electric motor drive for an application.

PROGRAMMING CONSIDERATIONS

A *parameter* is a property of an electric motor drive that is programmed or adjusted. Electric motor drive performance, motor specifications, the information that is displayed, and the functions of the drive inputs and outputs are some of the parameters that are programmed. A small, basic electric motor drive can contain 50 parameters, while large, sophisticated drives contain over 200. Regardless of the manufacturer, electric motor drives share many common parameters. The names of parameters may vary, but the functions performed are the same; for example, acceleration time and ramp-up time are different parameter names for the same function. Numbers assigned to a parameter and location in the menu structure also vary based on the manufacturer.

Electric motor drives are shipped with factory settings, or defaults, for most parameters. Default parameters are the most conservative and most frequently used parameter values and create the least amount of risk to equipment and personnel. The default values are identified in the instruction manual of an electric motor drive and function properly for most drive applications. Parameters for motor nameplate data are not factory set and must be set in the field by a technician.

When programming an electric motor drive, technicians must only adjust parameters that are related to a particular drive application. Changing a value of one parameter can effect another parameter. Certain parameters

can be changed "on the fly" (during electric motor drive and motor operation) and are identified in the electric motor drive instruction manual. It is uncommon that all parameters need to be adjusted from the default settings. A written record or printout of the parameter settings must be stored in a safe location for future reference.

Parameter Menu Format

Parameter menu formatting varies based on the manufacturer. Some manufacturers list parameters in numerical order and other manufacturers arrange parameters by file and group based on the functions performed. Some electric motor drives assign a number to a parameter, in addition to a file and group designation. See Figure 10-1.

Advanced Assembly Automation Inc.
An electric motor drive application may be too complex for factory default settings to be used.

PARAMETER MENU FORMATTING

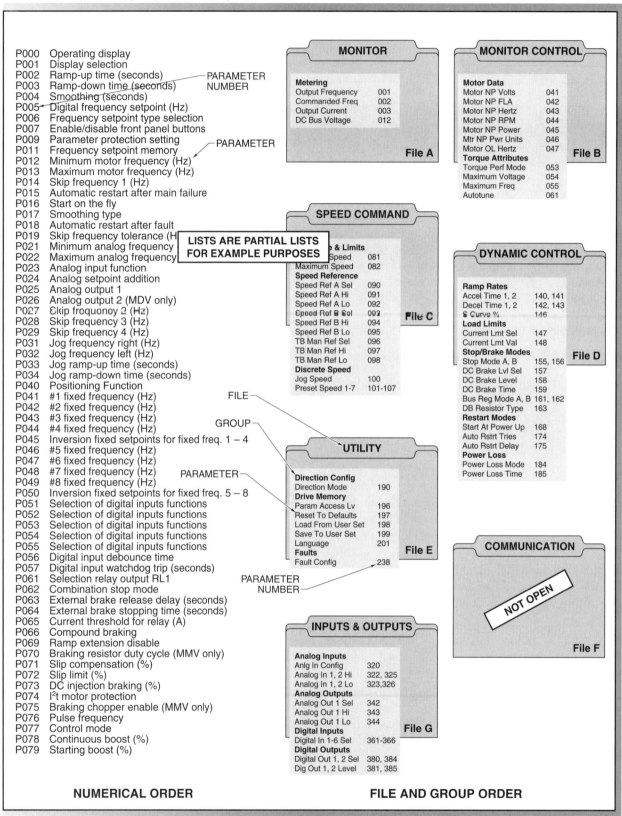

P000 Operating display
P001 Display selection
P002 Ramp-up time (seconds) —— PARAMETER NUMBER
P003 Ramp-down time (seconds)
P004 Smoothing (seconds)
P005 Digital frequency setpoint (Hz)
P006 Frequency setpoint type selection
P007 Enable/disable front panel buttons
P009 Parameter protection setting —— PARAMETER
P011 Frequency setpoint memory
P012 Minimum motor frequency (Hz)
P013 Maximum motor frequency (Hz)
P014 Skip frequency 1 (Hz)
P015 Automatic restart after main failure
P016 Start on the fly
P017 Smoothing type
P018 Automatic restart after fault
P019 Skip frequency tolerance (H
P021 Minimum analog frequency
P022 Maximum analog frequency
P023 Analog input function
P024 Analog setpoint addition
P025 Analog output 1
P026 Analog output 2 (MDV only)
P027 Skip frequency 2 (Hz)
P028 Skip frequency 3 (Hz)
P029 Skip frequency 4 (Hz)
P031 Jog frequency right (Hz)
P032 Jog frequency left (Hz)
P033 Jog ramp-up time (seconds)
P034 Jog ramp-down time (seconds)
P040 Positioning Function
P041 #1 fixed frequency (Hz)
P042 #2 fixed frequency (Hz)
P043 #3 fixed frequency (Hz)
P044 #4 fixed frequency (Hz)
P045 Inversion fixed setpoints for fixed freq. 1 – 4
P046 #5 fixed frequency (Hz)
P047 #6 fixed frequency (Hz)
P048 #7 fixed frequency (Hz)
P049 #8 fixed frequency (Hz)
P050 Inversion fixed setpoints for fixed freq. 5 – 8
P051 Selection of digital inputs functions
P052 Selection of digital inputs functions
P053 Selection of digital inputs functions
P054 Selection of digital inputs functions
P055 Selection of digital inputs functions
P056 Digital input debounce time
P057 Digital input watchdog trip (seconds)
P061 Selection relay output RL1
P062 Combination stop mode
P063 External brake release delay (seconds)
P064 External brake stopping time (seconds)
P065 Current threshold for relay (A)
P066 Compound braking
P069 Ramp extension disable
P070 Braking resistor duty cycle (MMV only)
P071 Slip compensation (%)
P072 Slip limit (%)
P073 DC injection braking (%)
P074 I²t motor protection
P075 Braking chopper enable (MMV only)
P076 Pulse frequency
P077 Control mode
P078 Continuous boost (%)
P079 Starting boost (%)

NUMERICAL ORDER

LISTS ARE PARTIAL LISTS FOR EXAMPLE PURPOSES

MONITOR

Metering
Output Frequency	001
Commanded Freq	002
Output Current	003
DC Bus Voltage	012

File A

MONITOR CONTROL

Motor Data
Motor NP Volts	041
Motor NP FLA	042
Motor NP Hertz	043
Motor NP RPM	044
Motor NP Power	045
Mtr NP Pwr Units	046
Motor OL Hertz	047

Torque Attributes
Torque Perf Mode	053
Maximum Voltage	054
Maximum Freq	055
Autotune	061

File B

SPEED COMMAND

& Limits
Speed	081
Maximum Speed	082

Speed Reference
Speed Ref A Sel	090
Speed Ref A Hi	091
Speed Ref A Lo	092
Speed Ref B Sel	093
Speed Ref B Hi	094
Speed Ref B Lo	095
TB Man Ref Sel	096
TB Man Ref Hi	097
TB Man Ref Lo	098

Discrete Speed
Jog Speed	100
Preset Speed 1-7	101-107

File C

DYNAMIC CONTROL

Ramp Rates
Accel Time 1, 2	140, 141
Decel Time 1, 2	142, 143
S Curve %	146

Load Limits
Current Lmt Sel	147
Current Lmt Val	148

Stop/Brake Modes
Stop Mode A, B	155, 156
DC Brake Lvl Sel	157
DC Brake Level	158
DC Brake Time	159
Bus Reg Mode A, B	161, 162
DB Resistor Type	163

Restart Modes
Start At Power Up	168
Auto Rstrt Tries	174
Auto Rstrt Delay	175

Power Loss
Power Loss Mode	184
Power Loss Time	185

File D

FILE
GROUP
PARAMETER

UTILITY

Direction Config
Direction Mode	190

Drive Memory
Param Access Lv	196
Reset To Defaults	197
Load From User Set	198
Save To User Set	199
Language	201

Faults
Fault Config	238

File E

PARAMETER NUMBER

COMMUNICATION

NOT OPEN

File F

INPUTS & OUTPUTS

Analog Inputs
Anlg In Config	320
Analog In 1, 2 Hi	322, 325
Analog In 1, 2 Lo	323, 326

Analog Outputs
Analog Out 1 Sel	342
Analog Out 1 Hi	343
Analog Out 1 Lo	344

Digital Inputs
Digital In 1-6 Sel	361-366

Digital Outputs
Digital Out 1, 2 Sel	380, 384
Dig Out 1, 2 Level	381, 385

File G

FILE AND GROUP ORDER

Figure 10-1. The parameter menu lists the parameters in numerical order or by files and groups based on parameter function.

Electric motor drive manufacturers produce drive models ranging from small, basic models to large, complex models. Manufacturers tend to use the same parameter menu formatting across various models produced and add parameters for more complex models. Technicians do not need to learn entirely new types of parameter formatting when working with different models from the same manufacturer.

Programs used by electric motor drives contain macros. A *macro* is a parameter that contains predefined values for a group of parameters. Macros are specific to a certain type of load or application, such as pumps, fans, or conveyors. Electric motor drive manufacturers set values contained in the macros for optimum performance of specific applications. For example, macros for a conveyor application contain acceleration and deceleration times specifically suited for a certain design of conveyor. Macros simplify programming and enhance electric motor drive performance.

Programming Devices

A number of different programming devices are available for electric motor drives. Programming devices include integral keypads, keypads with clear text displays, PLCs, and PCs with electric motor drive programming software. Programming devices allow a technician to view, monitor, and edit parameters, and control an electric motor drive with start, stop, reverse, up, and down buttons. The devices differ in the amount of information displayed, ease of use, cost, and whether or not remote programming is possible. An electric motor drive chosen for an application determines the appropriate type of programming device to be used. A single electric motor drive controlling an HVAC fan is easily programmed with a keypad. A multidrive conveyor line covering a large area is best programmed with a PC that connects to all electric motor drives via a network. The programming device used determines how a technician views and adjusts parameters.

Keypads. A keypad with an LED display is the standard programming device for most electric motor drives. See Figure 10-2. An integral keypad with LED display is used to program and operate an electric motor drive locally. LED displays show parameter and performance information as numbers. An electric motor drive manual is necessary to know what parameter a number represents, and what numerical selections or options for a parameter are available. The number, type, and function of the buttons on a keypad vary. Several buttons serve dual functions to reduce the size of the keypad. Buttons that are common to most keypads are the following:

- **I** Start or RUN button
- **O** Stop or OFF button
- ⌒ Controls the direction of rotation
- **△** Up button (dual function: increase speed or increase a numerical value)
- **▽** Down button (dual function: decrease speed or decrease a numerical value)

KEYPAD CONTROLS

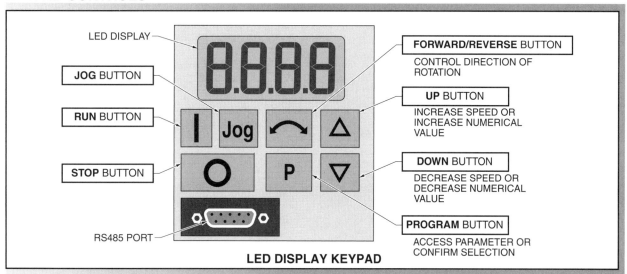

Figure 10-2. A keypad with an LED display is the standard programming device for electric motor drives.

A keypad with LED display is best suited for simple electric motor drive applications that do not require extensive programming and that cannot be remotely programmed.

Clear Text Displays. A clear text display provides more functional versatility than an LED display. Clear text displays have a keypad and an LCD screen that displays lines of text and numbers. See Figure 10-3. Clear text displays can be integral with an electric motor drive, or stand-alone devices that attach to a drive. Stand-alone models may be operated remotely and/or networked to more than one electric motor drive.

CLEAR TEXT KEYPAD CONTROL

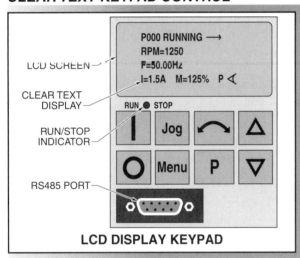

LCD DISPLAY KEYPAD

Figure 10-3. Clear text displays add a MENU button to the keypad and provide more information by providing lines of text.

Clear text displays simplify electric motor drive programming by allowing technicians to view the parameter information without constantly referring to an electric motor drive manual. The parameter, parameter number, and parameter settings are shown on clear text displays in alphanumeric formatting. In addition to programming, clear text displays provide the following:

• operational information (output voltage, speed)
• diagnostic information (fault codes)
• configuration options (display contrast, display language)

Clear text displays provide some of the functionality of a personal computer (PC) without the complications. The ability to store and transfer programs permits easy

program updating and storing of multiple sets of drive parameters in nonvolatile memory. Clear text displays are used to transfer sets of drive parameters between electric motor drives. Clear text displays are well suited to applications that have multiple electric motor drives of the same model, but lack a network.

Personal Computers and Software. Software programs developed by electric motor drive manufacturers allow a PC to communicate with a drive, or with multiple drives provided the drives are connected to a network. A special interface cable is used between the serial port of a PC and the communication port (serial port) on an electric motor drive to allow communication. Software allows a technician to program, monitor, and control an electric motor drive from a remote location with a PC. The use of a PC to control an electric motor drive has all the features of a clear text display with the advantage of a large screen and a user-friendly graphical interface. See Figure 10-4.

PROGRAMMING WITH A PERSONAL COMPUTER

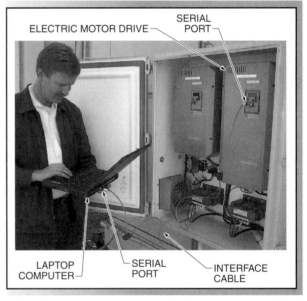

Figure 10-4. The serial port of a PC is used to communicate with an electric motor drive.

When connecting a PC to an electric motor drive, the following steps should be followed:

1. Connect the interface cable to the electric motor drive and to the serial port of the PC, also known as COM 1. Make sure the PC is OFF when connecting the cable to avoid damage to the serial port.

2. Start the PC and make sure the serial port is not dedicated to another device. The serial port must be free for use with the electric motor drive.

3. Load the electric motor drive software and follow the configuration instructions. The software helps a technician navigate through the setup process.

4. Assign an address to the electric motor drive and enable the drive for serial communication with a keypad. When serial communication is enabled, it may not be possible to control the electric motor drive from a keypad or from other inputs.

5. Establish communication with the electric motor drive so programming, controlling, and monitoring of the electric motor drive can be completed. Electric motor drive configuration information must be saved for future reference.

Programmable Logic Controllers. A *programmable logic controller (PLC)* is a solid state control device that is programmed to automatically control an industrial process or machine. As the use of automation in industrial applications increases, the need for communication between electric motor drives and PLCs increases. Many electric motor drive manufacturers also manufacture PLCs and other related industrial automation equipment. PLC programming software has been developed to allow PLCs and electric motor drives of the same manufacturer to communicate over a network. PLC software contains special instructions specifically designed to control an electric motor drive. Special instructions read and write parameters for electric motor drives. An application requiring complex coordination between inputs and outputs of electric motor drives is well suited to a PLC. See Figure 10-5.

PLC CONTROL OF ELECTRIC MOTOR DRIVES

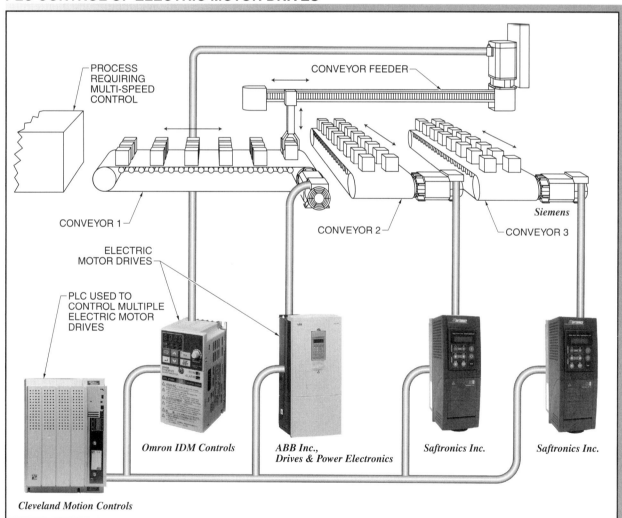

Figure 10-5. PLCs are used to control and synchronize the speed and movement of equipment on a production line.

Parameter Programming

When programming, a technician should always have the electric motor drive instruction manual available. See Figure 10-6. The instruction manual typically provides the following information:

- parameter name and description
- parameter number or designation
- parameter range of values
- parameter units of measure
- parameter default settings
- parameter warnings
- parameter interactions
- "on the fly" adjustability

TECH FACT

Parameters must not be changed while an electric motor drive is running unless the consequences of the change are fully understood.

INSTRUCTION MANUAL SYSTEM PARAMETERS

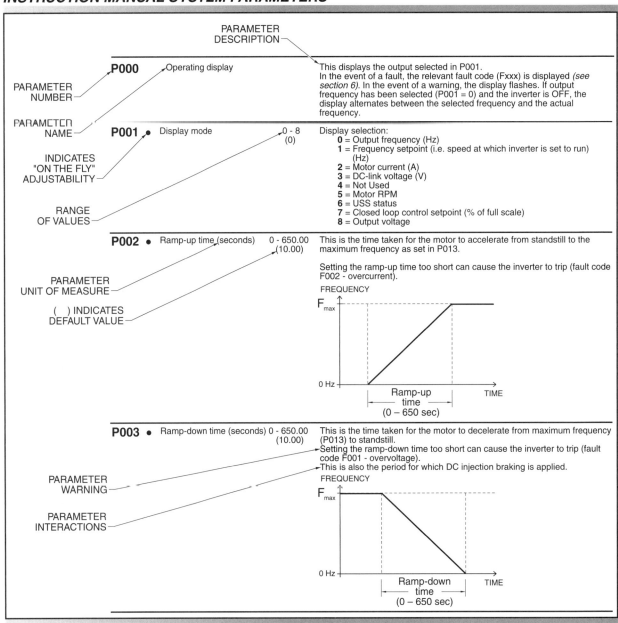

Figure 10-6. An electric motor drive instruction manual provides detailed information on programming parameters.

Parameter values that are adjustable are classified into the following three categories:

- A range of numbers that represent selections, such as numbers representing what can be shown on an electric motor drive LED display.
- A number that determines the state of a bit. If the bit is 0, the option is OFF or the condition is false. If the bit is 1, the option is ON or the condition is true.
- A number that represents a value. See Figure 10-7.

TECH FACT

Typically, the number of parameters increases as the size (horsepower) of an electric motor drive increases. Parameters for motor nameplate data must be entered for each electric motor drive application.

BASIC PARAMETERS

Basic parameters are a group of parameters that are adjusted in most electric motor drive installations. It is not always necessary to adjust all of the basic parameters. The number, type, and formatting of basic parameters vary between manufacturers, but most manufacturers group basic parameters together for ease of programming. Basic parameters include display modes, speed references, input modes, frequency setpoints, stop modes, control modes, acceleration time, deceleration time, minimum motor frequency, maximum motor frequency, and motor nameplate data.

Display Mode

The display mode allows a technician to adjust electric motor drive-related information that is shown on the LED or LCD display. The display mode is adjusted to provide useful information to a technician for startup and troubleshooting of electric motor drives. The following are common properties shown in display mode:

- electric motor drive output frequency (in Hz)
- electric motor drive frequency setpoint (in Hz)
- electric motor drive output voltage (in V)
- electric motor drive DC bus voltage (in V)
- motor current (in A)
- motor speed (in rpm)
- communication status of an electric motor drive networked with other devices

---

ADJUSTABLE PARAMETER VALUES

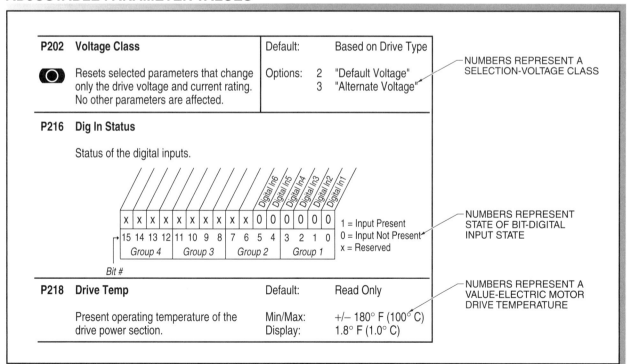

Figure 10-7. Parameter numerical values can represent a selection, the state of a bit, or a value.

Speed Reference

Speed reference (frequency source) is a signal that informs an electric motor drive of the speed at which to operate. The speed reference parameter determines the source of the speed reference signal for an electric motor drive whether the speed reference is internal or external. Internal speed references for an electric motor drive are obtained from other parameters, such as a frequency setpoint. External speed references are provided by an external analog signal, such as a potentiometer providing a 0 VDC–5 VDC signal, 0 VDC–10 VDC signal, or 4 mA–20 mA signal to an electric motor drive.

Input Mode

An *input mode (operating mode)* is a display mode that determines how an electric motor drive is controlled during starting and stopping. The operating mode parameter must be adjusted for each application. Common input modes to an electric motor drive include keypads, two-wire controls, three-wire controls, analog signals, potentiometers, and serial communication.

- A *keypad* is an input control device for electric motor drives using LED or LCD displays.
- *Two-wire control* is an input control for an electric motor drive requiring two conductors to complete a circuit.
- *Three-wire control* is an input control for an electric motor drive requiring three conductors to complete a circuit. The three-wire control used is similar to the three-wire control used with magnetic motor starters.

In 1939, a 3300 HP synchronous frequency changer converted 13,200 V, 60 Hz power to power with a voltage between 6600 V and 13,200 V, and frequency between 25 Hz and 60 Hz to control motor speed and torque.

- An analog signal is a type of input signal to an electric motor drive that can be either varying voltage or varying current.
- A *potentiometer* is an input control device that sends various resistance values to an electric motor drive.
- *Serial communication* is a communications port that uses a D-shell connector to connect an electric motor drive with other drives, PLCs, or PCs.

Not all electric motor drives have speed reference and input mode as adjustable parameters. Certain electric motor drives use one parameter to determine the speed reference and input mode parameters. Some electric motor drives allow more than one input mode to be used together, such as a keypad and analog signal to control an electric motor drive simultaneously.

Frequency Setpoints

A *frequency setpoint* is a parameter that sets the frequency at which an electric motor drive operates when speed reference is internal. The frequency setpoint parameter is measured in hertz (Hz) to ensure safety. The default setting is usually a low frequency, such as 5 Hz or 10 Hz.

Stop Modes

A *stop mode* is a parameter that determines how an electric motor drive stops when it receives a stop command. Two common stop modes used in electric motor drives include ramp stop and coast stop. See Figure 10-8.

Ramp Stop. The *ramp stop* method of stopping a motor is a stopping method in which the frequency applied to a motor is reduced, which decelerates the motor to a stop. The ramp stop method brings a motor to a smooth, controlled stop and is used when it is necessary to stop a load in a predetermined time or location. The length of time an electric motor drive takes to stop a motor is controlled by the deceleration time parameter. Ramp stop is usually the default setting of an electric motor drive.

Coast Stop. The *coast stop* method of stopping a motor is a stopping method in which the electric motor drive shuts OFF the voltage to the motor, allowing the motor to coast to a stop. When using the coast stop method, the electric motor drive does not have any control of the motor after a stop command is entered. The length of time a motor takes to stop depends on the motor type and type of load connected to the motor. The coast stop mode has no time control for stopping or specific location control.

STOP MODES

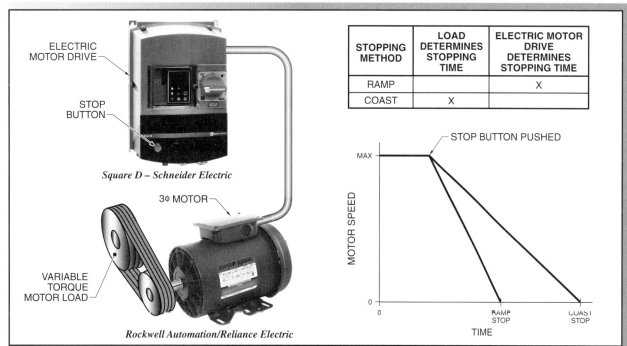

STOPPING METHOD	LOAD DETERMINES STOPPING TIME	ELECTRIC MOTOR DRIVE DETERMINES STOPPING TIME
RAMP		X
COAST	X	

Figure 10-8. Typical stop modes include ramp stop and coast stop.

Control Modes

A *control mode (volts-per-hertz pattern)* is a motor control parameter that determines the relationship between the voltage and frequency an electric motor drive outputs to a motor. The control mode parameter allows a single electric motor drive to be used with various loads and to determine the voltage and frequency delivered to a motor. There are multiple control modes that can be used by an electric motor drive including volts-per-hertz, quadratic volts-per-hertz, open-loop vector control, and closed-loop vector control.

Volts-Per-Hertz. *Volts-per-hertz* is a control mode that provides a linear voltage ratio to the frequency of a motor from 0 rpm to base speed. Volts-per-hertz mode is used for constant torque loads, such as conveyors. See Figure 10-9.

Quadratic Volts-Per-Hertz. *Quadratic volts-per-hertz* is a control mode that provides a nonlinear voltage to frequency ratio. The quadratic volts-per-hertz curve matches the torque requirements of variable torque loads.

Open-Loop Vector Control. *Open-loop vector control (sensorless control)* is a control mode that uses complex mathematical formulas to control the flux-producing and torque-producing currents to an AC motor. Open-loop vector control allows an AC motor to have torque

characteristics that are almost identical to those of a DC motor. The formulas calculate motor speed because there are no sensors (tachometer or encoder) to provide feedback on motor speed to the electric motor drive itself. Open-loop vector control is not a practical control at low speeds since the formulas cannot accurately calculate speed below a frequency of 5 Hz.

VOLTS-PER-HERTZ CONTROL MODES

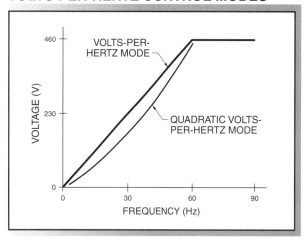

Figure 10-9. Constant torque is provided by the volts-per-hertz mode. Variable torque is provided by the quadratic volts-per-hertz mode.

Closed-Loop Vector Control. *Closed-loop vector control (vector control)* is a control mode that allows an AC motor to have torque characteristics identical to a DC motor. Closed-loop vector control is similar to open-loop vector control, but a tachometer or encoder provides motor speed information to the electric motor drive. Closed-loop vector control can provide 100% torque to a motor at 0 Hz speed.

Acceleration Time

Acceleration time (ramp-up time) is a motor control parameter that determines the length of time an electric motor drive takes to accelerate a motor from a standstill (0 rpm) to maximum motor speed (max rpm). Acceleration time is measured in seconds, with the common default setting at 10 sec. When acceleration time is too short, an electric motor drive draws excessive current, which blows fuses or causes overcurrent faults. Increasing the acceleration time reduces the likelihood of blowing fuses or causing overcurrent faults and decreases the mechanical stress on the motor and driven load. See Figure 10-10.

Deceleration Time

Deceleration time (ramp-down time) is a motor control parameter that determines how long the length of time an electric motor drive takes to decelerate a motor from maximum motor frequency speed to a standstill. Deceleration time is measured in seconds with 10 sec as the common default setting. When deceleration time is too short, the voltage regenerated from the motor is excessive, causing the electric motor drive to have a DC bus overvoltage fault. Increasing the deceleration time reduces the likelihood of an overvoltage fault and decreases the mechanical stress on the motor and driven load.

GE Motors & Industrial Systems
Pumps controlled by electric motor drives use the ramp-up time function to prevent overstressing the motor and to avoid instantaneous pressure surges in the piping system.

ACCELERATION AND DECELERATION TIME

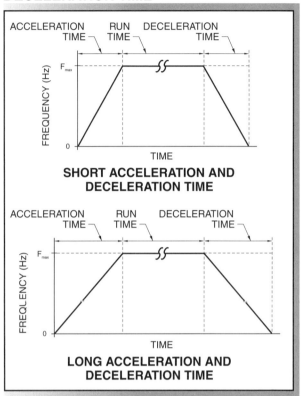

SHORT ACCELERATION AND DECELERATION TIME

LONG ACCELERATION AND DECELERATION TIME

Figure 10-10. Acceleration time and deceleration time change the slope of the acceleration and deceleration angles. Longer times are less stressful to the motor and the driven load.

Minimum Motor Frequency

Minimum motor frequency sets the lowest frequency an electric motor drive outputs to a motor to control speed. Minimum motor frequency is measured in hertz, with the default setting being 0 Hz.

Maximum Motor Frequency

Maximum motor frequency sets the highest frequency an electric motor drive outputs to a motor. Maximum motor frequency is measured in hertz, and 60 Hz is the common default setting. Electric motor drives are capable of outputting frequencies above 60 Hz, the base speed for 3ϕ AC motors. Output frequencies of 400 Hz are created by some electric motor drives.

TECH FACT

Licensed engineers must be consulted before operating a motor above base speed.

Motor Nameplate Data

Motor nameplate data is data that consists of parameters related to motor specifications. See Figure 9-11. An electric motor drive uses nameplate data to calculate drive performance (volts/hertz pattern) and fault protection (overload protection). Motor nameplate data must be entered for each application because motors from different manufacturers that have the same horsepower rating may have different nameplate values and the horsepower rating of a drive may not match the rating of the motor. For example, a 5 HP drive may be used to power a 2.5 HP motor. *Note:* The nameplate current of a motor cannot exceed the output current rating of the drive. Nameplate data and other information that must be programmed into an electric motor drive include the following:

- Motor nameplate frequency is the frequency required by a motor to achieve base speed. The default value is 60 Hz.
- Motor nameplate speed is the maximum speed, in rpm, at which a motor should be rotated.
- Motor nameplate current is the maximum current, in amps, that a motor must use. Full load amps (FLA) and full load current (FLC) are the same as motor nameplate current.
- Motor nameplate voltage is the voltage, in volts, required by a motor to achieve maximum torque.
- Motor nameplate power rating is the power rating listed on a motor nameplate in either horsepower

(HP) or kilowatts (kW). A conversion from horsepower to kilowatts may be necessary if an electric motor drive and motor use different units of measure (1 HP = 746 W).

- Motor magnetizing current is the current a motor draws, in amps, with no load when running at nameplate voltage and frequency. When this value is not specifically listed on the motor nameplate, run the motor at its nameplate voltage and frequency with no load connected. Measure the current with a true-rms clamp-on ammeter and enter this current value as the motor magnetizing current.
- Motor stator resistance is the resistance, in ohms, of the stator between any two phases. The measurement is taken at the electric motor drive with the motor connected to the circuit and power OFF.

ADVANCED PARAMETERS

Advanced parameters are used in more complex electric motor drive applications. An advanced parameter adds extra functions to an electric motor drive for greater motor control. Extra functions provide maintenance information, adjust the functions of input and output terminals, and provide enhanced acceleration and deceleration methods. As the size and cost of electric motor drives increase, the number and type of advanced parameters also increase. In many electric motor drive applications, advanced parameters are not

MOTOR NAMEPLATE DATA

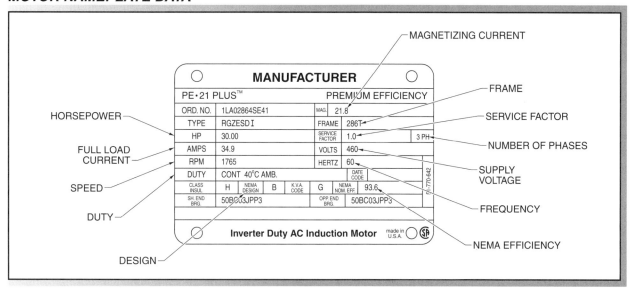

Figure 10-11. Motor nameplate data must be entered for each electric motor drive installation because the information is specific to that motor.

adjusted from the original default settings. Advanced parameters include electric motor drive information, parameter protection and keypad controls, functionality of inputs and outputs, enhanced acceleration and deceleration methods, and enhanced performance features.

ABB Inc., Drives & Power Electronics
Remote clear text display keypads are used to remotely control specific electric motor drives.

Electric Motor Drive Information Parameters

An *electric motor drive information parameter* is a parameter that provides information used for periodic maintenance and troubleshooting. A variety of information can be accessed by various electric motor drive parameters. Some information is permanently stored and cannot be changed, such as the software version in use. Other information is stored in memory and updated as new events occur, such as fault history and maintenance information.

Software Version. Electric motor drive software version is an advanced parameter used on many drives. Typically the software version is stored as a number, such as 2.02. When contacting an electric motor drive manufacturer for technical assistance, technicians are required to know the software version, drive model number, and the drive serial number. Two electric motor drives of the same make and model often have different software versions, depending on their date of manufacture. Electric motor drive manufacturers make revisions to software over the life of a particular drive model.

Fault History. A fault history advanced parameter stores the most recent electric motor drive faults. Fault history is a common advanced parameter used in many electric motor drives, with the number of faults stored varying from drive to drive. A fault code or a brief description of the fault is shown, depending on the type of display

(LED or LCD). See Figure 10-12. The fault history provides useful information for troubleshooting purposes and is especially helpful when a fault occurs and a technician is not present to document the fault.

Maintenance Information. Maintenance information is contained in one or more files in electric motor drives. Maintenance information files include the following:
• Number of hours an electric motor drive has been powered up
• Number of hours the internal cooling fans have run
• Value of the DC bus capacitors

The internal cooling fans and DC bus capacitors are electric motor drive components that require periodic replacement. Serious damage occurs to an electric motor drive when a cooling fan or DC bus capacitor fails. Cooling fan and capacitor malfunctions are avoided when maintenance parameters and replacement interval information contained in an electric motor drive manual are properly used.

TECH FACT

Most clear text displays can be attached or removed from electric motor drives that are energized but OFF.

Parameter Protection and Keypad Controls

Parameter protection and keypad controls are a group of parameters that control the functionality of a keypad and limit access to parameters along with resetting the drive to factory default settings. Often an electric motor drive is located where it is accessible to anyone. Personnel unfamiliar with an electric motor drive should not attempt to operate or program a drive. Unauthorized personnel attempting to operate or program an electric motor drive may cause malfunctions, damage to the drive, and/or personal injury.

Parameter Protection. *Parameter protection* is a parameter that limits access to electric motor drive parameters. Parameter protection has many levels of protection and often requires a password to change a parameter. The password is programmed at startup and should be stored in a secure location. The default setting for parameter protection is typically the most restrictive setting.

There is a hierarchy of access of parameters from most to least restrictive access levels. Read-only permits viewing a parameter, but does not allow altering of the parameter. Limited adjustment allows viewing of all parameters but only permits changing of a limited number of parameters. Unlimited access allows viewing and changing of all parameters.

FAULT HISTORY CODE NUMBERS

F001	Overvoltage	Check whether supply voltage is within the limits indicated on the rating plate. Increase the ramp down time (P003). Check whether the required braking power is within the specified limits.
F002	Overcurrent	Check whether the motor power corresponds to the inverter power. Check that the cable length limits have not been exceeded. Check motor lead and motor for short circuits and ground faults. Check whether the motor parameters (P081 – P086) correspond with the motor being set. Check the stator resistance (P089). Increase the ramp-up time (P002). Reduce the boost set in P078 and P079. Check whether the motor is obstructed or overloaded.
F003	Overload	Check whether the motor is overloaded. Increase the maximum motor frequency if a motor with high slip is used.
F005	Inverter overtemperature (internal PTC)	Check that the ambient temperature is not too high. Check that the air inlet and outlet are not obstructed. Check that the integral fan is working.
F006	Motor overtemperature	Check that the motor is not overloaded. Check that P087 has not been set to 1 without a PTC connected.
F008	USS protocol timeout	Check the serial interface. Check the settings of the bus master and P091 – P093. Check whether the timeout interval is too short (P093).
F010	Initialization fault/Parameter loss	Check the entire parameter set. Set P009 to '0000' before power down.
F011	Internal interface fault	Switch power OFF and switch ON again.
F012	External trip (PTC)	Check if motor is overloaded.
F013	Program fault	Switch power OFF and switch ON again.
F018	Auto-restart after fault	Automatic restart after fault (P018) is pending. **WARNING: The electric motor drive may start at any time.**
F030	PROFIBUS link failure	Check the integrity of the link.
F031	Option module to link failure	Check the integrity of the link.
F033	PROFIBUS configuration error	Check the PROFIBUS configuration.
F036	PROFIBUS module watchdog trip	Replace PROFIBUS module.
F074	Motor overtemperature by I²t calculation	Check that the motor does not exceed the value set in P083.
F106	Parameter fault P006	Parameterize fixed frequency(ies) and/or motor potentiometer on the digital inputs.
F112	Parameter fault P012/P013	Set parameter P012 < P013.
F151 – F153	Digital input parameter fault	Check the settings of digital inputs P051 to P053.
F188	Automatic calibration failure	Motor not connected to electric motor drive – connect motor. If the fault persists, set P088 = 0 and then enter the stator resistance of the motor onto P089 manually.
F201	P006 = 1 while P201 = 2	Change parameter P006 and/or P201.
F212	Parameter fault P211/P212	Set parameter P211 < P212.
F231	Output current measurement imbalance	Check motor cable and motor for short-circuits and earth faults.

FAULT CODE NUMBER — FAULT CAUSE — CORRECTIVE ACTION(S)

Figure 10-12. Electric motor drive manufacturers provide lists of fault codes, including possible causes and corrective actions.

Keypad Controls. *Keypad control* is a parameter that disables some or all of the buttons on an electric motor drive keypad. The start button, forward/reverse button, and jog button are buttons that can be disabled by the keypad control. See Figure 10-13. Most electric motor drive manufacturers do not allow stop buttons to be disabled because a safety hazard is created. The default setting for keypad control is for all buttons to be enabled.

KEYPAD CONTROLS

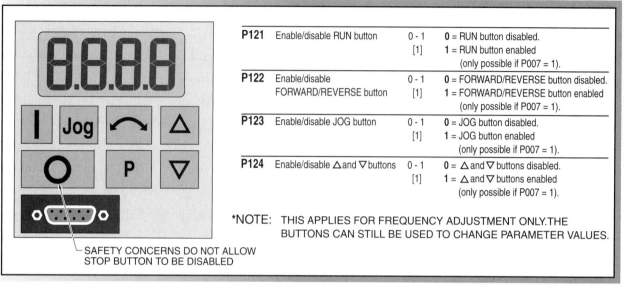

P121	Enable/disable RUN button	0 - 1 [1]	**0** = RUN button disabled. **1** = RUN button enabled (only possible if P007 = 1).
P122	Enable/disable FORWARD/REVERSE button	0 - 1 [1]	**0** = FORWARD/REVERSE button disabled. **1** = FORWARD/REVERSE button enabled (only possible if P007 = 1).
P123	Enable/disable JOG button	0 - 1 [1]	**0** = JOG button disabled. **1** = JOG button enabled (only possible if P007 = 1).
P124	Enable/disable △ and ▽ buttons	0 - 1 [1]	**0** = △ and ▽ buttons disabled. **1** = △ and ▽ buttons enabled (only possible if P007 = 1).

*NOTE: THIS APPLIES FOR FREQUENCY ADJUSTMENT ONLY. THE BUTTONS CAN STILL BE USED TO CHANGE PARAMETER VALUES.

SAFETY CONCERNS DO NOT ALLOW STOP BUTTON TO BE DISABLED

Figure 10-13. Keypad control can be used to disable certain buttons. It is unsafe for keypad control to allow a stop button to be disabled.

Reset to Factory Default. The reset to factory default parameter allows a technician to return all electric motor drive parameters to factory default settings. The factory default settings for all electric motor drive parameters are contained in an electrically erasable programmable read-only memory (EEPROM) chip. When troubleshooting an electric motor drive problem, and parameter misadjustment is suspected, resetting the parameters to default is the recommended action. The default settings are the most conservative settings and work for most electric motor drive and motor applications.

Functionality of Inputs and Outputs

The functionality of electric motor drive inputs and outputs is adjusted using parameters. The number, type, and function of inputs and outputs vary with the size of an electric motor drive and software. Larger electric motor drives have more inputs and outputs, with greater programming versatility. Typically, each input and output has its own parameter setting.

Electric motor drive inputs and outputs are either digital or analog signals. Digital inputs and outputs are either ON or OFF and analog inputs and outputs have many states that vary across a range of values.

Digital Inputs. Digital input devices such as pushbuttons, selector switches, relay contacts, and outputs of a programmable logic controller are connected to control terminal strips. The functionality of each individual terminal is controlled by the digital input parameter. The various functions that can be assigned to each terminal are jog, start, stop, and fault reset. See Figure 10-14.

Digital Outputs. Digital outputs range from a single set of contacts to multiple sets of relay contacts. The functionality of the digital output contacts is controlled by the digital output parameter. The various functions that can be assigned to each set of contacts include inverter is running, inverter frequency is zero, inverter fault indication, and motor overcurrent. The digital outputs are used to turn ON other pieces of equipment or activate panel indicator lights.

Analog Inputs. Analog input signals are connected to an electric motor drive at control terminal strips. The signals provide speed references to an electric motor drive from other sources, such as a potentiometer and temperature references from a temperature control panel. The operation of each analog input is controlled by analog input parameters. Analog inputs can be set for variable voltage or variable current signals of 0 VDC – 10 VDC, 2 VDC – 10 VDC, 0 mA – 20 mA, or 4 mA – 20 mA. The parameter setting must match the input signal type for proper electric motor drive operation.

PROGRAMMING DIGITAL INPUTS

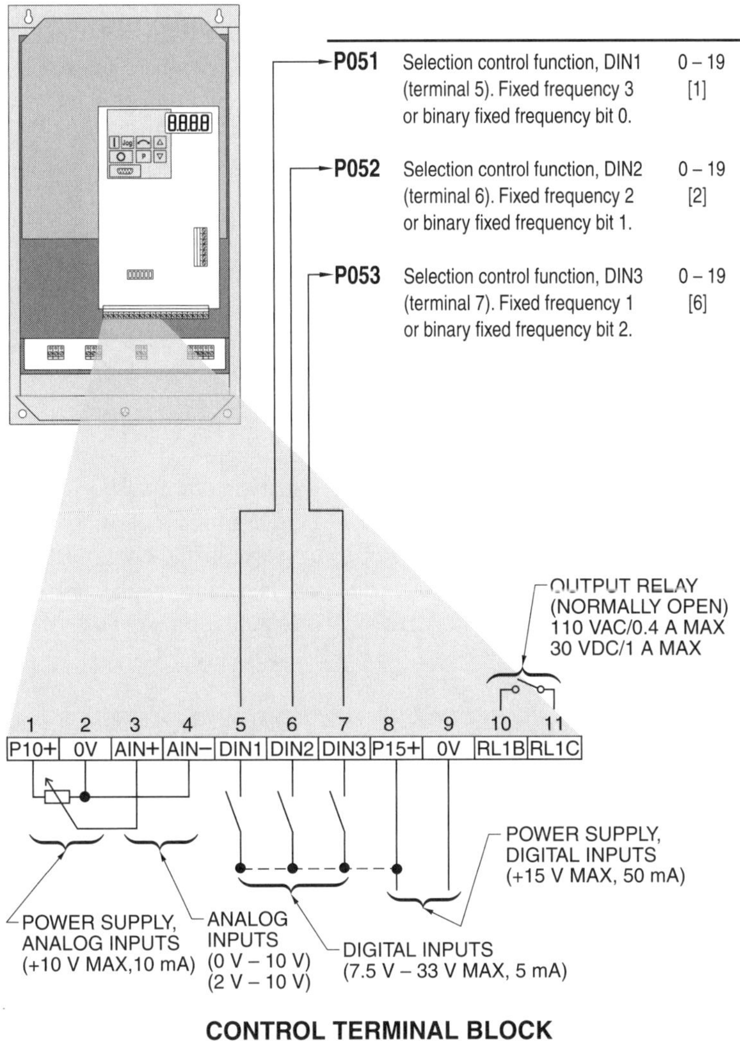

P051 Selection control function, DIN1 0 – 19
(terminal 5). Fixed frequency 3 [1]
or binary fixed frequency bit 0.

P052 Selection control function, DIN2 0 – 19
(terminal 6). Fixed frequency 2 [2]
or binary fixed frequency bit 1.

P053 Selection control function, DIN3 0 – 19
(terminal 7). Fixed frequency 1 [6]
or binary fixed frequency bit 2.

Value	Function of P051 to P053	Function, low state	Function, high state
0	Input Disabled		
1	ON right	OFF	ON right
2	ON left	OFF	ON left
3	Reverse	Normal	Reverse
4	OFF2	OFF2	ON
5	OFF3	OFF3	ON
6	Fixed frequencies 1 - 3	OFF	ON
7	Jog right	OFF	Jog right
8	Jog left	OFF	Jog left
9	Remote operation	Local	Remote
10	Fault code reset	OFF	Reset on rising edge
11	Increase frequency	OFF	Increase
12	Decrease frequency	OFF	Decrease
13	Disable analog input (setpoint is 0.0 Hz)	Analog on	Analog disabled
14	Disable the ability to change	Enabled	Disabled
15	Enable DC brake	OFF	Brake on
16	Do not use		
17	Binary fixed frequency control (fixed frequencies 1 – 7)	OFF	ON
18	As 6, but input high will also request RUN	OFF	ON
19	External trip/PTC	Yes (F012)	No

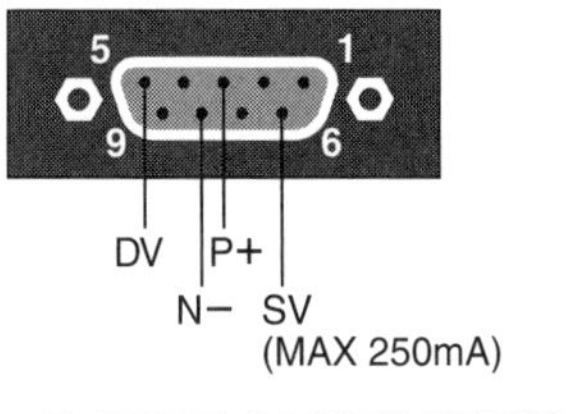

OUTPUT RELAY
(NORMALLY OPEN)
110 VAC/0.4 A MAX
30 VDC/1 A MAX

1	2	3	4	5	6	7	8	9	10	11
P10+	0V	AIN+	AIN–	DIN1	DIN2	DIN3	P15+	0V	RL1B	RL1C

POWER SUPPLY,
DIGITAL INPUTS
(+15 V MAX, 50 mA)

POWER SUPPLY,
ANALOG INPUTS
(+10 V MAX,10 mA)

ANALOG
INPUTS
(0 V – 10 V)
(2 V – 10 V)

DIGITAL INPUTS
(7.5 V – 33 V MAX, 5 mA)

CONTROL TERMINAL BLOCK

5 1
9 6

DV P+
N– SV
(MAX 250mA)

RS485 D-TYPE CONNECTION

Figure 10-14. The functionality of digital inputs is controlled by parameters. Parameter 051 is DIN 1, digital input 1.

Analog Outputs. Electric motor drive analog output signals are used as reference inputs to other pieces of equipment or instrumentation. An analog output parameter controls the type of signal (voltage or current) and what the signal represents. An analog output signal can represent numerous electric motor drive properties, such as inverter output frequency, motor current, motor rpm, or DC-bus voltage.

Enhanced Acceleration and Deceleration Methods

The electric motor drive, motor, and load determine the acceleration and deceleration method to be used for an application. Many electric motor drive applications simply require a drive to be ramped up to setpoint and ramped down to stop. Some applications require the use of enhanced acceleration and deceleration methods. The enhanced methods should only be used when necessary and with a thorough understanding of the application. Enhanced acceleration and deceleration methods include boost, skip frequency, S-curve, DC injection braking, and dynamic braking.

Boost. Boost covers two separate but closely related parameters, start boost and continuous boost. Certain electric motor drive applications require extra starting torque at low speeds, while other applications require extra torque to reach base speed. Start boost provides extra torque at startup by initially applying a higher voltage. Continuous boost provides extra torque by applying a higher voltage to reach base speed, with voltage not exceeding motor nameplate voltage. Both parameters alter the volts-per-hertz curve to provide more torque and current. See Figure 10-15. Start boost and continuous boost are measured in volts or percent of rated motor current. There is some interaction between the two parameters. The default settings vary amongst manufacturers.

Skip Frequency. *Skip frequency* is a parameter that prevents an electric motor drive from operating within a particular frequency range. When certain motor-driven loads are operated at particular speeds (frequencies), mechanical resonance causes unwanted noise or destructive vibration. The suppressed frequency range is bypassed and operation is not possible within the suppressed range. See Figure 10-16.

The skip frequency parameter consists of a skip frequency and a skip frequency bandwidth. Skip frequency is an objectionable frequency. Skip frequency bandwidth is the skipped frequency with plus and minus bands where the electric motor drive does not operate. The default setting for skip frequency is 0 Hz and is not enabled.

S-Curve. *S-curve (smoothing)* is a parameter that changes the acceleration and deceleration profile from a ramp to an S-curve slope. S-curves provide smooth, nonjerky acceleration and deceleration and are used in applications where sudden starts and stops are undesirable. S-curve parameter times (sec) are adjustable and are additive with acceleration and deceleration time. The default setting for an S-curve is 0 sec, and is not enabled. See Figure 10-17.

In 1939, flywheel motor-generator sets supplied variable-frequency AC to control the speed of steel mill motors.

> **TECH FACT**
>
> *Names for parameters differ from one electric motor drive manufacturer to another. For example, acceleration time is also called ramp-up time.*

VOLTS-PER-HERTZ-BOOST

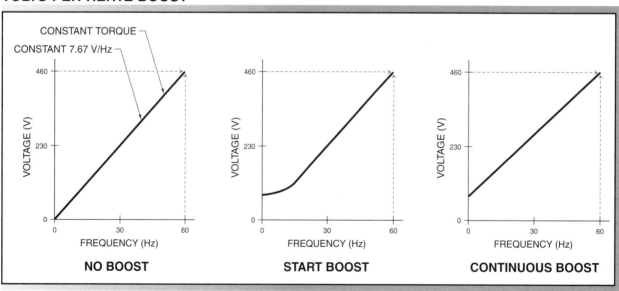

Figure 10-15. Start boost applies higher voltage initially and then follows the constant torque volts-per-hertz pattern. Continuous boost applies higher voltage from zero to base speed.

SKIP FREQUENCY

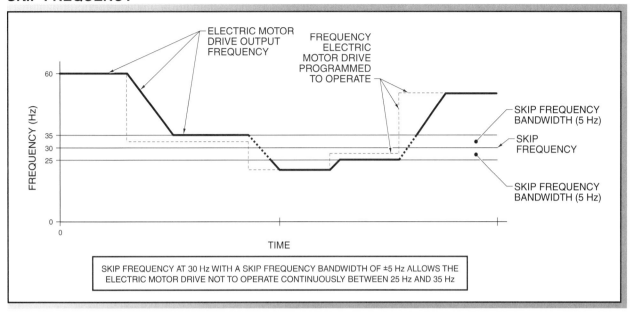

Figure 10-16. Skip frequency prevents stationary operation of an electric motor drive in a specific frequency range for noise reduction purposes.

ACCELERATION AND DECELERATION S-CURVE

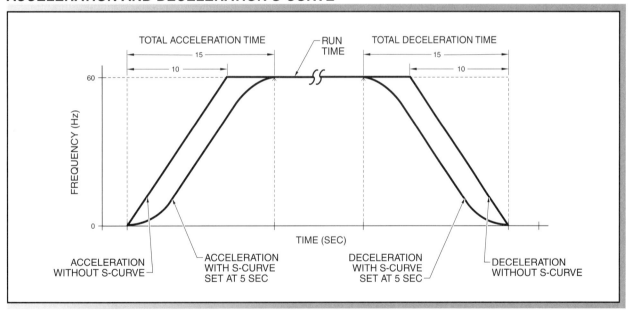

Figure 10-17. S-curve is used when sudden starts and stops of a motor are undesirable.

DC Injection Braking. *DC injection braking* is a deceleration method which brings a motor to a smooth, quick stop and can be used to hold a motor shaft stationary for brief periods of time. AC voltage is removed from the motor, and DC voltage from an electric motor drive is applied to the motor. A DC voltage creates a fixed magnetic field in the stator of a motor, which induces a magnetic field in the rotor. The fixed magnetic field causes opposite poles to align between the stator and rotor, stopping motor rotation. See Figure 10-18.

ENHANCED DECELERATION METHODS

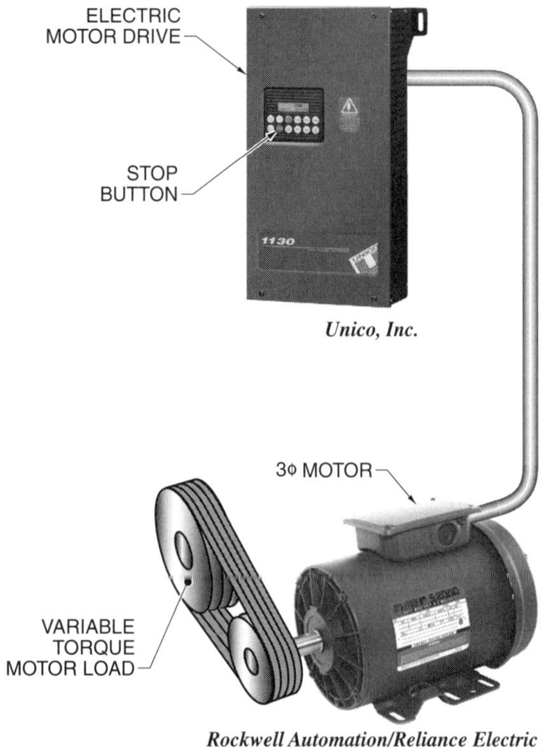

STOPPING METHOD	LOAD DETERMINES STOPPING METHOD	ELECTRIC MOTOR DRIVE DETERMINES STOPPING TIME
RAMP		X
COAST	X	
DC INJECTION BRAKING		X
S-CURVE		X
DYNAMIC BRAKING		X

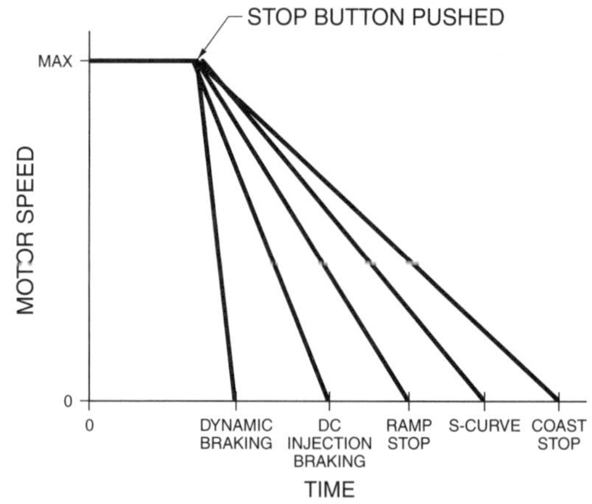

Unico, Inc.

Rockwell Automation/Reliance Electric

Figure 10-18. Enhanced deceleration methods include dynamic braking, DC injection braking, and S-curve.

The DC injection braking parameter has two components, time and braking level. Time (seconds) adjusts the length of time DC voltage is applied when the stop command is given. Braking level (volts or percent) adjusts the amount of DC voltage applied to the stator. The higher the voltage, the greater the braking force. The default setting for DC injection braking time is 0 sec, and is not enabled.

Dynamic Braking. *Dynamic braking* is a motor deceleration method that brings a motor to a smooth, quick stop, similar to DC injection braking. Unlike DC injection braking, dynamic braking requires a braking resistor and braking controller, and cannot hold a motor shaft stationary. Once an electric motor drive removes AC power from a motor, the inertia of the load continues to rotate the motor, causing the motor to act like a generator. The motor generates voltage back into the inverter, through the fly-back diodes, and to the braking resistor connected across the DC bus. The braking resistor serves as a load, dissipating the electrical energy as heat and slowing the

rotation of the motor. A braking module controls the flow of electricity to the resistor to avoid overheating the resistor. Some electric motor drives require more than one resistor, depending on the size of the motor. See Figure 10-19.

The dynamic braking parameter typically has enable and duty cycle components. Enable allows the user to enable or disable the dynamic braking feature, with disable as the default setting. Duty cycle adjusts the length of time a braking resistor is ON in relation to the length of time the braking resistor is OFF. Duty cycle controls the braking module and is measured in percent.

> **⚠ CAUTION**
>
> *DC injection braking causes motor conductors and windings to overheat. Frequent use of DC injection braking results in conductor and motor failure or fire, with risk of minor or moderate personnel injury.*

BRAKING RESISTOR AND CONTROLLER CIRCUIT

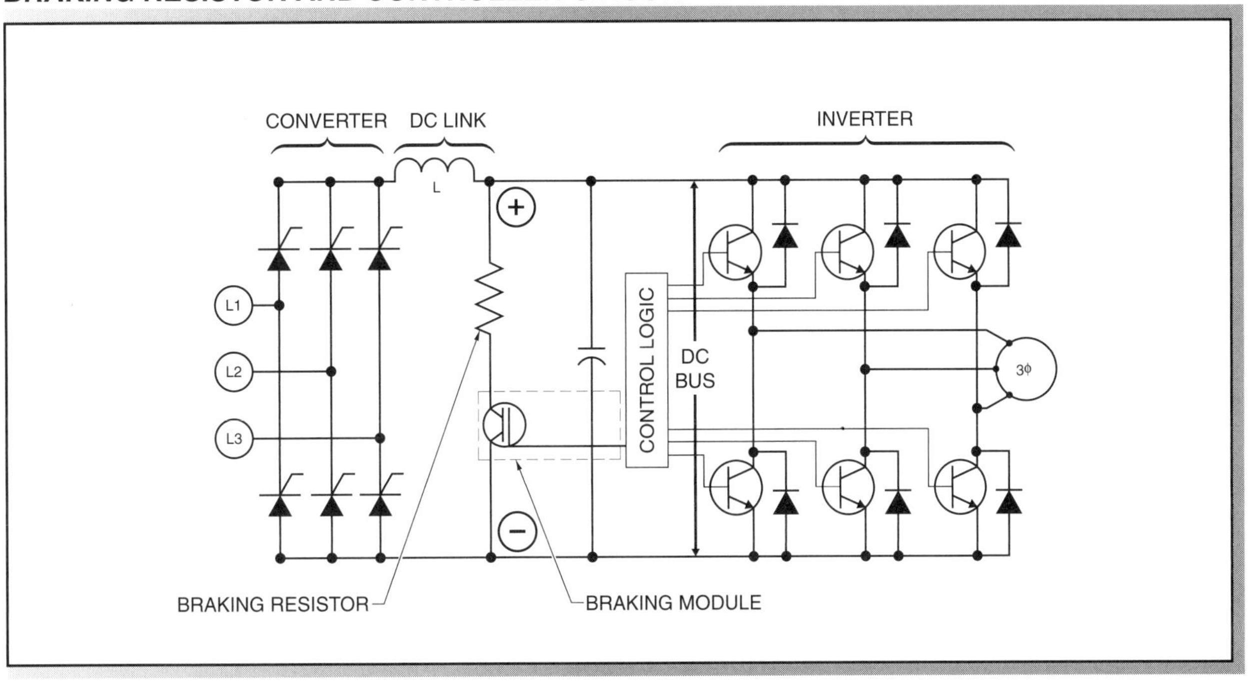

Figure 10-19. Dynamic braking utilizes one or more resistors and a controller to rapidly decelerate the driven load and protect the electric motor drive.

Enhanced Performance Features

Enhanced performance features provide greater flexibility and make electric motor drives easier to use for technicians. Certain electric motor drive applications require enhanced performance features. A thorough understanding of electric motor drive applications, parameters, and safety considerations is necessary before using enhanced performance features.

Automatic Calibration. The *automatic calibration (automatic tuning)* parameter is a parameter that fine tunes an electric motor drive to the characteristics of a motor for optimum performance. Automatic calibration involves different functions depending on the electric motor drive manufacturer. Automatic calibration may require an electric motor drive and motor to run with all loads disconnected, including removing sheaves from the motor shaft. An alternative method is to have the electric motor drive calculate the stator resistance without running the motor.

Reverse Inhibit. The *reverse inhibit* parameter is a parameter that prevents an electric motor drive from running a motor in reverse. In some applications running a motor in reverse damages the driven load. The parameter is either disabled or enabled, with disabled as the default setting.

Flying Start. *Flying start* is a parameter that allows an electric motor drive to lock in on the speed of a motor and ramp the motor up from that speed to setpoint. Certain types of high-inertia loads cause motors to spin without power. Applying power to a spinning motor from an electric motor drive causes a large flow of current, and the drive trips on an overcurrent fault. Flying start is useful when an electric motor drive is restarted after a brief power outage while the load is still spinning. The parameter is either disabled or enabled, with disabled as the default setting.

Automatic Restart after a Power Outage. *Automatic restart after a power outage* is a parameter that allows an electric motor drive to start automatically when power is once again present on the supply lines. Typically, electric motor drives must be restarted from a keypad after a power outage, even if the start inputs are still closed. Restarting several electric motor drives in a large manufacturing facility after a power outage causes unwanted downtime. Automatic restart after a power outage presents safety issues, since an electric motor drive and driven load can start unexpectedly. The automatic restart after a power outage parameter is either disabled or enabled, with disabled as the default setting.

Automatic Restart after a Fault. *Automatic restart after a fault* is a parameter that allows an electric motor drive to start automatically, provided the start inputs are still closed and the fault has cleared. Typically, an electric motor drive must be restarted from a keypad after a drive fault, even if the start inputs are still closed. It is possible the faulted electric motor drive is located miles from the maintenance shop, requiring time to restart manually. The number of restart attempts may be fixed or adjustable via other parameters. Automatic restart after a fault presents safety issues, since an electric motor drive and driven load can start unexpectedly. The automatic restart after a fault parameter is either disabled or enabled, with disabled as the default.

PWM Frequency (Carrier Frequency). The *PWM frequency parameter* is a parameter that allows the PWM frequency (the carrier frequency) of a drive to be adjusted. The parameter has a range of values such as 2000 Hz to 16,000 Hz. Motors will operate more quietly at higher PWM frequencies. At higher PWM frequencies, it may be necessary to derate the drive output due to switching losses. If quiet operation is not absolutely necessary, lower PWM frequencies are advised. The default PWM frequency settings vary between manufacturers.

Name_____ Date _____

True-False

T   **F**   **1.** Parameters for motor nameplate data are not factory-set and must be set in the field by a technician. *pg 269*

T   F   **2.** Clear text displays do not simplify electric motor drive programming. *pg 272*

T   F   **3.** When programming, a technician should always have an electric motor drive instruction manual present. *pg 274*

T   **F**   **4.** When acceleration time is too long, an electric motor drive draws excessive current, which can cause overcurrent faults. *short* *pg 278*

T   **F**   **5.** In many electric motor drive applications, advanced parameters are adjusted from the original default settings. *pg 280*

T   F   **6.** Most electric motor drive manufacturers do not allow stop buttons to be disabled because this creates a safety hazard. *pg 281*

T   F   **7.** When troubleshooting an electric motor drive problem, and parameter misadjustment is suspected, resetting the parameters to default is the recommended action. *pg 282*

T   **F**   **8.** Start boost and continuous boost do not alter the volts/hertz curve. *pg 284*

T   F   **9.** Applying power to a spinning motor from an electric motor drive causes a large current flow, and the drive trips on an overcurrent fault.

T   F   **10.** Enhanced acceleration and deceleration methods do not include ramp stop. *pg 283*

T   F   **11.** A programmable logic controller (PLC) is a solid state control device that is programmed to automatically control an industrial process or machine. *pg 273*

T   **F**   **12.** The coast stop method of stopping a motor is a stopping method in which the voltage applied to a motor is reduced, which decelerates the motor to a stop. *pg 276*

T   F   **13.** Full load amps and full load current are the same as motor nameplate current.

T   **F**   **14.** *Skip Frequency* S-curve (smoothing) is a parameter that prevents an electric motor drive from operating within a particular frequency range. *pg 284*

T   F   **15.** Speed reference is a signal that informs an electric motor drive of the speed at which to operate. *pg 275*

Completion

parameter   **1.** ___ are properties of an electric motor drive that are programmed or adjusted. *pg 269*

Keypad   **2.** A(n) ___ with an LED display is the standard programming device for electric motor drives. *pg 271*

_____Display mode_____ **3.** The ___ allows a technician to adjust electric motor drive-related information that is shown on the LED or LCD display. *pg 275*

Quadratic voltz-per-hertz **4.** The ___ curve matches the torque requirements of variable torque loads. *pg 277*

_____short_____ **5.** When deceleration time is too ___, the electric motor drive can have a DC bus overvoltage fault. *pg 278*

motor magnetizing current **6.** ___ is the current a motor draws with no load when running at nameplate voltage and frequency. *pg 279*

DC injection Breaking **7.** ___ is a deceleration method which brings a motor to a smooth, quick stop and can be used to hold a motor shaft stationary for brief periods of time. *pg 285*

reverse inhibit **8.** The ___ parameter prevents an electric motor drive from running a motor in reverse. *pg 287*

Carrier frequency **9.** The PWM frequency parameter allows the ___ of an electric motor drive to be adjusted. *pg 288*

_____Digital_____ **10.** ___ input devices such as pushbuttons, selector switches, and outputs of a programmable logic controller are connected to control terminal strips. *pg 282*

Multiple Choice

_____C_____ **1.** Acceleration time and ___ time are different parameter names for the same function.
A. deceleration   C. ramp-up
B. ramp-down   D. S-curve *pg 278*

_____B_____ **2.** When connecting a PC to an electric motor drive, the interface cable is connected to the serial port of the PC, also known as ___.
A. CD-ROM   C. LPT 1
B. COM 1   D. USB 1 *pg 272*

_____A_____ **3.** When using the ___ method of deceleration, an electric motor drive does not have any control of the motor after a stop command is entered.
A. coast stop   C. ramp stop *pg 276*
B. dynamic braking   D. S-curve

_____A_____ **4.** ___ control is a control mode that allows an AC motor to have torque characteristics identical to a DC motor.
A. Closed-loop vector   C. Quadratic volts/hertz *pg 278*
B. Open-loop vector   D. Volts/hertz

_____D_____ **5.** ___ is/are a parameter(s) that limit(s) access to electric motor drive parameters.
A. Acceleration time   C. Fault history
B. Advanced parameters   D. Parameter protection *pg 280*

Name_____ Date _____

PARAMETER DEFINITION

AU1- Parameter that automatically sets the acceleration and deceleration time to match the load condition

AU2- Parameter that automatically adjusts the torque boost when starting torque requirements are abnormally high. Setting 1 is used for 50 Hz motors and when standard automatic boost is preferred, setting 2 for 60 Hz motors without any feedback (tachometer) and standard automatic boost is preferred, setting 3 for 60 Hz motors without feedback and adjustable (through an extended drive parameter setting) boost is preferred

AU3- Parameter automatically sets the electric motor drive to better match the operating condition such as high ambient temperature

FMC- An electric motor drive allows a scope meter or ammeter to be connected directly to the drive. The parameter sets the electric motor drive output at the meter connection point to match the meter type being connected, scope meter or ammeter

SMS- Parameter used to set options as follows:

1.   Sets some factory default settings to match standard 50 Hz motor applications
2.   Sets some factory default settings to match standard 60 Hz motor applications
3.   Sets all parameter settings back to listed factory default settings
4.   Clears past fault history
5.   Clears accumulative operating time

FRS- Parameter that sets motor direction when motor is started from front panel

ACC- Parameter that sets the time required to accelerate from 0.1 Hz to the maximum frequency (MFS parameter setting)

DAC- Parameter that sets the time required to decelerate from maximum frequency (MFS parameter setting) to 0 Hz

MFS- Parameter that sets the output frequency to a maximum value

BFS- Parameter that sets the base operating frequency to match the nameplate rated frequency of the motor

CMS- Parameters that determine where an electric motor drive can be started and stopped. Panel is the front of the electric motor drive and terminal strip is where external inputs and outputs are connected to control motor starting, stopping, and direction

FMS- Parameter that determines where motor frequency is controlled. Panel is the front of the electric motor drive and terminal strip is where external inputs and outputs are connected to control motor speed

F300- Parameter that sets the pulse width modulation carrier frequency

F307- Parameter used to maintain a fixed output voltage from the electric motor drive, even when the input supply voltage momentarily drops

F600- Parameter that sets the motor overload protection (such as the overloads on magnetic motor starters). The electric motor drive is programmed as a percent of the drive's rated current output. The nameplate rated current of the motor is used to determine the setting

PARAMETER DEFINITION (CONTINUED)

F603- Parameter that selects a method for emergency stopping of a motor (when the electric motor drive type includes an emergency stop connection on the drive terminal strip). When option 0 is selected, the motor will coast to a stop based on the load connected to the motor. When option 1 is selected, the motor will stop based on the programmed deceleration time (parameter DAC). When option 2 is selected, parameter F604 must be programmed for a braking time

F604- Parameter used when parameter F603 is set to option 2

F900- Parameter used to select the serial communication bit transfer rate

TITLE	PARAMETERS	ADJUSTABLE RANGE	DEFAULT VALUE
AU1	Automatic acceleration and deceleration	0: NO 1: YES	0
AU2	Automatic torque boost*	0: NO 1: AUTOMATIC 50 Hz MOTOR 2: SENSORLESS VECTOR CONTROL 3: SENSORLESS VECTOR CONTROL AUTOMATIC TUNING	0
AU3	Automatic environment setting*	0: NO 1: AUTOMATIC 50 Hz MOTOR 2: AUTOMATIC 60 Hz MOTOR	0
FMC	FM terminal function selection	0: FREQUENCY METER 1: OUTPUT CURRENT METER	0
SMS	Standard mode selection*	1: 50 Hz STANDARD 2: 60 Hz STANDARD 3: DEFAULT SETTING 4: CLEARING LOG ERRORS 5: CLEARING ACCUMULATED OPERATION TIME	3
FRS	Forward/reverse selection (panel)	0: FORWARD 1: REVERSE	0
ACC	Acceleration time #1 (sec)	0.1 – 3600	10.0
DAC	Deceleration time #1 (sec)	0.1 – 3600	10.0
MFS	Maximum frequency (Hz)*	30.0 – 320.0	80.0
BFS	Base frequency (Hz)	25.0 – 320.0	60.0
CMS	Command mode selection	0: TERMINAL BLOCK, 1: PANEL	1
FMS	Frequency setting mode selection	0: TERMINAL BLOCK, 1: PANEL	1
F300	PWM carrier frequency (kHz)	2.2 – 12.0	12.0
F305	Overvoltage stall protection	0: ENABLED, 1: DISABLED	0
F306	Output voltage adjustment (%)	0 – 120	100
F307	Line voltage compensation*	0: NO, 1: YES	0
F600	Motor overload protection level (%)	10 – 100	100
F603	Emergency stop selection*	O: COAST TO STOP 1: STOPPING AFTER DECELERATION 2: STOPPING AFTER EMERGENCY DC BRAKING	0
F604	Emergency DC inject time (sec)	0.0 – 20.0	10
F800	Communication speed	0: 1200 bps 1: 2400 bps 2: 4800 bps 3: 9600 bps	3

* Parameters cannot be changed while running.

Activity 10-1. Programming Parameters–Application 1

An electric motor drive is used to control a bottling in-feed conveyor system. The electric motor drive output is rated for 460 VAC, 15 HP, and 25 A. Since the bottles are not yet filled, a longer acceleration and deceleration time are required to prevent the bottles from tipping over every time the conveyor is started and stopped. The operator manually controls the conveyor from the panel of the electric motor drive.

Determine the parameter title and setting based on the following application requirements and conditions. Assume that the electric motor drive was set to factory default settings at the time of installation and Parameter SMS is set to #3 (default setting). When a controlling parameter does not require changing from the default setting, leave at default setting value.

- Acceleration time is to be 30 seconds.

    _____ **1.** Parameter title is ___.   _____ **2.** Parameter setting is ___.

- Deceleration time is to be 30 seconds.

    _____ **3.** Parameter title is ___.   _____ **4.** Parameter setting is ___.

- The motor is not to operate above the nameplate speed rating.

    _____ **5.** Parameter title is ___.   _____ **6.** Parameter setting is ___.

- The motor should not be overloaded. The electric motor drive is to automatically stop the conveyor when the motor exceeds the nameplate current rating during operation.

    _____ **7.** Parameter title is ___.   _____ **8.** Parameter setting is ___.

- The bottles are light because the bottles are not yet filled and thus no additional torque boost is required when starting the conveyor.

    _____ **9.** Parameter title is ___.   _____ **10.** Parameter setting is ___.

ELECTRIC MOTOR DRIVE   SHIELDED POWER CABLE
CABLE TRAY   CABLE TRAY DROP
CONVEYOR MOTOR

Siemens

○	**MANUFACTURER**		○
PE·21 PLUS™		PREMIUM EFFICIENCY	
ORD. NO.	1LA02864SE41	MAG. 21.8	
TYPE	RGZESDI	FRAME 286T	
HP	15.00	SERVICE FACTOR 1.0	3 PH
AMPS	19.5	VOLTS 460	
RPM	1765	HERTZ 60	
DUTY	CONT 40°C AMB.	DATE CODE	
CLASS INSUL H	NEMA DESIGN B	K.V.A. CODE G	NEMA NOM. EFF. 93.6
SH. END BRG.	50BC03JPP3	OPP END BRG. 50BC03JPP3	
○	**Inverter Duty AC Induction Motor** made in U.S.A. ○ ⑤		

51-770-642

APPLICATION 1

Activity 10-2. Programming Parameters–Application 2

An electric motor drive is used to control a large conveyor system moving heavy boxes. The electric motor drive output is rated for 460 VAC, 50 HP, and 70 A. Since the boxes are heavy and hard to tip, the acceleration and deceleration times are not important. The operators control the conveyor from any one of several STOP/START pushbutton stations located along the conveyor system.

Determine the parameter title and setting based on the following application requirements and conditions. Assume that the electric motor drive was set back to factory default settings at the time of installation and Parameter SMS is set to #3 (default settings). When a controlling parameter does not require changing from the default setting, leave at default setting value.

1. A remote instrument is connected to the electric motor drive to monitor the system. The instrument set-up software displays the following screen. Check the Baud Rate of the instrument to match the electric motor drive.

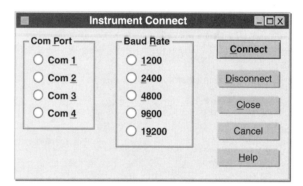

- The motor is not to operate at more than half nameplate rated speed.

_____ **2.** Parameter title is ___.   _____ **3.** Parameter setting is ___.

- The START/STOP pushbuttons must control the motor.

_____ **4.** Parameter title is ___.   _____ **5.** Parameter setting is ___.

- Motor speed will still be controlled at the electric motor drive panel.

_____ **6.** Parameter title is ___.   _____ **7.** Parameter setting is ___.

- The motor should never be overloaded. The drive is to automatically stop the conveyor when the motor exceeds the nameplate current rating during operation.

_____ **8.** Parameter title is ___.   _____ **9.** Parameter setting is ___.

- The boxes are heavy and an additional standard torque boost is required when starting the conveyor.

_____ **10.** Parameter title is ___.   _____ **11.** Parameter setting is ___.

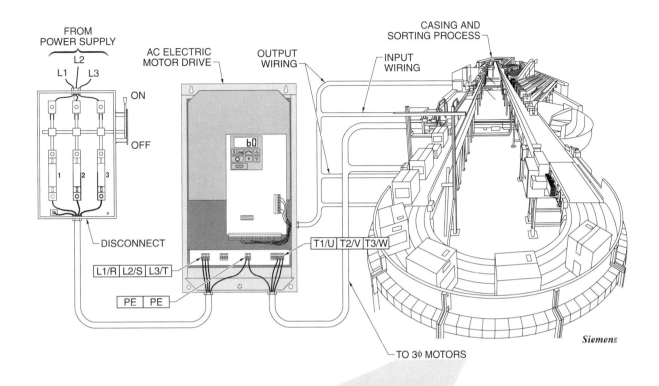

APPLICATION 2

Activity 10-3. Programming Parameters–Application 3

An electric motor drive is used to control a conveyor system. The electric motor drive is rated for 460 VAC, 10 HP, and 15 A.

Determine the parameter title and setting based on the following application requirements and conditions. Assume that the electric motor drive was set back to factory default settings at the time of installation and Parameter SSM is set to #3 (default setting). When a controlling parameter does not require changing from the default setting leave at default setting value.

- The motor should never be overloaded. The electric motor drive is to automatically stop the conveyor when the motor exceeds the nameplate current rating during operation.

_____ 1. Parameter title is ___.

_____ 2. Parameter setting is ___.

- The operator is to control the speed of the motor with an external potentiometer located at a control station.

_____ 3. Parameter title is ___.

_____ 4. Parameter setting is ___.

- The operator is to control the motor with a three-position selector switch (Forward/Stop/Reverse) located at the operator station.

_____ 5. Parameter title is ___.

_____ 6. Parameter setting is ___.

- The conveyor can operate at a speed 10% higher than nameplate rated speed.

_____ 7. Parameter title is ___.

_____ 8. Parameter setting is ___.

- To monitor motor speed, a frequency meter is connected to the drive and located at the operator station.

_____ 9. Parameter title is ___.

_____ 10. Parameter setting is ___.

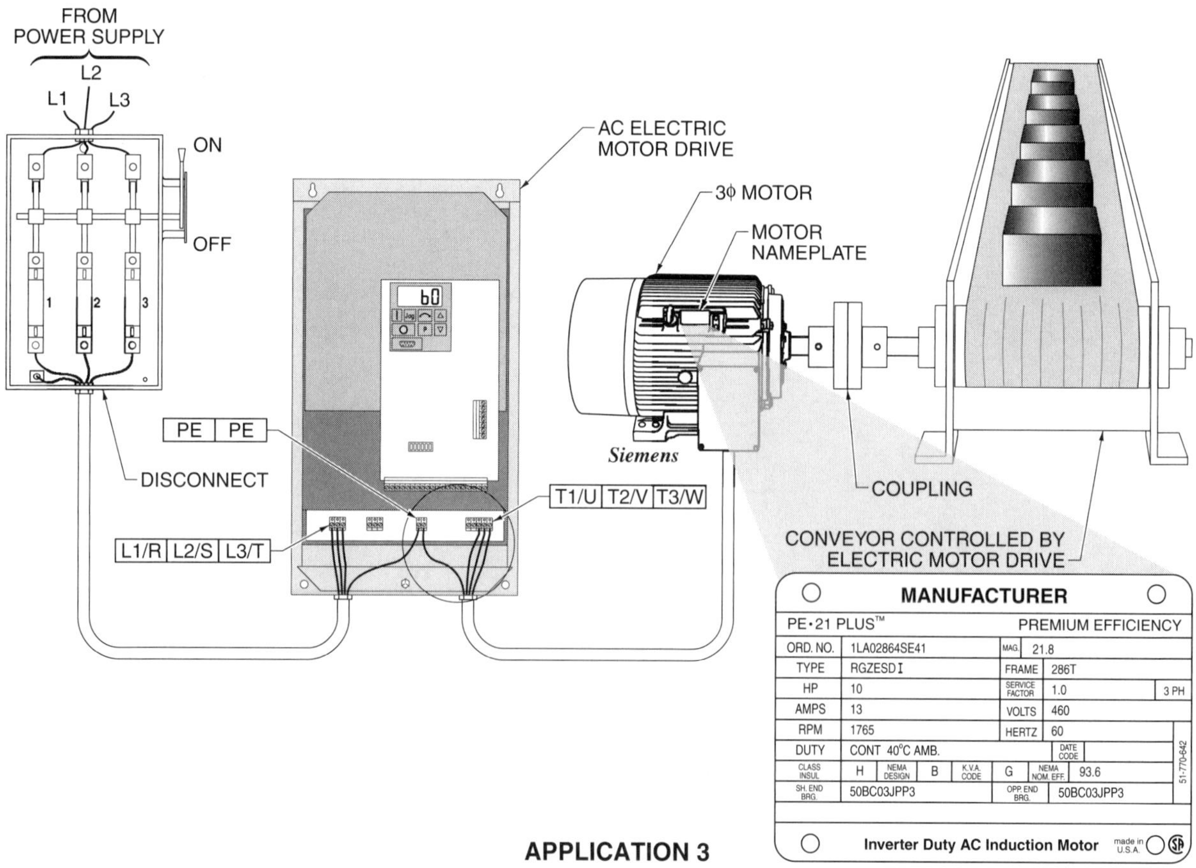

FROM POWER SUPPLY

L1 L2 L3

ON

OFF

AC ELECTRIC MOTOR DRIVE

3φ MOTOR

MOTOR NAMEPLATE

PE | PE

DISCONNECT

L1/R | L2/S | L3/T

Siemens

T1/U | T2/V | T3/W

COUPLING

CONVEYOR CONTROLLED BY ELECTRIC MOTOR DRIVE

○	MANUFACTURER		○
PE·21 PLUS™		PREMIUM EFFICIENCY	
ORD. NO.	1LA02864SE41	MAG. 21.8	
TYPE	RGZESDI	FRAME 286T	
HP	10	SERVICE FACTOR 1.0	3 PH
AMPS	13	VOLTS 460	
RPM	1765	HERTZ 60	
DUTY	CONT. 40°C AMB.	DATE CODE	
CLASS INSUL.	H NEMA DESIGN B K.V.A. CODE G	NEMA NOM. EFF 93.6	
SH. END BRG.	50BC03JPP3	OPP. END BRG. 50BC03JPP3	

○ Inverter Duty AC Induction Motor made in U.S.A. ○ (SP)

51-770-642

APPLICATION 3

11 Electric Motor Drive Test Tools

A variety of test tools are used when working with electric motor drives and electric motors. The most common test tools are digital multimeters, clamp-on meters, and megohmmeters. The features and operation of test tools vary between manufacturers and tool models, but the basic principles of operation remain the same. It is essential that technicians working on electric motor drives and electric motors understand how to safely and effectively use test tools.

TEST TOOL CONSIDERATIONS

Special consideration is required when selecting and using test tools with electric motor drives. Electric motor drives are nonlinear loads on the facility electrical system. The current input of CSI, PWM, and VVI drives is a nonsinusoidal waveform. A *nonsinusoidal waveform* is a waveform that has a distorted appearance when compared with a pure sine waveform. The electric motor drive voltage and current outputs are also nonsinusoidal waveforms. See Figure 11-1.

Digital multimeters (DMMs) and clamp-on ammeters fall into two categories according to how they measure voltage and current waveforms: average-responding and true-rms. An average-responding meter provides the correct reading for a sine wave (linear load) but an incorrect reading for a nonsinusoidal waveform (nonlinear load). A true-rms meter provides the correct reading regardless of the waveform. See Figure 11-2. True-rms meters are labeled true rms.

Meter bandwidth is a consideration when taking measurements on electric motor drives. *Bandwidth* is the range of frequencies to which the meter can respond. Average-responding and true-rms multimeters and clamp-on ammeters have a bandwidth rating. Electric motor drives have more than one frequency present at the output of the drive, the frequency of the voltage supplied to the motor and the PWM frequency (carrier frequency). Depending on the bandwidth, the high frequency

switching of the inverter-section semiconductors can cause an inaccurate reading. Two different meter models of the same measurement category (average-responding or true-rms) can have different readings when measuring the same voltage or current if they have different bandwidth capabilities. Some test tool manufacturers have designed test tools to measure low-frequency motor voltage to eliminate the discrepancy.

A true-rms DMM is used to measure voltage present between phases of the power supply.

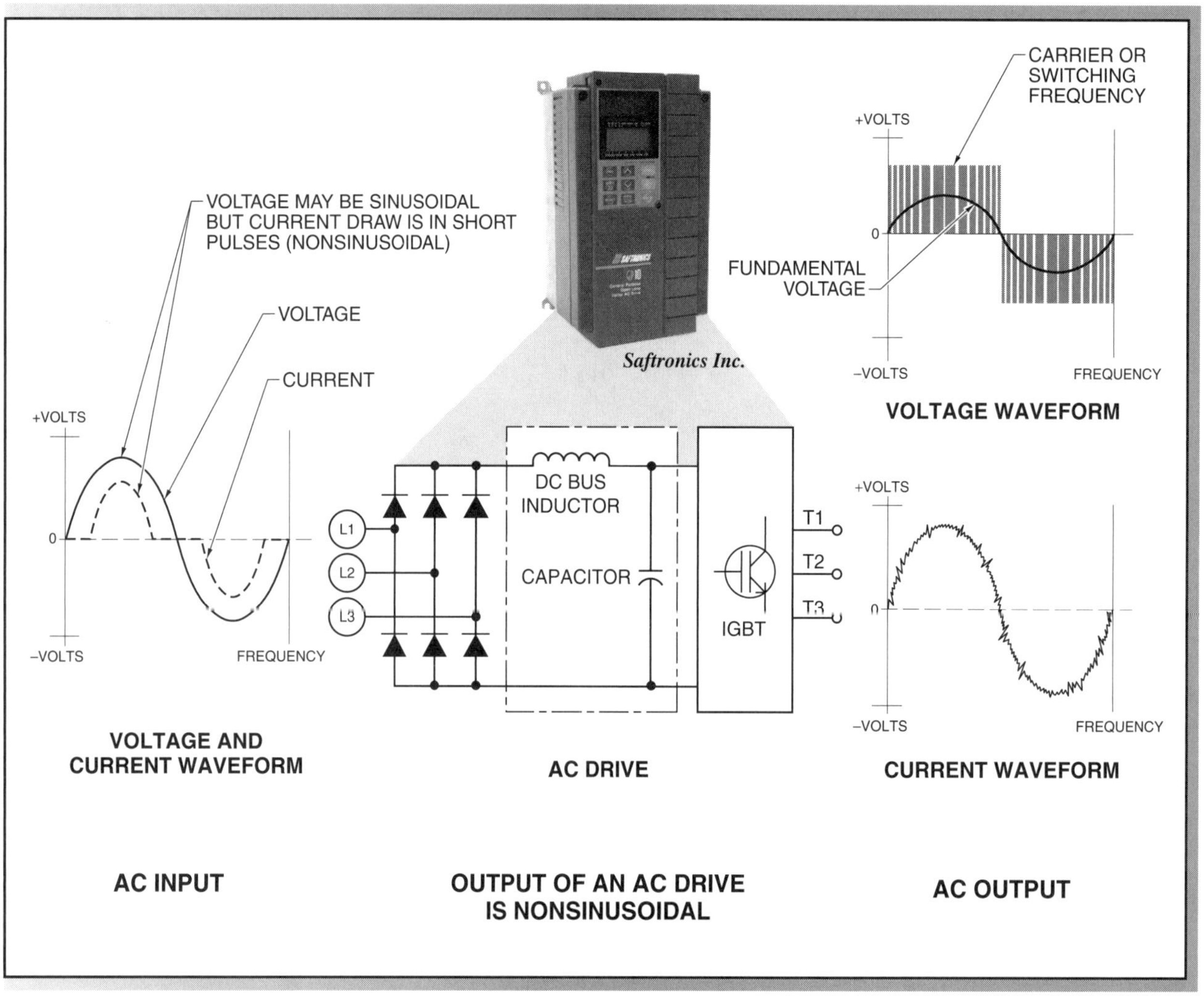

VOLTAGE AND CURRENT OUTPUT WAVEFORMS

VOLTAGE MAY BE SINUSOIDAL BUT CURRENT DRAW IS IN SHORT PULSES (NONSINUSOIDAL)

VOLTAGE

CURRENT

CARRIER OR SWITCHING FREQUENCY

FUNDAMENTAL VOLTAGE

VOLTAGE WAVEFORM

VOLTAGE AND CURRENT WAVEFORM

DC BUS INDUCTOR

CAPACITOR

IGBT

AC DRIVE

CURRENT WAVEFORM

AC INPUT

OUTPUT OF AN AC DRIVE IS NONSINUSOIDAL

AC OUTPUT

Saftronics Inc.

Figure 11-1. In nonlinear loads, the instantaneous current is not proportional to the instantaneous voltage.

⚠ DANGER

- *Do not use a meter if the meter or test leads have been damaged. Before using a meter, inspect the case by checking for cracks or missing plastic.*
- *Do not operate a meter in areas where explosive gas, vapor, or dust is present.*
- *Do not use a meter if any meter operation works abnormally.*
- *Remove the test leads from a meter before opening the battery door.*
- *Do not operate a meter with the battery door or portions of the cover loosened or removed.*
- *Use the proper terminals, function, and range when taking any measurement.*

ELECTRICAL SAFETY STANDARDS

An *electrical safety standard* is a document that provides information to reduce safety hazards that occur when using electrical test equipment such as DMMs, clamp-on meters, and megohmmeters. The *International Electrotechnical Commission (IEC)* is an organization that develops international safety standards for electrical equipment.

Voltage surges within a power distribution system can be a safety hazard. A *voltage surge* is a higher-than-normal voltage that temporarily exists in one or more power lines. Voltage surges vary in voltage level and duration. A common type of voltage surge is a transient voltage. A *transient voltage* is a high-energy, high-voltage, short-duration spike in an electrical system. Transient voltages typically exist for a short period of time and are very erratic. Transient voltages are produced by lightning strikes, unfiltered electrical equipment, or when large high-current loads are switched ON or OFF. Transient voltages of 1000 V are found on 120 V power lines, but transient voltages can reach several thousand volts on 480 V and 600 V power lines.

When a large motor (100 HP or more) is turned OFF, a transient voltage moves through the power distribution system. A DMM connected to a point along the system where the high transient voltage is present can have an arc created inside the meter. The arc causes a high-current short in the power distribution system even after the original high transient voltage has dissipated. The high-current short can turn into an arc blast. An *arc blast* is an explosion that occurs when the surrounding air becomes ionized and conductive. See Figure 11-3.

The amount of current flow and damage from a transient voltage depends on the specific location of the transient voltage within the power distribution system. High-energy transient voltages are weakened or dampened as they travel through the impedance (AC resistance) of the system and the system grounds. All power distribution systems have fuses and circuit breakers to

limit current flow and transient voltages. The current rating (size) of fuses and circuit breakers decreases the further away from the main distribution panel the fuses and circuit breakers are located and the less likely that transient voltages can cause damage.

Circuit breakers and fuses are connected in series with a circuit to protect the circuit from overcurrents, shorts, and transient voltages.

DMM AC VOLTAGE MEASUREMENT

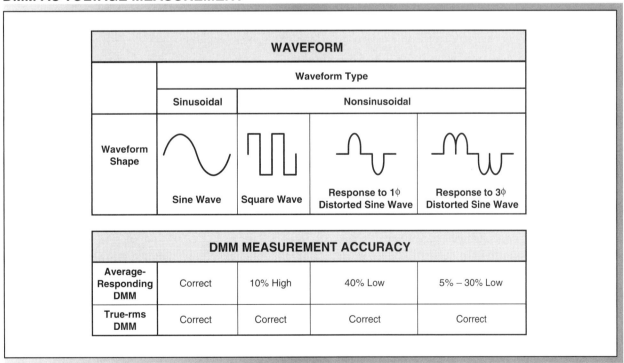

WAVEFORM			
	Waveform Type		
	Sinusoidal	**Nonsinusoidal**	
Waveform Shape	Sine Wave	Square Wave	Response to 1ϕ Distorted Sine Wave	Response to 3ϕ Distorted Sine Wave

DMM MEASUREMENT ACCURACY			
Average-Responding DMM	Correct	10% High	40% Low	5% – 30% Low
True-rms DMM	Correct	Correct	Correct	Correct

Figure 11-2. A true-rms DMM is required for measurement accuracy when measuring nonsinusoidal voltage or current waveforms.

TRANSIENT VOLTAGE

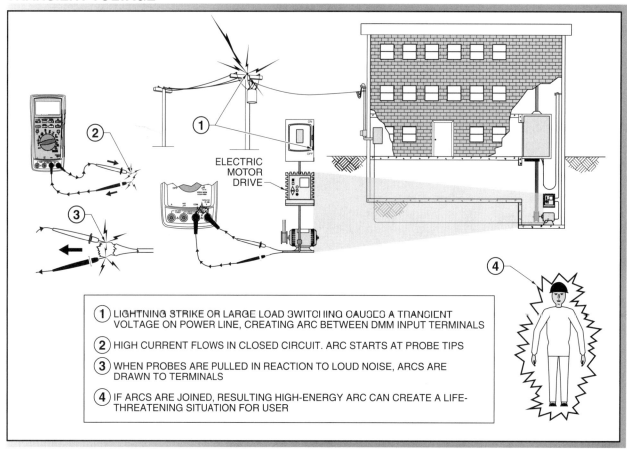

1 LIGHTNING STRIKE OR LARGE LOAD SWITCHING CAUSES A TRANSIENT VOLTAGE ON POWER LINE, CREATING ARC BETWEEN DMM INPUT TERMINALS

2 HIGH CURRENT FLOWS IN CLOSED CIRCUIT. ARC STARTS AT PROBE TIPS

3 WHEN PROBES ARE PULLED IN REACTION TO LOUD NOISE, ARCS ARE DRAWN TO TERMINALS

4 IF ARCS ARE JOINED, RESULTING HIGH-ENERGY ARC CAN CREATE A LIFE-THREATENING SITUATION FOR USER

Figure 11-3. When taking measurements in an electrical circuit, transient voltages can cause electrical shock and/or damage to equipment.

Overvoltage Installation Categories

The IEC 1010 standard divides DMM applications into four overvoltage installation categories. These categories are designated CAT I, CAT II, CAT III, and CAT IV, and determine what magnitude of transient voltage a DMM or electrical appliance must withstand. A DMM or electrical appliance operated at voltages below 600 V and used in a CAT III environment must withstand a 6000 V transient voltage without causing an arc. DMMs or electrical appliances operated at voltages between 600 V and 1000 V must withstand an 8000 V transient voltage without causing an arc. Higher CAT numbers indicate an electrical environment with higher power, larger short-circuit current, and higher transient voltages. DMMs designed to the CAT III standard are resistant to higher transient voltages than DMMs designed to the CAT II standard. DMMs and other test equipment designed to the IEC 1010 standard offer a high level of protection from transient voltages.

TECH FACT

To obtain accurate measurements when working on an electric motor drive circuit, true-rms meters must be used to measure the input and output current, and the output voltage of electric motor drives.

Within an IEC 1010 standard category, a higher voltage rating denotes the ability to withstand higher transient voltages. A CAT III–1000 V rated DMM has better protection than a CAT III–600 V rated DMM. A CAT III–600 V rated DMM has better transient protection compared to a CAT II–1000 V rated DMM. A DMM that can withstand a transient voltage may be damaged, but an arc does not start and no arc blast occurs. A DMM is chosen based on the IEC overvoltage installation category first and voltage second. See Figure 11-4.

IEC 1010 0VERVOLTAGE INSTALLATION CATEGORIES

Category	In Brief	Examples
CAT I	Electronic	• Protected electronic equipment • Equipment connected to (source) circuits in which measurements are taken to limit transient overvoltages to an appropriately low level • Any high-voltage, low-energy source derived from a high-winding resistance transformer such as the high-voltage section of a copier
CAT II	1φ receptacle-connected loads	• Appliances, portable tools, and other household and similar loads • Outlets and long branch circuits • Outlets at more than 30′ (10 m) from CAT III source • Outlets at more than 60′ (20 m) from CAT IV source
CAT III	3φ distribution, including 1φ commercial lighting	• Equipment in fixed installations, such as switchgear, electric motor drives, and motors • Bus and feeder in industrial plants • Feeders and short branch circuits and distribution panel devices • Lighting systems in larger buildings • Appliance outlets with short connections to service entrance
CAT IV	3φ at utility connection, any outdoors conductors	• Refers to the origin of installation, where low-voltage connection is made to utility power • Electric meters, primary overcurrent protection equipment • Outside and service entrance, service drop from pole to building, run between meter and panel • Overhead line to detached building

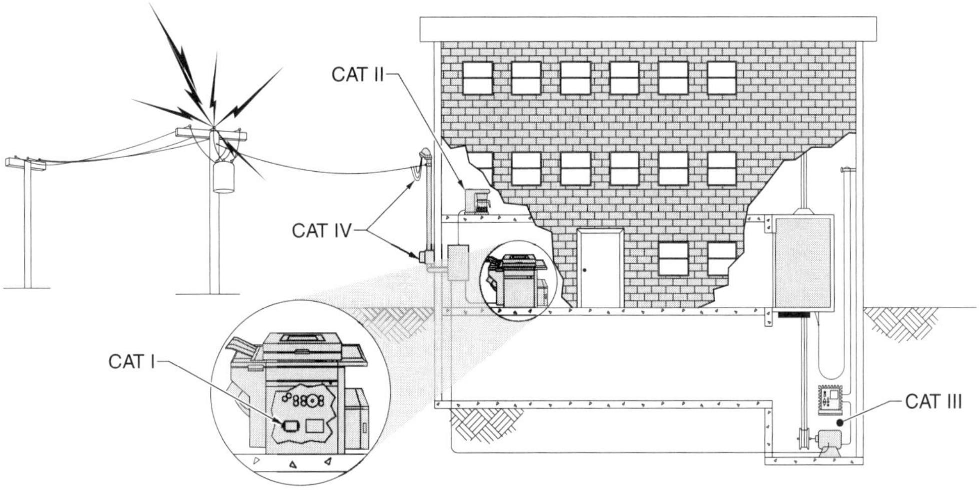

Figure 11-4. The new IEC 1010 standard categorizes the applications in which a DMM may be used in four overvoltage installation categories.

TECH FACT

When changing the ambient temperature of a DMM significantly, 30 minutes or more must be allowed for the meter to stabilize to the new ambient temperature before taking any readings. The specified accuracy of a DMM is valid within a small, 73° F (23° C) temperature range.

The IEC sets standards but does not test or inspect for compliance. A DMM with a symbol or listing number of an independent testing lab such as Underwriters Laboratories, Inc. (UL®), European Commission (CE), Canadian Standards Association (CSA), or other recognized testing organization indicates compliance with the IEC 1010 standard. A manufacturer can claim to "design to" a standard with no independent verification. To be UL®-listed, CE-listed, or CSA-certified, a manufacturer must employ the services of the approval/listing agency to test that the product complies with the standard. See Figure 11-5.

BACK OF DMM

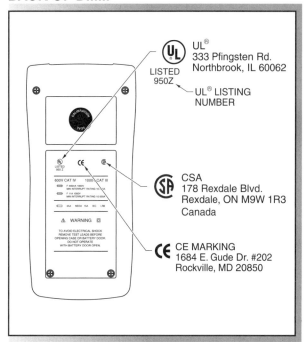

Figure 11-5. A symbol and listing number indicates compliance with the IEC 1010 standard.

DIGITAL MULTIMETER USAGE

A *digital multimeter (DMM)* is a test tool used to measure two or more electrical values. Electrical measurements on electric motor drives are commonly taken using a DMM. DMMs are used in residential, commercial, and industrial electrical equipment installation and maintenance applications. Electrical measurements are taken from exposed electric components that are normally enclosed. Safe work habits and personal protective equipment are required to prevent electrical shock when working with exposed electrical circuits.

DMM Safety Precautions

DMMs are used to measure electrical values such as voltage, current, resistance, and frequency in electrical and electronic circuits. Each DMM has specific features and limits. The user's manual details applications and specific DMM specifications, features, operating procedures, and safety precautions.

Conditions can change quickly as voltage and current levels vary in individual electrical circuits, requiring safe work habits. Following are some of the general safety precautions to be followed when using a DMM:

- Work on de-energized circuits whenever possible. Use proper lockout/tagout procedures.
- Assume all electrical components in a circuit are energized when taking electrical measurements.
- Check test leads for frayed or broken insulation. The DMM should have double-insulated test leads, recessed input jacks, and finger shrouds.
- Ensure that the test leads are connected properly. Test leads that are not connected to the correct jacks for the meter setting are dangerous. Use DMMs that are self-protected with a high-energy fuse.
- Ensure that the selector switch is set to the proper range and function before applying test leads to a circuit. A DMM set to the wrong function can be damaged.
- Start with the highest range when measuring unknown values. Using a range too low can damage the DMM.
- Connect the black (common) meter lead to the circuit, then connect the red (voltage) lead to the circuit when taking a measurement. When finished, remove the red meter lead, and then remove the black meter lead.
- Avoid holding a meter by hand to minimize personal exposure to the effects of transient voltages. Hang or rest the meter when taking a measurement.
- Use DMMs that conform to the IEC 1010 category to which the DMM was designed. To measure 480 V on an electric motor drive, a DMM rated at CAT III-600 V or CAT III-1000 V is used.
- Avoid taking measurements in humid or damp locations.
- Use recommended personal protective equipment when measuring high voltages and currents.
- Ensure that there are no atmospheric hazards such as flammable dust or vapors in the area.
- Keep one hand in your pocket when working on a live circuit to reduce the chance of an electric shock passing through the heart and lungs.
- Check a DMM by measuring a known (energized) voltage source before taking a measurement on an unknown voltage source. After taking the measurement on an unknown voltage source, recheck the DMM by again measuring the known voltage source. Checking a DMM prevents a blown fuse or other malfunction from giving a false reading of an energized circuit.

Digital Displays

A *digital display* is a display that shows numerical values using light-emitting diodes (LEDs) or liquid crystal displays (LCDs). Digital multimeters display readings as exact numerical values. Numerical values are displayed digitally on a DMM and help eliminate errors that occur when reading an analog meter.

Errors occur when reading a digital display if prefixes, symbols, and decimal points are not properly interpreted. The number displayed and the position of the decimal point determine the exact value of a digital display. The selected range determines the placement of the decimal point. Typical voltage ranges on a DMM are 3 V, 30 V, and 300 V or 6 V, 60 V, and 600 V. Always check the DMM manual for the specific ranges available. See Figure 11-6.

If the range is not set high enough, the DMM display reads "OL" (overload). An *autoranging DMM* is a meter that automatically adjusts to a higher range setting if the range is not high enough. Accurate readings are best obtained by using the range that provides the best resolution without overloading.

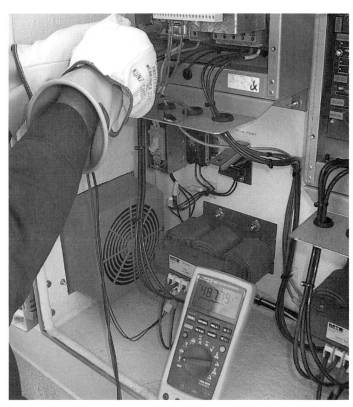

DMM dual displays typically have a 50,000-count primary display for high-resolution measurement readings, and a 51 segment bar graph analog display.

DMM RANGE SETTINGS

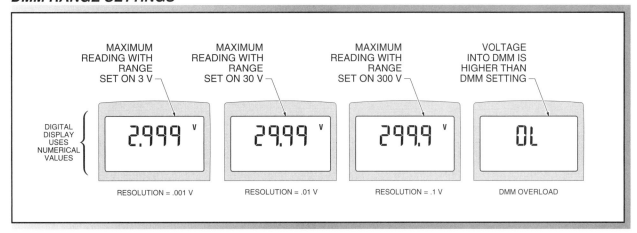

Figure 11-6. The setting of the range on a DMM determines the placement of the decimal point.

Fluke Corporation

A DMM is used to measure AC voltage, DC voltage, resistance, continuity, Fahrenheit and Celsius temperature, AC current, DC current, and test capacitors and diodes.

Bar Graph

A *bar graph* is a graph composed of segments that function as an analog pointer. Most digital displays include a bar graph. The displayed bar graph segments increase as the measured value increases and decrease as the measured value decreases. Reversing the position of the test leads eliminates the negative sign displayed at the beginning of a bar graph. A *wrap-around bar graph* is a bar graph that displays a fraction of the full range on the graph. The pointer wraps around and starts over when the limit of the bar graph is reached. See Figure 11-7.

A bar graph reading is updated 30 times per second. A digital display is updated four times per second. The bar graph is used when fast-changing signals cause the digital display to flash or when there is a change in the circuit that is too rapid for the digital display to show. Mechanical relay contacts may bounce open when exposed to vibration, causing rapid circuit changes. Contact bounce is displayed on the DMM by the movement of one or more bar graph segments as the contact opens then closes.

BAR GRAPHS

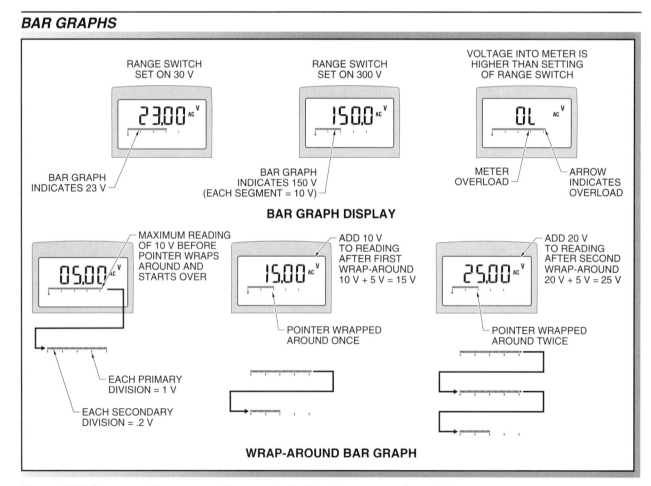

Figure 11-7. A bar graph is composed of segments that function as an analog pointer.

Ghost Voltages

Ghost voltage is a voltage reading on a DMM that is not connected to an energized circuit. Ghost voltage appears as changing numbers on the DMM display. The sensitivity of a meter to magnetic fields allows ghost voltages to occur. They are created by magnetic fields generated by current-carrying conductors, fluorescent lighting, and energized electrical equipment, and enter a DMM through the test leads. When not connected to a circuit, a DMM's test leads act as antennas for stray voltages. Ghost voltages do not damage DMMs or other meters. See Figure 11-8.

Avoid confusing ghost voltage readings with stable readings taken during a measurement. Ghost voltage are misread as circuit voltage when a DMM is connected to a circuit believed to be energized. A circuit that is not energized can also act as an antenna for stray voltages. To ensure a true circuit voltage reading, a DMM should be connected to a circuit long enough to obtain a stable reading.

GHOST VOLTAGE

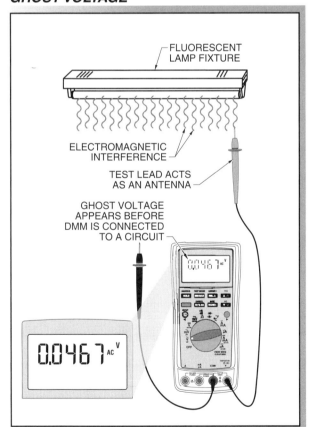

Figure 11-8. Ghost voltage is a voltage reading that appears as changing numbers on the display when a DMM is not connected to an energized circuit.

AC Voltage Measurement Procedures

AC voltage measurement procedures may vary slightly with different DMMs. Some DMMs have advanced features which provide convenience and specific measurement capabilities. To measure AC voltage between a phase conductor (hot) and ground or a phase conductor (hot) and neutral, connect the black (common) test lead to the circuit, then connect the red (voltage) lead to the circuit. After taking the AC voltage measurement, remove the red test lead, and then remove the black test lead. If the red test lead is connected to a phase conductor first, the black test lead is hot before the lead is connected to the circuit. See Figure 11-9.

To measure AC voltages with a DMM, apply the procedure:

1. Plug the black test lead into the common jack.
2. Plug the red test lead into the voltage jack.
3. Set the function switch to AC voltage. Set the range to the highest AC voltage setting if voltage in the circuit is unknown. Single setting DMMs power up in the autorange mode, which automatically selects a measurement range based on the voltage present.
4. Connect the test leads to the circuit.
5. Read the voltage measurement displayed on the DMM.
6. Turn the DMM OFF to prevent battery drain.

When taking AC voltage measurements, the circuit may have to be manually energized.

AC VOLTAGE MEASUREMENT

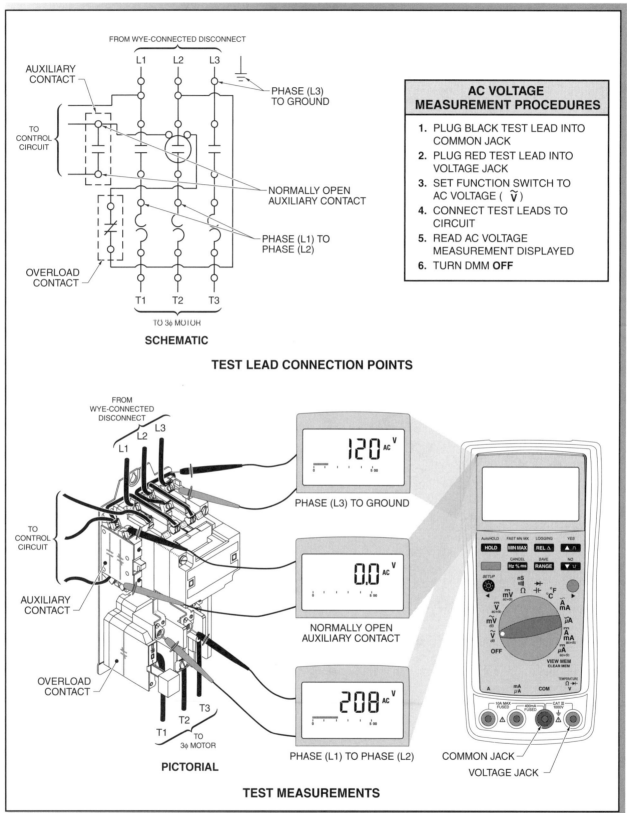

Figure 11-9. The DMM test leads do not have to match polarity when connected to an AC circuit.

DC Voltage Measurement Procedures

All DC voltage sources have positive and negative terminals, which establish polarity in a circuit. *Polarity* is the positive (+) or negative (–) electrical state of an object. All points in a DC circuit have positive and negative polarity. The DMM test leads should match the polarity of the DC voltage test points (red test lead to positive test point and black test lead to negative test point); if they do, a positive DC voltage measurement is displayed. If the DMM test leads do not match the polarity of the DC voltage point being tested (red test lead connected to negative test point and black test lead to positive test point), a negative sign appears to the left of the DC voltage measurement displayed. See Figure 11-10.

To measure DC voltages with a DMM, apply the procedure:

1. Plug the black test lead into the common jack.
2. Plug the red test lead into the voltage jack.
3. Set the function switch to DC voltage. If the DMM includes more than one DC setting, select the highest setting. For example, some DMMs include a VDC setting and an mVDC setting.
4. Connect the test leads to the circuit. Connect the black test lead to the negative polarity test point (circuit ground) first, and then connect the red test lead to the positive polarity test point. Reverse the test leads if a negative sign (–) appears to the left of the measurement displayed.
5. Read the voltage measurement displayed on the DMM.
6. Turn the DMM OFF to prevent battery drain.

⚠ WARNING

Before taking a resistance or continuity measurement, verify that the power is OFF, and that all capacitors in the circuit are discharged. Taking a resistance or continuity measurement on an energized circuit may result in the meter exploding, causing death or serious personal injury.

Resistance Measurement Procedures

DMMs are used to measure the amount of resistance in a component or circuit that is not energized. Resistance measurements are normally taken to determine the condition of a component or circuit. Resistance is displayed on the DMM in ohms (Ω), kilohms (kΩ), or megohms (MΩ). A reading of OL (overload) indicates infinite resistance or an open state. Components designed to insulate, such as rubber or plastic, have a very high resistance. Components designed to conduct, such as switch contacts, have a very low resistance. Other components such as the coil of a motor starter or a solenoid have fixed resistance values. See Figure 11-11.

To measure resistance with a DMM, apply the procedure:

1. Turn power to the circuit OFF, and lockout/tagout disconnect.
2. Isolate the component under test.
3. Plug the black test lead into the common jack.
4. Plug the red test lead into the resistance jack.
5. Set the function switch to Resistance mode. The DMM should display OL and the Ω symbol when the DMM is in the Resistance mode.
6. Connect the test leads across the component under test. Ensure the contact between the test leads and the component is good.
7. Read the resistance measurement displayed on the DMM.
8. Turn the DMM OFF to prevent battery drain.

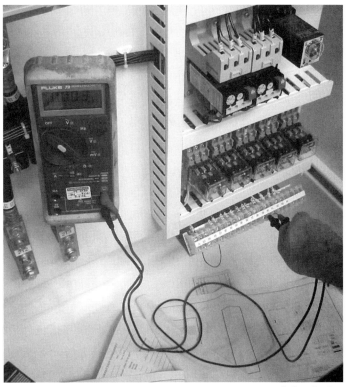

Saftronics Inc.
Resistance measurements of a circuit are affected by temperature, humidity, and quality of test lead connection.

DC VOLTAGE MEASUREMENT

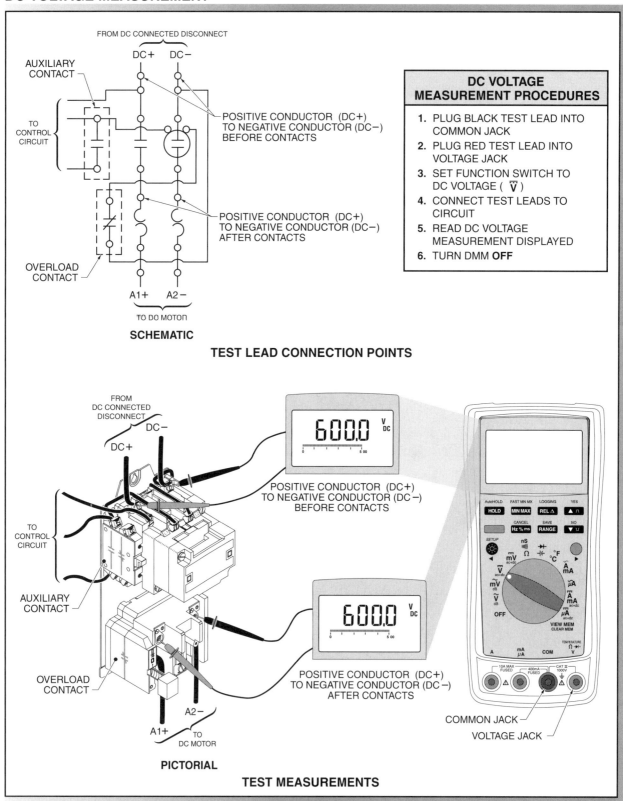

Figure 11-10. The DMM test leads are attached to the circuit when measuring DC voltage by connecting the black test lead to the negative polarity test point and then connecting the red test lead to the positive polarity test point.

RESISTANCE MEASUREMENT

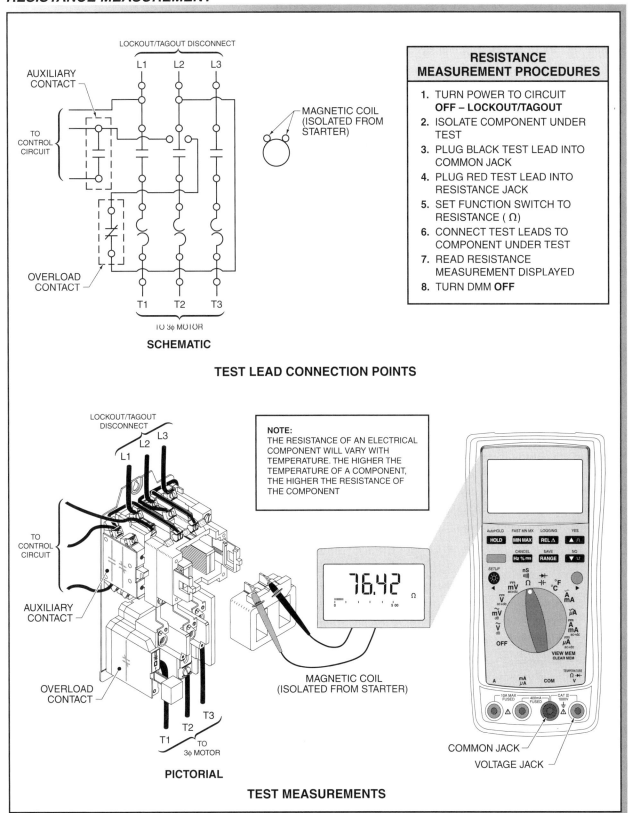

Figure 11-11. Resistance measurements are taken with the circuit de-energized.

Fluke Corporation

Scopemeters® are used to troubleshoot electric motor drives, motors, transformers, and analog and digital control circuits.

Continuity Test Procedures

Continuity is the presence of a complete path for current flow. A closed switch that is in good condition has continuity. An open switch does not have continuity. The Continuity Test mode on a DMM can be used to test electrical connections and components such as switches. The DMM emits an audible response (beep) when there is a complete path. Indication of a complete path can be used to determine the condition of a component. See Figure 11-12.

To test for continuity with a DMM, apply the procedure:

1. Turn power to the circuit OFF, and lockout/tagout disconnect.
2. Isolate the component under test.
3. Plug the black test lead into the common jack.
4. Plug the red test lead into the resistance jack.
5. Set the function switch to Continuity Test mode as required on the DMM. On most DMMs, the Continuity Test mode and Resistance mode share the same function switch position. The sign should appear in the DMM display. The DMM may still display OL and Ω.
6. Press the Continuity button if required. The sign appears in the DMM display.

7. Connect the test leads across the component under test.
8. If there is a complete path (continuity), the DMM beeps. If there is no continuity (open circuit), the DMM does not beep.
9. Turn the DMM OFF to prevent battery drain.

Testing Fuses Using Voltage Measurements

Fuses connected to AC or DC voltage sources are tested using a DMM set to measure voltage. There are two methods to test a fuse by measuring voltage: measuring voltage across an individual fuse, or measuring voltage across two fuses. See Figure 11-13. After a blown fuse is identified, remove the blown fuse as well as the other two phase fuses and verify condition by using a DMM set to measure resistance.

To test fuses with a DMM using voltage measurement, apply the procedure:

1. Gain access to the fuse or fuses to be tested.
2. Plug the black test lead into the common jack.
3. Plug the red test lead into the voltage jack.
4. Set the function switch to AC or DC voltage. Set the range to the highest setting if voltage in the circuit is unknown.
5. Verify correct supply phase-to-ground measurements.
6. Verify correct supply phase-to-phase measurements.
7. Connect the black test lead to the load side and the red test lead to the line side of an individual fuse. A blown fuse indicates source voltage.
8. An alternate step 7 is to connect a black test lead to the load side of a fuse, and connect the red test lead to the line side of another fuse. The fuse that is under test is the fuse with the black test lead on the load side. A blown fuse indicates 0 V or less than phase-to-phase voltage.
9. Repeat step 7 or 8 for each of the remaining fuses.
10. Turn the DMM OFF to prevent battery drain.

TECH FACT

To remove a fuse, use an insulated fuse puller to remove the top (line side) of the fuse first and then remove the bottom (load side) of the fuse. If the bottom of the fuse were removed first, the bottom of the fuse would be energized (hot) until the top of the fuse was removed. Never remove a fuse from a circuit that is supplying current to a load. To install a fuse, use an insulated fuse puller to insert the bottom of the fuse (load side) first and then insert the top of the fuse (line side).

CONTINUITY TESTING

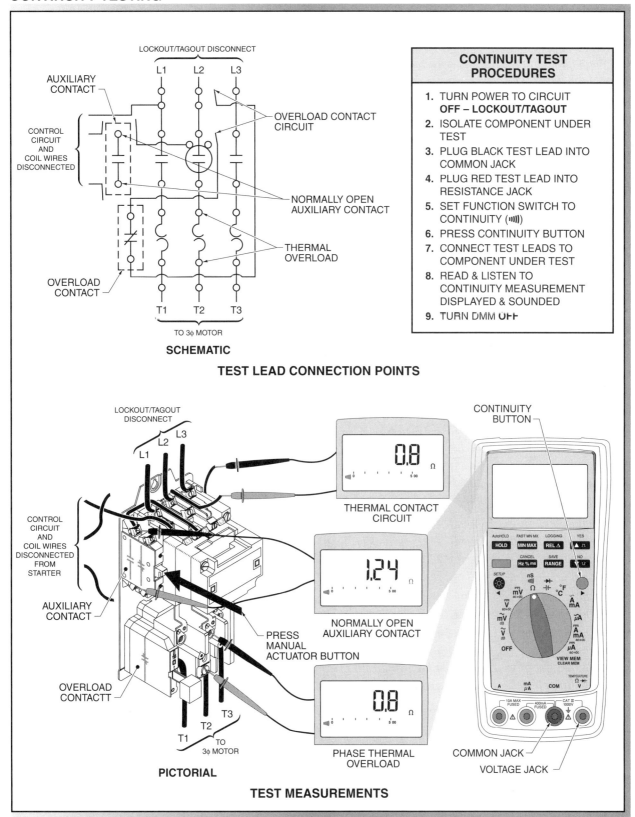

CONTINUITY TEST PROCEDURES

1. TURN POWER TO CIRCUIT **OFF – LOCKOUT/TAGOUT**
2. ISOLATE COMPONENT UNDER TEST
3. PLUG BLACK TEST LEAD INTO COMMON JACK
4. PLUG RED TEST LEAD INTO RESISTANCE JACK
5. SET FUNCTION SWITCH TO CONTINUITY (•))))
6. PRESS CONTINUITY BUTTON
7. CONNECT TEST LEADS TO COMPONENT UNDER TEST
8. READ & LISTEN TO CONTINUITY MEASUREMENT DISPLAYED & SOUNDED
9. TURN DMM OFF

Figure 11-12. The DMM beeps if there is continuity in the component or circuit.

FUSE TESTING – MEASURING VOLTAGE

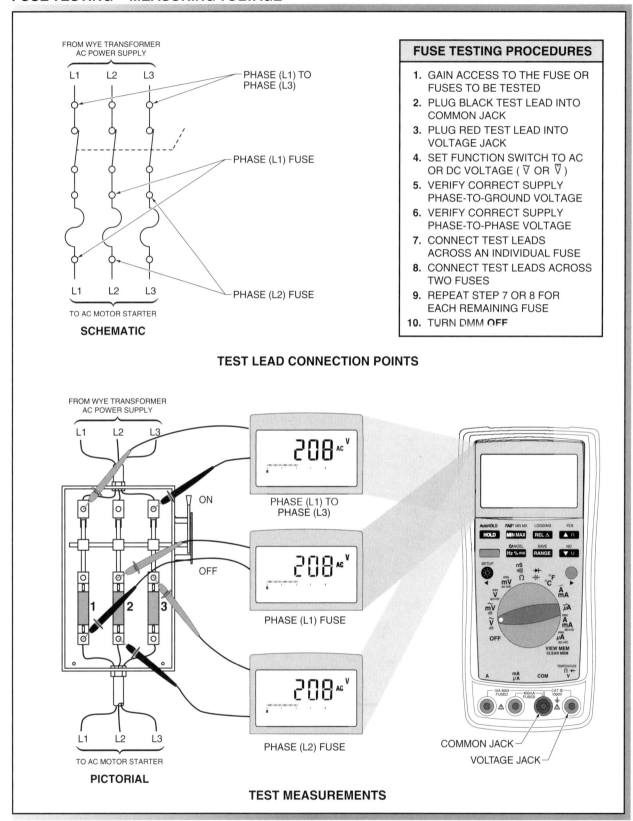

FUSE TESTING PROCEDURES

1. GAIN ACCESS TO THE FUSE OR FUSES TO BE TESTED
2. PLUG BLACK TEST LEAD INTO COMMON JACK
3. PLUG RED TEST LEAD INTO VOLTAGE JACK
4. SET FUNCTION SWITCH TO AC OR DC VOLTAGE ($\overline{V}$ OR $\widetilde{V}$)
5. VERIFY CORRECT SUPPLY PHASE-TO-GROUND VOLTAGE
6. VERIFY CORRECT SUPPLY PHASE-TO-PHASE VOLTAGE
7. CONNECT TEST LEADS ACROSS AN INDIVIDUAL FUSE
8. CONNECT TEST LEADS ACROSS TWO FUSES
9. REPEAT STEP 7 OR 8 FOR EACH REMAINING FUSE
10. TURN DMM OFF

Figure 11-13. Fuses can be tested while in an energized circuit, using a DMM set to measure voltage.

Testing Fuses Using Resistance Measurements

Fuses removed from a circuit can be tested using a DMM set to measure resistance. The resistance of a tested fuse must equal the resistance of a new fuse. Fuses from each phase must have equal resistance to achieve electrical balance between supply lines. See Figure 11-14.

To test fuses with a DMM using resistance, apply the procedure:

1. Turn power to the circuit OFF, and lockout/tagout disconnect.
2. Remove all fuses for testing.
3. Plug the black test lead into the common jack.
4. Plug the red test lead into the resistance jack.
5. Set the function switch to Resistance mode.
6. Connect the test leads to the fuse under test.
7. Read the resistance measurement displayed on the DMM.
8. Repeat steps 6 and 7 for each of the remaining fuses.
9. Turn the DMM OFF to prevent battery drain.

TECH FACT

Digital multimeters can be used to measure AC or DC by inserting the meter in series with the circuit under test. Most DMMs limit the amount of current that can be measured to 10 A. Since it is not safe or practical in many cases to insert the meter in series with the circuit under test, current clamp accessories are available. The current clamp accessories plug directly into the DMM and have moveable jaws that open to encircle the conductor under test.

Testing Diodes

A diode is tested by measuring the voltage drop across the diode when it is forward-biased. A good diode has a voltage drop across it when it is forward-biased and is allowing current to flow. The DMM Diode Test mode produces a small voltage between the test leads. The DMM then displays the voltage drop when the test leads are connected across a diode. The Diode Test mode can be used to test an individual diode or a diode that is an integral part of a bridge rectifier. The Diode Test mode can be used to test a diode that is removed from a circuit or part of a circuit, as long as all power to the circuit is OFF. See Figure 11-15.

To test a diode with a DMM, apply the procedure:

1. Turn power to the circuit OFF, lockout/tagout disconnect.
2. Ensure there is no voltage present in the circuit from charged capacitors.
3. Plug the black test lead into the common jack.

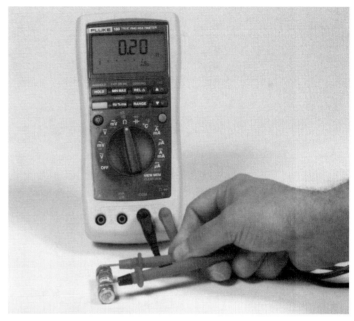

Accurately measuring resistance of a fuse requires that the fuse be laid on a nonconductive material with fingers kept behind the test lead guard.

4. Plug the red test lead into the resistance jack.
5. Set the DMM function switch to Diode Test mode.
6. Connect the black test lead to the cathode and the red test lead to the anode.
7. Read and record the measurement displayed on the DMM.
8. Connect the black test lead to the anode and the red test lead to the cathode.
9. Read and record the measurement displayed.
10. Turn the DMM OFF to prevent battery drain.

A forward-biased diode that is good displays a voltage drop ranging from .2 VDC to .8 VDC for the most commonly used silicon diodes. Some germanium diodes have a voltage drop ranging from .2 VDC to .3 VDC. The DMM displays OL when a good diode is reverse-biased. The OL reading indicates the diode is functioning as an open switch.

A diode has two failure modes: shorted or opened. Shorted is the most common failure mode. A shorted diode has the same voltage drop reading in both directions (0.0 VDC). An open diode does not allow current flow in either direction. The DMM then displays OL in both directions when the diode is opened.

⚠ CAUTION

Remove power and ensure capacitors are discharged prior to removing a diode or diode package for testing.

FUSE TESTING – MEASURING RESISTANCE

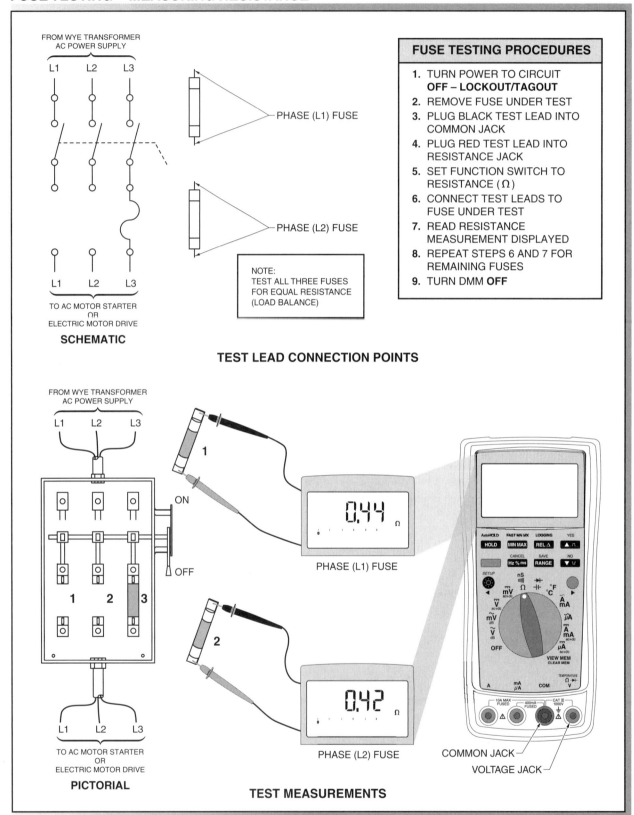

SCHEMATIC

FROM WYE TRANSFORMER AC POWER SUPPLY

L1 L2 L3

L1 L2 L3

TO AC MOTOR STARTER
OR
ELECTRIC MOTOR DRIVE

PHASE (L1) FUSE

PHASE (L2) FUSE

NOTE:
TEST ALL THREE FUSES
FOR EQUAL RESISTANCE
(LOAD BALANCE)

TEST LEAD CONNECTION POINTS

FUSE TESTING PROCEDURES

1. TURN POWER TO CIRCUIT **OFF – LOCKOUT/TAGOUT**
2. REMOVE FUSE UNDER TEST
3. PLUG BLACK TEST LEAD INTO COMMON JACK
4. PLUG RED TEST LEAD INTO RESISTANCE JACK
5. SET FUNCTION SWITCH TO RESISTANCE (Ω)
6. CONNECT TEST LEADS TO FUSE UNDER TEST
7. READ RESISTANCE MEASUREMENT DISPLAYED
8. REPEAT STEPS 6 AND 7 FOR REMAINING FUSES
9. TURN DMM **OFF**

PICTORIAL

FROM WYE TRANSFORMER AC POWER SUPPLY

L1 L2 L3

ON
OFF

1 2 3

L1 L2 L3

TO AC MOTOR STARTER
OR
ELECTRIC MOTOR DRIVE

PHASE (L1) FUSE

PHASE (L2) FUSE

COMMON JACK
VOLTAGE JACK

TEST MEASUREMENTS

Figure 11-14. Fuses can be tested with the fuse removed from the circuit using a DMM set to measure resistance.

DIODE TESTING

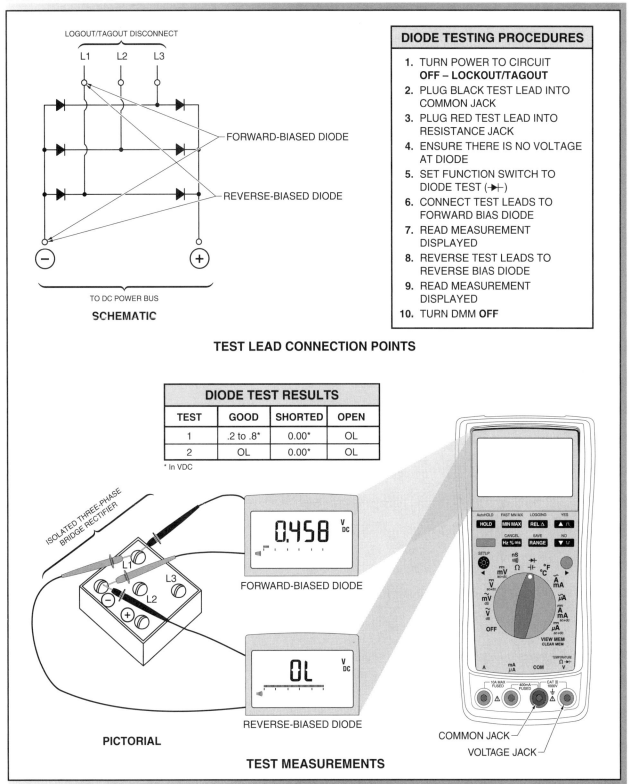

LOGOUT/TAGOUT DISCONNECT

L1 L2 L3

FORWARD-BIASED DIODE

REVERSE-BIASED DIODE

(−) (+)

TO DC POWER BUS

SCHEMATIC

DIODE TESTING PROCEDURES

1. TURN POWER TO CIRCUIT **OFF – LOCKOUT/TAGOUT**
2. PLUG BLACK TEST LEAD INTO COMMON JACK
3. PLUG RED TEST LEAD INTO RESISTANCE JACK
4. ENSURE THERE IS NO VOLTAGE AT DIODE
5. SET FUNCTION SWITCH TO DIODE TEST (→►⊢)
6. CONNECT TEST LEADS TO FORWARD BIAS DIODE
7. READ MEASUREMENT DISPLAYED
8. REVERSE TEST LEADS TO REVERSE BIAS DIODE
9. READ MEASUREMENT DISPLAYED
10. TURN DMM **OFF**

TEST LEAD CONNECTION POINTS

DIODE TEST RESULTS

TEST	GOOD	SHORTED	OPEN
1	.2 to .8*	0.00*	OL
2	OL	0.00*	OL

* In VDC

ISOLATED THREE-PHASE BRIDGE RECTIFIER

L1 L3 L2 (−) (+)

0.458 V DC

FORWARD-BIASED DIODE

OL V DC

REVERSE-BIASED DIODE

PICTORIAL

COMMON JACK
VOLTAGE JACK

TEST MEASUREMENTS

SEE APPENDIX FOR ⊕ TO ⊖ DIAGRAM

Figure 11-15. A diode is tested using the DMM Diode Test mode to measure voltage drop across the diode when it is forward-biased.

CLAMP-ON METER USAGE

A *clamp-on meter* is a meter that measures current in a circuit by measuring the strength of the magnetic field around a single conductor. If two conductors are enclosed in the jaws of a clamp-on meter, the magnetic fields of the two conductors cancel and an incorrect reading is obtained. Clamp-on meters are available for measuring both AC and DC current or AC current only. Clamp-on meters commonly include features that also allow for the measurement of AC voltage, DC voltage, frequency, and resistance. Some clamp-on meters have advanced features such as MIN/MAX and INRUSH. See Figure 11-16.

CLAMP-ON METER

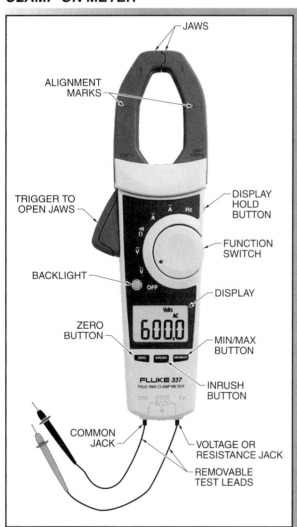

Fluke Corporation

Figure 11-16. Clamp-on meters are available with a variety of functions and features.

A clamp-on meter reading may vary depending upon the relative location of the conductor within the jaws. The jaws of a clamp-on meter have alignment marks. To obtain the most accurate reading, position the conductor within the jaws at the intersection of the alignment marks. If the conductor is positioned elsewhere within the jaws, the reading is not within the accuracy specification of the meter. A clamp-on meter reading may vary if the load is varying, as with a motor that powers a conveyor moving varying amounts of product.

TECH FACT

A true-rms clamp-on meter must be used when taking current measurements on the line or load side of an AC drive and when taking voltage measurements on the line or load side of the AC drive to obtain accurate readings.

Clamp-on Meter Safety Precautions

Clamp-on meters are used to measure electrical values such as current, frequency, voltage, and resistance in electrical and electronic circuits. Each clamp-on meter has specific features and limits. The clamp-on meter manual details meter applications, specifications, features, proper operating procedures, and safety precautions.

In addition to the general safety precautions required when using DMMs, the following precautions are required when using clamp-on meters:

- Before using the clamp-on meter, always refer to the user's manual for proper operating instructions, safety instructions, and limits.
- Never use a meter whose insulating protection has been impaired. Return the meter to the manufacturer for repair.
- Use extreme caution when clamping around uninsulated conductors or bus bars.
- If the clamp-on meter has test leads, remove the test leads and any input signals from the meter terminals before measuring current.

AC Current Measurement Procedures

Current is the quantity of electrons flowing through an electrical circuit. Current is measured in amperes. Current (amperage) measurements are taken to determine the amount of circuit loading. The advantage of a clamp-on meter is that current readings are taken without opening a circuit. See Figure 11-17.

AC CURRENT MEASUREMENT

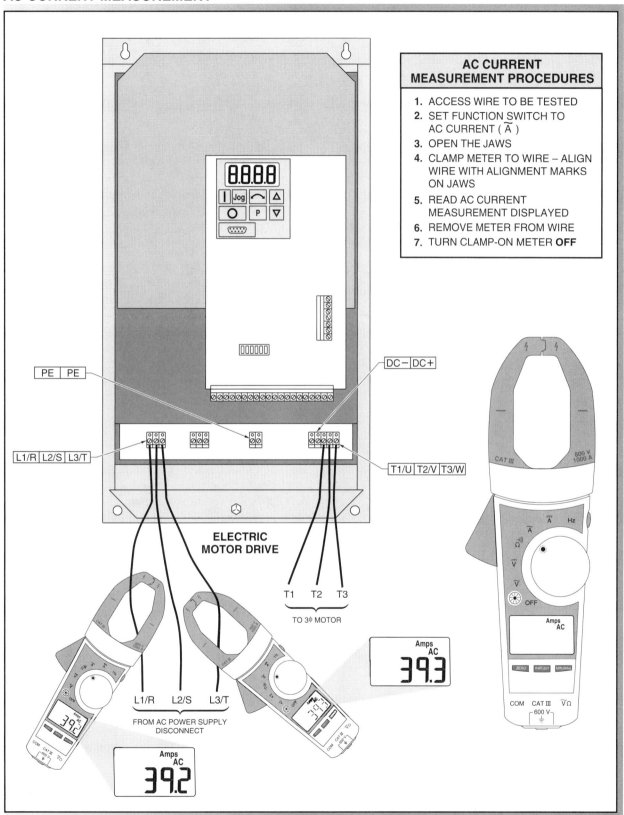

Figure 11-17. Clamp-on meters are used to measure current in AC circuits.

To measure AC current with a clamp-on meter, apply the procedure:

1. Determine if the clamp-on meter range is high enough to measure the maximum current that may exist in the test circuit. If the clamp-on meter range is not high enough, select a clamp-on meter with a higher range.
2. Access the wire to be tested.
3. Set the function switch to the correct AC current setting (600 A, 200 A, etc.). Select a setting greater than the highest possible circuit current if there is more than one current position or if the circuit current is unknown.
4. Open the jaws by pressing against the trigger.
5. Enclose one conductor in the jaws. Ensure that the jaws are completely closed and the conductor is positioned at the intersection of the alignment marks before taking the reading.
6. Read the current measurement displayed on the clamp-on meter.
7. Remove the clamp-on meter from the wire.
8. Turn the clamp-on meter OFF to prevent battery drain.

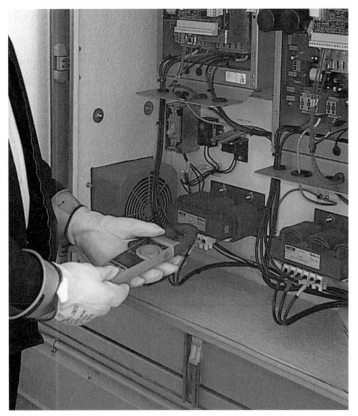

Clamp-on meters accurately measure the current through one conductor centered in the jaws.

A majority of clamp-on meters power up in the Autorange mode, which automatically selects a measurement range based on the voltage present.

DC Current Measurement Procedures

Clamp-on meters that measure DC current have an arrow embossed on the unit adjacent to the jaws. The arrow indicates the direction of meter polarity, with the arrow pointing toward the positive polarity. If the arrow points toward the negative polarity, a negative sign (–) is displayed next to the current value. See Figure 11-18.

To measure DC current with a clamp-on meter, apply the procedure:

1. Determine if the clamp-on meter range is high enough to measure the maximum current that may exist in the test circuit. If the clamp-on meter range is not high enough, select a clamp-on meter with a higher range.
2. Access the wire to be tested.
3. Set the function switch to the correct DC current setting (600 A, 200 A, etc.). Select a setting greater than the highest possible circuit current if the circuit current is unknown.
4. Push the zero button to set the meter display to zero.
5. Open the jaws by pressing against the trigger.
6. Enclose one conductor in the jaws. Ensure that the jaws are completely closed and the conductor is positioned at the intersection of the alignment marks before taking the reading. The arrow embossed on the meter points towards the positive polarity.
7. Read the current measurement displayed on the clamp-on meter.
8. Remove the clamp-on meter from the wire.
9. Turn the clamp-on meter OFF to prevent battery drain.

Frequency Measurement Procedures

Some clamp-on meters have the ability to measure frequency. Frequency measurements are only made on AC voltage circuits. The frequency measurement is displayed in hertz (Hz). The frequency of incoming electrical power to sensitive equipment such as PLCs or electric motor drives is measured to ensure that the frequency is within specification. The frequency of the electric motor drive output current is measured to calculate the speed of a motor. See Figure 11-19.

DC CURRENT MEASUREMENT

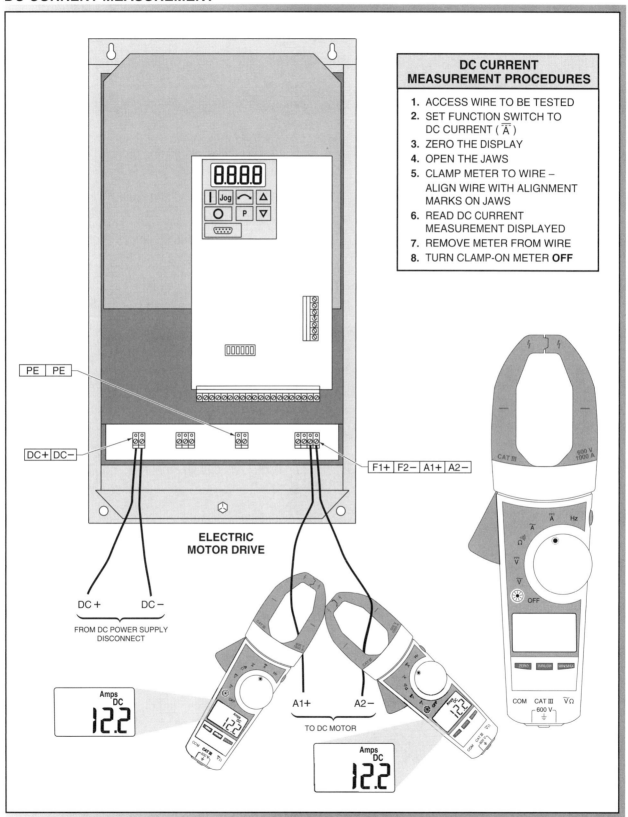

Figure 11-18. Clamp-on meters are used to measure current in DC circuits.

FREQUENCY MEASUREMENT

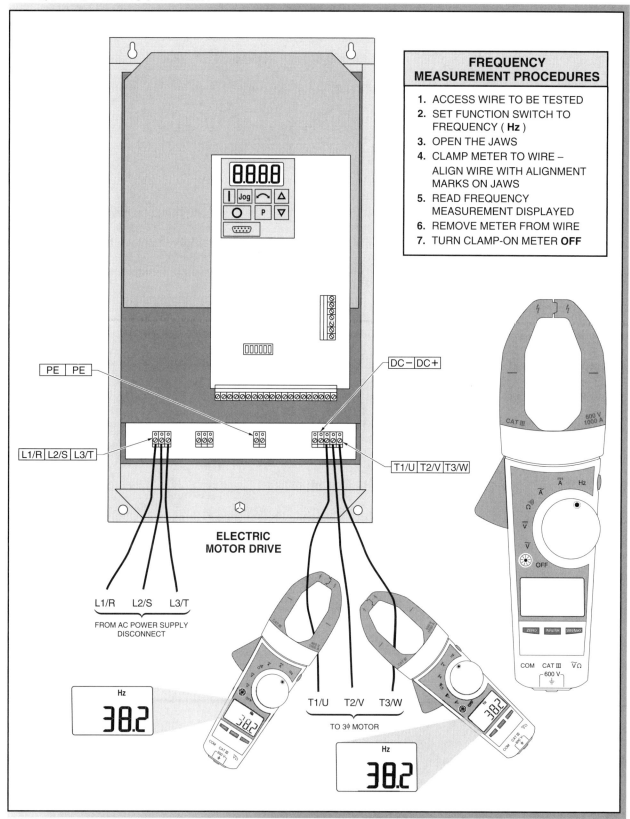

Figure 11-19. Clamp-on meters are used to measure frequency in AC circuits.

To measure frequency with a clamp-on meter, apply the procedure:

1. Determine if the clamp-on meter range is high enough to measure the maximum hertz that may exist in the test circuit. If the clamp-on meter range is not high enough, select a clamp-on meter with a higher range.
2. Access the wire to be tested.
3. Set the function switch to frequency.
4. Open the jaws by pressing against the trigger.
5. Enclose one conductor in the jaws. Ensure that the jaws are completely closed and the conductor is positioned at the intersection of the alignment marks before taking the reading.
6. Read the frequency measurement displayed on the clamp-on meter.
7. Remove the clamp-on meter from the wire.
8. Turn the clamp-on meter OFF to prevent battery drain.

⚠ CAUTION

In a 1φ electrical circuit, third-180 Hz harmonics create electrical system voltage distortions that cause overheating and equipment malfunctions. In a 3φ electrical circuit, mainly fifth-300 Hz and seventh-420 Hz harmonics create electrical system voltage distortions that cause overheating and equipment malfunctions.

Voltage Measurement Procedures

AC or DC voltage measurement procedures may vary slightly with different clamp-on meters. Some clamp-on meters have advanced features which provide convenience and specific measurement capabilities. The clamp-on meter test leads do not have to match any polarity when connected to an AC voltage. The AC voltage measurement displayed is positive, no matter which way the test leads are connected. When connecting the test leads, connect the black (common) test lead to the circuit, then connect the red (voltage) test lead to the circuit. To remove the test leads, remove the red test lead, then remove the black test lead. The clamp-on meter test leads should match the polarity of the DC voltage under test. If the DMM test leads do not match the polarity of the DC voltage being tested, a negative sign appears to the left of the DC voltage measurement displayed. See Figure 11-20.

TECH FACT

A 3φ, 60 Hz industrial circuit with harmonics created by nonlinear loads (electric motor drives) typically has the neutral conductor(s) oversized by 200%.

To measure AC or DC voltages with a clamp-on meter, apply the procedure:

1. Plug the black test lead into the common jack.
2. Plug the red test lead into the voltage jack.
3. Set the function switch to AC or DC voltage. Set the range to the highest AC or DC voltage setting if the voltage in the circuit is unknown.
4. Connect the test leads to the circuit.
5. Read the voltage measurement displayed on the clamp-on meter.
6. Turn the clamp-on meter OFF to prevent battery drain.

Resistance Measurement Procedures

Clamp-on meters measure the amount of resistance in a component or circuit that is not energized. Resistance measurements are normally taken to indicate the condition of a component or circuit. Resistance is displayed on the clamp-on meter as ohms (Ω) or kilohms (kΩ). A reading of OL (overload) indicates infinite resistance or an open. See Figure 11-21.

To measure resistance with a clamp-on meter, apply the procedure:

1. Turn power to the circuit OFF, and lockout/tagout disconnect.
2. Isolate the component under test.
3. Plug the black lead into the common jack.
4. Plug the red test lead into the resistance jack.
5. Set the function switch to Resistance mode. The clamp-on meter should display OL and the Ω symbol when the DMM is in the Resistance mode.
6. Connect the test leads across the component under test. Ensure the contact between the test leads and the component is good. Body contact with the metal ends of the test leads affects the resistance measurements.
7. Read the resistance measurement displayed on the clamp-on meter.
8. Turn the clamp-on meter OFF to prevent battery drain.

VOLTAGE MEASUREMENT

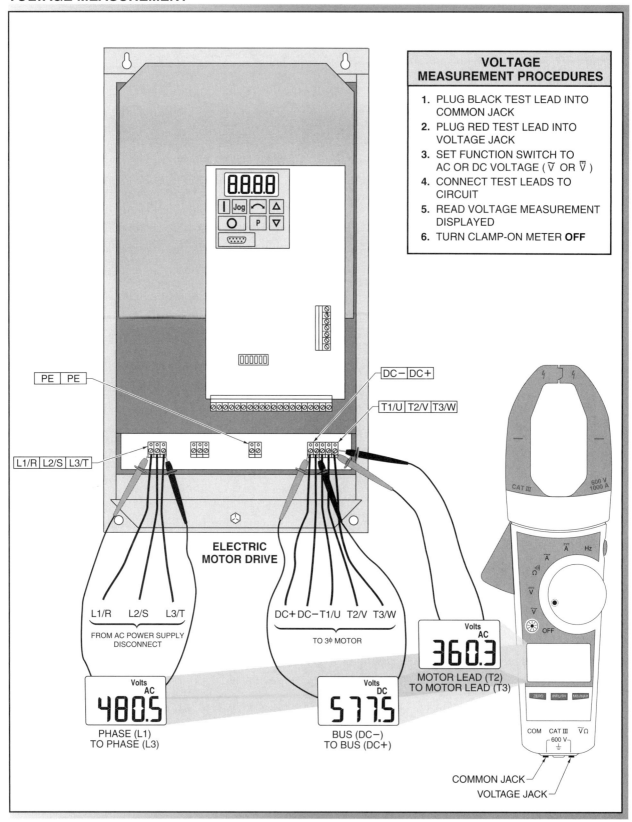

Figure 11-20. Clamp-on meters are used to measure voltage in AC and DC voltage circuits.

RESISTANCE MEASUREMENT

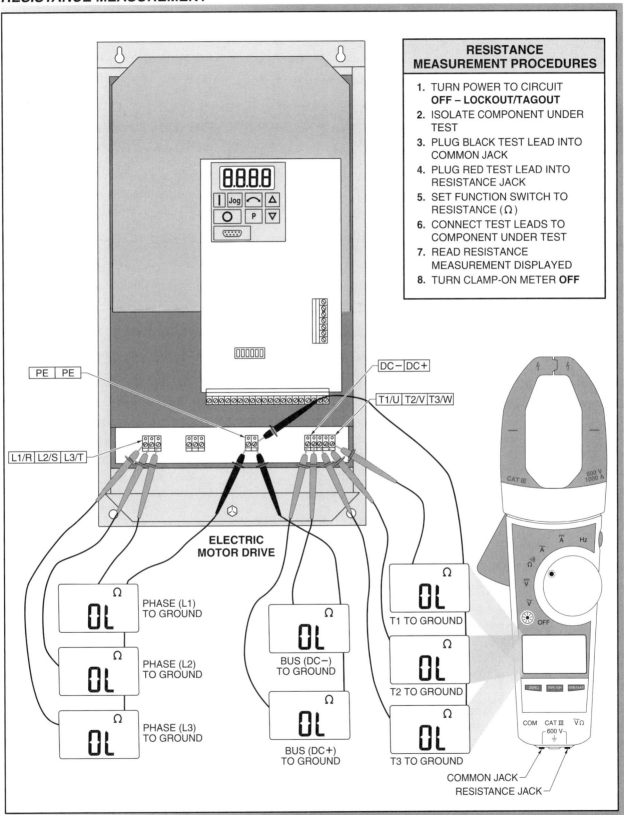

RESISTANCE MEASUREMENT PROCEDURES

1. TURN POWER TO CIRCUIT **OFF – LOCKOUT/TAGOUT**
2. ISOLATE COMPONENT UNDER TEST
3. PLUG BLACK TEST LEAD INTO COMMON JACK
4. PLUG RED TEST LEAD INTO RESISTANCE JACK
5. SET FUNCTION SWITCH TO RESISTANCE (Ω)
6. CONNECT TEST LEADS TO COMPONENT UNDER TEST
7. READ RESISTANCE MEASUREMENT DISPLAYED
8. TURN CLAMP-ON METER **OFF**

PE | PE

DC− | DC+

T1/U | T2/V | T3/W

L1/R | L2/S | L3/T

ELECTRIC MOTOR DRIVE

Ω
OL
PHASE (L1) TO GROUND

Ω
OL
PHASE (L2) TO GROUND

Ω
OL
PHASE (L3) TO GROUND

Ω
OL
BUS (DC−) TO GROUND

Ω
OL
BUS (DC+) TO GROUND

Ω
OL
T1 TO GROUND

Ω
OL
T2 TO GROUND

Ω
OL
T3 TO GROUND

COMMON JACK
RESISTANCE JACK

Figure 11-21. Clamp-on meters are used to measure resistance in de-energized circuits.

MEGOHMMETER USAGE

A *megohmmeter* is a high-resistance-range meter that is used to measure the insulation integrity of individual conductors, motor windings, transformer windings, etc. The megohmmeter imposes a high DC voltage into the insulation being tested (conductor or motor winding) and measures the leakage current. The megohmmeter reading is displayed in megohms, which are calculated using Ohm's law.

Megohmmeters come in a variety of styles. Most megohmmeters have a selector switch to choose the appropriate test voltage. The megohmmeter display can be an analog display or a digital display. The megohmmeter power source can be a hand crank, battery powered, or 120 VAC. Some models have dual power sources (hand crank or 120 VAC). See Figure 11-22.

Good insulation has a high resistance reading. Poor insulation has a low resistance reading. Insulation can be damaged by moisture, oil, dirt, excessive heat, excessive cold, corrosive vapors, aging, and vibration. The ideal megohmmeter reading is infinite resistance between the conductor or motor winding being tested, and ground. Infinite resistance is depicted by the symbol, (∞). Often the megohmmeter reading is less than infinite resistance. Typically, megohmmeters should read a minimum of 1 MΩ of resistance for every 1000 V of insulation rating. Wire used in 480 VAC or 240 VAC distribution systems has a rating of 600 V. For testing purposes consider the wire to have 1000 V insulation. The stator winding insulation for inverter duty motors has a rating of approximately 1500 V. For testing purposes consider the stator winding to have 2000 V insulation.

Megohmmeter Safety Precautions

Megohmmeters are used to measure resistance in conductors and motor windings. Each megohmmeter has specific features and limits. The user's manual details specific applications and specifications, features, proper operating procedures, and safety precautions.

In addition to the general safety precautions required when using DMMs or clamp-on meters, the following precautions are required when using megohmmeters:

- Before using the megohmmeter, always refer to the user's manual for proper operating instructions, safety instructions, and limits.
- Ensure that the conductors or windings to be tested are de-energized. Never use a megohmmeter on an energized circuit.
- A high voltage is developed across the test leads of a megohmmeter. Do not touch the metal ends of the test leads during a test.

- The technician must ensure that the other end of the conductor or winding under test is clear of ground and personnel. The megohmmeter voltage is present at the other end of the conductor. The megohmmeter voltage is also present at the motor leads not being tested because of the internal winding connections. This voltage can injure personnel.
- The conductor or winding under test develops a capacitive charge from the voltage applied by the megohmmeter. After the measurement is complete, the capacitive charge must be discharged. Follow the megohmmeter's instructions.
- Never use a megohmeter on conductons that are connected to a drive. The voltage from the megohmeter can damage the drive. Always disconnect the conductors before testing.

Insulation Spot-Test Procedures

An *insulation spot-test* is a test that checks the insulation integrity of the stator windings and load conductors of a motor. An insulation spot-test is taken when a motor is placed in service and every six months thereafter. The test should also be taken after a motor has maintenance performed or has been rewound. See Figure 11-23.

To perform an insulation spot-test using a megohmmeter, apply the procedure:

1. Plug the black test lead into the common jack.
2. Plug the red test lead into the line or resistance jack.
3. Select the test voltage level. The test voltage is typically set higher than the voltage rating of the insulation under test in order to stress the insulation. The 1000 V setting is normally used for motors and conductors operating at 480 VAC. If the megohmmeter does not have a 1000 V setting, use the voltage setting closest to but not greater than 1000 V.
4. Connect the black test lead of the megohmmeter to a grounded surface.
5. Connect the red test lead of the megohmmeter to one of the motor winding leads or an individual conductor.
6. Apply the test voltage for 60 sec. Record the megohmmeter reading. Record the lowest reading on an insulation spot test graph if all readings are above the minimum acceptable reading. The lowest reading is used because a motor or a set of feeder conductors is only as good as its weakest point.
7. Discharge the circuit under test.
8. Repeat steps 5, 6, and 7 for the remaining motor winding leads or individual conductors.
9. Turn the megohmmeter OFF to prevent battery drain.
10. Interpret the readings taken.

MEGOHMMETER STYLES

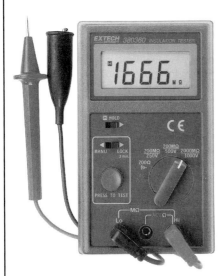

Extech Instruments

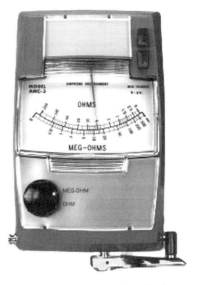

Advanced Test Products

Fluke Corporation

Fluke Corporation

**TESTING TRANSFORMER
INSULATION**

Fluke Corporation

**TESTING WIRE
RUN INSULATION**

Fluke Corporation

**TESTING ELECTRIC
MOTOR INSULATION**

Figure 11-22. Megohmmeters are available in different designs with a variety of functions and features.

INSULATION INTEGRITY MEASUREMENT

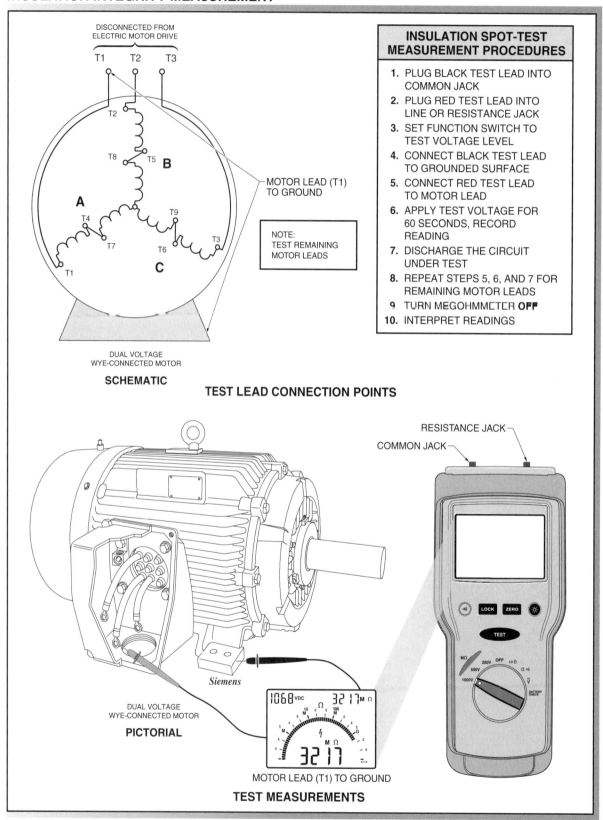

Figure 11-23. An insulation spot-test checks motor insulation over the life of a motor.

Megohmmeter readings must be interpreted. See Figure 11-24. A motor installed outdoors and tested two days in a row can have two different readings depending on the weather (foggy conditions one day would result in low MΩ and sunny conditions the next day would result in high MΩ). In general, megohmmeter readings are the most useful when taken semiannually over a period of years. A sudden drop in the megohmmeter readings of a motor (100 MΩ to 2 MΩ over a six-month period) is an indication of a problem, even if the reading is above the accepted value. A large difference between megohmmeter readings (L1 = 20 MΩ, L2 = 21 MΩ, and L3 = 1 MΩ) is an indication of a problem.

The cause of the low megohmmeter readings must be determined. The cause can be moisture, dirt, or damaged insulation. Typically, low megohmmeter readings require the motor or conductors to be repaired or replaced. The repaired or replaced items need to be tested with the megohmmeter before they are placed into service.

TECH FACT

Environmental conditions such as temperature and humidity affect the readings taken by a megohmmeter.

MEGOHMMETER READING INTERPRETATION

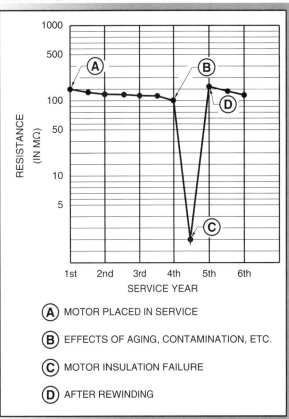

Figure 11-24. Megohmmeter readings should be charted for proper interpretation.

Name_____ Date _____

True-False

T (F) 1. A true-rms meter gives the correct reading regardless of the waveform. *pg 297*

T (F) 2. Voltage surges vary only in voltage level. *pg 298*

(T) F 3. Digital multimeters (DMMs) designed to the category (CAT) III standard are resistant to higher transient voltages than DMMs designed to the CAT II standard. *pg 300*

T (F) 4. Ghost voltage is a voltage on a DMM that is connected to an energized circuit.

(T) F 5. The DMM test leads should match the polarity of the DC voltage test points. *not pg 308*

T (F) 6. A disadvantage of a clamp on ammeter is that current readings are taken without opening the circuit.

T (F) 7. An open switch that is in good condition has continuity.

(T) F 8. Insulation can be damaged by moisture, oil, dirt, excessive heat, excessive cold, corrosive vapors, aging, and vibration. *closed pg 324*

(T) F 9. A motor installed outdoors and tested two days in a row with a megohmmeter can have different readings depending on the weather. *pg 327*

T (F) 10. A good diode has a voltage drop across it when it is reversed-biased and is allowing current to flow. *forward*

(T) F 11. When taking measurements using a DMM, avoid holding the meter by hand to minimize exposure to the effects of transient voltages. *pg 302*

T (F) 12. Clamp-on ammeters are used to measure only current in an electrical circuit. *pg 323*

T F 13. A megohmmeter is a high-resistance-range meter that is used to test the insulation integrity of individual conductors, motor windings, transformer windings, etc. *pg 324*

(T) F 14. Continuity is the presence of a complete path for current to flow.

T (F) 15. Environmental conditions such as temperature and humidity do not affect the readings taken by a megohmmeter. *pg 327*

Completion

nonsinusoidal waveform 1. A(n) ___ is a waveform that has a distorted appearance when compared with a pure sine waveform. *pg 297*

compliance 2. The IEC sets standards but does not test or inspect for ___. *pg 301*

double-insulated 3. A DMM should have ___ test leads, recessed input jacks, and finger shrouds.

Contact Bounce 4. ___ is displayed on the DMM by the movement of one or more bar graph segments as the contact opens then closes. *pg 304*

Cancel 5. If two conductors are enclosed in the jaws of a clamp-on ammeter, the magnetic fields of the two conductors ___ and an incorrect reading is obtained. *pg 316*

AC voltage 6. Frequency measurements are only made on ___ circuits. *pg 318*

Shorted 7. ___ is the most common failure mode of a diode. *pg 313*

forward-Biased 8. A(n) ___ diode that is good displays a voltage drop ranging from .2 VDC to .8 VDC for the most commonly used silicon diodes. *pg 313*

Arrow 9. Clamp-on ammeters that measure DC current have a(n) ___ embossed on the unit adjacent to the jaws. *pg 318*

arc Blast 10. A(n) ___ is an explosion that occurs when the surrounding air becomes ionized and conductive. *pg 299*

Multiple Choice

_D_ 1. The IEC 1010 standard divides DMM applications into ___ overvoltage installation category(ies).
    A. one   C. three *pg 300*
    B. two   D. four

_C_ 2. A(n) ___ DMM automatically adjusts to a higher range setting if the range is not high enough.
    A. automatic   C. autoranging *pg 303*
    B. autoloading   D. true-rms

_C_ 3. A bar graph of a DMM is updated ___ times per second.
    A. 4   C. 30 *pg 304*
    B. 10   D. 60

_C_ 4. On a DMM, a reading of ___ indicates infinite resistance or an open state.
    A. 0 Ω   C. OL *pg 321*
    B. 100 Ω   D. 1 MΩ

_D_ 5. A(n) ___ reading may vary depending upon the relative location of the conductor within the jaws.
    A. megohmmeter   C. voltmeter
    B. ammeter   D. clamp-on ammeter
pg 316

Name _____ Date _____

Activity 11-1. Overvoltage Installation Categories

Determine the overvoltage installation category (CAT I to Cat IV) for each troubleshooting application. The category indicates the required CAT rating of the meter.

_____ **1.** The overvoltage installation category when troubleshooting the START/STOP pushbutton station connected to the electric motor drive at Location 1 is CAT ___.

_____ **2.** The overvoltage installation category when troubleshooting the lamp ballast at Location 2 is CAT ___.

_____ **3.** The overvoltage installation category when troubleshooting the circuit breakers at the main switchboard of the facility is CAT ___.

_____ **4.** The overvoltage installation category when troubleshooting the line reactors and load reactors of an electric motor drive inside the motor control center is CAT ___.

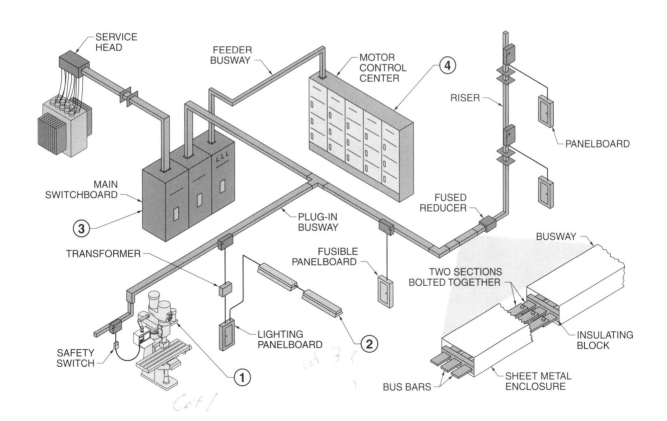

Activity 11-2. Troubleshooting Fuses

Using the meter readings, determine whether the fuse is good, bad, or questionable. A fuse is questionable when the meter readings do not provide enough information to determine whether the fuse is definitely good or bad.

_____ **1.** Fuse 1 is ___.

_____ **2.** Fuse 2 is ___.

_____ **3.** Fuse 3 is ___.

_____ **4.** Fuse 1 is ___.

_____ **5.** Fuse 2 is ___.

_____ **6.** Fuse 3 is ___.

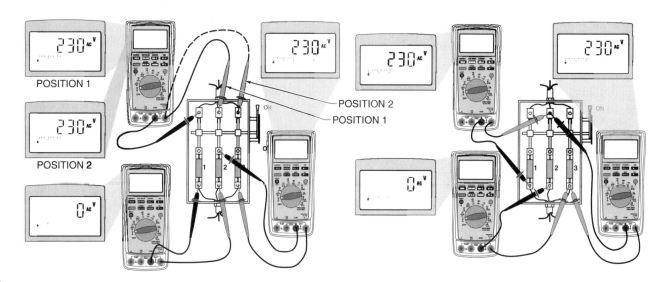

Activity 11-3. Troubleshooting Circuit Breakers

Using the meter readings, determine whether the circuit breaker is good, bad, or questionable. A circuit breaker is questionable when the meter readings do not provide enough information to determine whether the circuit breaker is definitely good or bad.

_____ **1.** The circuit breakers in Disconnect 1 are ___.

_____ **2.** The circuit breakers in Disconnect 2 are ___.

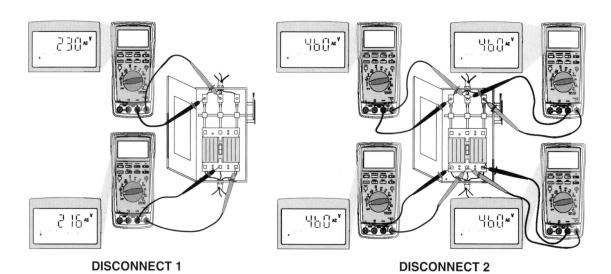

DISCONNECT 1   **DISCONNECT 2**

Activity 11-4. Meter Setting and Connections

1. Set Meter 1 selector switch to test the fuses in the disconnect. Connect and set Meter 1 to test Fuse 1. Connect and set Meter 2 to measure the DC bus voltage of the electric motor drive. Set Meter 3 to measure the output current of the electric motor drive.

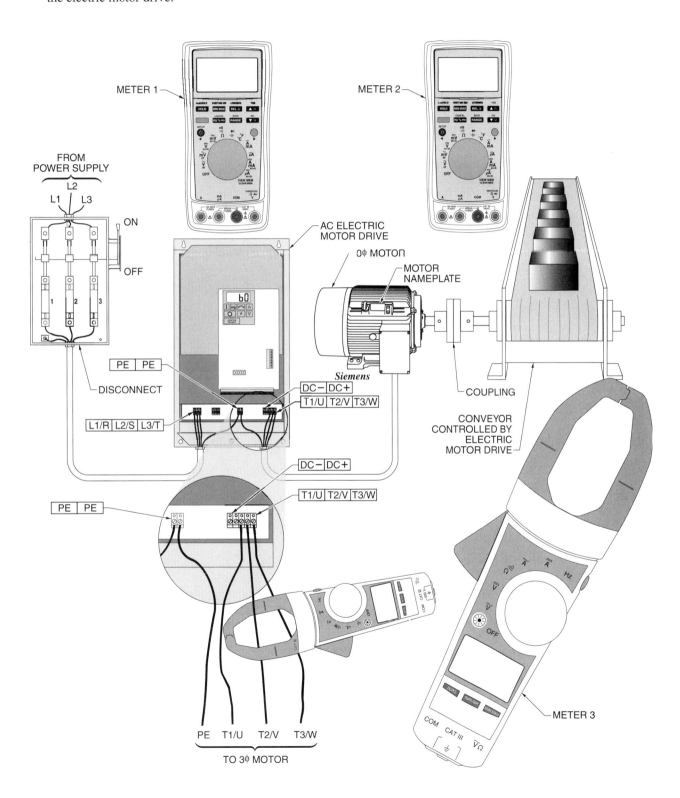

Activity 11-5. Insulation Spot-Test Measurements

A 230 V, 3φ motor has provided years of service. Semiannual insulation spot-test measurements were taken.

1. Develop an Insulation Spot-Test Graph from the readings. Mark the point at which the motor requires service.

INSULATION SPOT-TEST	
TEST TAKEN	RESISTANCE READING (IN MΩ)
TEST 1	400
TEST 2	350
TEST 3	325
TEST 4	250
TEST 5	225
TEST 6	225
TEST 7	200
TEST 8	3
TEST 9	2

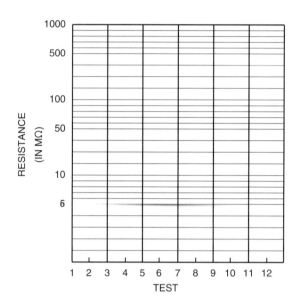

INSULATION SPOT-TEST GRAPH

12 Electric Motor Drive Start-up Procedures

The start-up procedure of an electric motor drive consists of systematic checks. The start-up procedure uses initial checks, secondary checks, and final checks to verify that an electric motor drive is fully operational. During the start-up procedure, the electric motor drive, electric motor, and driven load are checked for correct installation and the parameters of the drive are adjusted. Following the proper start-up procedures ensures that an electric motor drive and electric motor are installed correctly and are fully operational, in the least amount of time and with minimal problems.

INITIAL CHECKS

Initial checks are performed on an electric motor drive, motor, and driven load. All power to the electric motor drive must be OFF during the initial checks.

To perform the initial checks, apply the procedure:
1. Turn disconnect OFF and lockout/tagout disconnect.
2. Verify that no voltage is present at the electric motor drive terminals using a digital multimeter (DMM). Test for voltage at the power terminal strips and the control terminal strips.

A foreign control voltage present on any control terminal strip must be turned OFF. A *foreign control voltage* is a voltage that originates outside the electric motor drive, such as voltage from a control panel or sensor that interfaces with the drive. See Figure 12-1.

Initial Motor Inspection

Technicians performing an electric motor drive start-up must inspect the motor, the driven load, and the alignment between the motor and load. A technician may not have installed the motor and it is possible that the motor was installed incorrectly or was damaged after installation. Often one trades person installs and aligns a motor as another trades person performs the electrical work required. The exterior of a motor must be clean and the area adjacent to the motor must be free of debris to ensure proper ventilation for motor cooling.

Technicians must check that the voltage, current, horsepower, and service factor ratings of a motor are compatible with the electric motor drive. A motor should be inverter rated, NEMA MG-1, Section IV, Part 31 compliant. Contact the motor manufacturer or licensed engineer before powering a motor that is not NEMA MG-1, Section IV, Part 31 compliant with an electric motor drive. Motor nameplate data must be recorded in the electric motor drive manual or other type of recordkeeping at the time of installation. Motor nameplate information is used in the start-up procedure and in the troubleshooting procedures.

Initial motor inspection requires technicians to inspect the motor, the driven load, and the alignment between the motor and load, and ensure that the area is clean and free of debris.

ELECTRIC MOTOR DRIVE NO-VOLTAGE TESTING POSITIONS

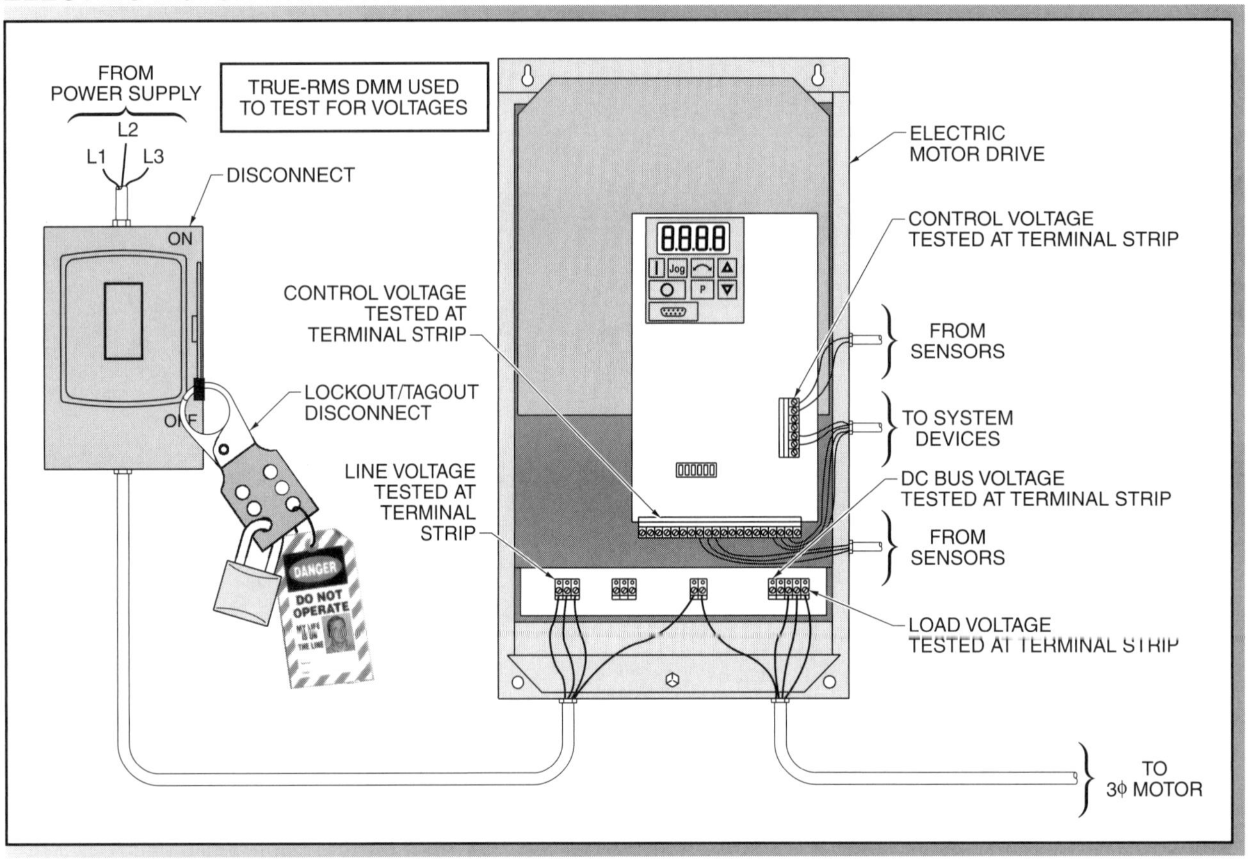

Figure 12-1. Electric motor drive voltage is tested at the line power terminal strip, load power terminal strip, DC bus terminal strip, and the control terminal strips.

In addition to verifying and recording motor data, motor connections must be checked. With the motor termination box open, technicians must verify the following:

- The motor connections are correct for the supply voltage; for example, 480 VAC connections are used when an electric motor drive is supplied with 480 VAC.
- The motor connections are properly insulated with no bare conductors exposed.
- The load conductors and motor leads have not been damaged.

After inspecting the motor connections and making any necessary corrections, replace the motor termination box cover, exercising care not to pinch any of the conductors.

A motor must be correctly aligned with the driven load and securely fastened in place. Improper alignment between a motor and driven load causes excessive current draw, premature motor and driven load bearing failure, and premature wear and damage to the driven load, and can prevent the motor from rotating

the driven load. A motor that is not securely fastened in place causes alignment and vibration problems. See Figure 12-2.

Motors are connected to loads by a variety of methods using couplings, gears, belts, and chains. Checks must be made that the motor shaft and driven load shaft rotate freely. A rough movement in the rotation of either the motor shaft or the driven load indicates a problem. A rough movement causes problems similar to alignment problems. When possible, the motor must be disconnected from the driven load. Rotating the motor and driven load may be impossible because of the method used to connect the motor to the load, the location of the motor and load, or the physical size of the motor and load.

TECH FACT

An improperly installed motor can cause an electric motor drive to malfunction.

MOTOR MECHANICAL INSPECTION

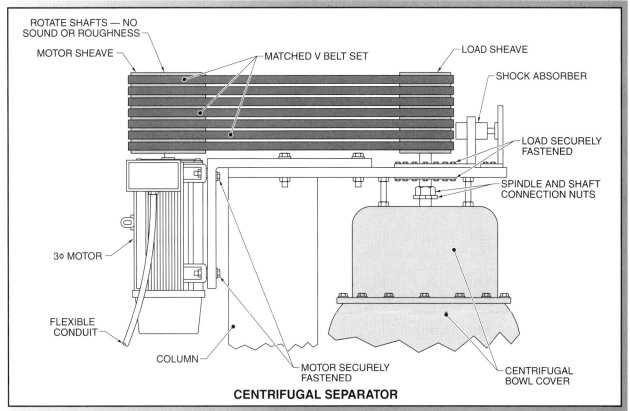

Figure 12-2. A motor and load must be secured in place and correctly aligned, and the shafts must be rotated to detect mechanical problems.

Initial Motor Testing

Motor testing involves testing the integrity of the motor winding insulation and conductor insulation. Moisture, oil, dirt, excessive heat, excessive cold, corrosive vapors, and vibrations damage motor insulation and conductor insulation. Insulation integrity is tested by using a megohmmeter to perform an insulation spot-test. Good insulation has a high resistance reading. Poor insulation has a low resistance reading.

An ideal megohmmeter reading is infinite (∞) resistance between a conductor and ground or a motor winding and ground. Often, a megohmmeter reading is less than infinite resistance. A rule of thumb is that for every 1000 V of insulation rating, there is 1 MΩ of resistance. Wire used in 480 VAC or 240 VAC distribution systems has a rating of 600 V. For testing purposes the wire is considered to have 1000 V insulation. The stator winding insulation for inverter duty motors is rated at approximately 1500 V. NEMA MG-1, Section IV, Part 31 defines an inverter duty motor as having insulation able to withstand voltage peaks of 3.1 times the rated voltage.

For testing purposes, the winding insulation is considered to have 2000 V insulation. The minimum megohmmeter reading when testing load conductors and motor windings having 2000 V insulation is 2 MΩ. The motor windings and load conductors are simultaneously tested because the megohmmeter readings are taken from the electric motor drive end of the conductors.

Conductor and Motor Winding Insulation Spot-Test. During the conductor and motor winding insulation spot-test, a megohmmeter is connected between the load conductors and ground, and a test voltage is applied. Never use a megohmmeter on conductors that are connected to a drive. Always disconnect conductors before testing. See Figure 12-3.

To perform a conductor and motor winding insulation spot-test, apply the procedure:

1. Set the megohmmeter to the selected test voltage level. The test voltage is typically set higher than the voltage rating of the insulation under test in order to stress the insulation. The 1000 V setting is typically used for motors and conductors operating at 480 VAC.

Some megohmmeters do not have a 1000 V setting and use a voltage setting close to 1000 V, but not exceeding 1000 V.
2. Plug the black test lead into the common jack.
3. Plug the red test lead into the resistance jack.
4. Connect the black lead of the megohmmeter to a grounded surface.

5. Connect the red lead of the megohmmeter to one of the individual conductors.
6. Apply the test voltage for 60 sec. Record the megohmmeter reading.
7. Discharge the circuit under test.
8. Repeat steps 5, 6, and 7 for the remaining load conductors.

MOTOR CONDUCTOR AND WINDING INSULATION SPOT-TEST

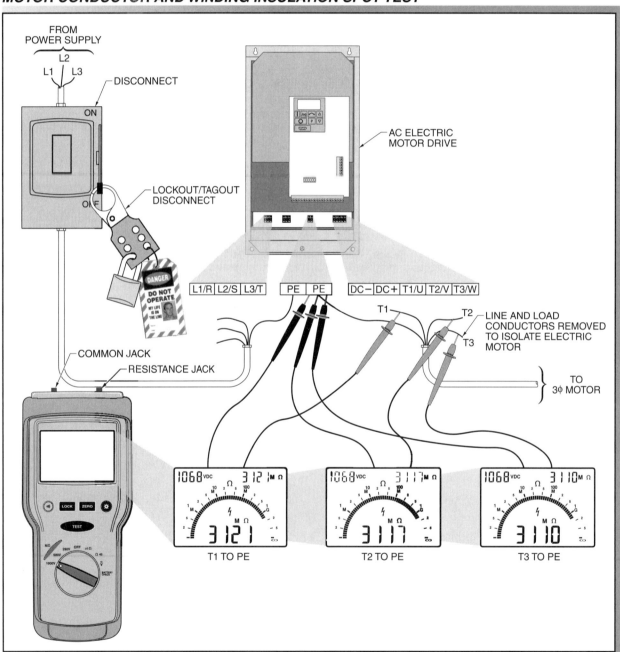

Figure 12-3. A conductor and motor winding insulation spot-test connects a megohmmeter between a load conductor and ground and stresses the insulation with a voltage.

The megohmmeter readings must be at least 2 MΩ and be relatively close to one another, such as T1 to ground measuring 10 MΩ, T2 to ground measuring 9.7 MΩ, and T3 to ground measuring 9.8 MΩ. A reading less than 2 MΩ, or a large difference between the readings (greater than 50%), is cause for additional testing.

Additional testing is necessary to find the cause of low reading(s) during the conductor and motor winding insulation spot-test. The cause of a low reading(s) may be the load conductors from the electric motor drive to the motor, or the motor windings. See Figure 12-4.

To isolate the cause of the low reading(s), apply the procedure:

1. Open the motor termination box. Disconnect the load conductors from the motor leads. Do not disconnect the connections between windings.
2. Use the conductor and motor winding insulation spot-test procedure to test the load conductors. Technicians must ensure that the load conductors are disconnected at the electric motor drive and that all conductors are clear of ground and personnel. Megohmmeters produce voltages that can injure personnel or damage an electric motor drive.
3. Use the conductor and motor winding insulation spot-test procedure to test the motor windings. Technicians must ensure that the motor leads not being tested are clear of ground and personnel. Megohmmeter voltage is present at the motor leads not being tested because of internal winding connections of the motor.

   When the cause of a low reading has been determined, the load conductors or the motor requires repair or replacement. When repairs are complete, the load conductors and motor windings must be rechecked with the megohmmeter.
4. Reconnect the load conductors to the motor leads and replace the motor termination box cover.

> **⚠ WARNING**
> *Technicians must consult OSHA Standard 1910.147,* **The Control of Hazardous Energy,** *for industry standards on lockout/tagout.*

Initial Electric Motor Drive Checks

The initial checks of an electric motor drive consist of a visual inspection, a check for unwanted grounds, and an electric motor drive component test. The checks are used to find problems caused by shipping, careless installation practices, or faulty components. Electric motor drive checks can prevent a minor problem from causing serious damage to a drive, a driven load, or personnel.

Omron IDM Controls

Initial electric motor drive check consists of a visual inspection, a check for unwanted grounds, and a drive component test.

Visual Inspection. A thorough visual inspection must be made of the interior of an electric motor drive for packing and shipping materials that are still in place, tools and hardware remaining from the installation, debris from the installation, and damage to components. To visually inspect an electric motor drive, apply the procedure:

- Remove all packing and shipping materials, tools, and hardware.
- Vacuum the interior of the electric motor drive with an ESD safe vacuum even if there is no debris visible. See Figure 12-5. Debris such as metal filings from holes drilled in the drive enclosure can fall into an electric motor drive during the installation process. The debris can cause grounds or shorts that damage the electric motor drive. Special attention must be paid to ensure that debris is not blocking the free flow of air across the electric motor drive's heat sink.
- Contact the manufacturer of the electric motor drive if damaged components are found.

Technicians must record nameplate information from an electric motor drive on the Electric Motor Drive Record Sheet. The information recorded is the electric motor drive model number, serial number, input voltage, input current, output current, and horsepower rating. A *model number* is a number that identifies the design of a drive and how the drive is to be used. A *serial number* is a number that identifies when a drive was manufactured, what modifications were made to the design, and what software is used. *Input voltage* is the voltage supplied to an electric motor drive. *Input current* is the current required by an electric motor drive to avoid sine wave flat-topping. *Output current* is the current sent to a motor to cause rotation; output current must be equal to or greater than the motor nameplate current. Horsepower ratings are not used to compare electric motor drives and motors because two motors with the same horsepower rating may have different motor nameplate current values. An electric motor drive may have different output current values for constant torque loads and variable torque loads. Also, electric motor drives have different output current values for different pulse width modulated (PWM) frequencies.

INSULATION SPOT-TESTS

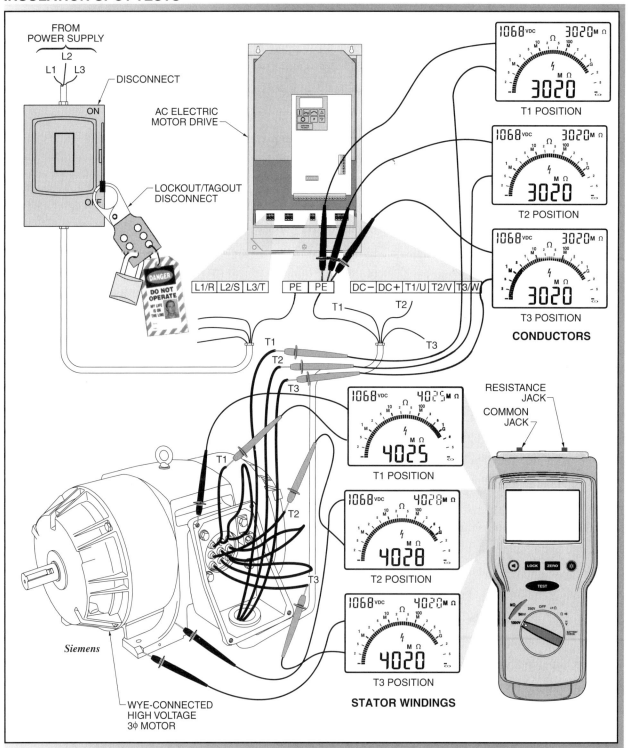

Figure 12-4. An insulation spot-test identifies a problem as either with the conductors to a motor or the motor windings.

PREPARING ELECTRIC MOTOR DRIVE FOR POWER

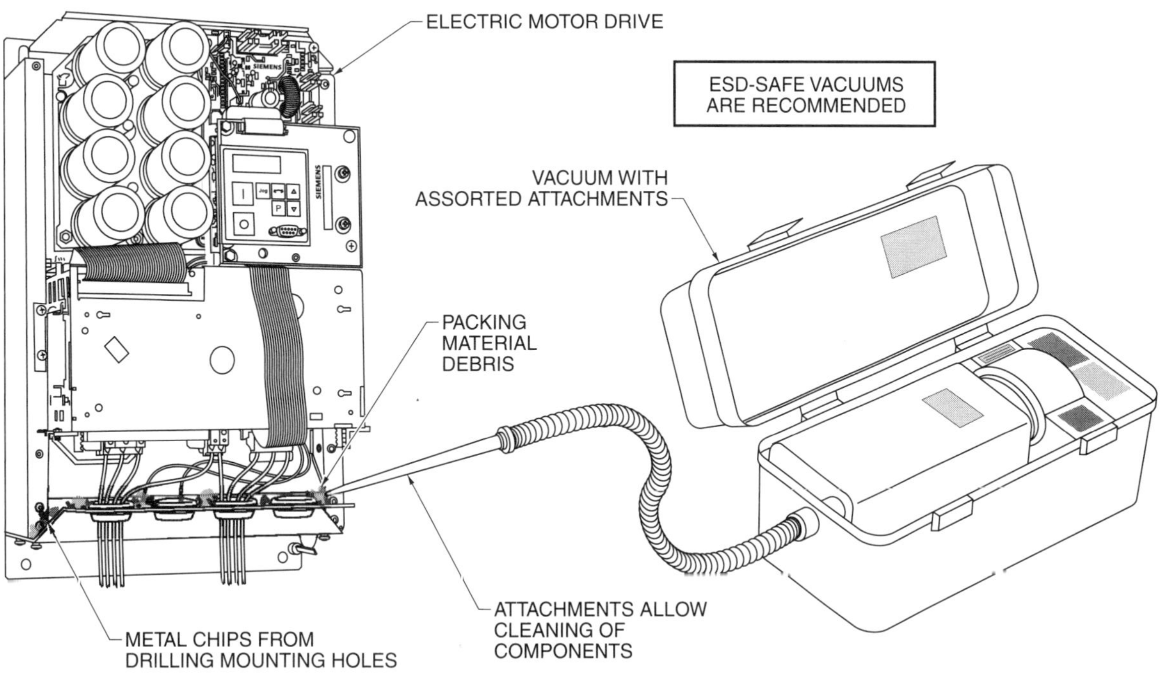

Figure 12-5. An ESD-safe vacuum with accessories is used to safely remove debris from an electric motor drive.

Ground Testing. Technicians must check for unwanted grounds in an electric motor drive. A ground in the converter section (rectifier), DC bus section, or inverter section indicates a problem with an electric motor drive. Faulty components or conductive debris are possible causes. The tests are accomplished at the power terminal strip of an electric motor drive using a DMM set to measure resistance. See Figure 12-6.

To check for electric motor drive grounds, apply the following procedure:

1. Ensure that the power is OFF at the local disconnect and that the line conductors are disconnected from the terminal strip of the electric motor drive.
2. Remove the load conductors from the terminal strip of the electric motor drive to isolate the drive.
3. Test the converter section of the electric motor drive. Place the DMM common probe on the ground terminal and touch the positive probe to L1 (R), then L2 (S), and then L3 (T). Each DMM resistance measurement must read OL, indicating no grounds.
4. Test the DC bus section of the electric motor drive. Place the DMM common on the ground terminal and touch the positive probe to DC+ and then DC–. Each DMM resistance measurement must read

OL, indicating no grounds. Some electric motor drive manufacturers denote the DC bus with B+ and B–. The B+ and B– indicate the location where a dynamic braking resistor is or may be installed. Make sure that B+ and B– go directly to the DC bus and not to a switching transistor. The DC bus section test is invalid if B+ and B– do not go directly to the DC bus.

5. Test the inverter section of the electric motor drive. Place the DMM common probe on the ground terminal and touch the positive probe to T1 (U), then T2 (V), and then T3 (W). Each DMM resistance measurement must read OL, indicating no grounds.
6. If a reading other than OL is obtained, a ground exists and the ground must be eliminated before the electric motor drive can be energized.
7. Reconnect the line conductors and load conductors to the electric motor drive.

> **⚠ CAUTION**
> *Applying power to an electric motor drive that has an unwanted ground may result in personal injury and/or damage to the drive.*

ELECTRIC MOTOR DRIVE GROUND TESTING

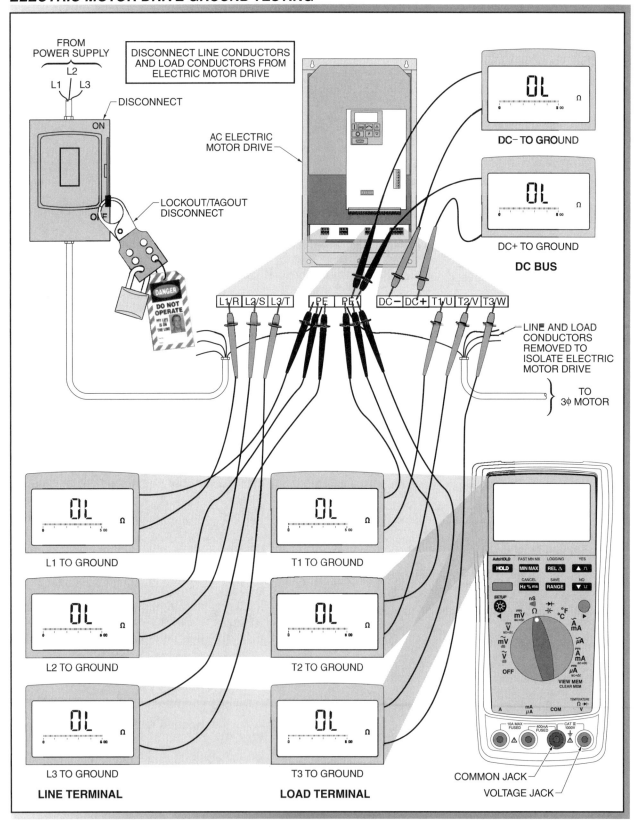

Figure 12-6. A DMM set to measure resistance is used to check for grounds in an electric motor drive.

Initial Electric Motor Drive Component Testing. Electric motor drive component testing uses the diode test mode of a DMM to test the semiconductor devices in the converter section and the inverter section of an electric motor drive. The diode test mode places a voltage across and a current through the semiconductor device. A good semiconductor reads 0.2 VDC to 0.8 VDC when forward-biased and OL when reverse-biased. A shorted semiconductor reads 0.0 VDC in both directions. An open semiconductor reads OL in both directions.

The following items must be considered when testing the converter and inverter sections of an electric motor drive:

- Ensure that power is OFF and that the line conductors are disconnected from the terminal strip of the electric motor drive.
- Remove the load conductors from the terminal strip of the electric motor drive to isolate the drive.
- Perform component tests at the power terminal strips of the electric motor drive.
- When a semiconductor is reverse-biased by a DMM, the DC bus capacitors are charged. A few seconds are required for the capacitors to charge. The voltage value increases (assuming the device is good) and an OL reading appears. When the capacitors are charged, all measurements are taken in which an OL reading is anticipated.
- A DMM may not be able to fully charge the large capacitors found in high-horsepower electric motor drives.
- When semiconductors are paralleled in the converter and inverter sections of high-horsepower electric motor drives, tests indicate which group of semiconductors is defective, not which device is defective.
- Reconnect the line conductors after the tests are completed.
- Some electric drive manufacturers denote the DC bus with B+ and B−. B+ and B− are used to indicate the location where a dynamic braking resistor can be installed. Make sure B+ and B− go directly to the DC bus and not to a switch transistor. Converter and inverter tests are invalid if B+ and B− do not go directly to the DC bus.

Converter Section Testing. Typically, the converter section consists of full-wave rectifiers or bridge rectifiers that are made up of diodes. See Figure 12-7.

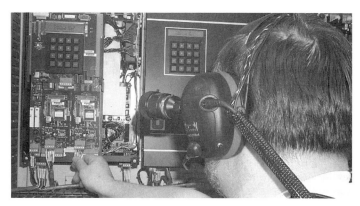

Unico, Inc.

Special vision aids can be required when performing component tests on micro drives.

To test the converter section, apply the procedure:
1. Verify that the disconnect is OFF and the disconnect is locked out/tagged out.
2. Use a DMM set to diode test mode and follow the Diode Test Matrix to test the converter section.
3. When a DMM reading does not match the matrix, a problem exists with a semiconductor.
4. A diode that reads 0.0 VDC when forward-biased and reverse-biased is shorted and must be replaced.
5. A diode that reads OL when forward-biased and reverse-biased is open and must be replaced.

Inverter Section Testing. Typically, the inverter section consists of transistors and fly-back diodes. See Figure 12-8.

To test the inverter section, apply the following procedure:
1. Use a DMM set to diode test mode and follow the IGBT Test Matrix to test the inverter section.
2. When a DMM reading does not match the matrix, a problem exists with a semiconductor.
3. A transistor that reads 0.0 VDC when forward-biased and reverse-biased is shorted and must be replaced.
4. A transistor that reads OL when forward-biased and reverse-biased is open and must be replaced.

When testing the inverter section transistors, a DMM is actually testing the fly-back diodes. The fly-back diodes and transistors are in the same package or module. Because of the close physical proximity, when one component fails the other typically fails.

SECONDARY CHECKS

A *secondary check* is a check that is performed between the initial checks and final checks of an electric motor drive. Secondary checks focus on the electric motor drive wiring. Line wiring, load wiring, and control wiring of an electric motor drive are inspected and errors that occurred during installation are discovered and corrected. Power is applied to the electric motor drive during secondary checks, allowing parameters to be adjusted. An electric motor drive is not started (run) during the secondary checks.

CONVERTER SECTION DIODE TESTING

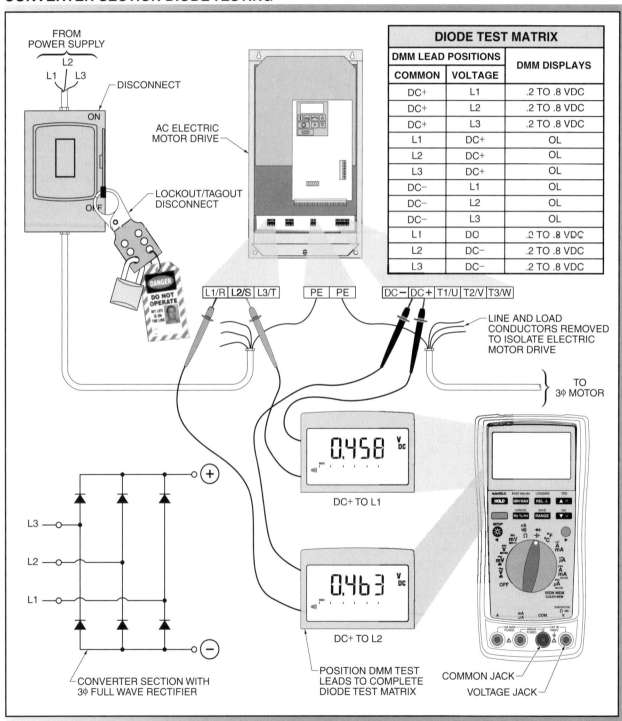

DIODE TEST MATRIX		
DMM LEAD POSITIONS		DMM DISPLAYS
COMMON	VOLTAGE	
DC+	L1	.2 TO .8 VDC
DC+	L2	.2 TO .8 VDC
DC+	L3	.2 TO .8 VDC
L1	DC+	OL
L2	DC+	OL
L3	DC+	OL
DC−	L1	OL
DC−	L2	OL
DC−	L3	OL
L1	DC−	.2 TO .8 VDC
L2	DC−	.2 TO .8 VDC
L3	DC−	.2 TO .8 VDC

Figure 12-7. A DMM set to the diode check mode is used to test the diodes in the converter section of an electric motor drive.

INVERTER SECTION DIODE AND TRANSISTOR TESTING

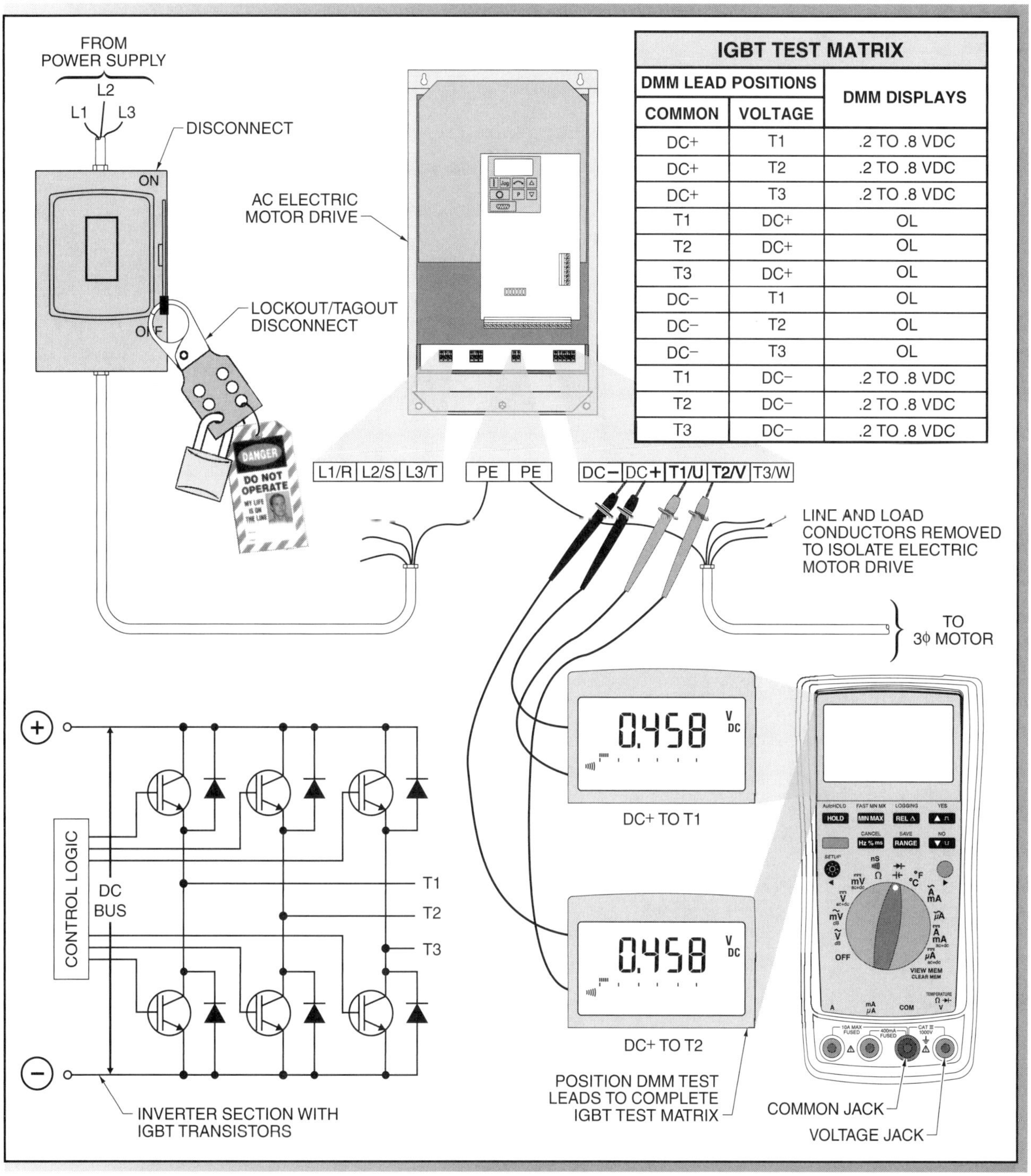

IGBT TEST MATRIX		
DMM LEAD POSITIONS		DMM DISPLAYS
COMMON	VOLTAGE	
DC+	T1	.2 TO .8 VDC
DC+	T2	.2 TO .8 VDC
DC+	T3	.2 TO .8 VDC
T1	DC+	OL
T2	DC+	OL
T3	DC+	OL
DC−	T1	OL
DC−	T2	OL
DC−	T3	OL
T1	DC−	.2 TO .8 VDC
T2	DC−	.2 TO .8 VDC
T3	DC−	.2 TO .8 VDC

Figure 12-8. A DMM set to the diode check function is used to test the transistors and fly-back diodes in the inverter section of an electric motor drive.

TECH FACT

Verify that all connections are tight before applying power to an electric motor drive.

Power Wiring

The power wiring consists of the conductors feeding the electric motor drive (line conductors) and the conductors feeding the motor from the drive (load conductors).

Technicians must verify that the line voltage is within the operating range of the electric motor drive. Consult the electric motor drive manual for the allowable voltage range. See Figure 12-9. To check line voltage, apply the following procedure:

1. Remove the lockout/tagout from disconnect.
2. Open the cover of the disconnect.
3. Set the DMM for AC voltage.
4. Measure the incoming line voltage, L1 to L2, L1 to L3, and L2 to L3.

POWER SUPPLY VOLTAGE TESTING

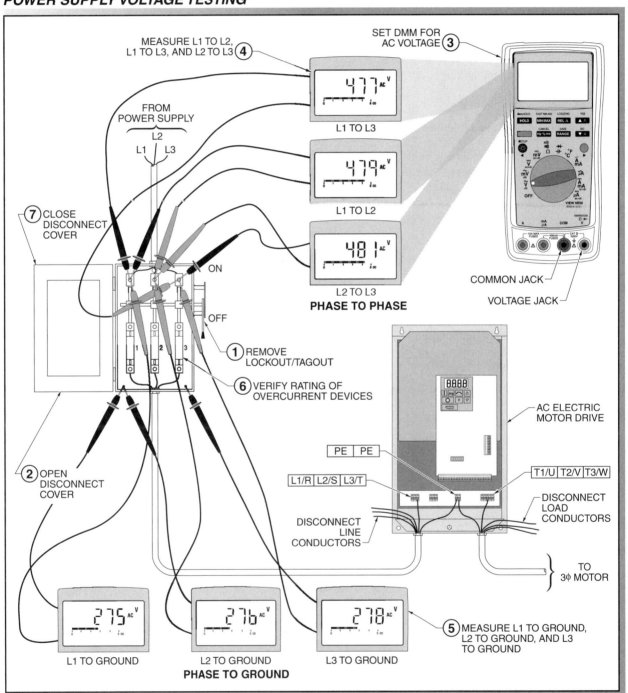

Figure 12-9. Power supply voltage to an electric motor drive is measured on the line side of the disconnect with a DMM.

5. Measure the incoming line voltage to ground to verify that the system is electrically symmetrical with respect to ground, L1 to GRD, L2 to GRD, and L3 to GRD.
6. When a disconnect also incorporates an overcurrent device, verify that the disconnect has the correct type and rated overcurrent devices.
7. Close the disconnect cover and reinstall the lockout/tagout.

Technicians must verify that the line conductors and grounds terminate at the correct positions on an electric motor drive and that all wire connections are tight. Technicians must also verify that all conduit connections are tight. See Figure 12-10.

Control Wiring

Control wiring consists of inputs and outputs (all wiring excluding power wiring) connected to an electric motor drive. The control wiring drawings or engineered drawings show where the control wiring must be connected.

Technicians must verify that the control wiring of an electric motor drive matches the drawings.

To verify the control wiring, apply the procedure:
1. Verify that control wires terminate at the correct locations.
2. Verify that terminal screws are tight and that strands of wire are not sticking out from underneath the terminals.
3. Verify that the AC, DC, analog, and serial communication control wiring of an electric motor drive are separated from each other as much as possible. Wires that have to cross must cross at a 90° angle to minimize the effects of electromagnetic interference (EMI).
4. Verify that the shield and drain wires of twisted-pair cables are only grounded at the electric motor drive.

TECH FACT

Electric motor drive DIP switches or jumpers must be set. Record settings on the Electric Motor Drive Record Sheet.

VERIFYING WIRE CONNECTIONS

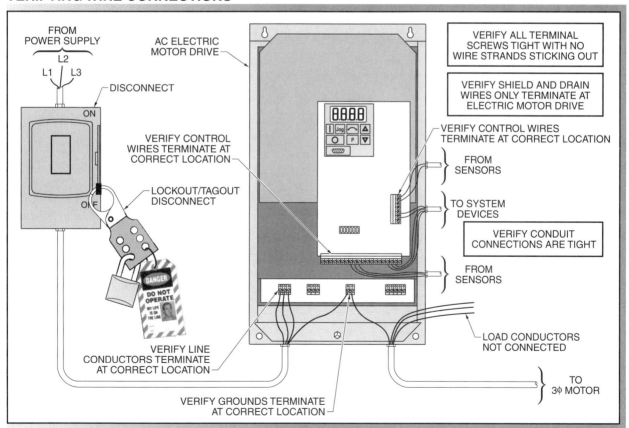

Figure 12-10. The line conductors, grounds, and control wiring must be tight and must terminate at the correct positions on an electric motor drive.

Secondary Electric Motor Drive Checks

Secondary electric motor drive checks consist of setting DIP switches, adjusting overloads, applying power to the drive, and adjusting drive parameters. The electric motor drive is not run during secondary drive check procedures but the drive is energized. The secondary electric motor drive check procedures are for typical installations. Manufacturer instructions and all applicable state and local codes must be followed at all times.

An electric motor drive may contain DIP switches or jumpers that select various functions, such as an analog input signal configuration. The DIP switches and jumpers are set for a specific application. Record all settings on the Electric Motor Drive Record Sheet for future reference. See Figure 12-11.

Overload relays must be externally installed in a circuit when an electric motor drive is not approved to provide motor overload protection, multiple motors are powered from a single drive, or a bypass contactor is used. Overload relays are either adjustable over a range of values or require a specific heater element. Overload(s) must be sized and set for the nameplate current of a motor in order to properly protect the motor. See Figure 12-12.

When DIP switches, jumpers, wiring, and overload protection are correct, apply power to the electric motor drive. See Figure 12-13. To apply power to an electric motor drive, apply the following procedure:

1. Verify that the load conductors are disconnected from the terminal strip of the electric motor drive.
2. Install the covers of an electric motor drive because a major electric motor drive failure can lead to an explosion when the drive is energized.
3. Remove the lockout/tagout from disconnect.
4. Stand to the side of the disconnect and electric motor drive when energizing, in case of a major failure. Turn the local disconnect ON. Do not push the START (I) button. The electric motor drive LED display or clear text display activates. The electric motor drive cooling fan(s) may or may not start when power is applied, depending on the drive model. If the fans do not start at this point, technicians must check that the fans start when the START (I) button is pushed. If there are any loud noises, smoke, or explosions, immediately turn the local disconnect OFF. Refer to the troubleshooting matrices for the electric motor drive and motor.

DIP SWITCH SETTINGS

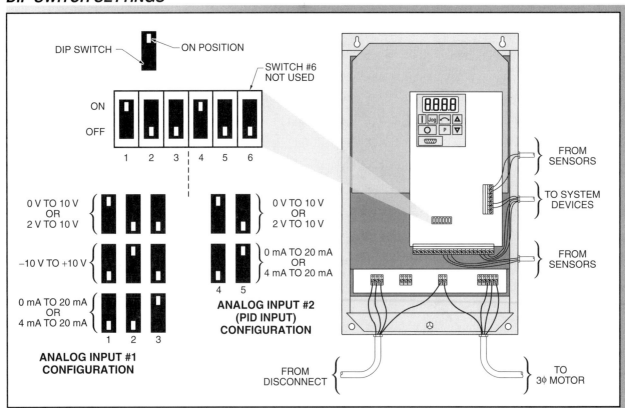

Figure 12-11. DIP switches or jumpers are used on electric motor drives to select various functions for a specific application.

MOTOR STARTERS AND OVERLOAD RELAYS

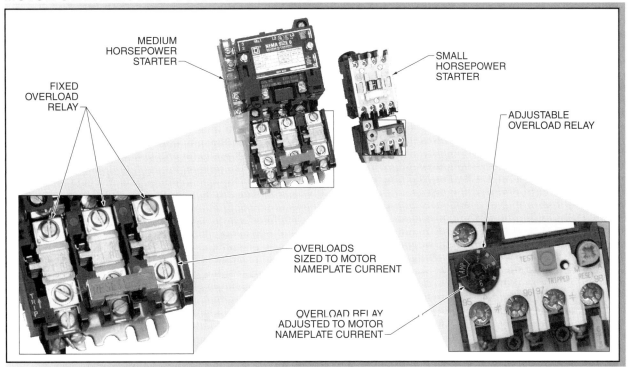

Figure 12-12. Overload relays must be sized for the nameplate current of a motor and are available in a variety of configurations.

APPLYING POWER TO AN ELECTRIC MOTOR DRIVE

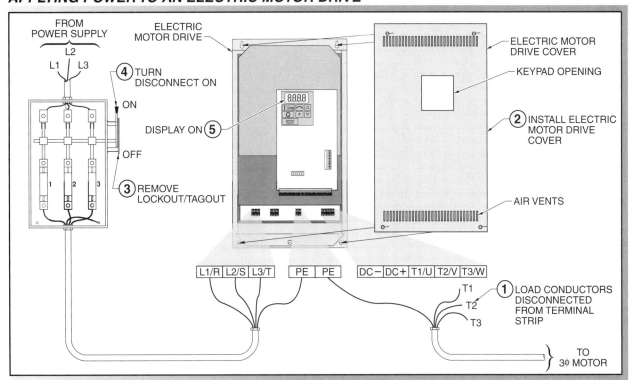

Figure 12-13. Power is applied to an electric motor drive after DIP switches, jumpers, wiring, and overload protection are set or configured correctly.

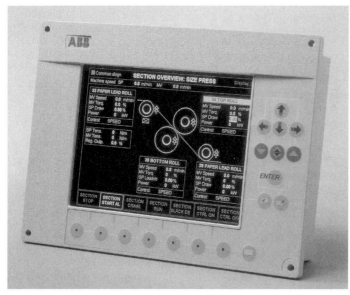

ABB Inc. Drives & Power Electronics
Some electric motor drive manufacturers have designed special devices to aid in parameter programming.

Parameter Programming

When an electric motor drive is energized, parameters are adjusted to the specific motor application. Any parameter values that are changed from the default settings must be recorded on the Electric Motor Drive Record Sheet. A record of the parameters that have been changed is useful in the event of an electric motor drive problem or failure. Additional changes to parameter values may be necessary during other procedures. The following items must be considered when programming electric motor drive parameters:

- Verify the parameter software version, if applicable. Record the software version number on the Electric Motor Drive Record Sheet.
- Program the appropriate parameters with the motor nameplate data collected during the motor inspection. For multi-motor installations, the stator resistance must remain at the default value and the motor nameplate current must be set to the maximum value. The motor nameplate current is set to maximum value because external overload relays are protecting the individual motors.
- Change parameters from default settings when specifically required for an application. When there is no set of parameter specifications, the parameter default settings should be used. Default settings are typically the safest option for many applications.
- When a licensed engineer or qualified person has provided a set of parameter specifications, program the electric motor drive with the provided specifications.

- Program the reverse inhibit parameter to prevent a motor from running in reverse when running in reverse damages the motor or driven load.
- Program the flying start parameter to allow an electric motor drive to synchronize with the speed of a motor to prevent overcurrent faults when the load is turning the motor before power is applied.
- Program the input mode to keypad, allowing the electric motor drive to be controlled by the integral keypad.
- Program speed reference to internal.
- Program the display mode to show electric motor drive output frequency in hertz (Hz).

Secondary Electric Motor Drive Test

A *secondary electric motor drive test* is a test that verifies the basic functionality of a drive. When the control mode is sensorless vector control or closed-loop vector control, it may not be possible to run certain electric motor drives with the motor disconnected. When possible, technicians must change the control mode to constant torque or variable torque in order to perform the secondary electric motor drive test. See Figure 12-14.

To test an electric motor drive, apply the following procedure:

1. Do not use two-way radios around electric motor drives, especially with the drive covers removed. Electric motor drives are susceptible to radio frequency interference (RFI) from two-way radios. Using two-way radios in close proximity to an electric motor drive can result in the drive running unexpectedly.
2. Stand to the side of the disconnect and electric motor drive when pushing the start button, in case of a major drive failure. Push the START (I) button. The cooling fan(s) must start if they did not start when power was applied. The LED display must ramp up to a low speed, such as 5 Hz. If the LED display shows 0 Hz, push the RAMP UP (↑) button until 5 Hz is shown.
3. Push the RAMP UP (↑) button until 60 Hz is shown on the LED display.
4. Program the display mode to show electric motor drive output voltage. The electric motor drive output voltage must be approximately the same as the drive input voltage at 60 Hz (assuming a 60 Hz supply), such as 480 VAC displayed when the input voltage is 485 VAC.
5. Push the STOP (O) button. The voltage must decrease to 0 VAC.
6. Turn the disconnect OFF and lockout/tagout disconnect.

7. Wait for the DC bus capacitors to discharge. Do not manually discharge the capacitors by shorting + to –. Remove the electric motor drive cover. Use a DMM to verify that the AC line voltage is not present. Use a DMM to verify that the DC bus capacitors have discharged. Do not rely on the DC bus charge LEDs because LEDs can burn out, giving a false indication.

8. Connect the load conductors (conductors to the motor) to the power terminal strip, T1 (U), T2 (V), and T3 (W). Reinstall the electric motor drive cover.

9. Remove the lockout/tagout from disconnect.

10. Stand to the side of the disconnect and electric motor drive when energizing, in case of a major drive failure. Turn the local disconnect ON. Do not push the START (I) button.

FINAL CHECKS

Final checks are checks used to limit the source of potential problems during start-up of an electric motor drive. Each successive test eliminates an element as a potential problem and adds another element to be tested. When problems occur during the final checks, refer to the troubleshooting procedures for the electric motor drive and motor. Items to be considered during the final checks include the following:

- The integral keypad is used to control an electric motor drive with the motor disconnected from the load.
- The integral keypad is used to control an electric motor drive with the motor connected.
- The electric motor drive and motor are tested to the specific motor application.
- The bypass contactor is tested.

SECONDARY ELECTRIC MOTOR DRIVE TEST

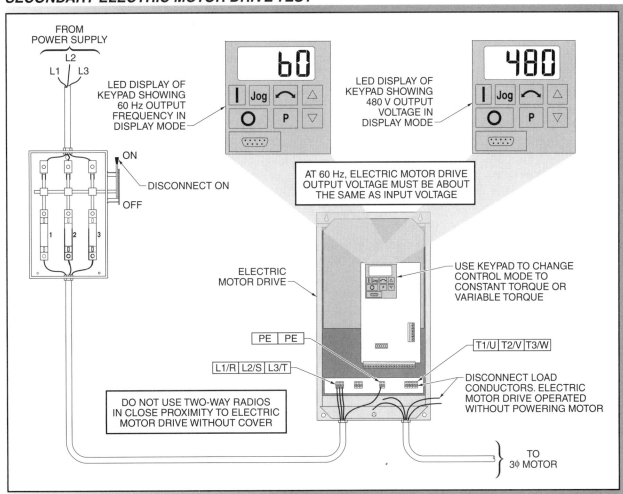

Figure 12-14. The functionality of an electric motor drive is tested by running the drive without the motor conductors connected.

Some electric motor drive models have an automatic calibration (automatic tuning) parameter. The automatic calibration parameter fine-tunes an electric motor drive to the characteristics of the motor being used for optimum performance. Automatic calibration may require the motor and electric motor drive to run with the driven load disconnected. When it is impossible to run the electric motor drive and motor with the driven load disconnected, automatic calibration must be skipped.

TECH FACT

Running an electric motor drive with the motor disconnected is possible with many drive models, but is not possible with drives programmed for open-loop vector control or closed-loop vector control.

Final Electric Motor Drive and Motor Test

The final electric motor drive and motor test verifies that a drive and motor function together properly to power a driven load. Parameter changes are not required because the keypad is controlling the electric motor drive. Return the control mode to its original setting. See Figure 12-15.

To test the electric motor drive and motor, apply the following procedure:

1. Disconnect the driven load from the motor to establish correct motor rotation if running the driven load in reverse may cause damage.
2. For the safety of personnel and equipment, a technician must monitor and control an electric motor drive while another technician monitors the motor. Do not start the electric motor drive before verifying that personnel are not at risk.
3. Program the display mode to show electric motor drive output frequency.
4. Stand to the side of the disconnect and electric motor drive when pushing the start button, in case of a major drive failure. Push the START (I) button. Verify the direction of rotation of the motor. If the direction is incorrect, push the STOP (O) button and proceed to step 5. If the direction is correct, proceed to step 7.
5. Turn disconnect OFF and lockout/tagout disconnect.
    a. Wait for the DC bus capacitors to discharge. Do not manually discharge the capacitors by shorting + to –. Remove the electric motor drive cover. Use a DMM to verify that the AC line voltage is not present. Use a DMM to verify that the DC bus capacitors have discharged. Do not rely on the DC bus charge LED(s).

    b. Interchange any two load conductors at the power terminal strip to reverse the direction of rotation. The direction of rotation of a motor cannot be changed by interchanging two line conductors to an electric motor drive because the line conductors feed the converter section, which produces DC bus voltage. Reinstall the electric motor drive cover.
    c. Remove lockout/tagout from disconnect.
6. Stand to the side of the disconnect and electric motor drive when energizing, in case of a major drive failure. Turn disconnect ON.
7. Increase the speed of the motor to 60 Hz via the RAMP UP (↑) button. The motor must accelerate smoothly to 60 Hz. Record the frequency at which any unusual noise or vibration occurs. Unusual noises or vibrations can indicate alignment problems or can require the use of the skip frequency parameter to avoid unwanted mechanical resonance.
8. Remove the electric motor drive cover. Exercise extreme caution and use the appropriate personal protective equipment.
9. Measure and record the current in each of the load conductors using a true-rms clamp-on ammeter. Current readings are taken at 60 Hz because the motor nameplate current is based on 60 Hz.
10. Reinstall the electric motor drive cover.
11. Decrease the speed of the motor to 0 Hz using the RAMP DOWN (↓) button. The motor must decelerate smoothly to 0 Hz. Any unusual noises or vibrations must be recorded and the frequency of the occurrence recorded.
12. Push the STOP (O) button.

Compare the current readings taken in step 9. The current readings must be equal or very close, for example T1 = 9.5 A, T2 = 9.4 A, and T3 = 9.6 A. A slight difference between readings is due to a slight difference in the impedance of the load conductors or the motor windings, the position of the wire in relation to the clamp-on ammeter jaws, or slight variation in the load. The phase-to-phase output voltages (L1 to L2, L1 to L3, and L2 to L3) of an electric motor drive are equal because the inverter section creates the voltage. The impedance of the load conductors and the impedance of the motor coils must be equal when in good condition. With the impedances equal, the current in each of the load conductors must be equal or very close. There is a problem with the load conductors or motor when the current readings are not equal or close.

Compare the current readings taken in step 9 to the motor nameplate current. The readings indicate an underloaded motor. An overloaded motor is not acceptable. An *underloaded motor* is a motor that has an amperage reading of 0% to 95% of nameplate current. An underloaded motor indicates that the motor load is not at its maximum, such as a conveyor that is moving without product. A *fully loaded motor* is a motor that has an amperage reading of 95% to 105% of nameplate current. An *overloaded motor* is a motor that has a current rating greater than 105% of nameplate rating. A motor without a load should draw less than the motor nameplate current.

ELECTRIC MOTOR DRIVE AND MOTOR FINAL TEST

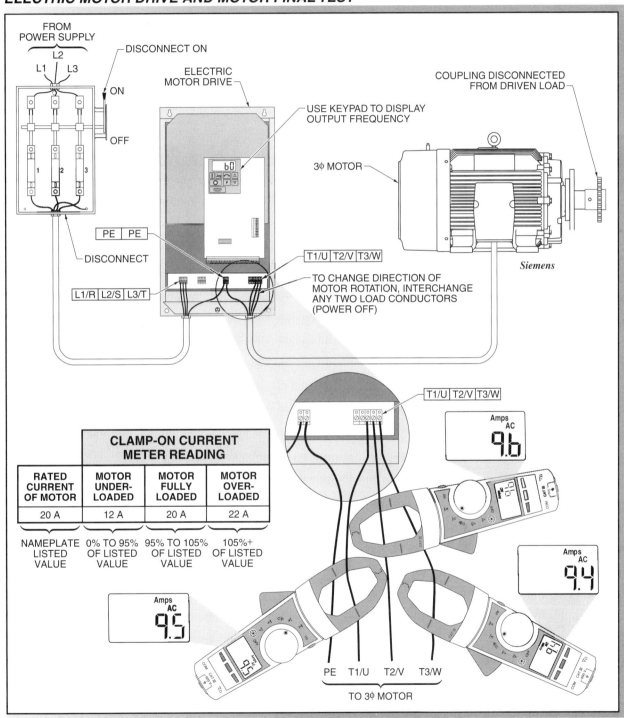

CLAMP-ON CURRENT METER READING			
RATED CURRENT OF MOTOR	MOTOR UNDER-LOADED	MOTOR FULLY LOADED	MOTOR OVER-LOADED
20 A	12 A	20 A	22 A
NAMEPLATE LISTED VALUE	0% TO 95% OF LISTED VALUE	95% TO 105% OF LISTED VALUE	105%+ OF LISTED VALUE

Figure 12-15. To compare the measured current reading with the nameplate current rating of a motor, the motor must be operated at a speed of 60 Hz.

Final Electric Motor Drive, Motor, and Driven Load Test

The final electric motor drive, motor, and driven load test verifies that the drive, motor, and driven load function properly when operated as designed, such as being controlled by a 4 mA to 20 mA signal from an HVAC control system. The complexity of the electric motor drive application may require more than one person to complete the final test. Another technician or qualified person that is familiar with the process the electric motor drive is controlling must assist the start-up technician. The technician or qualified person must be able to verify that the electric motor drive is working correctly in relation to the entire process. The technician or qualified person must also be able to make suggestions for optimizing the electric motor drive, such as reducing the speed of a mixing motor in a batch process that is controlled by an electric motor drive. See Figure 12-16.

To test the electric motor drive, motor, and load, apply the following procedure:

1. Program the input mode and speed reference of the electric motor drive to match the specific application.
2. For the safety of personnel and equipment, a technician must monitor and control an electric motor drive while another technician monitors the motor and driven load during the test. Do not start the electric motor drive before verifying that personnel are not at risk.
3. Stand to the side of the disconnect and electric motor drive when pushing the start button, in case of a major drive failure. Push the START (I) button.
4. Monitor the electric motor drive, the motor, and the driven load under full load conditions.
5. Verify that the electric motor drive application works properly. Adjust parameters as required to optimize the performance of the drive and the process it controls.
6. Verify the functionality of all inputs and outputs connected to the electric motor drive, such as a digital output used as an alarm.
7. Remove the electric motor drive cover. Dangerous voltage levels exist when the drive cover is removed and the drive is energized. Exercise extreme caution and use the appropriate personal protective equipment.
8. Measure and record the current in each of the line conductors using a true-rms clamp-on ammeter. Verify that the input current rating of the drive is not exceeded.
9. Measure and record the current in each of the load conductors using a true-rms clamp-on ammeter. Verify that the motor is not overloaded.
10. Reinstall the electric motor drive cover.
11. Enable parameter protection to prevent unauthorized personnel from adjusting parameters.
12. Record any parameter values that have been changed from the default settings on the Electric Motor Drive Record Sheet. Store all records in a safe location.

TECH FACT

Line voltage applied to the load terminals of an electric motor drive severely damages the drive.

Bypass Contactor Test

The *bypass contactor test* is a test that verifies that the bypass contactor works and that the motor rotates in the correct direction when the bypass contactor is energized. See Figure 12-17.

To test the bypass contactor, apply the following procedure:

1. If the electric motor drive is running, push the STOP (O) button.
2. Verify that the input line voltage cannot be applied to the load terminals of the electric motor drive through the output contactor.
3. Turn the bypass selector switch to BYPASS to start the motor. Verify the direction of rotation of the load. If the direction is incorrect, turn the bypass selector switch to OFF, and proceed to step 4. If the direction is correct, proceed to step 6.
4. Turn disconnect OFF and lockout/tagout disconnect.
   a. Using a DMM, verify that the AC line voltage is not present at the line side of the bypass contactor.
   b. Interchange any two load conductors on the load side of the bypass contactor.
   c. Remove the lockout/tagout from disconnect.
5. Stand to the side of the disconnect when energizing, in case of a major failure. Turn the disconnect ON.
6. Turn the bypass selector switch to DRIVE. Do not start the electric motor drive.
7. If the bypass contactor has automatic controls to initiate the bypass function, the automatic controls must be tested to ensure proper function. Testing the automatic controls may require removing power from the electric motor drive.
8. Stand to the side of the disconnect and electric motor drive when pushing the start button, in case of a major drive failure. Push the START (I) button. The electric motor drive, motor, and driven load must run as designed.

ELECTRIC MOTOR DRIVE, MOTOR, AND DRIVEN LOAD FINAL TEST

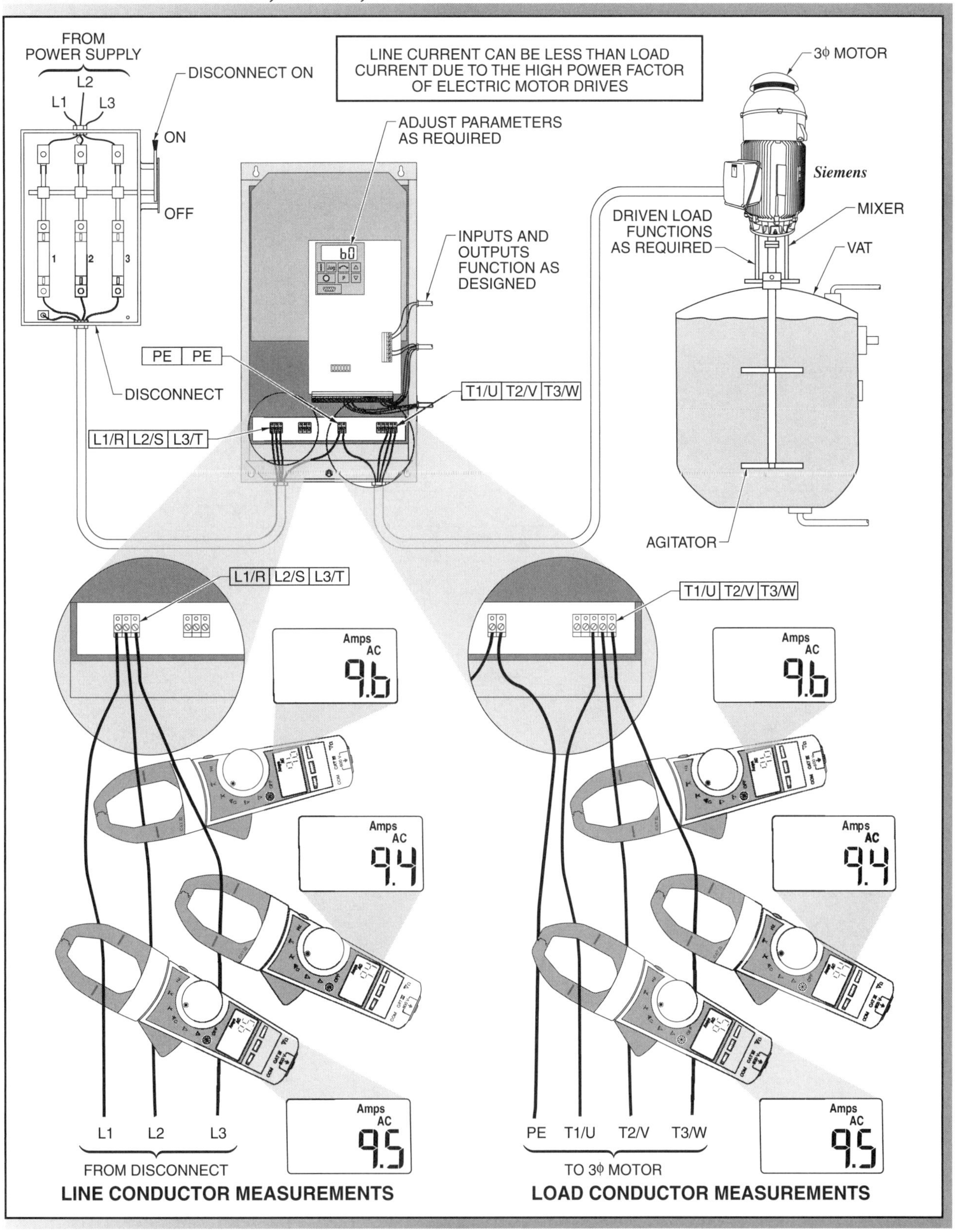

Figure 12-16. The electric motor drive, motor, and driven load test verifies that a 4 mA – 20 mA signal from a mixing process provides proper speed reference to the electric motor drive.

BYPASS CONTACTOR TEST

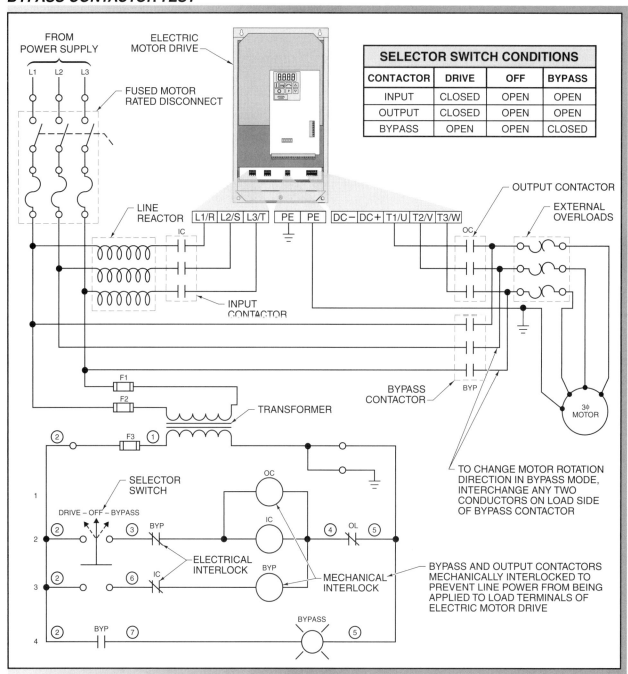

Figure 12-17. The operation of a bypass contactor must be tested before the start-up of an electric motor drive is complete.

Name_____ Date _____

True-False

T F **1.** Motor nameplate data must be recorded in the electric motor drive manual or other type of recordkeeping at the time of installation. *p) 337*

T **(F) 2.** Good insulation has a low resistance reading.

(T) F **3.** As part of the visual inspection of an electric motor drive, the interior of the electric motor drive should be vacuumed with an ESD safe vacuum. *p) 339*

(T) F **4.** When semiconductors are paralleled in the converter and inverter sections of electric motor drives, tests indicate which group is defective, not which device is defective. *p) 343*

T **(F) 5.** An electric motor drive is started during the secondary checks. *p) 344*

T **(F) 6.** When verifying the control wiring, the technician must verify that the shield and drain wires of twisted-pair cables are ~~not~~ grounded at the electric motor drive. *p) 347*

T **(F) 7.** Electric motor drives are ~~not~~ susceptible to radio frequency interference (RFI) from two-way radios. *p) 350*

(T) F **8.** When the control mode is sensorless vector control or closed-loop vector control, it may not be possible to run certain electric motor drives with the motor disconnected. *p) 352*

(T) F **9.** The direction of rotation of a motor ~~cannot~~ be changed by interchanging two line conductors to an electric motor drive because the line conductors feed the converter section, which produces DC bus voltage. *p) 352*

(T) F **10.** An overloaded motor is a motor that has an amperage reading greater than 105% of nameplate current. *p) 353*

(T) F **11.** Typically, the converter section of an electric motor drive consists of full-wave rectifiers or bridge rectifiers that are made up of diodes.

(T) F **12.** A rule of thumb during initial motor testing is that for 1000 V of insulation rating, there is 1 MΩ of resistance. *p) 337*

T **(F) 13.** The control wiring consists of line conductors feeding the electric motor drive and load conductors feeding the motor from the drive.

T **(F) 14.** A model number is a number that identifies when a drive was manufactured, what modifications were made to the design, and what software is used.

(T) F **15.** During the final electric motor drive and motor test, the current readings of each of the load conductors must be equal or very close.

358 ELECTRIC MOTOR DRIVE INSTALLATION AND TROUBLESHOOTING

Completion

foreign control   1. A(n) ___ voltage originates outside the electric motor drive, such as voltage from a control panel. *pg 335*

PWM   2. Electric motor drives have different output current values for different ___ frequencies. *pg 340*

transistors   3. Typically, the inverter section of an electric motor drive consists of ___. *pg 343*

DIP   4. An electric motor drive may contain ___ switches that select various functions, such as an analog input signal configuration. *pg 348*

Default   5. ___ parameter settings are typically the safest option for many applications. *pg 350*

Parameters   6. When an electric motor drive is energized, ___ are adjusted to the specific application. *pg 350*

Automatic tuning   7. The ___ parameter fine-tunes an electric motor drive to the characteristics of the motor being used for optimum performance. *pg 352*

Underloaded   8. A(n) ___ motor is a motor with an amperage reading of 0% to 95% of nameplate current. *pg 352*

Load   9. The bypass and output contactors are mechanically interlocked to prevent line power from being applied to the ___ terminals of an electric motor drive during a bypass contactor test. *pg 354*

Resistance   10. Ground testing is performed at the power terminal strip of an electric motor drive using a DMM set to measure ___. *pg 341*

Multiple Choice

_C_   1. Wire used in 480 VAC or 240 VAC distribution systems has a rating of ___ V.
     A. 240   C. 600
     B. 480   D. 1000   *pg 337*

_A_   2. Horsepower ratings are not used to compare electric motor drives and motors because two motors with the same horsepower rating may have different nameplate ___.
     A. current   C. slip   *pg 340*
     B. frequency   D. speed

_B_   3. Secondary checks focus on the ___.
     A. electric motor drive   C. load   *pg 344*
     B. electric motor drive wiring   D. motor

_A_   4. Overload relays must be sized and set for the nameplate ___ of a motor in order to protect the motor.
     A. current   C. speed   *pg 348*
     B. horsepower   D. voltage

_D_   5. A(n) ___ clamp-on ammeter is used to record the current in each of the electric motor drive's load conductors
     A. analog   C. digital   *pg 354*
     B. average-responding   D. true-rms

Name_____ Date _____

Activity 12-1. Electric Motor Drive Installation Steps

An electric motor drive must be installed in logical steps to ensure safe and proper operation.

Using the circuit diagram as a guide, place the seven basic installation steps in order of operation.

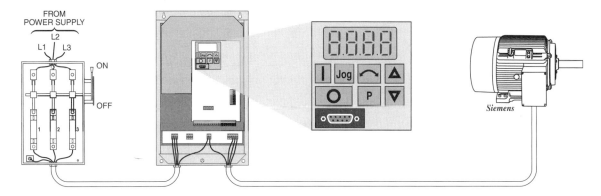

_____ 1. Step 1

_____ 2. Step 2

_____ 3. Step 3

_____ 4. Step 4

_____ 5. Step 5

_____ 6. Step 6

_____ 7. Step 7

A. Connect the electric motor drive power supply. Verify that the incoming power is the correct type and level for the electric motor drive. Lockout/tagout the incoming power disconnect before making any connections and follow all safety requirements.

B. Secure the work order for installing the electric motor drive and motor. Read the instruction manuals to make sure the drive and motor are compatible and that all manufacturer recommendations are understood. Check to make sure that all required tools, meters, and safety equipment are in working order.

C. Connect motor to the electric motor drive. Verify that the motor is the correct type and size and is compatible with the electric motor drive and application, and is properly mounted.

D. Check the electric motor drive and motor environment. Verify that the ambient temperature, installation altitude, and relative humidity are within ratings and the electric motor drive does not require derating.

E. Make a preliminary check of the electric motor drive program. Turn ON power to the electric motor drive before the motor is connected. Reset the drive back to factory default settings and program the known drive settings into the drive, such as all the motor nameplate data and application requirements.

F. Turn power ON and test the electric motor drive and motor. Turn ON and test the motor to verify that the electric motor drive and motor operate as required. Make any final parameter changes.

G. Install the electric motor drive per manufacturer recommendations, which include installing the electric motor drive with proper clearances and sufficient cooling, adding proper warning labels, and ensuring a secure and safe mounting.

Activity 12-2. Wiring Electric Motor Drive Electromechanical Outputs

For many electric motor drive and motor applications, the required devices include a fused (or circuit breaker) disconnect switch, an electric motor drive, a motor, and any required control devices (pressure switch, potentiometer, level switch, etc.). After the initial startup and observations (testing and taking measurements of the system) additional components and changes may be required. Additional components that may be required include input line reactors to reduce problems in the electrical distribution system, output line reactors to protect the motor, input (electromagnetic interference) EMI filters to reduce noise placed on the power lines, and dynamic braking resistors to produce additional braking (stopping) torque for some applications.

Using the manufacturer specification data and circuit diagram, list the model number, part number, and price for each electric motor drive and accessory. Select the electric motor drive accessories that most closely match the incoming power supply and driven load, without undersizing any component. Electric motor drive size, load reactors, and braking resistors are based on the motor used. Line reactors and EMI filters are based on electric motor drive input ratings.

ELECTRIC MOTOR DRIVE SPECIFICATION

MODEL	PRICE*	MOTOR RATING†	DRIVE INPUT‡	DRIVE OUTPUT§	MODEL	PRICE*	MOTOR RATING†	DRIVE INPUT‡	DRIVE OUTPUT§
100–01	225.00	½/0.4	1φ, 115–120/12.6	3φ, 230/2.5	203–03	300.00	3/2.2	3φ, 208–230, 12.5	3φ, 230/10
200–01	225.00	½/0.4	1φ, 208–230/6.3	3φ, 230/2.5	403–03	400.00	3/2.3	3φ, 440–480, 7	3φ, 460/5
300–01	200.00	½/0.4	3φ, 208–230/2.9	3φ, 230/2.5	205–04	450.00	5/3.7	3φ, 208–230, 19.6	3φ, 230/17
101–02	250.00	1/0.75	1φ, 208–230, 11.5	3φ, 230/11.5	405–04	500.00	5/4	3φ, 440–480, 8.5	3φ, 460/8.2
201–02	250.00	1/0.75	3φ, 208–230, 6.3	3φ, 230/5	206–06	625.00	7.5/5.5	3φ, 202–230, 31	3φ, 230/8.25
401–02	300.00	1/0.8	3φ, 440–480, 4.2	3φ, 460/3	406–06	700.00	7.5/5.5	3φ, 440–480, 14	3φ, 460/13
103–03	325.00	3/2.2	1φ, 208–230, 27	3φ, 230/10	407–07	825.00	10/7.5	3φ, 440–480, 20.6	3φ, 460/18

* in $
† in HP_{max}/kW_{max}
‡ in V/A
§ in V/A_{max}

LINE REACTORS

PART NUMBER	PRICE*	RATED HP_{MAX}	PHASE	RATING†	IMPEDANCE‡	INDUCTANCE§	PART NUMBER	PRICE*	RATED HP_{MAX}	PHASE	RATING†	IMPEDANCE‡	INDUCTANCE§
LNR–01	80.00	0.5	1	200–230/7	3	6.5	LNR–15	160.00	7	3	200–230/35	3	0.8
LNR–02	95.00	1	1	200–230/15	3	3.0	LNR–21	100.00	1	3	430–460/5	3	4.5
LNR–03	140.00	3	1	200–230/30	3	2.5	LNR–21	110.00	3	3	430–460/8	3	4.0
LNR–11	100.00	0.5	3	200–230/4	3	6.5	LNR–21	125.00	5	3	430–460/10	3	3.0
LNR–12	100.00	1	3	200–230/7	3	3.0	LNR–21	140.00	7.5	3	430–460/16	3	2.5
LNR–13	125.00	3	3	200–230/14	3	1.3	LNR–21	150.00	10	3	430–460/23	3	1.5
LNR–14	140.00	5	3	200–230/22	3	1.0							

* in $
† in V/A_{max}
‡ in %
§ in mH

LOAD REACTORS

PART NUMBER	PRICE*	RATED HP$_{MAX}$	PHASE	RATING†	IMPEDANCE‡	INDUCTANCE§	PART NUMBER	PRICE*	RATED HP$_{MAX}$	PHASE	RATING†	IMPEDANCE‡	INDUCTANCE§
LDR–X1	100.00	0.5	3	200–230/4	3	6.5	LDR–X6	100.00	1	3	430–460/5	3	4.5
LDR–X2	100.00	1	3	200–230/7	3	3.0	LDR–X7	110.00	3	3	430–460/8	3	4.0
LDR–X3	125.00	3	3	200–230/14	3	1.3	LDR–X8	125.00	5	3	430–460/10	3	3.0
LDR–X4	140.00	5	3	200–230/22	3	1.0	LDR–X9	140.00	7.5	3	430–460/16	3	2.5
LDR–X5	160.00	7	3	200–230/35	3	0.8	LDR–X10	150.00	10	3	430–460/23	3	1.5

* in $
† in V/A$_{max}$
‡ in %
§ in mH

BRAKING RESISTORS

RESISTOR NUMBER CYCLE FOR 208–240 V RATED DRIVE OUTPUT	PRICE*	MOTOR HP$_{MAX}$	RESISTANCE†	POWER‡	DUTY§
BR2–0	35.00	0.5	200	80	10
BR2–1	40.00	1	200	100	10
BR2–2	45.00	2	100	300	10
BR2–3	50.00	3	70	350	10
BR2–5	55.00	5	40	400	10
BR2–7.5	60.00	7.5	30	500	10

* in $
† in Ω
‡ in W
§ in %

BRAKING RESISTORS

RESISTOR NUMBER CYCLE FOR 430–480 V RATED DRIVE OUTPUT	PRICE*	MOTOR HP$_{MAX}$	RESISTANCE†	POWER‡	DUTY§
BR4–0	45.00	0.5	750	80	10
BR4–1	55.00	1	400	300	10
BR4–2	60.00	2	300	350	10
BR4–3	70.00	3	200	400	10
BR4–5	80.00	5	150	450	10
BR4–7.5	90.00	7.5	125	500	10
BR4–10	100.00	10	100	1000	10

* in $
† in Ω
‡ in W
§ in %

EMI FILTERS

PART NUMBER	PRICE*	DRIVE INPUT VOLTAGE†	POWER SUPPLY‡	MAXIMUM CURRENT§	PART NUMBER	PRICE*	DRIVE INPUT VOLTAGE†	POWER SUPPLY‡	MAXIMUM CURRENT§
EMI–11	50.00	110–120	1	10	EMI–32	70.00	208–240	3	20
EMI–12	60.00	110–120	1	20	EMI–33	80.00	208–240	3	30
EMI–21	55.00	208–240	1	10	EMI–34	90.00	208–240	3	40
EMI–22	65.00	208–240	1	20	EMI–41	75.00	430–480	3	10
EMI–23	75.00	208–240	1	30	EMI–42	85.00	430–480	3	20
EMI–24	85.00	208–240	1	40	EMI–43	95.00	430–480	3	30
EMI–31	60.00	208–240	3	10	EMI–44	105.00	430–480	3	40

* in $
† in V
‡ in φ
§ in A

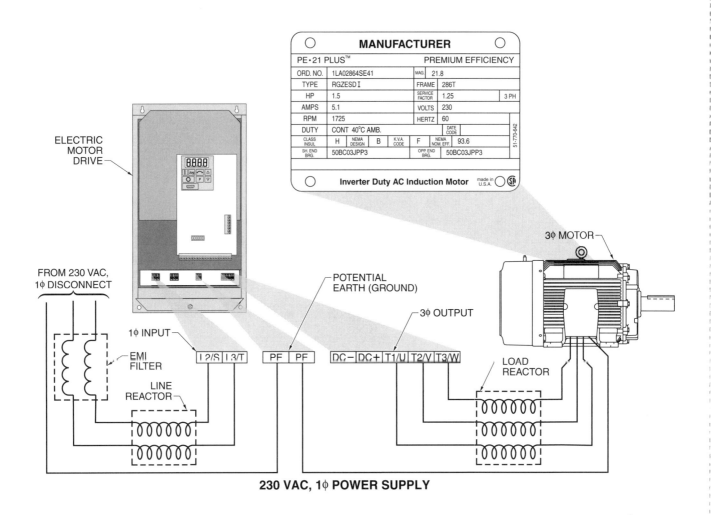

230 VAC, 1φ POWER SUPPLY

- Electric motor drive

  _____ **1.** Model and part number

  _____ **2.** Price

- EMI filter

  _____ **3.** Model and part number

  _____ **4.** Price

- Line reactor

  _____ **5.** Model and part number

  _____ **6.** Price

- Braking resistor

  _____ **7.** Model and part number

  _____ **8.** Price

- Load reactor

  _____ **9.** Model and part number

  _____ **10.** Price

13 Electric Motor Drive Troubleshooting

Troubleshooting is a procedure that consists of systematic testing designed to identify and correct a problem. Technicians must gather information about a specific problem and utilize tests to identify the problem and develop an appropriate solution. Following established troubleshooting procedures ensures that problems are identified and quickly corrected, minimizing equipment and production downtime.

TROUBLESHOOTING SAFETY

Troubleshooting electric motor drives and electrical equipment is inherently dangerous. Troubleshooting typically involves removing covers from electric motor drives, exposing internal parts with dangerous voltages present. The motor and driven load can be running, exposing personnel to machine hazards. Unexpected events occur during troubleshooting procedures such as an electric motor drive stopping unexpectedly due to a drive undervoltage fault. The following safety guidelines must be observed at all times when troubleshooting electric motor drives:

- Only qualified personnel must troubleshoot electric motor drives.
- Always refer to recommendations and instructions of the manufacturer and applicable federal, state, and local regulations. Failure to follow manufacturer recommendations can result in serious physical injury and/or property damage.
- Technicians must understand the machinery and the process that an electric motor drive controls, plus the consequences of starting, stopping, and running a drive—for example, running an electric motor drive can cause product to fall off a conveyor.
- Avoid using two-way radios around electric motor drives, especially with the drive covers removed. Electric motor drives are susceptible to two-way radio frequency interference (RFI). Using two-way radios in close proximity to an electric motor drive can result in the drive running unexpectedly.
- Flame resistant clothing and personal protective clothing including electrical gloves with cover gloves shall be used based upon the incident energy exposure for the flash hazard of a specific task. Insulating matting is used to provide maximum insulation from electrical shock.
- After a problem is identified, technicians must make every effort to find the cause of the problem.
- After a problem is corrected, a technician must verify that the electric motor drive and motor operate as designed for the application.

When troubleshooting electric motor drives, proper personal protective equipment must be worn and safety procedures practiced at all times.

INITIAL TROUBLESHOOTING STEPS

Initial troubleshooting steps consist of gathering information and inspecting an electric motor drive and application. The cause of a problem can often be found during the initial troubleshooting steps. The problem is not always the electric motor drive. Technicians need to troubleshoot all elements of an electric motor drive application including the incoming power, motor, load, and the drive.

TECH FACT

Troubleshooting procedures used to test an electric motor drive are similar to the start-up procedures for a drive. A systematic approach to troubleshooting ensures that problems are located quickly and equipment downtime is kept to a minimum.

Gathering Information

The initial task of a technician is to gather information about the electric motor drive application problem. Technicians are sent to unfamiliar locations to troubleshoot electric motor drive application problems without the aid of engineering or maintenance shop records. Machine operators and other technicians are valuable sources of information about the electric motor drive application. Questions technicians should ask in order to gather useful troubleshooting information include the following:

- What function was the electric motor drive performing when it failed, such as accelerating, decelerating, or running at speed?
- Did the electric motor drive display a fault code or error message? If so, what was the fault code?
- How long has the problem been occurring?
- Does the problem occur all the time, at a particular time, or randomly?
- Is the problem linked to a time of day, a specific event, or a specific process?
- Has anyone worked on the electric motor drive or motor recently? If so, what was done and who did it? Did the problem start after the work was finished?
- Have there been any changes to the load, system, or electric motor drive programming recently?

Technicians must obtain all appropriate electric motor drive manuals and programming parameters. Electric motor drive manuals include installation, operation, and troubleshooting procedures. The manuals also contain drive schematics, fault code explanations, and parameter descriptions. Troubleshooting an electric motor drive without the manuals is extremely difficult. Electric motor drive parameters are saved as hard copies, or as electronic files that are downloaded to clear text display units or personal computers (PCs). Electric motor drive parameter record systems guarantee that electric motor drive parameters are not lost or destroyed when a drive is reset to factory default settings.

Inspecting an Electric Motor Drive Application

After gathering information, technicians must inspect the electric motor drive application. An inspection allows technicians to become familiar with the physical layout and operation of an application. Inspections typically yield clues as to the cause of an electric motor drive application problem. See Figure 13-1. To inspect an electric motor drive application, apply the procedure:

1. Verify that all power disconnects are ON.
2. Access the fault history of the electric motor drive for information on possible causes and record the software version number.
3. Inspect the electric motor drive for physical damage and signs of overheating or fire.
4. Record the electric motor drive nameplate model number, serial number, input voltage, input current, output current, and horsepower rating.
5. Inspect the exterior of the motor and the area adjacent to the motor for debris to ensure proper ventilation to cool the motor.
6. Verify that the motor power rating corresponds to the electric motor drive power rating.
7. Verify that the motor is correctly aligned with the driven load.
8. Verify that the coupling or other connection method between the motor and driven load is not loose or broken.
9. Verify that the motor and the driven load are securely fastened in place.
10. Verify that an object is not preventing the motor or load from rotating.
11. Determine if any special equipment is **required** to work on the electric motor **drive application.**

ELECTRIC MOTOR DRIVE APPLICATION INSPECTION POINTS

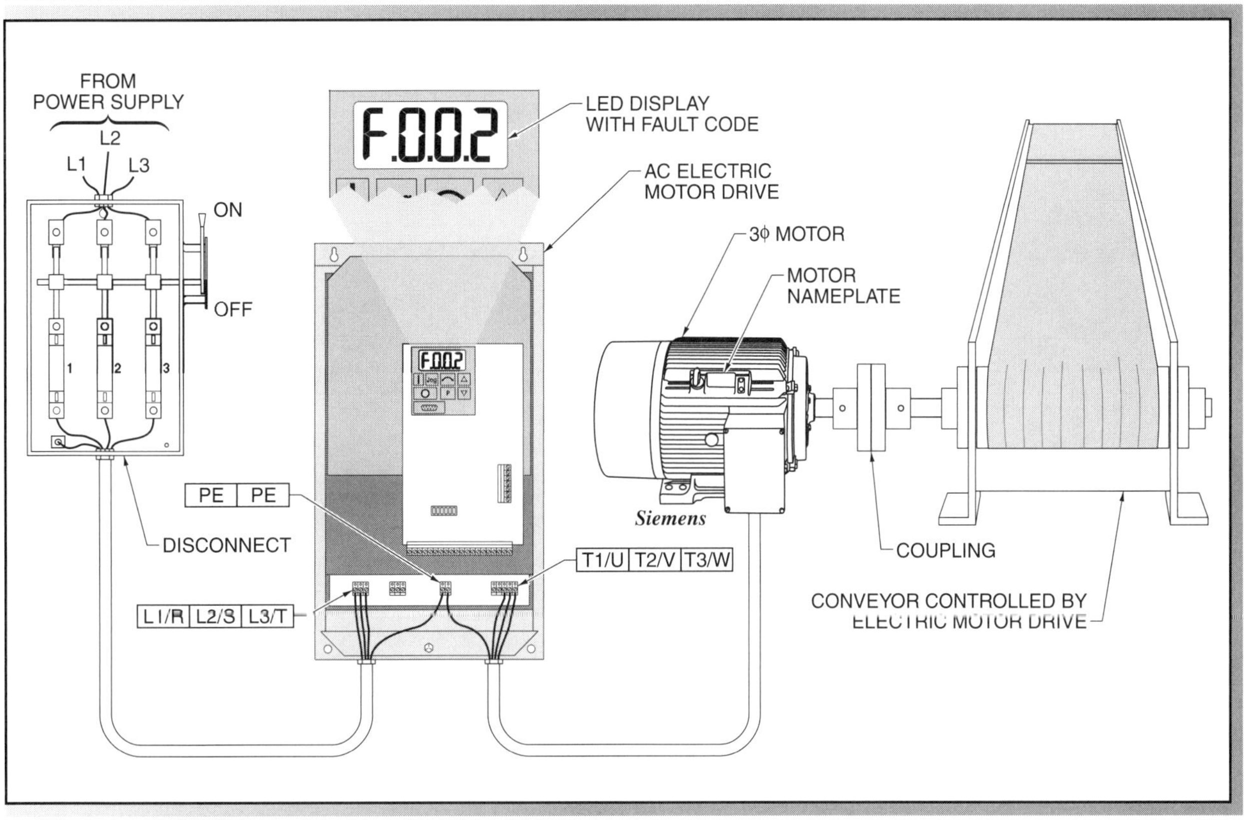

Figure 13-1. Electric motor drive information, motor information, and possible causes of the drive problem are obtained by a thorough visual inspection.

TROUBLESHOOTING TESTS AND SOLUTIONS

Troubleshooting tests are a series of tests designed to identify an electric motor drive application problem. Each element of an application must be tested in order to identify the specific problem. When initial troubleshooting steps do not identify the problem, other tests are typically performed in the following order: incoming power, electric motor drive, motor, and then load. During the troubleshooting process, examine the motor and the load together. When the problem cannot be found with a particular element, the technician must proceed to test other elements, identifying the problem through the process of elimination. Some tests used during an electric motor drive start-up process are used in the troubleshooting process. Performing the tests in an orderly manner identifies the problem in the least amount of time.

The solutions to electric motor drive problems are numerous and varied. Solutions include adjusting a parameter, repairing an electric motor drive, replacing a drive, and/or installing reactors and filters. As a rule of thumb, when repairs would cost 50% or more of the cost of replacing an electric motor drive or motor, replacement is more cost-effective. The decision to repair or replace equipment depends on several factors such as the availability of repair parts, the availability of a replacement unit, cost of downtime, cost of labor, and skill of technicians. When technicians are unable to solve a problem, technical support is available from equipment manufacturers.

Incoming Power Tests

In order for an electric motor drive to function properly, the incoming power must be ON, the power must be within the voltage operating range of the drive, and the power must have sufficient kVA capacity and be free of quality problems. Common incoming power problems are high input voltage, low input voltage, no input voltage, voltage unbalance, improper grounding, and harmonics. Momentary power problems also exist such as voltage sag or voltage swell. A *voltage sag* is a drop in voltage of not more than 10% below the normal rated line voltage lasting from 8 ms to 1 min. A *voltage swell* is an increase in voltage of not more than 10% above the normal rated line voltage lasting from 8 ms to 1 min.

Technicians must frequently test the incoming power supplies to electric motor drives. The tests are performed at the power terminal strip of electric motor drives using digital multimeters (DMMs) set to measure AC voltage. See Figure 13-2. To test the incoming power supply, apply the procedure:

1. Verify that all disconnects are ON and that fuses or circuit breakers are operational.
2. Verify that the line conductors from the local disconnect terminate at the correct spot on the electric motor drive.
3. Verify that the line conductors are shielded cable or are in separate metal conduits with no other conductors.
4. Verify that the line conductors are the proper AWG size.

MEASURING ELECTRIC MOTOR DRIVE LINE VOLTAGE

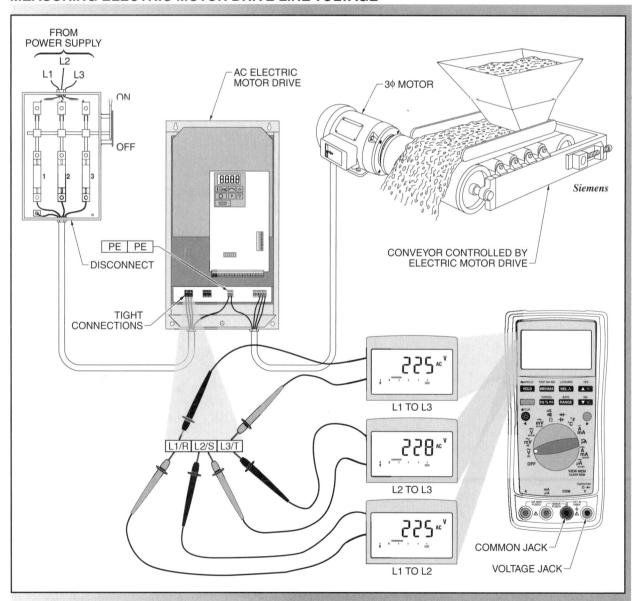

Figure 13-2. The incoming line power is checked at the power terminal strip of an electric motor drive.

5. Check that the grounding conductor is the proper AWG size and terminates at the correct position.
6. Check that all connections at the power supply terminal strip are tight.
7. Measure and record the line voltage with no load (electric motor drive not running), L1 to L2, L1 to L3, and L2 to L3. Verify that the voltage is within the operating range of the electric motor drive. When a measurement of no voltage (0 VAC) is present, technicians must perform additional electrical distribution system tests.
8. Measure and record the line voltage under full-load operating conditions, L1 to L2, L1 to L3, and L2 to L3. Compare full-load readings with the no-load readings from Step 7. A voltage difference greater than 3% between no-load and full-load indicates that the electric motor drive is not receiving sufficient capacity (kVA) or the drive is overloaded.
9. Use the readings from Step 7 to calculate voltage unbalance. A value greater than 2% is not acceptable. To calculate voltage unbalance, apply the formulas:

$$V_u = \frac{V_d}{V_a} \times 100$$

where

V_u = voltage unbalance (%)
V_d = largest voltage deviation from average (in V)
V_a = voltage average (in V)
100 = constant

For example, what is the voltage unbalance when L1 to L2 measures 451 V, L1 to L3 measures 466 V, and L2 to L3 measures 456 V with no load?

$$V_u = \frac{V_d}{V_a} \times 100$$

$$V_u = \frac{8}{458} \times 100$$

$$V_u = .017 \times 100$$

$$V_u = \mathbf{1.7\%}$$

When no voltage is found at Step 7, technicians must determine the cause. The local disconnect can be OFF, a fuse may be blown, or a circuit breaker may be tripped. See Figure 13-3. To identify the cause of a 0 VAC reading, continue with procedure:

10. If the local disconnect is OFF, turn ON. Stand to the side of the disconnect and electric motor drive when energizing, in case of a major failure. Return to Step 7.

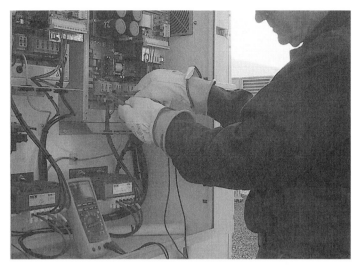

Testing the line voltage of an electric motor drive involves following safety procedures and systematic troubleshooting procedures, and performing math calculations.

11. If the local disconnect is ON, a technician must verify that voltage is present at the disconnect.
  a. Open the disconnect cover and measure the line voltage, L1 to L2, L1 to L3, and L2 to L3. If any of the measurements is 0 VAC or significantly less than the known line voltage, a problem exists in the electrical distribution system.
  b. Use a DMM to check fuses and circuit breakers. Verify that fuses or circuit breakers have the correct voltage rating, current rating, and trip characteristic for the electric motor drive. Replace any blown fuses or reset any tripped breakers. Do not remove or install fuses with the disconnect ON.
  c. Do not turn circuit breakers ON while the cover of the disconnect is open.
  d. Close the disconnect cover. Stand to the side of the disconnect and electric motor drive when energizing, in case of a major failure. Turn the local disconnect ON. Return to Step 7.
12. If a fuse blows or circuit breaker trips again when power is applied to an electric motor drive, turn the local disconnect OFF.
  a. Use a DMM to verify that the AC line voltage is not present at the electric motor drive power terminal strip. Disconnect the line conductors at the power terminal strip of the electric motor drive and insulate the conductors.
  b. Replace any blown fuses or reset any tripped breakers. Turn the local disconnect ON.
  c. If fuses do not blow or circuit breakers do not trip, the electric motor drive has a problem. If fuses blow or circuit breakers trip, there is a problem with the wiring to the electric motor drive.

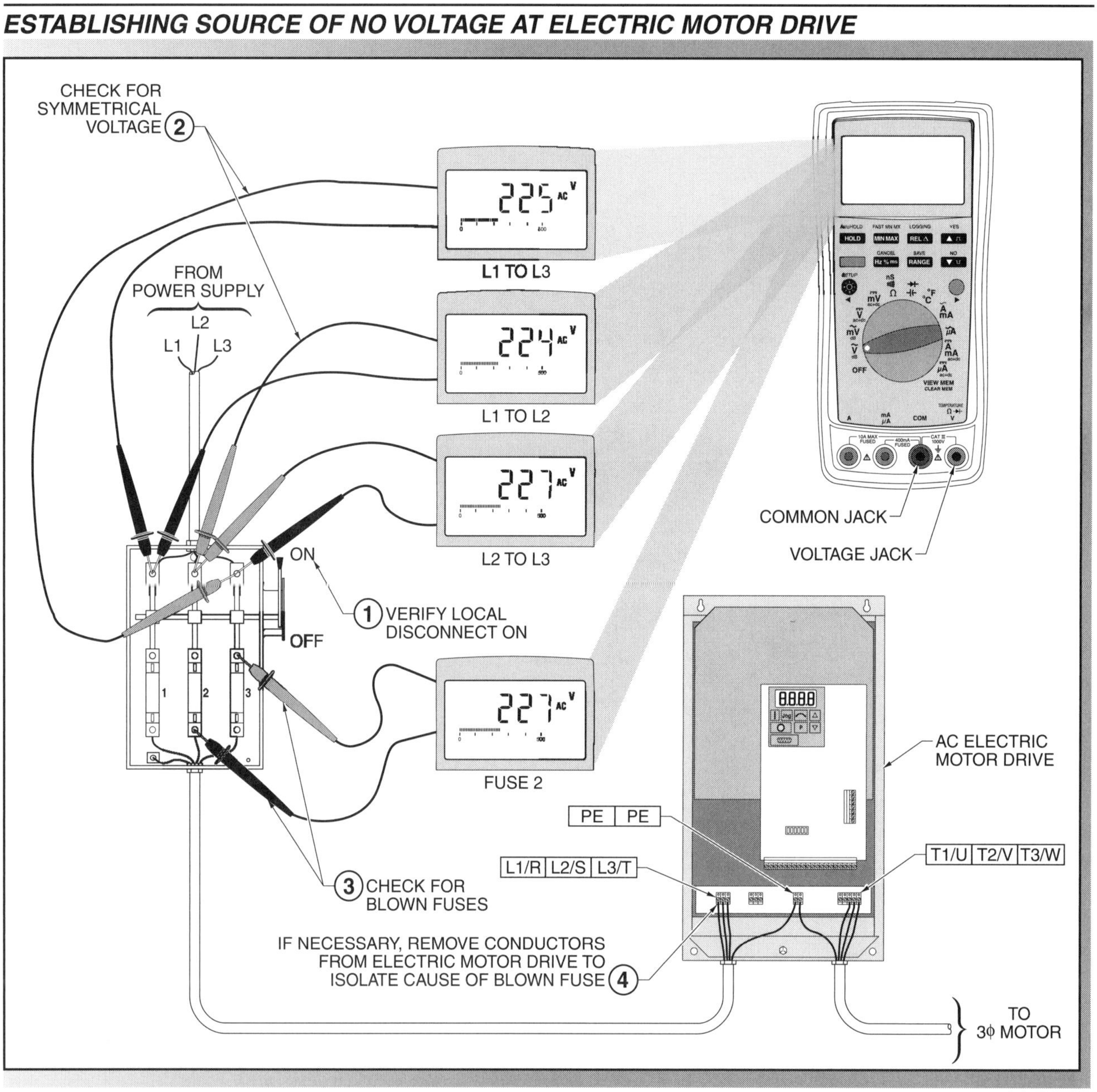

ESTABLISHING SOURCE OF NO VOLTAGE AT ELECTRIC MOTOR DRIVE

Figure 13-3. Electric motor drive line power can be interrupted by blown fuses, tripped circuit breakers, or problems with the power distribution system.

Incoming Power Solutions

When a problem with the incoming power is identified, the appropriate solution is applied. The solution may require modifications to the power source that supplies the electric motor drive, or troubleshooting the drive. The Incoming Power Troubleshooting Matrix should be consulted for solutions. Electric motor drives can be the cause or victim of harmonics and related problems. Input reactors limit the effects on electric motor drives of harmonics caused by other loads. Input reactors also limit the effects on the electrical distribution system of harmonics caused by electric motor drives. See Figure 13-4. See Appendix.

Electric Motor Drive Tests

When the incoming power is eliminated as the source of a problem, the electric motor drive is the next element to test. A series of tests is used to eliminate the source of possible problems within the electric motor drive application. The possible sources of problems are the electric motor drive, drive parameters, input and output devices, motor, and load. Common electric motor drive problems are component failure, incorrect parameter settings, and input and output devices that are incorrectly connected or that have failed. Motor problems and load problems can be mistaken for electric motor drive problems. Electric motor drive fault codes aid in identifying problems. Tests must be performed in the proper sequence to correctly identify a problem in the least amount of time.

⚠ EXPLOSION WARNING

Verify that the category and voltage rating of a DMM is correct for the application. Using a DMM with the improper category and voltage rating can cause DMM components to explode, resulting in death or serious personal injury. CAT IV 600 V meters are recommended.

Saftronics Inc.
Large horsepower electric motor drives can have line voltages of 600 V or higher.

INCOMING POWER TROUBLESHOOTING

INCOMING POWER TROUBLESHOOTING MATRIX			
SYMPTOM/FAULT CODE	**PROBLEM**	**CAUSE**	**SOLUTION**
ELECTRIC MOTOR DRIVE OVERVOLTAGE FAULTS. BLOWN CONVERTER (RECTIFIER) SEMICONDUCTOR	HIGH INPUT VOLTAGE/VOLTAGE SWELL	SWITCHING OF POWER FACTOR CORRECTION CAPACITORS	STOP SWITCHING POWER FACTOR CORRECTION CAPACITORS. INSTALL ELECTRIC MOTOR DRIVE ON ANOTHER FEEDER
		UTILITY SWITCHING TRANSFORMER TAPS FOR LOAD ADJUSTMENT	INSTALL A LINE REACTOR, OR INSTALL ELECTRIC MOTOR DRIVE ON ANOTHER FEEDER
		PROXIMITY TO LOW IMPEDANCE VOLTAGE SOURCE	INSTALL LINE REACTOR
		TRANSFORMER SECONDARY VOLTAGE IS HIGH	ADJUST TAPS ON THE TRANSFORMER
ELECTRIC MOTOR DRIVE OVERLOAD FAULTS	VOLTAGE UNBALANCE GREATER THAN 2%	UNBALANCE FROM UTILITY	CONTACT UTILITY
		SINGLE-PHASE LOADS DROPPING ON AND OFF THE SAME FEEDER AS THE ELECTRIC MOTOR DRIVE	INSTALL ELECTRIC MOTOR DRIVE ON SEPARATE FEEDER
HARMONICS	HARMONICS PRESENT ON ELECTRICAL DISTRIBUTION SYSTEM	ELECTRIC MOTOR DRIVE OR EXISTING NONLINEAR LOADS ARE POSSIBLE SOURCE	INSTALL LINE REACTOR. INSTALL HARMONIC FILTER IF NECESSARY
DRIVE DOES NOT TURN ON	NO INPUT VOLTAGE	INCORRECT FUSE OR CIRCUIT BREAKER	INSTALL CORRECT FUSE OR CIRCUIT BREAKER
		CONDUCTORS FEEDING THE ELECTRIC MOTOR DRIVE ARE SHORTED OR HAVE GROU	REPAIR OR REPLACE CONDUC

Figure 13-4. Problems with the incoming line power to an electric motor drive are found by following matrices.

Initial Test. An *initial test* is a test that verifies if an electric motor drive is operational. A partial failure of an electric motor drive is uncommon. Electric motor drives typically work or do not work. An electric motor drive set to factory default settings and controlled by an integral keypad is tested with the motor disconnected. At this point parameter settings, inputs, outputs, the motor, and the load are not tested as the source of the problem. When the control mode is sensorless vector control or closed-loop vector control, it may not be possible to run the vector control drive with the motor disconnected. When possible, technicians should change the control mode to constant torque or variable torque in order to perform electric motor drive tests. See Figure 13-5. To test an electric motor drive, apply the procedure:

1. If the electric motor drive is ON, push the STOP (O) button.
2. Turn disconnect OFF. Lockout/tagout disconnect.
3. Wait for the DC bus capacitors to discharge. Do not manually discharge the capacitors by shorting + to –. Remove the electric motor drive cover. Use a DMM to verify that the AC line voltage is not present. Use a DMM to verify that the DC bus capacitors have discharged. Do not rely on the DC bus charge LEDs because LEDs can burn out, giving a false indication.
4. Disconnect the load conductors from the electric motor drive power terminal strip. Note where motor wires are connected in order to maintain the correct rotation upon reconnection. Reinstall the electric motor drive cover.
5. Remove the lockout/tagout from the local disconnect.

Saftronics Inc.

Sometimes electric motor drives are removed from service so shop or benchtop troubleshooting can be performed.

TECH FACT

Before performing any test to locate an electric motor drive problem, technicians must first check the fault history of the drive and visually inspect the drive for possible causes.

6. Stand to the side of the disconnect and electric motor drive when energizing, in case of a major failure. Turn the local disconnect ON. Do not push the START (I) button. The electric motor drive LED display or clear text display activates. The electric motor drive cooling fan(s) may or may not start when power is applied, depending on the drive model. If the fans do not start, a technician must check that the fans start when the START (I) button is pushed. If there are any loud noises, smoke, or explosions, immediately turn the local disconnect OFF and proceed to electric motor drive component tests.
7. Record or download electric motor drive parameter values. Reset parameters to factory default settings.
8. Program the input mode to keypad to control the electric motor drive by the integral keypad.
9. Program speed reference to internal.
10. Program the display mode to show electric motor drive output frequency in hertz (Hz).
11. Stand to the side of the electric motor drive when pushing the START (I) button, in case of a major drive failure. Push the START (I) button. The cooling fan(s) should start, if cooling fan(s) did not start when power was applied. The LED display should ramp up to a low speed. If the LED display shows 0 Hz, push the RAMP UP (↑) button until 5 Hz is shown.
12. Push the RAMP UP (↑) button until 60 Hz is shown on the LED display.
13. Program display mode to show the electric motor drive output voltage. The electric motor drive output voltage should be approximately the same as the 60 Hz drive input voltage, such as 480 VAC displayed when the input voltage is 485 VAC.
14. Push the STOP (O) button. The voltage should decrease to 0 VAC.
15. If the electric motor drive performed without any problem, it is not the source of the problem. Proceed to the next test.
16. If the electric motor drive did not perform correctly, it has a problem. Proceed to Electric Motor Drive Component Tests.

ELECTRIC MOTOR DRIVE INITIAL TEST

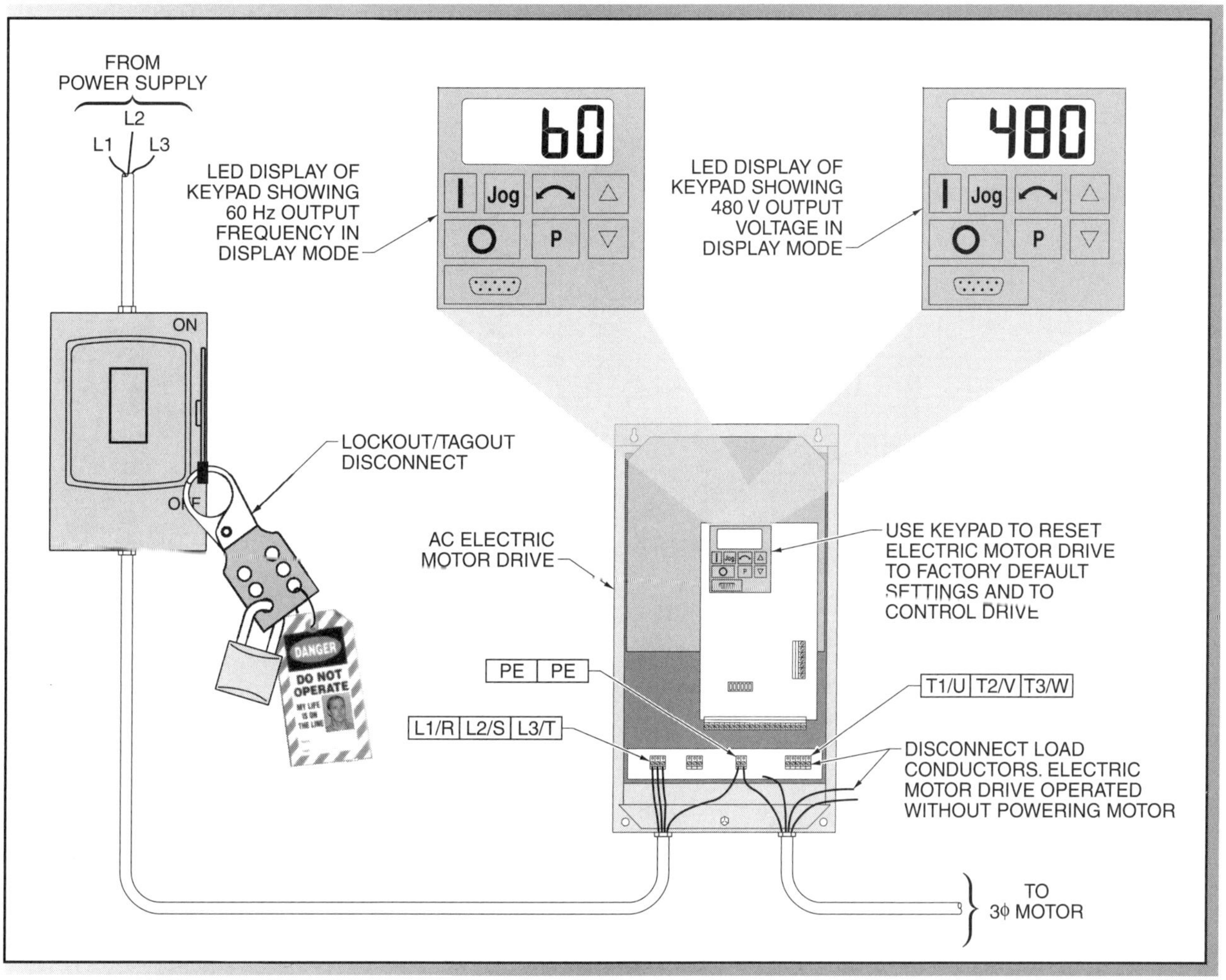

Figure 13-5. The functionality of an electric motor drive is tested by running the drive without the motor conductors connected.

Electric Motor Drive, Motor, and Load Test. An *electric motor drive, motor, and load test* is a test used to verify that a drive and motor function together properly to rotate the driven load. An electric motor drive set to factory defaults, and controlled by the integral keypad, is tested with the motor connected. At this point inputs and outputs are not tested as the source of the problem. If the control mode was changed to perform the electric motor drive test, return the control mode to its original setting. See Figure 13-6. To test an electric motor drive, motor, and load apply the procedure:

1. If the electric motor drive is ON, push the STOP (O) button.
2. Turn disconnect OFF. Lockout/tagout disconnect.
3. Wait for the DC bus capacitors to discharge. Do not manually discharge the capacitors by shorting + to –. Remove the electric motor drive cover. Use a DMM

to verify that the AC line voltage is not present. Use a DMM to verify that the DC bus capacitors have discharged. Do not rely on the DC bus, charge LED(s).

4. Reconnect the load conductors to their previous locations on the power terminal strip in order to maintain correct motor rotation because incorrect motor rotation causes damage in certain applications. Reinstall the electric motor drive cover.
5. Remove lockout/tagout from disconnect.
6. Stand to the side of the disconnect and electric motor drive when energizing, in case of a major failure. Turn disconnect ON. Do not push the START (I) button. The electric motor drive LED display or clear text display activates.
7. Program the appropriate parameters into the electric motor drive with motor nameplate data.

8. Program the display mode to show electric motor drive output frequency.

9. For the safety of personnel and equipment, technicians must monitor and control the electric motor drive as another technician monitors the motor and driven load during the test. Do not start the electric motor drive until a check is made that personnel are not at risk from the driven load.

ELECTRIC MOTOR DRIVE, MOTOR, AND LOAD TEST

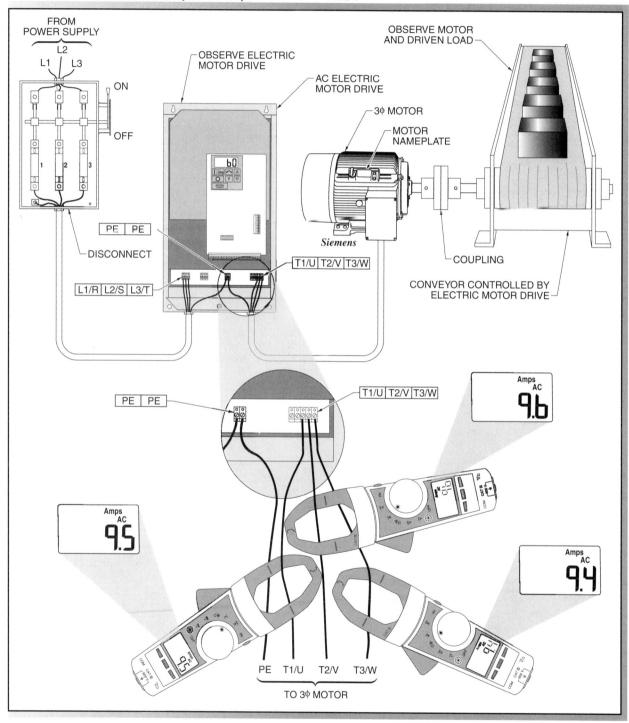

Figure 13-6. Using the integral keypad to control the electric motor drive, the drive, motor, and load are tested together to ensure proper control of the driven load.

10. Stand to the side of the disconnect and electric motor drive when energizing, in case of a major drive failure. Push the START (I) button. The LED display should ramp up to a low speed. If the LED display shows 0 Hz, push the RAMP UP (↑) button until 5 Hz is shown.

11. Increase the speed of the motor to 60 Hz using the RAMP UP (↑) button. The motor and driven load must accelerate smoothly to 60 Hz. Any unusual noises or vibrations must be recorded and the frequency of the occurrence recorded. Unusual noises or vibrations indicate alignment problems or require the use of the skip frequency parameter to avoid unwanted mechanical resonance.

12. Remove the electric motor drive cover. Dangerous voltage levels exist when the electric motor drive cover is removed and the drive is energized. Exercise extreme caution and use the appropriate personal protective equipment.

13. Measure and record the current in each of the three load conductors using a true-rms clamp-on ammeter. True-rms clamp-on ammeters are required because the current waveform of an electric motor drive is nonsinusoidal. Current readings are taken at 60 Hz because the motor nameplate current is based on 60 Hz.

    a. Current readings of the three load conductors must be equal or very close to each other—for example, T1 = 9.5 A, T2 = 9.4 A, and T3 = 9.6 A. A problem with the load conductors or motor is present if the current readings of the load conductors are not equal or very close.

    b. An *overloaded motor* is a motor that has a current reading greater than 105% of nameplate current rating. There is a problem with the motor or the load if the current readings are greater than 105% of the nameplate current rating.

    c. Reinstall the electric motor drive cover.

14. Decrease the speed of the motor to 0 Hz using the RAMP DOWN (↓) button. The motor and driven load must decelerate smoothly to 0 Hz. Any unusual noises or vibrations must be recorded and the frequency of the occurrence recorded. Unusual noises or vibrations indicate alignment problems, or can require the use of the skip frequency parameter to avoid unwanted mechanical resonance.

15. Push the STOP (O) button.

16. If the electric motor drive, motor, and load performed without any problems, the drive, motor, and load are not the source of the problem. Proceed to Electric Motor Drive Input and Output Test.

17. There is a problem if the electric motor drive, motor, and load did not perform correctly. The problem is with the electric motor drive parameters, motor, or load. The electric motor drive was eliminated as the problem in the Initial Test. Proceed to the Electric Motor Drive Solutions section.

⚠ WARNING

Dangerous voltage is exposed when the cover of an electric motor drive is removed. Conditions are present that could result in death or serious injury from electrical shock.

Electric Motor Drive Input and Output Test. An *electric motor drive input and output test* is a test used to verify that the inputs and outputs of a drive function properly when operated as designed. An electric motor drive programmed for a specific application, and controlled by inputs and outputs specific to the application, is tested with the motor connected. The complexity of an electric motor drive application can require more than one technician or a technician and qualified person to complete the test. A qualified person must be able to verify that the electric motor drive is working properly in relation to the entire process. The qualified person must also be able to make suggestions for optimizing an electric motor drive such as slowing down motors on a conveyor packaging process. See Figure 13-7. To test electric motor drive inputs and outputs, apply the procedure:

1. If the electric motor drive is ON, push the STOP (O) button.

2. Return the electric motor drive parameters to the application values copied or downloaded in Step 7 of the Initial Test.

3. For the safety of personnel and equipment, technicians must monitor and control the electric motor drive as another technician monitors the motor and driven load during the test. Do not start an electric motor drive until a check has been made that personnel are not at risk from the driven load.

4. Stand to the side of the electric motor drive when pushing the start button, in case of a major drive failure. Push the electric motor drive START (I) button.

ELECTRIC MOTOR DRIVE INPUT AND OUTPUT TEST

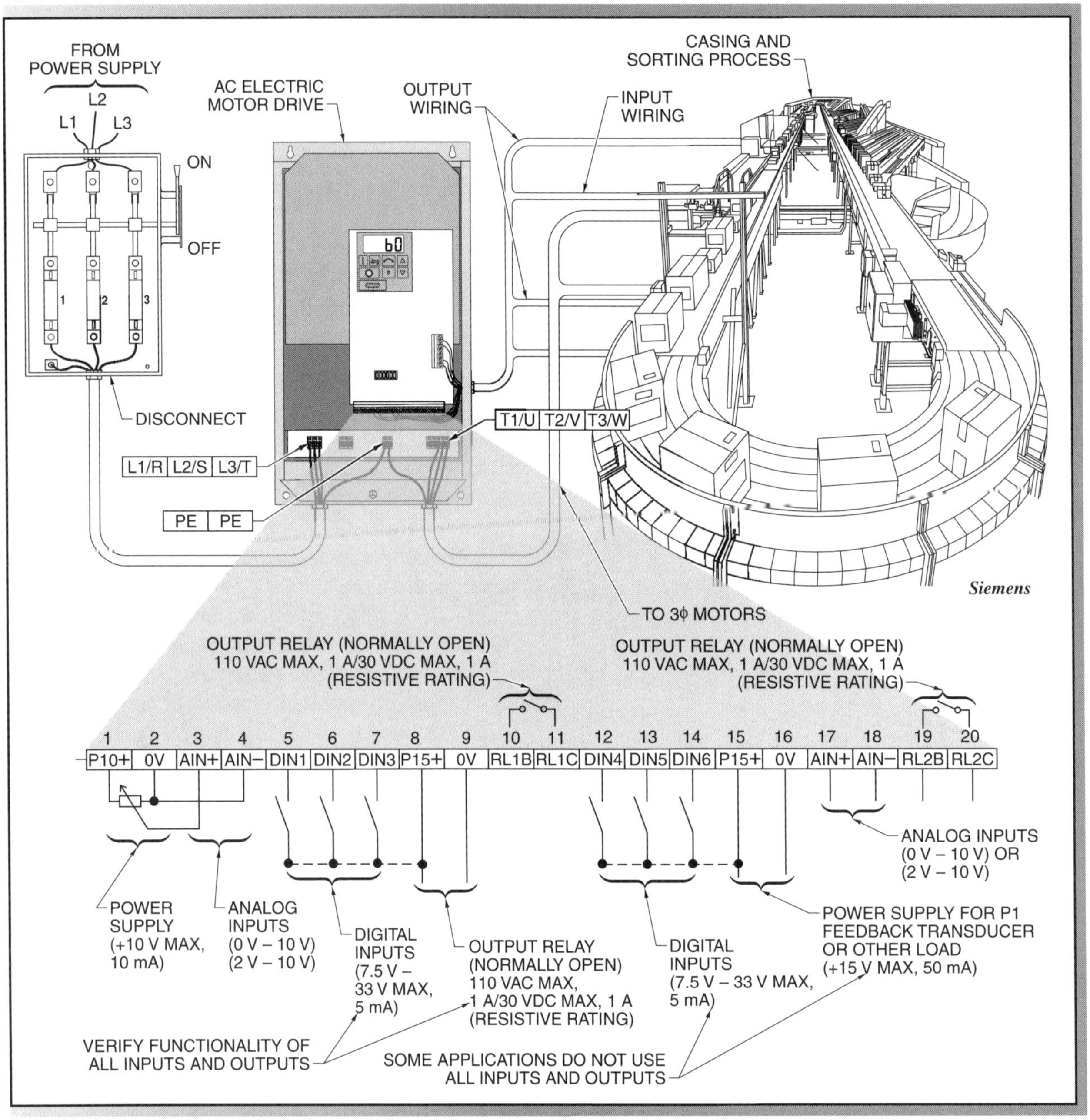

Figure 13-7. Analog and digital signals from process control systems provide speed reference to an electric motor drive. The electric motor drive is tested under normal application operating conditions.

5. Monitor the electric motor drive, motor, and driven load under full-load condition.

6. Verify that the electric motor drive application works properly. Adjust parameters as needed to optimize performance of the electric motor drive and the controlled process.

7. Verify the functionality of all inputs and outputs connected to the electric motor drive.

8. Remove the electric motor drive cover. Dangerous voltage levels exist when the electric motor drive cover is removed and the drive is energized. Exercise extreme caution and use the appropriate personal protective equipment.

9. Measure and record the current in each of the load conductors using a true-rms clamp-on ammeter. Verify that the motor is not overloaded.

10. Reinstall the electric motor drive cover.
11. Record or download the electric motor drive application parameter values. Store this information in a safe location.
12. Enable parameter protection to prevent unauthorized personnel from adjusting parameters.
13. If the electric motor drive, motor, load, inputs, and outputs performed without any problems, the inputs and outputs are not the source of the problem. The electric motor drive application is ready for use.
14. If the electric motor drive, motor, load, inputs, and outputs did not perform correctly, there is a problem. The problem is an electric motor drive parameter problem or an input and output problem. The electric motor drive, motor, and load were eliminated as possible problems in the electric motor drive, motor, and load test.

TECH FACT

Components of small electric motor drives cannot be replaced; the entire drive must be replaced.

Electric Motor Drive Component Tests

An *electric motor drive component test* is a test that identifies which components in a drive are defective. The DC bus capacitors, bus capacitor balancing resistors, electric motor drive cooling fan(s), converter semiconductors, and inverter semiconductors are tested. The tests consist of visual inspections, converter test, and an inverter test. The tests must be performed in sequence. An electric motor drive component test does not test every possible component that can fail. Technicians that are unable to identify a defective component have technical support available from electric motor drive manufacturers.

Visual Inspection. A thorough visual inspection of the electric motor drive interior can identify a defective component. See Figure 13-8. To inspect an electric motor drive interior, apply the procedure:
1. If the electric motor drive is ON, push the STOP (O) button.
2. Turn disconnect OFF. Lockout/tagout disconnect.
3. Wait for the DC bus capacitors to discharge. Do not manually discharge the capacitors by shorting + to –. Remove the electric motor drive cover. Use a DMM to verify that the AC line voltage is not present. Use a DMM to verify that the DC bus capacitors have discharged.

4. Inspect the DC bus capacitors and the capacitor balancing resistors for signs of overheating or fire. DC bus capacitors that are swollen and/or have a protruding pressure relief valve are defective.
5. Inspect the electric motor drive cooling fan(s). The electric motor drive cooling fan(s) must rotate freely by hand, and be noise-free. A noisy fan indicates a fan that is failing. Technicians must also verify that electric motor drive cooling fan(s) work when power is applied.

Converter and Inverter Tests. A *converter and inverter test* is a test used to verify correct operation of semiconductor components in the converter and inverter sections of an electric motor drive. A converter and inverter test uses the diode test mode of a DMM to test the semiconductor components. The diode test mode of a DMM places a voltage across a semiconductor component by passing a small current through the component. A good semiconductor reads about 0.2 VDC to 0.8 VDC when forward-biased and OL when reverse-biased. A shorted semiconductor reads 0 VDC in both directions. An open semiconductor reads OL in both directions. When testing the converter and inverter sections of an electric motor drive, note the following:
- Remove the line and load conductors from the terminal strips of an electric motor drive to isolate the drive.
- Converter and inverter tests are performed at the power terminal strips of an electric motor drive.
- Some electric motor drive manufacturers denote the DC bus with B+ and B–. B+ and B– are used to indicate the location where a dynamic braking resistor is installed or can be installed. Make sure that B+ and B– go directly to the DC bus and not to a switching transistor. Converter and inverter tests are invalid if B+ and B– do not go directly to the DC bus.
- When a semiconductor is reverse-biased by a DMM, the DC bus capacitors are charged. A few seconds are required for the DC bus capacitors to charge. The voltage value increases and the OL reading appears. When the DC bus capacitors are charged, take all remaining measurements in which an OL reading is anticipated.
- A DMM may not be able to charge large capacitors such as are found in higher horsepower electric motor drives.
- Semiconductors are paralleled in the converter and inverter sections of high-horsepower electric motor drives. Converter and inverter tests indicate which group of paralleled semiconductors are bad, but not the individual component.

ELECTRIC MOTOR DRIVE VISUAL INSPECTION

DC BUS CAPACITORS

BUS CAPACITOR BALANCING RESISTORS

DC BUS CAPACITOR BANK REMOVED FOR PRESSURE RELIEF VALVE INSPECTION

DC BUS CAPACITORS

PRESSURE RELIEF VALVE

CAPACITOR PLATE BACKSIDE

ELECTRIC MOTOR DRIVE WITHOUT COVER

SCREW TERMINALS

PRESSURE RELIEF VALVE

BULGE

TOP VIEWS

Siemens

COOLING FAN REMOVED FROM BEHIND PANEL FOR INSPECTION

COOLING FAN

PRESSURE RELIEF VALVE FLUSH WITH TOP OF CAPACITOR

PRESSURE RELIEF VALVE PROTRUDING FROM TOP OF CAPACITOR

FRONT VIEWS

GOOD CAPACITOR　　　　**BAD CAPACITOR**

Figure 13-8. A visual inspection of the interior of an electric motor drive can identify defective drive components.

Curtis Instruments, Inc.

Small capacitors and other components are difficult to visually inspect due to the density of components on electric motor drive printed circuit boards.

The converter section, typically consisting of diodes, is tested first. See Figure 13-9. To test the converter section, apply the procedure:

1. Verify disconnect OFF and that the disconnect is locked out/tagged out.
2. Use a DMM set to diode test mode and follow the Diode Test Matrix to test the converter section.
3. If the DMM display does not match the matrix, a problem exists with a semiconductor.
4. A diode that reads 0 VDC when both forward-biased and reverse-biased is shorted and requires replacement.
5. A diode that reads OL when forward-biased and reverse-biased is open, and must be replaced.

CONVERTER SECTION DIODE TESTING

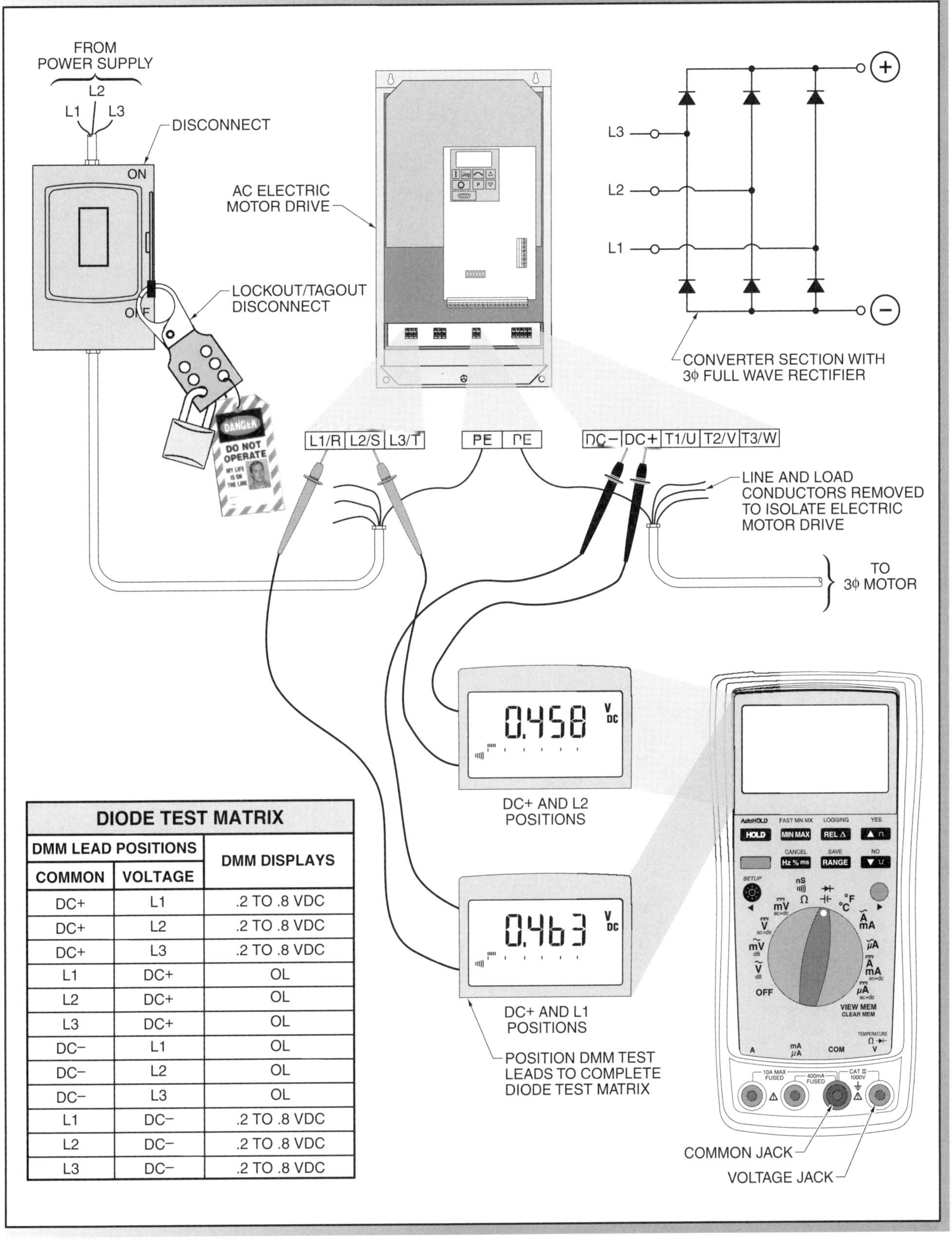

DIODE TEST MATRIX		
DMM LEAD POSITIONS		**DMM DISPLAYS**
COMMON	**VOLTAGE**	
DC+	L1	.2 TO .8 VDC
DC+	L2	.2 TO .8 VDC
DC+	L3	.2 TO .8 VDC
L1	DC+	OL
L2	DC+	OL
L3	DC+	OL
DC−	L1	OL
DC−	L2	OL
DC−	L3	OL
L1	DC−	.2 TO .8 VDC
L2	DC−	.2 TO .8 VDC
L3	DC−	.2 TO .8 VDC

Figure 13-9. A DMM set to the diode test function is used to test diodes in the converter section of an electric motor drive.

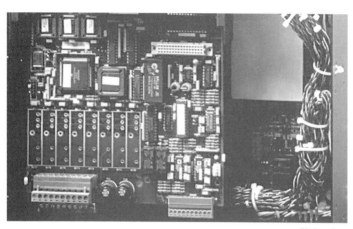

Unico, Inc.

Testing positions can be supplied for testing components of the converter section and inverter section of electric motor drives.

The inverter section, which typically consists of transistors, is tested. See Figure 13-10. To test the inverter section, apply the procedure:

1. Use a DMM set to diode test mode and follow the IGBT Test Matrix to test the inverter section.
2. If the DMM display does not match the matrix, a problem exists with a semiconductor.
3. A transistor that reads 0 VDC when forward-biased and when reverse-biased is shorted and must be replaced.
4. A transistor that reads OL when forward-biased and when reverse-biased is open and must be replaced.

When testing the inverter section transistors, the DMM is actually testing the fly-back diodes. The fly-back diodes and the transistors are in the same package (module). When either fails, the other fails because the fly-back diodes and transistors are in such close physical proximity.

Electric Motor Drive Solutions

When the problem with an electric motor drive is identified, the appropriate solution is applied. The solution requires changing a parameter value, replacing a component, or troubleshooting the motor and the load. See Figure 13-11. See Appendix. Items that must be followed when applying a solution to an electric motor drive are:

- Replace a soldered component where cost-effective. Typically, replacing multiple components is not cost-effective because on small horsepower and newer electric motor drives, all components are soldered in place using special equipment.
- Follow the instructions of the electric motor drive manufacturer when replacing a component.

- Exercise electrostatic discharge (ESD) precautions when working with circuit boards and components.
- Mark and record locations of wires and cables before disconnecting control cables and circuit boards in order to avoid mistakes during reconnection.
- Verify pin and socket alignment before reconnecting cables to circuit boards.
- Replace all DC bus capacitors when any DC bus capacitor is defective because good DC bus capacitors are the same age as the failed capacitor and may fail at any time. Also, the defective DC bus capacitor can cause other capacitors to be overstressed.
- Electrify spare DC bus capacitors periodically per electric motor drive manufacturer recommendations. Follow the instructions of the manufacturer regarding DC bus capacitor storage and replacement.
- Remove old heat sink compound when replacing semiconductors. Apply new heat sink compound before installing new semiconductors.
- Replace both semiconductors when one semiconductor of a paralleled pair is defective.

TECH FACT

An increase in heat sink temperature of an electric motor drive indicates a problem.

Motor and Load Tests

After the incoming power and the electric motor drive have been eliminated as sources of problems, the motor and the load are the next elements to test. A series of tests are used to eliminate the source of possible problems. Motor and load tests must be performed in sequence to identify a problem. Common motor and load problems are a grounded motor stator, defective motor bearings, and motor-to-load misalignment. Motor and load problems can require the assistance of the manufacturer to solve. Electric motor drive fault codes aid in identifying motor and load problems.

Insulation Spot-Test. An *insulation spot-test* is a test that checks the insulation integrity of the stator windings and load conductors of a motor. Megohmmeters are used to perform insulation spot-tests. Motor insulation is damaged by moisture, oil, dirt, excessive heat, excessive cold, corrosive vapors, and vibration. Good motor insulation has a high resistance reading. Poor motor insulation has a low resistance reading.

INVERTER SECTION DIODE TESTING

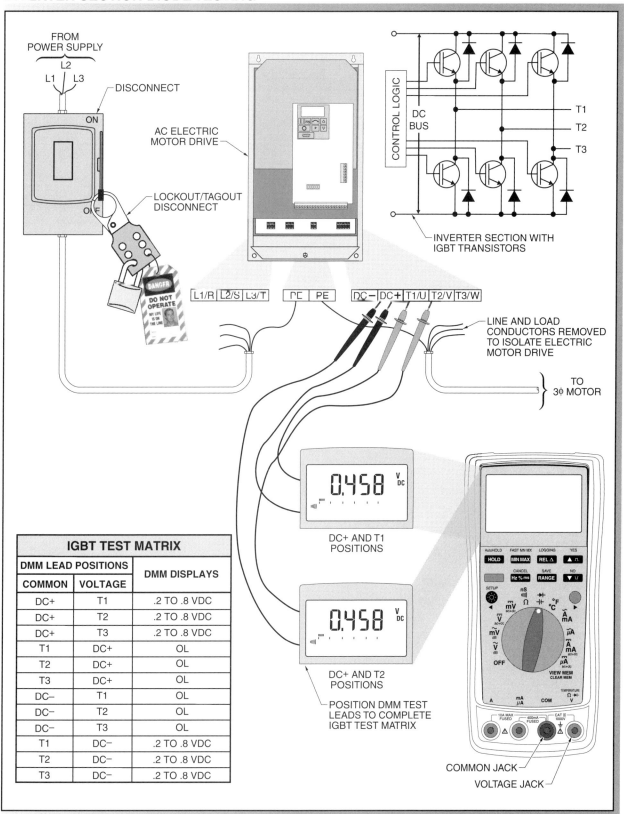

IGBT TEST MATRIX		
DMM LEAD POSITIONS		**DMM DISPLAYS**
COMMON	**VOLTAGE**	
DC+	T1	.2 TO .8 VDC
DC+	T2	.2 TO .8 VDC
DC+	T3	.2 TO .8 VDC
T1	DC+	OL
T2	DC+	OL
T3	DC+	OL
DC−	T1	OL
DC−	T2	OL
DC−	T3	OL
T1	DC−	.2 TO .8 VDC
T2	DC−	.2 TO .8 VDC
T3	DC−	.2 TO .8 VDC

Figure 13-10. A DMM set to the diode test function is used to test the transistors in the inverter section of an electric motor drive.

ELECTRIC MOTOR DRIVE TROUBLESHOOTING

ELECTRIC MOTOR DRIVE TROUBLESHOOTING MATRIX			
FAULTS			
SYMPTOM/FAULT CODE	**PROBLEM**	**CAUSE**	**SOLUTION**
ELECTRIC MOTOR DRIVE OVERVOLTAGE FAULT	ELECTRIC MOTOR DRIVE OVERVOLTAGE	DECELERATION TIME IS TOO SHORT	INCREASE DECELERATION TIME
		HIGH INPUT VOLTAGE (VOLTAGE SWELL)	*SEE INCOMING POWER TROUBLESHOOTING MATRIX*
		LOAD IS OVERHAULING MOTOR	ADD DYNAMIC BRAKING RESISTOR AND/OR INCREASE DECELERATION TIME
COMPONENT FAILURES			
ELECTRIC MOTOR DRIVE DOES NOT TURN ON. BLOWN FUSE OR TRIPPED BREAKER	DEFECTIVE CONVERTER SECTION (RECTIFIER SEMICONDUCTOR)	HIGH INPUT VOLTAGE (VOLTAGE SWELL)	REPLACE CONVERTER SECTION SEMICONDUCTOR OR REPLACE ELECTRIC MOTOR DRIVE. *SEE ALSO INCOMING POWER MATRIX*
		ELECTRIC MOTOR DRIVE COOLING FAN IS DEFECTIVE	REPLACE CONVERTER SECTION SEMICONDUCTOR AND COOLING FAN OR REPLACE ELECTRIC MOTOR DRIVE
PARAMETER PROBLEMS			
UNUSUAL NOISES OR VIBRATIONS WHEN ELECTRIC MOTOR DRIVE POWERING MOTOR	PARAMETERS INCORRECT	PARAMETER(S) INCORRECTLY PROGRAMMED	ADJUST SKIP FREQUENCY PARAMETER
	PROBLEM WITH MOTOR AND/OR LOAD	PROBLEM WITH MOTOR AND/OR LOAD	*SEE MOTOR AND LOAD TROUBLESHOOTING MATRIX*
INPUT AND OUTPUT PROBLEMS			
ELECTRIC MOTOR DRIVE DOES NOT OPERATE CORRECTLY WHEN INPUT MODE IS OTHER THAN KEYPAD, MOTOR AND LOAD ARE CONNECTED, AND DRIVE IS OPERATED AS DESIGNED	EXTERNALLY CONNECTED INPUTS AND OUTPUTS INCORRECT	INPUT(S) AND/OR OUTPUT(S) INCORRECTLY WIRED. INPUT OR OUTPUT DEVICES NOT FUNCTIONAL	TIGHTEN LOOSE WIRES AND/OR REPLACE NON-FUNCTIONAL OR INCORRECT DEVICES FOR APPLICATION
		PROBLEM WITH INPUTS THAT SUPPLY START, STOP, REFERENCE, OR FEEDBACK SIGNALS	CHECK INPUT SYSTEM FOR PROPER INPUT
	PARAMETERS INCORRECT	PARAMETERS INCORRECTLY PROGRAMMED	*SEE ELECTRIC MOTOR DRIVE PARAMETER PROBLEMS*
OPERATIONAL PROBLEMS			
MOTOR ROTATION INCORRECT WHEN POWERED BY ELECTRIC MOTOR DRIVE	INCORRECT PHASING	WIRING	INTERCHANGE TWO OF THE LOAD CONDUCTORS AT THE ELECTRIC MOTOR DRIVE LOAD TERMINAL STRIP
MOTOR ROTATION INCORRECT WHEN IN BYPASS MODE	INCORRECT PHASING	WIRING	INTERCHANGE TWO LINE CONDUCTORS AT DISCONNECT. *NOTE: ASSUMES ELECTRIC MOTOR DRIVE AND BYPASS SHARE COMMON FEED*

Figure 13-11. Problems with electric motor drives are found by following fault, component, parameter, input and output, and operational matrices.

The ideal megohmmeter reading is infinite resistance (∞ resistance) between the conductor or winding being tested, and ground. Megohmmeter readings of less than infinite resistance are common. A rule of thumb states that for every 1000 V of insulation rating, 1 MΩ of resistance should exist. Wires used in 240 VAC or 480 VAC distribution systems have a rating of 600 V. During insulation spot-tests, consider the wires to have 1000 V insulation. The stator winding insulation for inverter duty motors is rated at approximately 1500 V.

For insulation spot-tests of inverter duty motors, consider the winding insulation to have a 2000 V rating. Using the rule of thumb, the minimum megohmmeter reading when testing the load conductors and the motor windings is 2 MΩ. The motor windings and the load conductors are simultaneously tested because the megohmmeter readings are typically measured from the electric motor drive end of the motor conductors. See Figure 13-12.

CONDUCTOR AND LOAD INSULATION SPOT-TEST

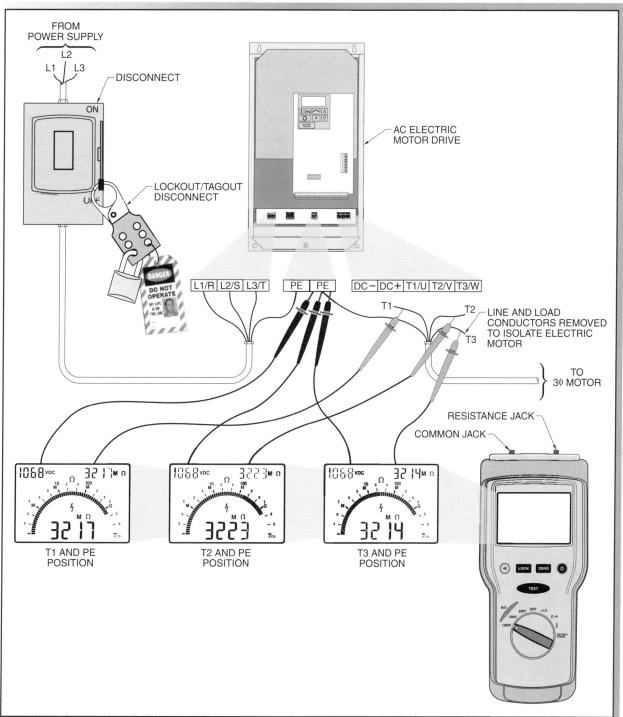

Figure 13-12. The load conductors must be disconnected from an electric motor drive before the conductors and motor windings are tested with a megohmmeter.

To perform a conductor and load insulation spot-test, apply the procedure:

1. When an electric motor drive is ON, push the STOP (O) button.
2. Turn disconnect OFF. Lockout/tagout disconnect.
3. Wait for the DC bus capacitors to discharge. Do not manually discharge the capacitors by shorting + to –. Remove the electric motor drive cover. Use a DMM to verify that AC line voltage is not present. Use a DMM to verify that the DC bus capacitors have discharged.
4. Disconnect the load conductors from the electric motor drive at the power terminal strip. Note where conductors were connected in order to maintain the correct rotation upon reconnection. Failure to disconnect load conductors from the electric motor drive results in damage to the drive due to voltage from the megohmmeter.
5. Technicians using megohmmeters must read and understand all instructions on megohmmeter usage. Particular attention must be given to the procedure for safely discharging a megohmmeter.
6. Set the megohmmeter to the selected test voltage level. The test voltage is typically set higher than the voltage rating of the insulation under test in order to stress the insulation. The 1000 V setting is typically used for motors and conductors operating at 480 VAC or less. Some megohmmeters do not have a 1000 V setting and use a voltage setting close to 1000 V, but not to exceed 1000 V.

7. Perform insulation spot-test.
8. Interpret the readings taken.

TECH FACT

Technicians using megohmmeters must read and understand all instructions and pay extra attention to safely discharging the megohmmeter. Touching the leads of a megohmmeter prior to discharging causes a condition that could result in death or serious injury from electrical shock.

Megohmmeter readings must be a minimum of 2 MΩ and be relatively close to each other, such as T1 to ground measuring 10 MΩ, T2 to ground measuring 9.7 MΩ, and T3 to ground measuring 9.8 MΩ. A reading less than 2 MΩ, or a large difference between the readings (greater than 50%), is cause for additional testing.

When additional testing is necessary, the cause of the low readings or large difference between readings must be determined. Possible causes are the load conductors from the electric motor drive to the motor or the motor windings. See Figure 13-13. To isolate the cause of the low readings or large difference between readings, apply the procedure:

1. Open the motor termination box. Disconnect the load conductors from the motor leads. Do not disconnect the connections between windings.
2. Perform the insulation spot-test to test the load conductors. Technicians must ensure that the load conductors are disconnected from the electric motor drive and are clear of ground and personnel. Megohmmeters produce voltages that can injure personnel and damage electric motor drives.
3. Perform the insulation spot-test to test the motor windings. Technicians must ensure that the motor leads not under test are clear of ground and personnel. Megohmmeter voltage is present at the motor leads not under test because of the internal winding connections inside a motor.
4. When the cause of the low readings or large difference between readings is determined, the motor and/or load conductors must be repaired or replaced.
5. The load conductors and motor windings must be insulation spot-tested with a megohmmeter after repair or replacement work has been performed.
6. Reconnect the load conductors to the motor leads and replace the motor termination box cover. Reconnect the load conductors to the appropriate positions on the electric motor drive terminal strip.

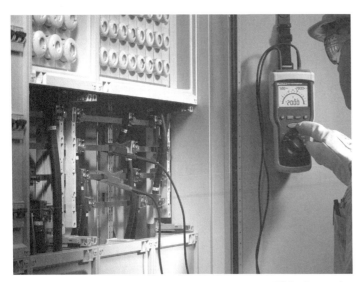

Fluke Corporation

A conductor and load insulation test performed on large horsepower electric motor drives requires that the megohmmeter leads be clamped onto the conductor bars.

INSULATION SPOT-TEST

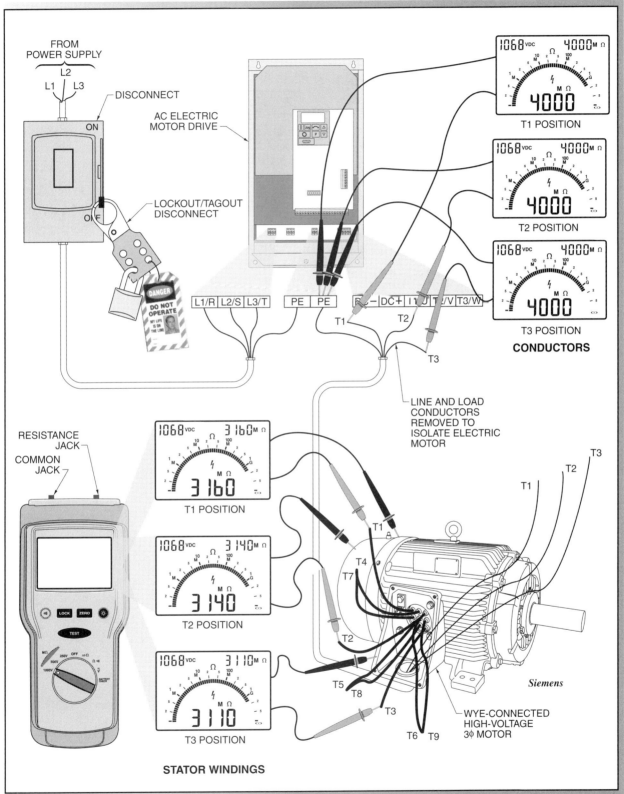

Figure 13-13. The load conductors must be disconnected from the electric motor drive and motor to isolate the source of a low conductor and motor winding megohmmeter reading.

Motor Mechanical Test. A *motor mechanical test* is a test that checks the mechanical operation of a motor. A mechanical problem with a motor results in electric motor drive faults. See Figure 13-14. To perform a motor mechanical test, apply the procedure:

1. Verify that the motor and driven load are aligned correctly for the type of coupling used.
2. Verify that the coupling connecting the motor and driven load is not loose or broken.
3. Verify that the motor is securely bolted in place.
4. Verify that an object is not preventing the motor from rotating.

5. Disconnect the motor from the driven load. Turn the motor shaft by hand. The shaft must rotate freely and not be noisy when rotated. A bind in the rotation of the motor shaft or noise when the shaft is rotating indicates a problem with the motor.

TECH FACT

Disconnecting the driven load from a motor can be required to determine whether the load or motor is causing electric motor drive faults.

MOTOR MECHANICAL TEST

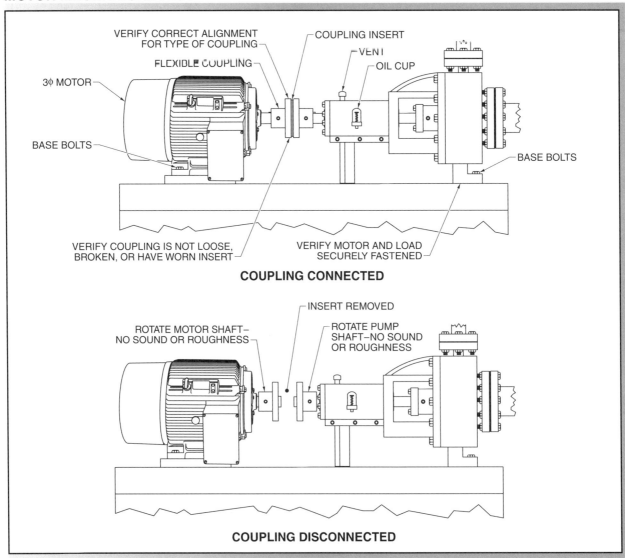

COUPLING CONNECTED

COUPLING DISCONNECTED

Figure 13-14. A motor mechanical test requires that the load be disconnected from the motor so that all aspects of the motor can be properly checked.

Motor Current Test. A *motor current test* is a test used to find hidden motor problems not found with the motor mechanical test. A possibility exists that a motor mechanical test did not detect a problem that is just starting to develop within a motor. The motor current test is designed to catch hidden problems in motors. A motor that is disconnected from the driven load draws less than its motor nameplate current when running at 60 Hz. See Figure 13-15. To perform a motor current test, apply the procedure:

1. Remove the lockout/tagout from the disconnect.
2. Stand to the side of the disconnect and electric motor drive when energizing, in case of a major failure. Turn disconnect ON.

MOTOR CURRENT TEST

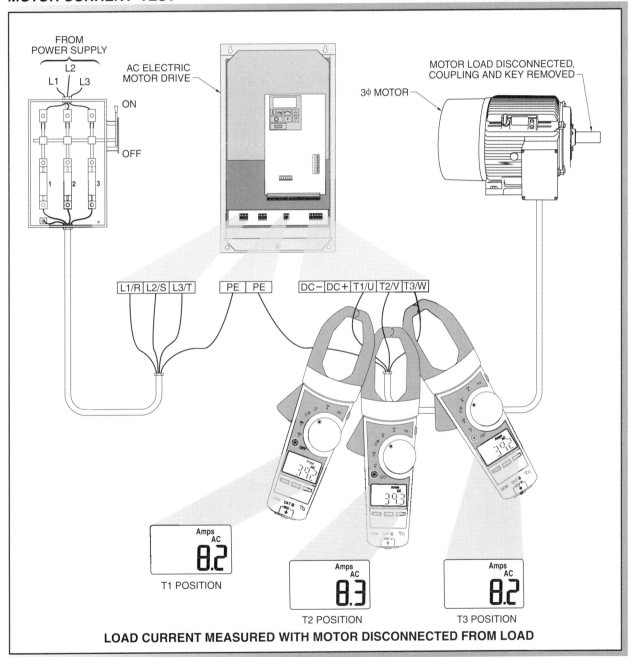

LOAD CURRENT MEASURED WITH MOTOR DISCONNECTED FROM LOAD

Figure 13-15. The load current of a motor is measured at the electric motor drive end of the load conductors with the load disconnected from the motor.

3. For the safety of personnel and equipment, a technician must monitor and control the electric motor drive as another technician monitors the motor during testing. Do not start an electric motor drive until a check has been made that personnel are not at risk from the driven load.

4. Program the display mode to show the electric motor drive output frequency.

5. Stand to the side of the electric motor drive when pushing the start button, in case of a major drive failure. Push the START (I) button.

6. Increase the speed of the motor to 60 Hz using the RAMP UP (↑) button.

7. Remove the electric motor drive cover. Dangerous voltage levels exist when the electric motor drive cover is removed and the drive is energized. Exercise extreme caution and use the appropriate personal protective equipment.

8. Measure and record the current in each of the load conductors with no load using a true-rms clamp-on ammeter. Current readings equal to or greater than the motor nameplate current indicate a problem with the motor. The current readings must be equal or very close to each other.

9. Reinstall the electric motor drive cover.

10. Push the STOP (O) button.

11. Turn disconnect OFF. Lockout/tagout disconnect.

12. If problems were not found using the insulation spot-test, the motor mechanical test, or the motor current test, the motor is not the source of the problem. The load is the problem.

13. If problems were found using the motor insulation test, the motor mechanical test, or the motor current test, the motor is the source of the problem. Proceed to the Motor and Load Solutions section.

Load Mechanical Test. A *load mechanical test* is a test that checks the mechanical operation of a load. A mechanical problem with a load results in electric motor drive faults. See Figure 13-16. To perform a load mechanical test, apply the procedure:

1. Check that the driven load is securely fastened in place.

2. Check that an object is not preventing the load from turning.

3. Turn the load by hand. The shaft should rotate freely and should not be noisy when rotated. A bind in the rotation of the load or noise when the load is rotated indicates a problem with the load. The physical size of the load, location of the load, or type of load may make it impossible to rotate the shaft.

⚠ WARNING

Lockout/tagout disconnect before touching motor shaft or shaft of driven load.

LOAD MECHANICAL TEST

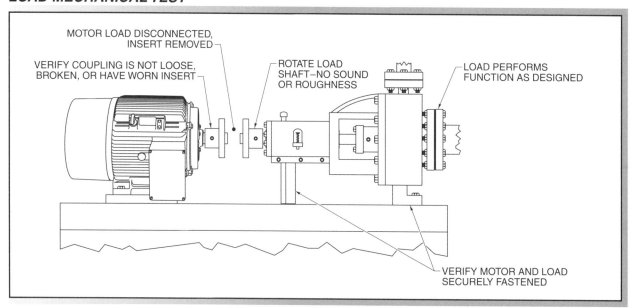

MOTOR LOAD DISCONNECTED, INSERT REMOVED

VERIFY COUPLING IS NOT LOOSE, BROKEN, OR HAVE WORN INSERT

ROTATE LOAD SHAFT—NO SOUND OR ROUGHNESS

LOAD PERFORMS FUNCTION AS DESIGNED

VERIFY MOTOR AND LOAD SECURELY FASTENED

Figure 13-16. A load mechanical test requires that the load be disconnected from the motor to properly check all aspects of the load.

Load Current Test. A possibility exists that a load mechanical test did not detect a problem that is just starting to develop. The load current test is designed to identify this type of problem. To perform a load current test, apply the procedure:

1. Reconnect the motor with the driven load. Check that the motor is correctly aligned with the driven load.
2. Remove lockout/tagout from disconnect.
3. Stand to the side of the disconnect and electric motor drive when energizing, in case of a major failure. Turn the disconnect ON.
4. For the safety of personnel and equipment, a technician must monitor and control an electric motor drive as another technician monitors the motor and driven load during the test. Do not start the electric motor drive until a check has been made that personnel are not at risk from the driven load.
5. Program the display mode to show electric motor drive output frequency.
6. Stand to the side of the electric motor drive when pushing the start button, in case of a major drive failure. Push the START (I) button.
7. Increase the speed of the motor to 60 Hz using the RAMP UP (↑) button.
8. Remove the electric motor drive cover. Dangerous voltage levels exist when the electric motor drive cover is removed and the drive is energized. Exercise extreme caution and use the appropriate personal protective equipment.
9. Measure and record the current in each of the load conductors using a true-rms clamp-on ammeter. Verify that the motor is not overloaded. The current readings must be equal or very close to each other.
10. Push the STOP (O) button.
11. Reinstall the electric motor drive cover.
12. If problems were not found using the load mechanical test or the load current test, the load is not the source of the problem.
13. If problems were found using the load mechanical test or the load current test, the load is the source of the problem. Proceed to Motor and Load Solutions section.

Motor and Load Solutions

When the problem with a motor and/or the load is identified, the appropriate solution is applied. The solution requires changing a parameter value, repairing the motor, or replacing the motor. A technician may require the assistance of other trades in order to correct a motor

and/or load problem. See Figure 13-17. See Appendix. Consult the Motor and Load Troubleshooting Matrix for solutions.

Points to remember when applying a solution to a motor and/or load problem include the following:

- Motors connected to an electric motor drive should be inverter rated or inverter duty, NEMA MG-1 Section IV Part 31 compliant.
- Improper installation of an electric motor drive and motor can result in motor problems and failures, such as long distances between an electric motor drive and a motor resulting in destructive voltage spikes at the terminals of the motor.
- Dirty motors cause excessive motor heating.
- Bearings fail from lack of lubrication or overlubrication.
- Rotor and/or stator damage results from overlubrication.
- The services of professional engineers may be required to solve load problems.

MAINTENANCE

After successful installation and startup, AC drives and AC motors do not require a large amount of maintenance. In most electric motor drive and motor applications, an annual maintenance inspection is all that is required. Turn power OFF to the electric motor drive during an annual maintenance inspection. An annual maintenance inspection must be scheduled so the inspection does not conflict with the process the electric motor drive controls. Problems found during an annual maintenance inspection must be corrected before the problems lead to electric motor drive failure and equipment downtime. Retorquing the power connection is common during the maintenance inspection process.

Saftronics Inc.
Troubleshooting of an electric motor drive ends when the application performs as required.

MOTOR AND LOAD TROUBLESHOOTING

MOTOR AND LOAD TROUBLESHOOTING MATRIX			
MOTOR AND LOAD PROBLEMS			
SYMPTOM/FAULT CODE	**PROBLEM**	**CAUSE**	**SOLUTION**
DRIVE OVERVOLTAGE FAULT	ELECTRIC MOTOR DRIVE OVERVOLTAGE	LOAD IS OVERHAULING MOTOR	CONTACT ELECTRIC MOTOR DRIVE MANUFACTURER *SEE ELECTRIC MOTOR DRIVE TROUBLESHOOTING MATRIX*
UNUSUAL NOISES OR VIBRATIONS WHEN ELECTRIC MOTOR DRIVE IS POWERING LOAD	PROBLEM WITH MOTOR AND/OR LOAD	MISALIGNMENT OF MOTOR AND LOAD	ALIGN MOTOR AND LOAD
		MOTOR AND/OR LOAD NOT SECURELY FASTENED IN PLACE	SECURELY FASTEN MOTOR AND LOAD TO BASE
		DEFECTIVE BEARING(S) IN MOTOR AND/OR LOAD	REPLACE DEFECTIVE BEARING(S) OR REPLACE MOTOR OR LOAD
		MOTOR AND ELECTRIC MOTOR DRIVE NOT PROPERLY SIZED FOR LOAD	CONTACT ELECTRIC MOTOR DRIVE MANUFACTURER
	PARAMETERS INCORRECT	PARAMETER(S) INCORRECTLY PROGRAMMED	*SEE ELECTRIC MOTOR DRIVE TROUBLESHOOTING MATRIX*
RELATED PROBLEMS			
ELECTRIC MOTOR DRIVE DOES NOT OPERATE CORRECTLY WHEN SET TO DEFAULT PARAMETERS, INPUT MODE IS KEYPAD WITH MOTOR AND LOAD CONNECTED	PROBLEM WITH MOTOR AND/OR LOAD	PROBLEM WITH MOTOR AND/OR LOAD	*SEE MOTOR AND LOAD PROBLEMS*
	PARAMETERS INCORRECT	PARAMETER(S) INCORRECTLY PROGRAMMED	*SEE ELECTRIC MOTOR DRIVE TROUBLESHOOTING MATRIX*

Figure 13-17. Problems with the motor or load are found by following motor, load, and related matrices.

Drive Maintenance Data

Technicians must take measurements and record information from an electric motor drive. The maintenance data collected is used to establish a baseline. See Figure 13-18. A significant deviation from the prior annual maintenance inspection data is a warning of imminent failure. Technicians must determine the cause of the deviations and take appropriate corrective actions. To collect maintenance data, apply the procedure:

• Record the electric motor drive model number, serial number, input voltage, input current, output current, and horsepower rating from the nameplate of the drive.
• Record motor nameplate data.
• Measure and record line voltages, L1 to L2, L1 to L3, and L2 to L3.
• Measure and record the DC bus voltage with the electric motor drive running at a constant speed. Unstable DC bus voltage when the electric motor drive is running at a constant speed is an indication of imminent failure or defective DC bus capacitors.
• Measure and record the currents in each line conductor using a true-rms clamp-on ammeter with the electric motor drive powering a motor at 60 Hz.
• Measure and record the currents in each load conductor using a true-rms clamp-on ammeter with the drive powering a motor at 60 Hz. The reading must be 105% or less than the current listed on the nameplate of the motor.
• Perform an insulation spot-test of the load conductors and motor windings. Record the lowest reading on an insulation spot-test graph if all readings are above the minimum acceptable reading.
• Record the heat sink temperature. An increase in heat sink temperature between annual inspections indicates a problem.

ELECTRIC MOTOR DRIVE RECORDS

ELECTRIC MOTOR DRIVE RECORD SHEET

ELECTRIC MOTOR DRIVE INFORMATION

Drive Name:		DRIVE LOCATION	
Drive Manufacturer:		Building:	Floor:
Drive Model:		System:	
Drive Serial Number:		Machine:	
Drive Horsepower:	Software Version:		

INPUT	OUTPUT	PARAMETERS CHANGED FROM DEFAULT	
Voltage:	Voltage:	P –	P –
Current:	Current:	P –	P –
Frequency:	Frequency:	P –	P –
Phase:	Phase:	P –	P –
KVA·	KVA:	P –	P –

MOTOR INFORMATION

Horsepower or Kilowatts:	Service Factor:
Voltage:	Current:
Speed:	Frequency:
Magnetizing Current:	Stator Resistance:
NEMA Design:	NEMA Efficiency:
Duty:	Frame:
Motor is NEMA MG-1 Section IV part 31 compliant	☐ YES ☐ NO

DRIVE MAINTENANCE INFORMATION

LINE VOLTAGE			LINE CURRENT	LOAD CURRENT
L1 to L2:	L1 to L3:	L2 to L3:	L1:	T1:
DC Bus Voltage:	Heat Sink temp:		L2:	T2:
			L3:	T3:

MEGOHMMETER–MOTOR AND LOAD CONDUCTOR TEST

T1 to Ground:	T2 to Ground:	T3 to Ground:

MAINTENANCE LOG

DATE	PROBLEM FOUND	SOLUTION

Figure 13-18. Electric motor drive maintenance records must include information about the drive, programming, motor, and electrical values found in the application.

Replacement Parts

Cooling fans and DC bus capacitors of electric motor drives are two components that have fixed life expectancies. Cooling fans and capacitors must be replaced periodically to avoid major electric motor drive failures. Always follow the recommended replacement intervals and procedures given by the electric motor drive manufacturer. Points to consider to maintain the cooling fan(s) of an electric motor drive include the following:

• The cooling fan must operate correctly. The fan must not be noisy and should rotate freely by hand when not powered. A fan that is noisy or does not rotate freely requires replacement.

• The fan housing, fan blades, and air intake must be free of dirt, dust, and obstructions. Replace the air intake filter per manufacturer schedule or when dirty.

• An increase in heat sink temperature indicates a fan is beginning to fail or that the intake filter is dirty.

• Typically fans are replaced every 3 years to 5 years. To maintain the DC bus capacitors of an electric motor drive, apply the procedure:

• Inspect the DC bus capacitors. DC bus capacitors that are swollen and/or have a protruding pressure relief valve are failing and must be replaced. If one DC bus capacitor is defective, replace all DC bus capacitors.

• When DC bus capacitors are replaced, also check the capacitor balancing resistors. Defective capacitor balancing resistors damage DC bus capacitors.

• DC bus capacitors must be replaced every 5 years to 10 years. Ambient temperatures and electric motor drive loading affect the life expectancy of capacitors.

• Some clean text displays have a battery. Typically the battery is used to power the clock function of the clean text display during a power outage. The batteries need to be replaced every 2 to 5 years depending on environmental conditions.

Visual Inspection

A thorough visual inspection of an electric motor drive and motor must be made periodically. Components are checked for wear and loose connections. To perform a visual inspection, apply the procedure:

• Inspect an electric motor drive for physical damage and signs of overheating.

• Inspect an electric motor drive heat sink to ensure that the heat sink is clean and air can flow freely across the heat sink.

• Inspect the input contactor, output contactor, and bypass contactor for loose connections, worn or damaged parts, and burned or pitted contacts. Power must be OFF and proper lockout/tagout procedures observed.

• Tighten all connections on the control terminal strip and power terminal strip per the recommendations of the manufacturer.

• Inspect the motor for physical damage and signs of overheating. Check that the exterior of the motor and the area adjacent to the motor are free of debris to ensure proper ventilation. Check that the motor and load are fastened securely in place, correctly aligned, and that the coupling method is not loose or broken.

Name_____ Date _____

True-False

T F   1. Electric motor drives are susceptible to two-way radio frequency interference (RFI).
Pg 363

T F   2. Tests used during an electric motor drive start-up process are never used in the troubleshooting process.

T F   3. Electric motor drives can be the cause or victims of harmonics and related problems.
Pg 368

T F   4. A partial failure of an electric motor drive is common.
uncommon   pg 370

T F   5. For the safety of personnel and equipment, the technicians must monitor and control the electric motor drive as another technician monitors the motor and driven load during a test.   *pg 372*

T F   6. The complexity of an electric motor drive application can require more than one technician or a technician and a qualified person to complete certain test procedures.
pg 373

T F   7. Converter and inverter tests indicate which group of paralleled semiconductors are bad, but not the individual component.   *Pg 375*

T F   8. When one semiconductor of a paralleled pair is defective, only the defective semiconductor should be replaced.   *replace both   pg 378*

T F   9. A motor that is disconnected from the driven load draws more than its nameplate current when running at 60 Hz.   *less   pg 385*

T F   10. As a rule of thumb, when repairs of an electric motor drive cost 50% or more of the cost of replacing the drive or motor, replacement is more cost-effective.   *pg 365*

T F   11. During a converter and inverter test, a good semiconductor reads OL when forward-biased.   *pg 375   reverse*

T F   12. An increase in heat sink temperature of an electric motor drive indicates a problem.
pg 390

T F   13. During testing of the incoming power supply of an electric motor drive, a voltage difference greater than 3% between no-load and full-load indicates that the electric motor drive does not have sufficient capacity (kVA) or the drive is overloaded.
pg 367

T F   14. When testing the converter section of an electric motor drive, a diode that reads OL when forward-biased and reverse-biased is shorted.   *open   pg 376*

T F   15. During a conductor and load insulation spot-test, the megohmmeter readings must be a minimum of 2 MΩ and be relatively close to each other.   *pg 382*

Completion

__Unexpected__ 1. ___ events occur during troubleshooting procedures such as an electric motor drive stopping due to a drive undervoltage fault. *pg 363*

__Voltage sag__ 2. A(n) ___ is a drop in voltage of not more than 10% below the normal rated line voltage lasting 8 ms to 1 min. *pg 365*

__true-rms__ 3. ___ clamp-on ammeters are required because the current waveform of an electric motor drive is nonsinusoidal. *pg 373*

__Disconnect__ 4. Technicians should stand to the side of the ___ and electric motor drive when energizing, in case of a major drive failure. *pg 373*

__swollen__ 5. DC bus capacitors that are ___ and/or have a protruding pressure relief valve are defective. *pg 375*

__insulation spot-test__ 6. A(n) ___ is a test used to check the integrity of the stator windings and load conductors of a motor. *pg 381*

__motor mechanical__ 7. A(n) ___ test is a test that checks the mechanical operation of a motor. *pg 384*

__Heat sink__ 8. An increase in ___ temperature between annual inspections indicates a problem. *pg 388*

__Input Reactors__ 9. ___ limit the effects on electric motor drives of harmonics caused by other loads and also limit the effects on the power distribution system of harmonics caused by the drive. *pg 368*

__Sensorless vector__ 10. When the control mode is ___ control, it may not be possible to run the electric motor drive with the motor disconnected.

Multiple Choice

__C__ 1. When inspecting an electric motor drive application, all of the following information should be recorded from the electric motor drive except the ___.
   A. input voltage   C. NEMA enclosure type   *pg 364*
   B. model number   D. serial number

__A__ 2. A voltage swell is an increase in voltage of not more than ___% above the normal rated line voltage lasting from 8 ms to 1 min.
   A. 10   C. 20   *pg 366*
   B. 15   D. 25

__B__ 3. An overloaded motor is a motor with a current reading greater than ___% of nameplate current rating.
   A. 103   C. 110   *pg 373*
   B. 105   D. 115

__B__ 4. The rule of thumb for megohmmeter readings states that for every ___ V of insulation rating, 1 MΩ of resistance should exist.
   A. 500   C. 1500   *pg 380*
   B. 1000   D. 2000

__B__ 5. ___ and DC bus capacitors of electric motor drives are two components that have fixed life expectancies.
   A. Diodes   C. Heat sinks
   B. Fans   D. SCRs

pg 390

13 Electric Motor Drive Troubleshooting — Activities

Name________________________________ Date ______________

ELECTRIC MOTOR DRIVE
CONTROL TERMINAL STRIP
1 2 3 4 5 6 7 8 9 10 11 12 13 14
POWER TERMINAL STRIP
L1/R L2/S L3/T PE PE DC− DC+ T1/U T2/V T3/W
2-WIRE CONTROL (ON/OFF)
FORWARD
REVERSE
FORWARD/REVERSE
START
STOP
3-WIRE CONTROL (START/STOP)
SPEED POTENTIOMETER
FAULT OUTPUT CONTACTS
DRIVE ERROR RESET

MANUFACTURER WIRING DIAGRAM

CONTROL TERMINAL STRIP
1 2 3 4 5 6 7 8 9 10 11 12 13 14
START
STOP
FORWARD
REVERSE
L1/R L2/S L3/T PE PE DC− DC+ T1/U T2/V T3/W
L1 L2 L3 G
IC
BYC
OC
T1 T2 T3
3φ
DRIVE OFF BYPASS
OLs
POWER ON
INPUT CONTACTOR
OUTPUT CONTACTOR
BYPASS CONTACTOR
DRIVE
BYPASS

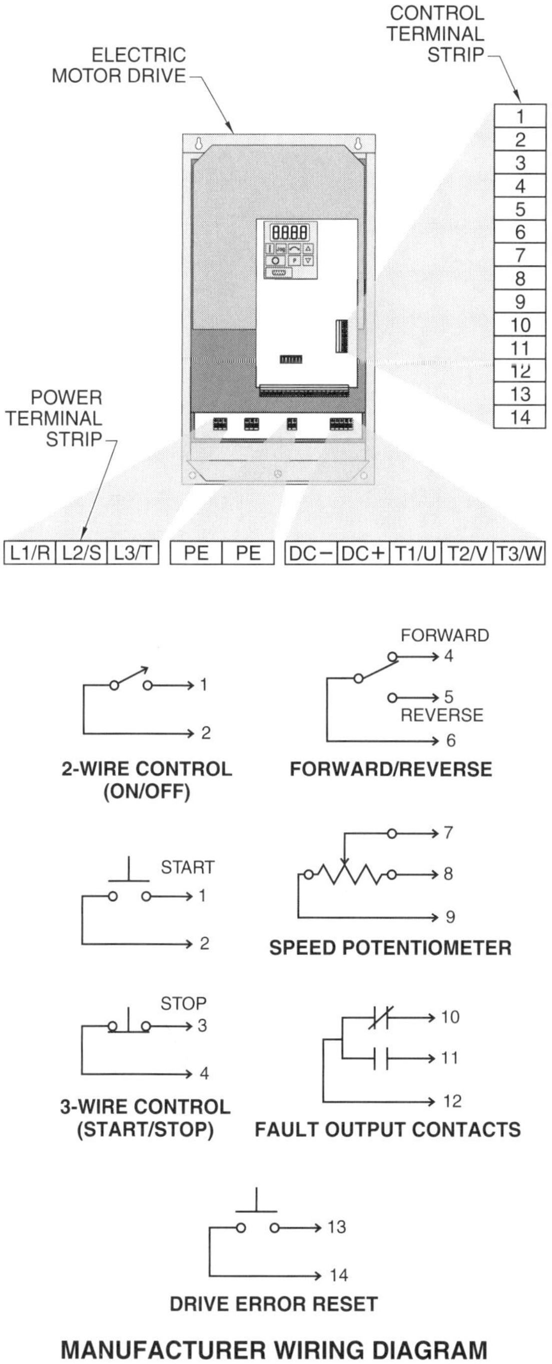

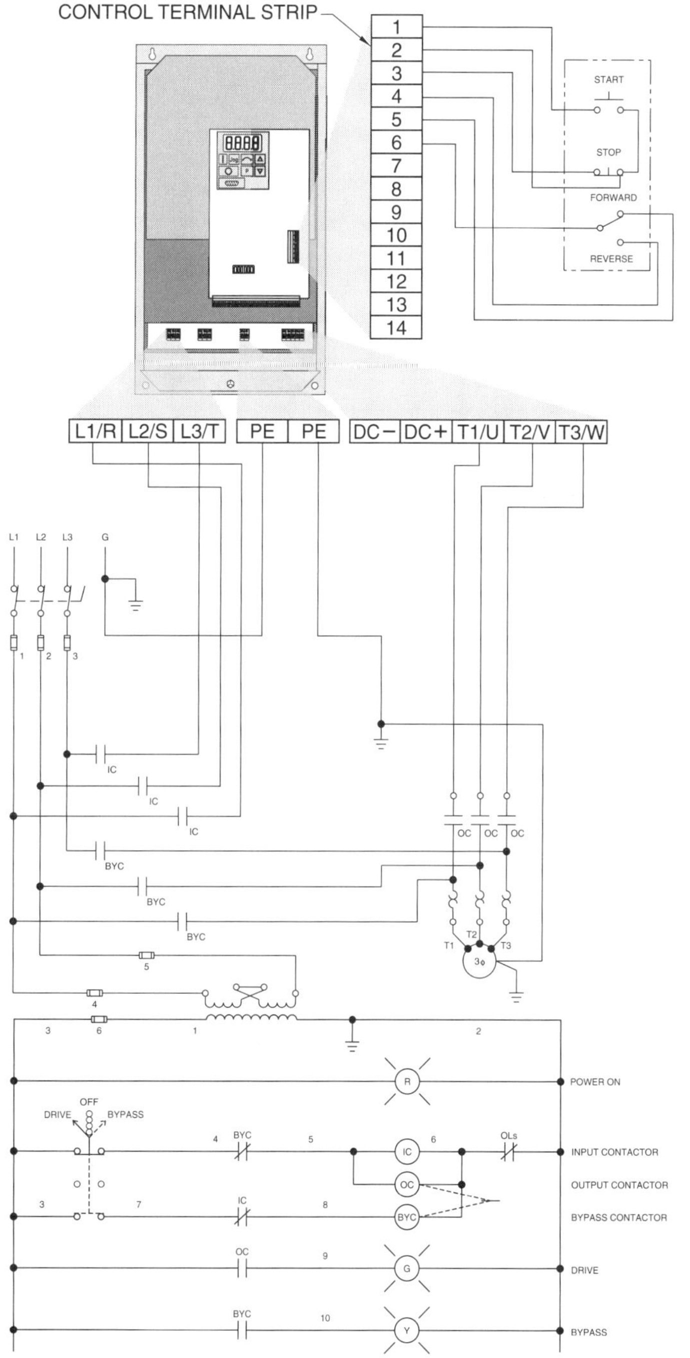

Activity 13-1. Troubleshooting Electric Motor Drive Power Circuits

Answer the questions based on the manufacturer wiring and line diagrams. The circuit is operating properly and has a 230 VAC power supply.

_____ 1. When the selector switch is in the OFF position, Meter 1 reads ___ V.

_____ 2. When the selector switch is in the DRIVE position and the motor is operating at full speed, Meter 2 reads ___ V.

_____ 3. When the selector switch is in the BYPASS position and the motor is operating at full speed, Meter 3 reads ___ V.

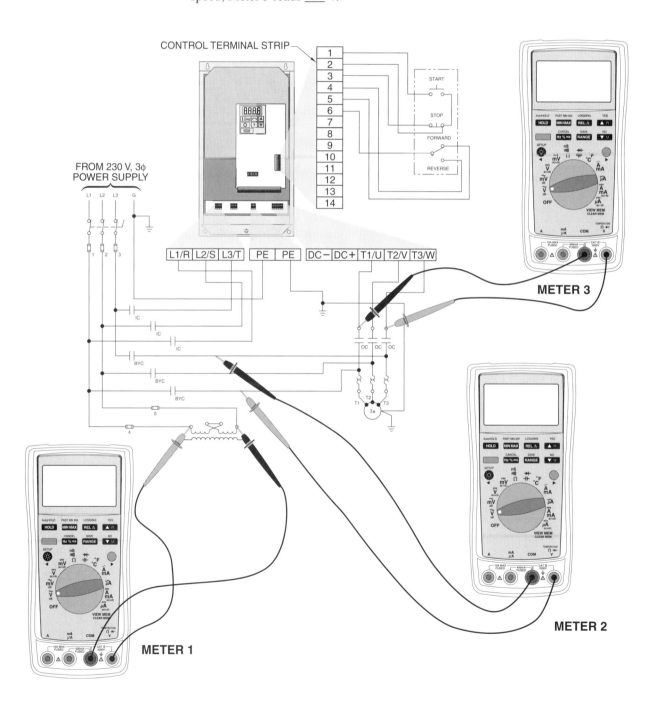

Activity 13-2. Troubleshooting Electric Motor Drive Control Circuits

Answer the questions based on the manufacturer wiring and line diagrams. The control circuit is 115 VAC.

_____ 1. The voltage reads ___ V on Meter 1 when in position 1.

_____ 2. The voltage reads ___ V on Meter 1 when in position 2 and when Fuse 6 is good.

_____ 3. The voltage reads ___ V on Meter 1 when in position 2 and when Fuse 6 is blown.

_____ 4. When the selector switch is in the DRIVE position and the motor is operating at full speed, Meter 2 reads ___ V.

_____ 5. Meter 2 reads 115 VAC, but the INPUT contactor and OUTPUT contactor are not energized. The problem is ___.

_____ 6. The selector switch is in the ___ position when Meter 3 reads 115 VAC.

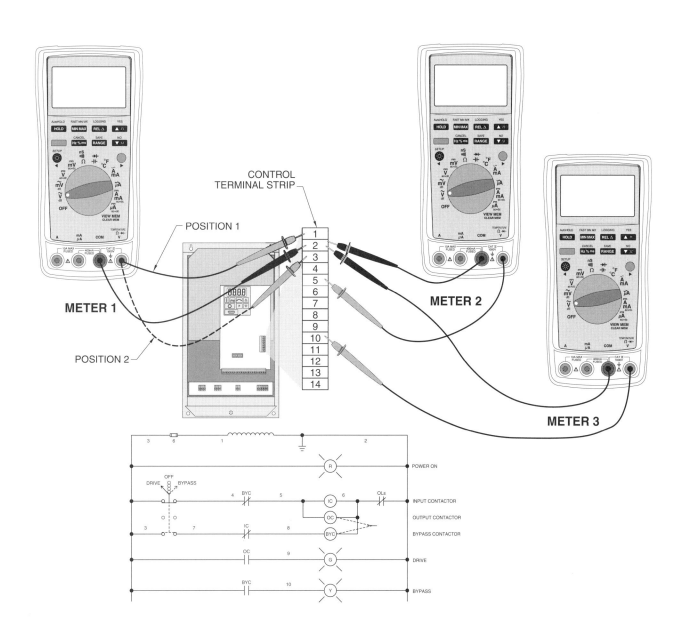

Activity 13-3. Troubleshooting the System

Connect the meters to the electric motor drives. The control circuit is 115 VAC.

1. Connect Ohmmeter 1 to test the normally open fault output contacts on the electric motor drive. Connect Ohmmeter 2 to test the normally closed fault output contacts on the electric motor drive.

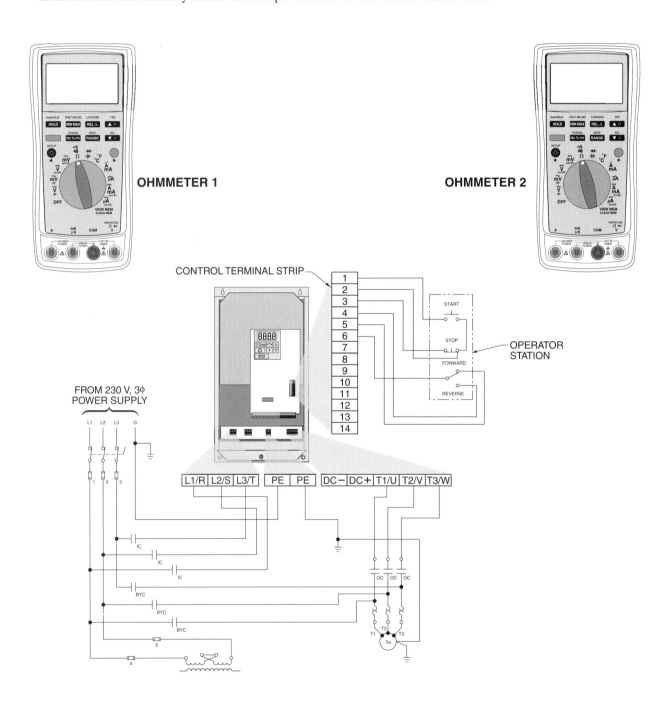

14 Electric Motor Drive Selection

Electric motor drives are used in a variety of applications and provide numerous benefits over conventional motor controls. Completely understanding an electric motor drive application and following selection guidelines is necessary when selecting a drive to reduce the possibility of the incorrect motor drive being selected for an application. When in doubt about selecting an electric motor drive for a specific application, contact the drive manufacturer or a licensed engineer.

AC DRIVE BENEFITS

Prior to the development of AC drives, DC drives and DC motors were the predominant form of variable speed control. The complexities of DC drives and the high maintenance requirements of DC motors were limiting factors in DC drive and DC motor use. The development of AC drives allowed the number of applications for AC variable speed control to increase. AC drives provide economic benefits in the form of energy savings when used with variable torque loads and a number of performance benefits with most applications. The use of AC drives is increasing each year. The benefits of AC drives include the following:

- easier installation and maintenance than DC drives
- less maintenance with AC motors than with DC motors
- significant energy savings using AC drives for pump and fan applications
- enhanced performance through a variety of advanced parameters such as skip frequency, which reduces unwanted mechanical resonances
- precise control of speed and torque of AC motors
- elimination of high starting currents and high transient voltages resulting from large motors turning ON and OFF because AC drives ramp up and ramp down motors
- reversing of motor rotation without the requirement of additional contactors

- communication capabilities with other automated equipment such as PLCs
- less mechanical wear and stress on motors because of slower than typical operation speeds.
- operation of multiple motors
- operation of motors above base speed
- ability to operate 3ϕ motors when supplied from a 1ϕ power source
- ability to serve as "conversion units" to operate equipment that requires a nonstandard voltage such as operating a machine that requires 380 VAC, 50 Hz power from an AC drive connected to 460 VAC, 60 Hz power

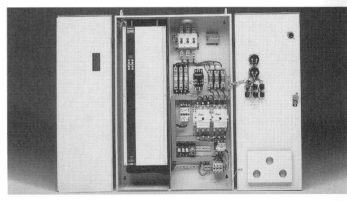

Danfoss Drives
AC drives provide economic benefits, precise control of speed and torque, and enhanced performance when advanced parameters are used.

Energy Savings with Variable Torque Loads

Almost 50% of all energy used by industry is consumed by variable torque loads (fans and most pumps) that are powered by 3ϕ constant-speed induction motors. See Figure 14-1. Not all pumps are variable torque loads; for example, reciprocating pumps are constant torque loads. AC drives are commonly installed on pumps and fans in HVAC applications because AC drives save energy and money. A reduction in the amount of energy used by pump and fan motors reduces the utility bills of a facility and also reduces the emissions of greenhouse gases from power plants.

The *Affinity Laws* are physics laws that cover the relationships between speed, flow, pressure, and horsepower for variable torque loads. See Figure 14-2. The Affinity Laws are as follows:

• Speed and flow are proportional.
• Pressure (head) varies as the square of speed.
• Power requirements vary as the cube of speed.

Often a system requires less than full output (flow) from a pump or fan. When an AC drive is installed to control a pump or fan motor, the speed of the motor can be varied to control the product flow rate allowing less energy to be used. When an AC drive is not installed to control a motor, the speed of the motor is constant and flow is adjusted by using throttling or other restrictive devices. Throttling valves used with pumps and inlet vanes or outlet dampers used with fans cause a motor to use more energy than a motor powered by a drive.

TECH FACT

An added benefit of using AC drives versus DC drives is that AC motors require less maintenance than DC motors. The brushes—which wear out and require replacement—the commutator, armature, and field of a DC motor must be inspected and serviced periodically. Failure to properly maintain a DC motor leads to motor failure and system downtime.

Electric motor drive manufacturers provide software programs to calculate the cost savings realized when using an AC drive with pumps and fans. Information such as motor horsepower, motor efficiency, cost of electricity, cost of the AC drive, and length of time the load is operated at different flow rates is entered into the program. The cost savings program calculates the annual cost savings and the payback period. In many cases the software is available free or as a download from the web site of the electric motor drive manufacturer.

Pump Applications. A *pump curve* is a graph that shows the relationship between flow and pressure for a particular pump at a single speed. A *system curve* is a graph that shows the relationship between the pressure and flow required by a system. A system curve takes into account system losses such as friction in the piping system. The intersection of the pump curve and the system curve is the point at which the pump and system operate. See Figure 14-3.

INDUSTRIAL USE OF MOTORS

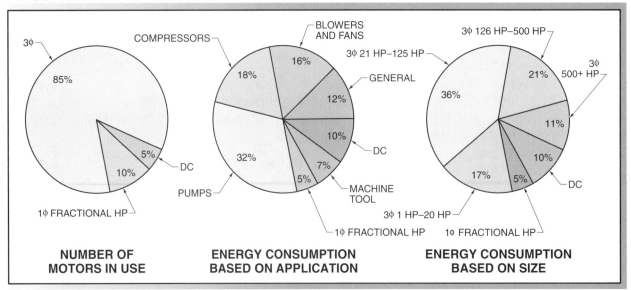

Figure 14-1. Three-phase motors are the most common motors used in industry. Pumps and fans powered by 3ϕ constant-speed induction motors account for almost 50% of all energy consumed.

to

AFFINITY LAWS FOR VARIABLE TORQUE LOADS

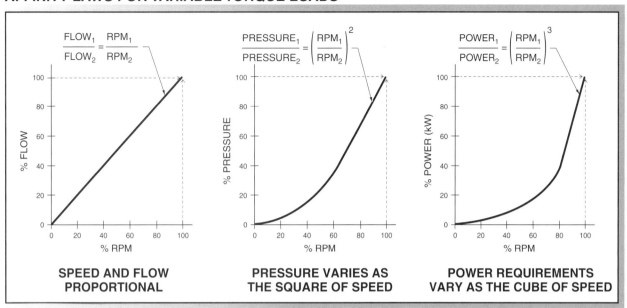

$$\frac{FLOW_1}{FLOW_2} = \frac{RPM_1}{RPM_2}$$

$$\frac{PRESSURE_1}{PRESSURE_2} = \left(\frac{RPM_1}{RPM_2}\right)^2$$

$$\frac{POWER_1}{POWER_2} = \left(\frac{RPM_1}{RPM_2}\right)^3$$

SPEED AND FLOW PROPORTIONAL

PRESSURE VARIES AS THE SQUARE OF SPEED

POWER REQUIREMENTS VARY AS THE CUBE OF SPEED

Figure 14-2. The Affinity Laws cover the relationships among speed, flow, pressure, and horsepower for variable torque loads.

ENERGY SAVINGS USING THROTTLING VALVE

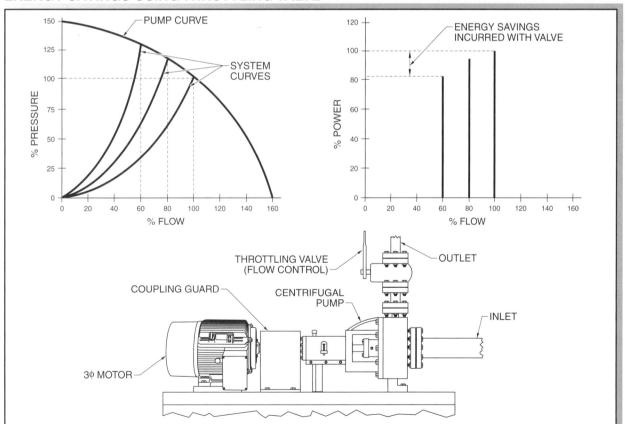

Figure 14-3. Reducing the flow of a pump with a valve results in nominal energy savings, but reducing the flow of a pump with an AC drive results in new pump curves and significant energy savings.

Changing the position of an outlet damper on an HVAC fan generates a new system curve for the particular flow rate.

When a throttling valve is used to control the flow of liquid out of a pump, a new system curve is generated for that specific flow rate. A throttling valve restricting the flow of liquid out of a pump decreases flow but increases pressure. As flow decreases, a minor energy savings occurs. When an AC drive is used to control the flow of liquid out of a pump, a new pump curve is generated. The AC drive controls the flow of liquid by controlling the speed of the pump. As the flow decreases, a substantial energy savings occurs because power varies as the cube of speed.

Fan Applications. A *fan curve* is a graph that shows the relationship between flow and pressure for a particular

fan at a single speed. The system curve takes into account system losses such as ductwork and restrictions. The intersection of the fan curve and the system curve is the point at which the fan and system operate. See Figure 14-4.

When an outlet damper is used to control the flow of air, a new system curve is generated for the particular flow rate. An outlet damper restricting the flow of air out of a fan decreases flow but increases pressure. As flow decreases, a minor energy savings occurs. When an inlet vane is used to control the flow of air, a new fan curve is generated. An inlet vane controls the amount of air going into a fan, resulting in reduced flow and reduced pressure. Inlet vane control yields greater energy savings than using outlet dampers. When an AC drive is used to control the flow of air, a new fan curve is also generated. An AC drive controls the flow of air by controlling the speed of the fan motor. As the flow decreases, a substantial energy savings occurs because power varies as the cube of speed.

ANALYSIS OF DRIVEN LOAD

The first step in selecting an electric motor drive is to analyze the load requirements on the motor. Loads are classified according to how torque and horsepower vary with speed. The three types of loads are constant torque, constant horsepower, and variable torque.

ENERGY SAVINGS USING INLET VANE

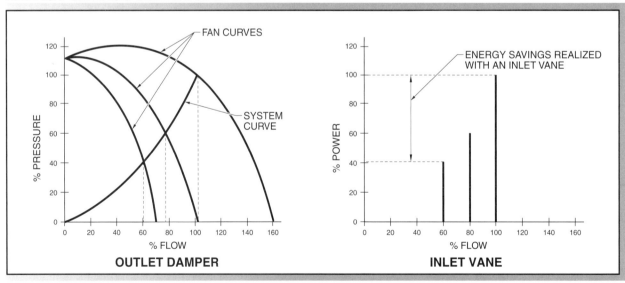

Figure 14-4. Reducing the flow of a fan with outlet dampers results in nominal energy savings, but reducing the flow of a fan with inlet vanes results in greater energy savings, and reducing the flow of a fan with an AC drive results in significant energy savings.

Constant torque loads are loads in which the torque required remains constant from 0 rpm to base speed. The horsepower required by constant torque loads increases linearly from 0 rpm to base speed, as with conveyors and extruders. See Figure 14-5.

Constant horsepower loads are loads with which the motor operates above base speed. The horsepower required remains constant across the speed range, as with center-driven winders and milling machines. The torque required by the load decreases as speed increases. See Figure 14-6.

Variable torque loads are loads in which the torque required increases with the speed of the load. A typical variable torque load has the torque equal to the speed squared and the horsepower equal to the speed cubed, as with centrifugal pumps and centrifugal fans. See Figure 14-7.

CONSTANT TORQUE LOAD ANALYSIS

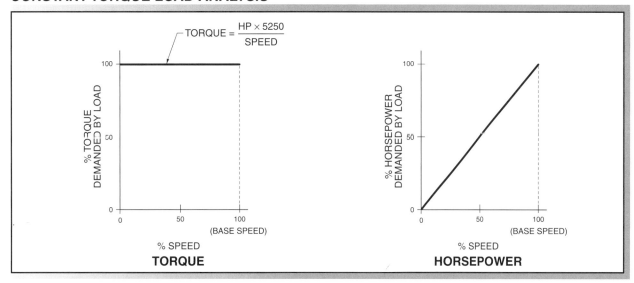

Figure 14-5. Torque remains constant from 0 rpm to base speed when a motor is connected to a constant torque load.

CONSTANT HORSEPOWER LOAD ANALYSIS

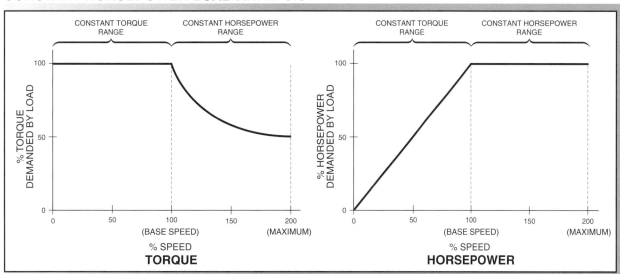

Figure 14-6. Motors are operated above base speed when connected to constant horsepower loads.

VARIABLE TORQUE LOAD ANALYSIS

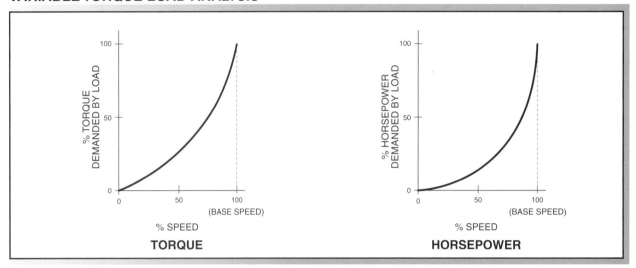

Figure 14-7. Torque is equal to the speed squared, and horsepower is equal to the speed cubed when a motor is connected to a variable torque load.

Load Considerations

Certain applications require an electric motor drive to be oversized because of load dynamics. Shock loads routinely overload an electric motor drive. A *shock (impact) load* is a load which varies from a fraction of rating (horsepower and torque) to several hundred percent of rating, as with coal crushers and wood chippers. Some applications require very high starting torque, frequent reversals, or short acceleration times. An electric motor drive manufacturer or a licensed engineer must be consulted regarding specific load applications. Tables of typical load characteristics are used as guidelines. See Figure 14-8.

MOTOR DATA

The second step in selecting an electric motor drive is to gather information about the motor the drive is controlling, such as the operating conditions and nameplate data of the motor. The conditions of operation for a motor vary from application to application. The motor must be compatible with the specific application. The following conditions must be verified to ensure motor compatibility:

- Altitude. Most motors are designed to operate below 3300 ft (1000 m). A motor that is operated above 3300 ft must be derated because thin air does not dissipate heat as quickly as air at sea level.
- Temperature. Most motors are designed to operate at an ambient temperature of 104° F (40° C). Motors operated at a higher ambient temperature must be derated.

TECH FACT

Electric motor drive manufacturers provide assistance in selecting the correct type and size drive for a specific application. Applications with shock loads can overload an electric motor drive, requiring that the drive be oversized by at least one size to ensure the drive can handle the load.

- Environment. Motor enclosures must be compatible with the environment in which a motor is mounted. Motors operated in environments with explosive dust or vapors or subject to washdowns require special enclosures.
- Speed. Constant horsepower applications require a motor to be operated above base speed. A motor must be capable of safely operating above base speed by handling the additional stresses on motor bearings, rotor, and winding insulation.
- Cooling. A motor that is operated below base speed has the cooling effect of the fan or rotor fins diminished. Motors operating at very low speeds and connected to constant torque loads may need to be derated or have an auxiliary blower installed to provide cooling. Typically, motors operating variable torque loads at speeds as low as 30% of base speed are not affected by the reduced cooling, since the load requires less torque at low speed.

TYPICAL LOAD CHARACTERISTICS

Load	Classification	Locked Rotor Torque	Load	Classification	Locked Rotor Torque
Agitators			Machines		
Liquid	*VT	Moderate	Boring	CT	Moderate
Slurry	*VT	Moderate	Bottling	CT	Moderate
Blowers			Milling	*CH	Moderate
Centrifugal	VT	Low	Mills		
Positive displacement	CT	Low (Unloaded)	Rolling	*CT	Moderate
Calenders	CT	Low	Rubber	*CT	Moderate
Card machines	CT	Moderate	Mixers		
Centrifuges	CT	Moderate	Chemical	CT	High
Chippers	*CT	High	Dough	CT	High
Compressors			Slurry	CT	High
Axial–centrifugal	VT	Low	Planers	CT	Moderate
Reciprocating	*CT	Moderate	Plows–conveyor	CT	Moderate
Rotary	CT	Moderate	Presses		
Conveyors			Printing	CT	Moderate
Belt	CT	Moderate	Punch	*CT	Moderate
Screw	*CT	High	Pullers–car	CT	Moderate
Shaker	*CT	Moderate	Pumps		
Cranes			Centrifugal	VT	Low
Bridge	CT	Moderate	Positive displacement	CT	Moderate
Trolley	CT	Moderate	Slurry	CT	High
Hoist	CT	Moderate	Roll benders	CT	Moderate
Crushers	*CT	High	Sanders	CT	Low
Drill presses	CH	Moderate	Saws	*CT	Moderate
Elevators	CT	Moderate	Shakers	*CT	High
Extruders	CT	Moderate	Shears	*CT	Low
Fans–centrifugal	VT	Low	Tension drives	CH	Moderate
Frames–spinning	CH	Low	Tool machines	CH	Moderate
Grinders	CH	Moderate	Walkways	CT	Low
Kilns	CT	High	Winches	CT	Moderate
Looms	CT	Moderate	Winders	CH	Moderate
Lathes	*CH	Moderate	Washers	CT	Moderate

* potential shock load
CT = constant torque
CH = constant horsepower
VT = variable torque

Figure 14-8. A load is classified by the specific application of the load.

Motor nameplate data is required to size an electric motor drive and determine if the motor is compatible with the drive. See Figure 14-9. The following conditions must be verified to ensure motor compatibility:

- Inverter rated motors. NEMA has developed a set of specifications (NEMA MG-1, Section IV, Part 31) for motors that are powered by electric motor drives. Contact the motor manufacturer to verify compatibility before powering a motor that is not NEMA MG-1, Section IV, Part 31 compliant from an electric motor drive.
- Current, horsepower, and voltage. The nameplate current, horsepower, and voltage must be recorded. The size of the electric motor drive is approximately determined by the horsepower of the motor, but the motor nameplate current is the actual determining factor for sizing a drive.

The output voltage of an AC drive must be compatible with the requirements of the motor.

- NEMA design. NEMA has established standards for classifying the speed-torque characteristics of motors (NEMA design designations A, B, C, D, and E). NEMA B design motors are the most commonly used motors with AC drives. When a NEMA B design motor is not used, contact the motor manufacturer regarding compatibility with electric motor drives.

ELECTRIC MOTOR DRIVE SELECTION

The final step in selecting an electric motor drive involves sizing the drive, determining the type of drive, determining the drive features required for the application, and ensuring the drive is compatible with the environment at the point of use. All steps must be followed when selecting an electric motor drive because the possibility for error or oversight is high. Failure to follow the recommended steps can result in an electric motor drive application that does not work properly or causes a catastrophic failure.

Electric Motor Drive Sizing

The size of an electric motor drive is based on the horsepower and nameplate current of the motor. The output current rating of an electric motor drive must be equal to or greater than the nameplate current of the motor. See Figure 14-10. Items to be considered when sizing an electric motor drive include the following:

- The horsepower rating of the motor must not exceed the horsepower rating of the electric motor drive.
- In multimotor applications, an electric motor drive is sized based on the sum of the nameplate currents of all the motors.
- An electric motor drive requires derating if the load torque exceeds 150% for a constant torque load or 110% for a variable torque load during starting or intermittently while running.
- An electric motor drive requires derating when the drive is connected to a 1φ power source or operated from a power source with reduced voltage.
- An electric motor drive requires derating if the drive is installed above 3300 ft (1000 m), is operated above normal temperature, is operated at a high

carrier frequency, controls shock loads, or controls loads that are difficult to start.

- For constant torque or constant horsepower loads, oversizing an electric motor drive by one size is a practical method to ensure a drive can handle a load without any problems—for example, using a 25 HP drive where a 20 HP unit has been determined to be appropriate. An electric motor drive that is not oversized can have a change in operating conditions, such as a load increasing slightly, that leads to drive failures or drive faults such as drive overcurrent faults.

TECH FACT

Electric motor drive installations must comply with the NEC® and all applicable municipal, state, and federal codes.

Electric Motor Drive Types

The four major types of electric motor drives are DC, AC inverter, AC open-loop vector, and AC closed-loop vector. DC drives and AC drives are typically used on any type of load, with DC drives used for constant torque or constant horsepower loads. Many AC drives can operate any type of load such as constant torque, constant horsepower, or variable torque loads. Some AC drives are designed for specific applications such as variable torque loads in HVAC applications. New electric motor drive installations typically use AC drives, but DC drives are still found in certain applications.

ELECTRIC MOTOR DRIVE SIZING DATA

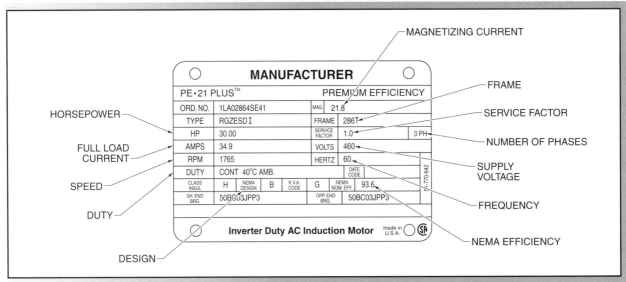

Figure 14-9. Motor nameplate data are required to size an electric motor drive for an application.

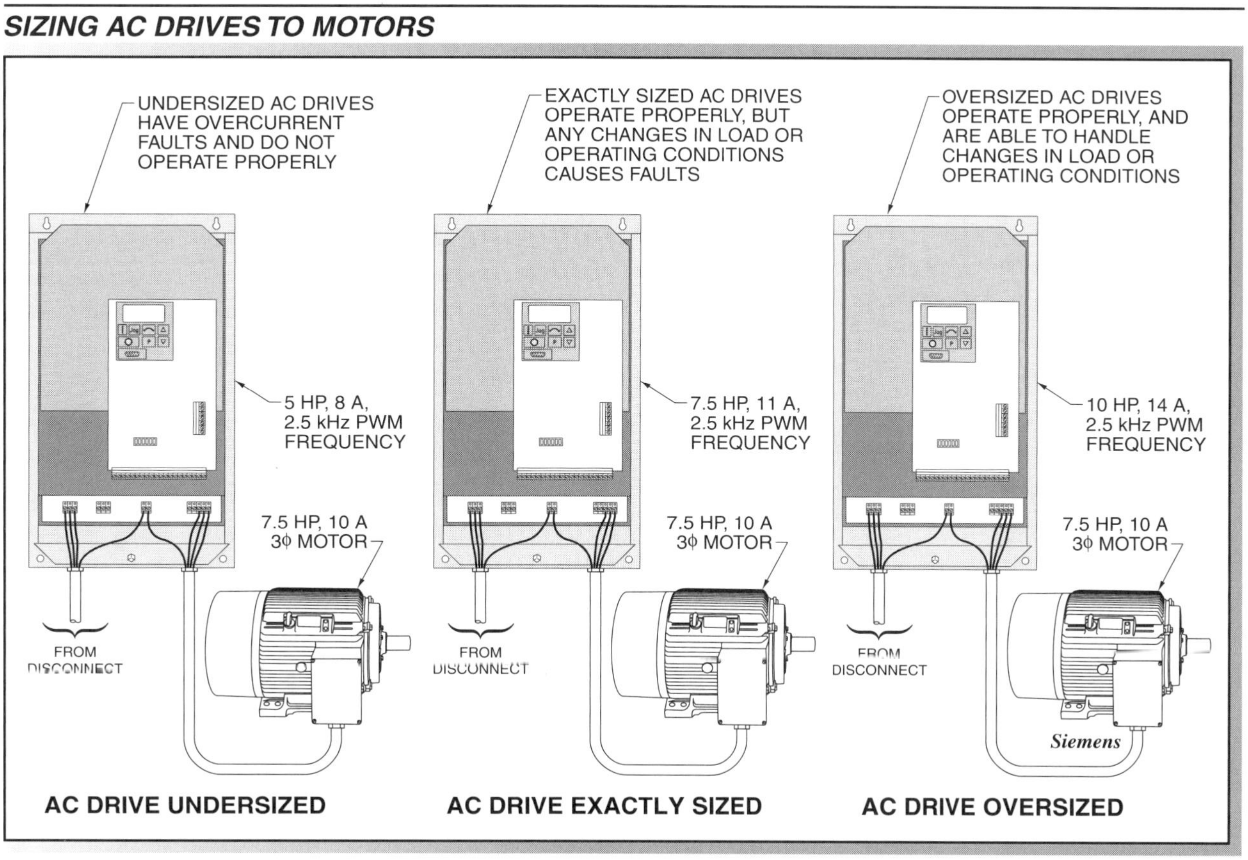

SIZING AC DRIVES TO MOTORS

Figure 14-10. The output current rating of an electric motor drive must be equal to or greater than the nameplate current of the motor.

The performance characteristics of an electric motor drive must be considered when selecting a type of drive for a specific application. See Figure 14-11. Speed range and speed regulation are two important characteristics of an electric motor drive and motor. *Speed range* is the minimum speed and maximum speed at which an electric motor drive or motor can operate under constant torque or variable torque conditions. Speed range is expressed as a ratio between base speed and minimum speed; for example, a drive with a speed range of 20:1 can operate a nameplate rated 1740 rpm motor anywhere between 1740 rpm and 87 rpm. Typically, the speed range of an electric motor drive is wider than the speed range of a motor because of motor cooling considerations. *Speed regulation* is the numerical measure of how accurately an electric motor drive can maintain the speed of a motor when the load changes. Speed regulation is expressed as the percentage of change in speed between no load and full load when compared to base speed.

Integrated Motor Drives. Integrated motor drives are a special type of electric motor drive. Integrated motor drives consist of an AC drive mounted directly on an AC motor. See Figure 14-12. Integrated motor drives are designed with protection against the heat and vibration developed by motors and are available up to 10 HP. The advantages of integrated motor drives are the elimination of a separate electric motor drive enclosure, the elimination of long conduit or cable runs between the drive and motor, and the elimination of reflected waves as the result of the elimination of long conductor runs between the drive and motor. Many integrated motor drives have communication capabilities that allow the integrated motor drive to be remotely controlled.

> ⚠ **WARNING**
>
> *Electric motor drives installed in hazardous locations without approved enclosures for the environment or drives that are of the wrong type or size for the application can cause fires and/or explosions, resulting in death or serious injury.*

DRIVE PERFORMANCE COMPARISONS				
FEATURE	DC	INVERTER	OPEN-LOOP VECTOR	CLOSED-LOOP VECTOR
Relative speed range	20:1	5:1 – 10:1	30:1	6000:1
Speed control	Yes	Yes	Yes	Yes
Torque control	Possible	No	Possible	Yes
Positioning	Possible	No	Possible	Possible
Open-loop speed regulation	±1% – 2% Base speed	±1% – 3% Base speed	±0.5% Base speed	—
Closed-loop speed regulation	±1% Set speed	—	—	±0.1% Set speed
Maintain full torque at zero speed	No	No	No*	Yes
Motor brushes required	Yes	No	No	No
Operate multiple motors from one drive	Yes	Yes	No	No

*(Only to a few Hz)

Figure 14-11. When selecting an electric motor drive, the performance characteristics of the drive must match the application.

INTEGRATED MOTOR DRIVES

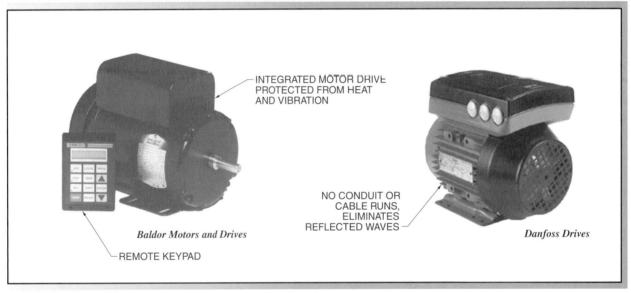

Baldor Motors and Drives

Danfoss Drives

Figure 14-12. Integrated motor drives are available in a variety of sizes and are protected from heat and vibrations created by the motor.

Electric Motor Drive Features

An *electric motor drive feature* is a function and/or accessory of a drive that is required for a specific application. Electric motor drive manufacturers have an assortment of drive models with a variety of features. The complexity of the application determines the number and type of features required by an electric motor drive. Purchasing an electric motor drive with features that are never used adds to the cost of a drive and can complicate the installation. Electric motor drive features that must be considered include:

• the number of digital inputs and outputs required for the application
• the number of analog inputs and outputs required for the application

• the communication capabilities required such as RS422, Profibus, DeviceNet, etc.
• special braking requirements such as dynamic braking
• bypass requirements for critical loads such as an exhaust fan that removes fumes from a parking garage
• supplemental overcurrent devices and overload protection required for multimotor applications

Electric Motor Drive Environment

Electric motor drives are used in a variety of environments. The environments range from clean, air-conditioned control rooms to remote oil wells in extremely hot and dirty locations. When selecting an electric motor drive, consideration must be given to the environment in which the drive is installed.

See Figure 14-13. Environment considerations in selecting an electric motor drive include the following:

- Avoid locations that subject the electric motor drive to excessive vibration or shock, corrosive atmospheres, and high humidity.
- Ensure that an electric motor drive enclosure is rated for the location; for example, use a NEMA 3R enclosure for outdoor installations.
- Ensure the motor is NEMA MG-1, Section IV, Part 31 compliant (inverter rated).
- Ensure that the ambient temperature range of the electric motor drive is not exceeded. Auxiliary cooling can be required in extremely hot locations, or auxiliary heating in extremely cold locations.
- Ensure the distance between an electric motor drive and motor does not exceed recommended lengths.
- Ensure that proper installation techniques are followed to minimize EMI and power quality problems; for example, install the line and load conductors in separate metal conduits.

ELECTRIC MOTOR DRIVE RETROFITS

In addition to new installations, many existing applications are retrofitted with electric motor drives. Applications that are retrofitted with electric motor drives include HVAC fans, HVAC pumps, cooling towers, domestic water pumping systems, conveyors, elevators, and printing presses. Typically, a retrofit involves replacing a magnetic motor starter with an AC electric motor drive or replacing a DC electric motor drive and motor with an AC electric motor drive and motor.

Electric Motor Drive Retrofit Benefits

Retrofitting an existing application with an electric motor drive produces many benefits. Benefits include energy savings, enhanced system performance, complete control of a system, decreased pollution, reduced maintenance due to less stress on the electrical and mechanical systems, reduced utility bills, and utility rebates. Electric motor drive retrofits require extra attention to

ELECTRIC MOTOR DRIVE ENVIRONMENT CONSIDERATIONS

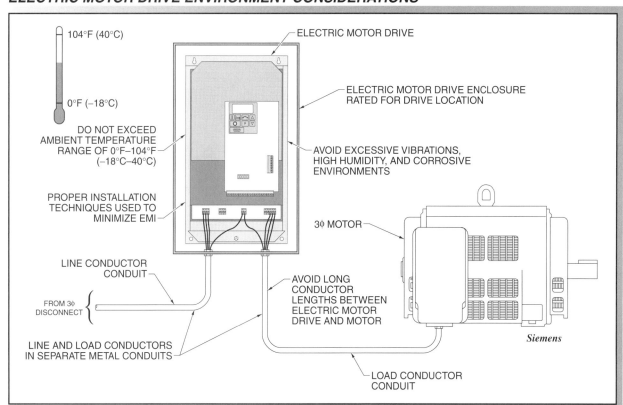

Figure 14-13. Electric motor drive environment is a major consideration when selecting a drive.

detail because information about the existing application may be missing or incomplete. In many applications, electric motor drive retrofits are more challenging than new installations. See Figure 14-14.

The primary benefit of retrofitting an existing application with an electric motor drive is energy savings. An electric motor drive can run a motor and its load at less than base speed. The reduction in speed results in a reduction in energy use, which translates into lower utility bills. The amount of energy saved depends on the type of load, torque, and horsepower. The Affinity Laws show that a large amount of energy can be saved when variable torque loads are powered by electric motor drives. According to the Affinity Laws, power requirements vary as the cube of speed. The most common variable torque loads include HVAC fans and HVAC pumps. In brief, customers receive a substantial savings on their utility bill by installing electric motor drives.

RETROFIT BENEFITS

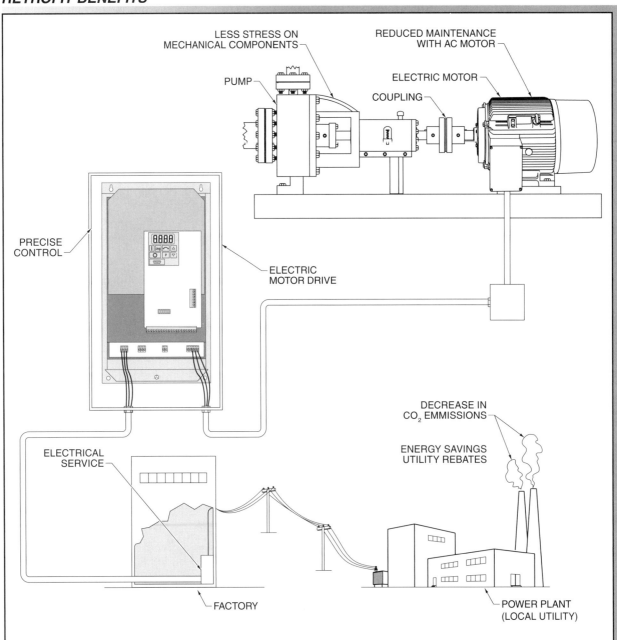

Figure 14-14. Retrofitting existing applications with electric motor drives results in numerous benefits.

Most electricity is generated by burning fossil fuel. Carbon dioxide is a by-product of burning fossil fuels such as coal, fuel oil, or natural gas. Carbon dioxide traps heat in the Earth's atmosphere. As more countries are becoming industrialized, the amount of carbon dioxide released into the atmosphere is increasing. The additional heat trapped in the atmosphere raises the surface temperature of the Earth, resulting in global warming. An important benefit related to electric motor drive energy savings is a reduction in the amount of carbon dioxide emitted into the atmosphere.

Another benefit of retrofitting an existing application with an electric motor drive is complete control of a system or process. Many existing installations that are retrofitted with an electric motor drive use magnetic motor starters with a single-speed or two-speed motor. The inability to vary the speed of a motor between 0 rpm and base speed limits the degree of control. An electric motor drive increases the degree of control because it allows the speed of the motor to be varied. Variable control can be used in applications such as conveyors, HVAC fans, and cooling towers.

Many existing cooling towers have two-speed motors. In order to maintain the temperature setpoint of the condenser water, the cooling tower motor is run at high or low speed. The temperature of the condenser water circulating between the cooling tower and the chiller rises and falls below the setpoint because the motor only has two speeds. This causes the chiller to operate in an inefficient manner. An electric motor drive retrofit allows the condenser water temperature to remain close to the temperature setpoint by varying the speed of the cooling tower motor. Consequently, the chiller operates more efficiently.

Retrofitting an existing application with an electric motor drive results in less stress on the electrical distribution system and the mechanical system. When a motor is started across the line by a magnetic motor starter, the motor can draw an inrush current exceeding six times the nameplate-rated current for an instant. This current can cause voltage fluctuations and peak power charges from the utility company. Also, when a motor is started across the line, the rapid acceleration from 0 rpm to base speed places additional stress on the mechanical elements such as the driven load, pulleys, belts, bearings, and rollers. When an electric motor drive is used to start a motor, the acceleration time can be lengthened to limit the amount of inrush current and reduce the mechanical stress due to rapid acceleration. The acceleration time (ramp-up time) is adjustable.

Retrofitting an existing application with an electric motor drive can result in reduced maintenance. For example, replacing a DC electric motor drive and motor with an AC electric motor drive and motor significantly reduces motor maintenance because 3ϕ AC motors do not have brushes, which eliminates the need for periodic inspection and replacement of the brushes. Replacing a magnetic motor starter with an electric motor drive reduces stress on the mechanical system, which reduces wear and provides a longer life for belts, bearings, pulleys, and rollers. The reduction in maintenance and associated costs vary depending on the application.

Utility Rebates. Utility rebates are monetary benefits received after retrofitting an application with an electric motor drive. The demand for electricity continues to increase as the population grows. The purpose of utility rebates is to encourage customers to reduce power demand. Rebates are available for many different items, such as energy-efficient lighting, solar power, electric motor drives, and premium efficient motors. A decreased demand for electricity decreases the amount of CO_2 released into the atmosphere.

The rebates and rebate process for electric motor drives vary among utilities. Some utilities provide a flat rate rebate based on the horsepower of the drive. Other utilities base the rebate on the annual savings in kilowatt-hours. Applications for rebates must be filled out by utility customers. The application contains customer information and project details, including location, cost, description, and estimated energy savings. Most electric motor drive manufacturers have free software programs that estimate energy savings and returns on investment (ROI) for HVAC fan and pump applications. The software generates an estimate based on the Affinity Laws. The actual savings is typically less due to power losses caused by reduced airflow in narrow ductwork and reduced water flow in piping and elbows. Energy savings software can be found on the manufacturers' web sites. See Figure 14-15.

TECH FACT

A human interface module (HIM) is a manually operated input control unit that includes programming keys, system operating keys, and normally a status display. The status display may be a liquid crystal display (LCD) or light emitting diode (LED) display. Liquid crystal displays or LED displays are used to present programming information, drive status conditions, and diagnostic data during system operation.

ENERGY SAVINGS SOFTWARE

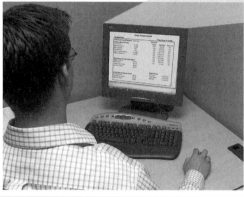

Energy Savings Estimator

Fan Application   Project Name: Air Handler

Total Annual Hours Of Operation: 8,736 Hours		Duty Cycle		
Operation/Motor/VFD Data		**% Flow**	**Time (Hrs)**	**Time (%)**
Cost per kWh:	12.00 ct.	100%	84.4 Hrs	1%
Motor Horse Power:	15.0 HP	90%	174.4 Hrs	2%
Motor Efficiency	95%	80%	786.2 Hrs	9%
Drive Efficiency	97%	70%	1,485.1 Hrs	17%
Power Company Incentive	18.0 $/HP	60%	2,096.6 Hrs	24%
Drive Cost	$2,325	50%	1,485.1 Hrs	17%
		40%	1,135.7 Hrs	13%
		30%	961.0 Hrs	11%
Annual Enegy Cost per Control Method		20%	524.2 Hrs	6%
No Speed Control	$12,382	10%	0.0 Hrs	0%
Drive	$2,716			
Outlet Damper Control	$8,997			
Inlet Vane Control	$6,757			

		Payback Period	
		No Control	0.234 Years
Annual Enegy Savings per Control		Outer Damper	0.355 Years
No Speed Control	$9,666	Inlet Vane	0.539 Years
Outlet Damper Control	$6,282		
Inlet Vane Control	$4,041		

Figure 14-15. Many electric motor drive manufacturers have free software programs that estimate energy savings and returns on investment for HVAC fan and pump applications.

Electric Motor Drive Retrofit Installations

When an existing application is retrofitted with an electric motor drive, additional issues need to be addressed. In retrofitted installations, detailed information may be missing or incomplete. Interface issues may exist with the current installation, and there may be special scheduling requirements. See Figure 14-16.

Missing or incomplete information, such as conductor length and motor compatibility, can present installation problems. The length of the conductors between the electric motor drive and motor must be kept to a minimum. Drive manufacturer requirements may vary.

Commonly, in retrofit installations, the drive is placed where the motor starter was located. However, in many existing installations, the distance from the motor starter to the motor is not evident, particularly if the conductors are run through a concrete floor or underground. As built drawings of the building can be consulted or the electric motor drive located adjacent to the motor. If a motor nameplate does not indicate that the motor is inverter-rated, the motor manufacturer should be contacted regarding compatibility. If a motor is not compatible with an electric motor drive, it should be replaced with an inverter-rated motor. A replacement motor may be eligible for a utility rebate.

SPECIAL CONSIDERATIONS

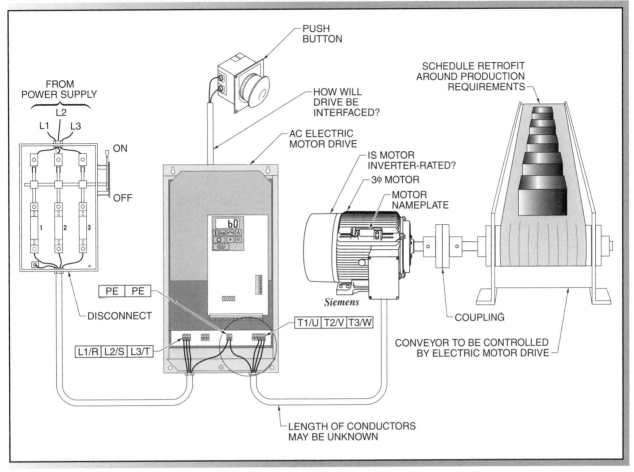

Figure 14-16. Retrofitting existing systems or processes with electric motor drives presents special installation considerations.

Specific issues involving the interface of the drive to the existing installation may arise when an existing installation is retrofitted with an electric motor drive. Issues may include how to start and stop the drive, how to control the speed of the drive, and special circumstances or requirements related to the existing installation. The existing installation must be thoroughly understood in order to ensure that the electric motor drive functions properly and the system works correctly.

Pushbuttons, selector switches, temperature control sensors, and building automation systems may be used to control a drive that controls the starting, stopping, and speed of a motor. In some applications, control components may need to be added, which may involve other trades. For example, a temperature sensor that provides an analog input (water temperature) to a drive may be added to the sump of a cooling tower. A welder must add the well for the temperature sensor to the cooling tower sump.

Some existing installations have special interface requirements, which need to be taken into consideration before starting the retrofit. Special requirements may include interface with a fire alarm/life safety system and interface with bypass contactors. Electric motor drives installed on HVAC fans typically interface with the building fire alarm/life safety system. The interface may include a duct smoke detector to shut down the drive or a fan control sequence to pressurize parts of a high rise building for smoke control in the event of a fire. Bypass contactors may be used on systems that cannot tolerate a shutdown due to a drive failure. A bypass contactor bypasses the drive and allows the motor to run at base speed until the drive is repaired.

Scheduling the retrofit of an existing system or process with an electric motor drive can present several challenges. Frequently, limited time periods are allowed to work on the systems. The customer's needs dictate that the system or process cannot be offline for extended periods of time. Technicians must complete the work in a short period of time or complete the work in phases. When a retrofit is done in phases, technicians perform preparatory work during time periods allowed by the customer, when the system can be offline. The system continues to run until the final conversion (cut over). The retrofit of certain systems can be scheduled for specific times of the year to correspond with the weather or with a slower production period. For example, the retrofit of a cooling tower is typically scheduled for fall or winter, when the cooling needs of a building are minimal.

Name _____ Date _____

True-False

T   F   1. Prior to the development of AC drives, DC drives and DC motors were the predominant form of variable speed control. *pg 397*

T   F   2. DC motors require less maintenance than AC motors. *pg 397*

T   F   3. AC drives are commonly installed on pumps and fans in HVAC applications because AC drives save energy. *pg 398*

T   F   4. In pump applications, an AC drive controls the flow of liquid by controlling the speed of the pump. *pg 400*

T   F   5. An outlet damper controls the amount of air going into a fan, resulting in reduced flow and reduced pressure. *pg 400*

T   F   6. A typical variable torque load has the torque equal to the speed squared and the horsepower equal to the speed cubed. *pg 401*

T   F   7. Most motors are designed to operate at an ambient temperature of 104°C. *pg 402 F*

T   F   8. Oversizing an electric motor drive by one size is a practical method to ensure a drive can handle a load without any problems. *pg 404*

T   F   9. Electric motor drives should not be installed in locations that subject an electric motor drive to excessive vibration, corrosive atmospheres, or high humidity. *pg 407*

T   F   10. Speed regulation is the numerical measure of how accurately an electric motor drive can maintain the speed of a motor when the load changes. *pg 405*

T   F   11. Constant horsepower applications of electric motor drives require a motor to operate below base speed. *above*

T   F   12. An electric motor drive requires derating if the load torque exceeds 150% for a constant torque load. *pg 404*

T   F   13. The horsepower rating of a motor must exceed the horsepower rating of the electric motor drive. *pg 404 not*

T   F   14. The monetary benefits received after retrofitting an application with an electric motor drive are called utility reactions. *pg 409 rebates*

T   F   15. Replacing a DC electric motor drive and motor with an AC electric motor drive and motor significantly reduces motor maintenance. *pg 409*

Completion

Contactors   1. A benefit of using AC drives is the reversing of motor rotation without additional ___.

constant torque   2. Not all pumps are variable torque loads; for example, reciprocating pumps are ___ loads.

Affinity laws   3. The ___ are physics laws that cover the relationships among speed, flow, pressure, and horsepower for variable torque loads.

Fan   4. A(n) ___ curve is a graph that shows the relationship between flow and pressure for a particular fan at a single speed.

Shock loads   5. ___ routinely overload an electric motor drive.

Motor current   6. The size of the electric motor drive is approximately determined by the horsepower of the motor, but the nameplate ___ is the actual determining factor for sizing a drive.

Speed range   7. ___ is the minimum speed and maximum speed at which an electric motor drive or motor can operate under constant torque or variable torque conditions.

Integrated motor drives   8. ___ consist of an AC drive mounted directly on an AC motor.

_CO_2_   9. An important benefit related to electric motor drive energy savings is a reduction in the amount of ___ emitted into the atmosphere.

energy savings   10. The primary benefit of retrofitting an application with an electric motor drive is ___.

Multiple Choice

_B_   1. Almost ___% of all energy used by industry is consumed by variable torque loads (fans and many pumps).
     A. 40   C. 70
     B. 50   D. 80

_A_   2. A pump curve is a graph that shows the relationship between ___ and pressure for a particular pump at a single speed.
     A. flow   C. speed
     B. horsepower   D. watts

_B_   3. The horsepower required by constant torque loads increases linearly from 0 rpm to ___ speed.
     A. average   C. maximum
     B. base   D. median

_C_   4. A motor that is operated above ___' must be derated because thinner air does not dissipate heat as quickly.
     A. 1100   C. 3300
     B. 2200   D. 4400

_B_   5. Retrofitting an existing application with an electric motor drive results in ___ air pollution.
     A. more   C. the same amount of
     B. reduced   D. the total elimination of

Name_____ Date _____

THREE-PHASE MOTOR

HP	RPM (60Hz)	Voltage	Catalog Number	List Price*
1	3450	230/460 - 60 Hz	M-1A	415
1	1725	230/460 - 60 Hz	M-1B	400
1½	3450	230/460 - 60 Hz	M-2A	450
1½	1725	230/460 - 60 Hz	M-2B	425
2	3450	230/460 - 60 Hz	M-3A	475
2	1725	230/460 - 60 Hz	M-3B	450
3	3450	230/460 - 60 Hz	M-4A	525
3	1725	230/460 - 60 Hz	M-4B	500
5	3450	230/460 - 60 Hz	M-5A	675
5	1725	230/460 - 60 Hz	M-5B	640
7½	3450	230/460 - 60 Hz	M-6A	775
7½	1725	230/460 - 60 Hz	M-6B	725
10	3450	230/460 - 60 Hz	M7A	900
10	1725	230/460 - 60 Hz	M-7B	850
15	3450	230/460 - 60 Hz	M-8A	1400
15	1725	230/460 - 60 Hz	M-8B	1350

* In $

ELECTRIC MOTOR DRIVE

HP	Input Voltage	Output Current		Catalog Number	List Price†
		Cont.*	Peak*		
1	230	2	8	D-LV-1	850
1	460	4	4	D-HV-1	900
2	230	7	14	D-LV-2	925
2	460	4	8	D-HV-2	975
3	230	10	20	D-LV-3	1150
3	460	5	10	D-HV-3	1275
5	230	16	32	D-LV-4	1300
5	460	8	16	D-HV-4	1360
7½	230	22	44	D-LV-5	1750
7½	460	11	22	D-HV-5	1840
10	230	28	56	D-LV-6	2400
10	460	14	28	D-HV-6	2525
15	230	42	84	D-LV-7	3000
15	460	21	42	D-HV-7	3200
20	230	55	100	D-LV-8	3700
20	460	27	54	D-HV-8	3850

* In A
† In $

MANUFACTURER ELECTRIC MOTOR DRIVE AND MOTOR CATALOG LISTINGS

Activity 14-1. Calculating Motor Torque

Torque is the force that produces rotation. A motor must produce enough torque to start and maintain load movement. The operating speed (in rpm), torque (in lb-ft), and horsepower (in HP) determine the work a motor produces. To determine the theoretical torque a motor can produce when motor horsepower and speed are known, apply the following formula:

$$T = \frac{HP \times 5252}{rpm}$$

where

T = torque (in lb-ft)

HP = horsepower

5252 = constant

rpm = revolutions per minute

Answer the questions using the Manufacturer Electric Motor Drive and Motor Catalog Listings and torque formula.

_____ 1. A catalog number M-1A motor will produce ___ torque output when connected to 230 V.

_____ 2. A catalog number M-1A motor will produce ___ torque output when connected to 460 V.

_____ 3. A catalog number M-1B motor will produce ___ torque output.

_____ 4. If the horsepower size of catalog number M-1A motor is doubled, the torque output of the larger motor is ___.

_____ 5. If the horsepower size of catalog number M-1A motor were increased ten times, the torque output of the larger motor is ___.

Activity 14-2. Calculating Motor Horsepower

Horsepower (HP) is a unit for measuring power. Horsepower equals 550 lb-ft/sec or 746 W electrically. To determine the theoretical horsepower of a motor when speed and torque are known, apply the following formula:

$$HP = \frac{rpm \times T}{5252}$$

where
HP = horsepower
rpm = revolutions per minute
T = motor torque (in lb-ft)
5252 = constant

Answer the questions using the Manufacturer Electric Motor Drive and Motor Catalog Listings and horsepower formula.

_____ 1. A customer requires a 230 V motor that can theoretically produce 17.5 lb-ft of torque at 1725 rpm. If the customer requests a motor that is rated 30% higher to ensure proper operation over time, what catalog motor would be recommended?

_____ 2. The recommended motor would cost ___.

_____ 3. If the customer wanted to know how much more the next larger size motor would cost (in case the application requirements are increased), what would the additional cost be?

_____ 4. The cost of an electric motor drive to operate the motor in question 1 is ___.

_____ 5. When using a 460 VAC electric motor drive and motor instead of 230 VAC drive and motor, the higher voltage equipment would ___ the installation cost.

Activity 14-3. Calculating Flow Speed When Speed Changes

Speed and flow are proportional. Changing motor speed changes the amount of output flow. Flow rate directly varies as speed varies. To determine flow rate, apply the following formula:

$$Q_{NEW} = Q_{OLD} \frac{rpm_{NEW}}{rpm_{OLD}}$$

where

Q_{NEW} = new flow rate (in CFM)
Q_{OLD} = original flow rate (in CFM)
rpm_{NEW} = new motor speed (in rpm)
rpm_{OLD} = original motor speed (in rpm)

Answer the questions using the Manufacturer Electric Motor Drive and Motor Catalog Listings and flow rate formula.

_____ 1. A heating, ventilation, and air-conditioning (HVAC) application has a catalog number M-7A motor. The motor drives a 1500 CFM blower, which moves heated air into a control room. A magnetic motor starter operates the motor and blower at full speed. The heat requirements change based on the control room usage and outside temperature. Replacing the magnetic motor starter with an electric motor drive that operates the motor at 862.5 rpm creates a power savings of ___%.

ACTIVITY 14-4—Air Handling Unit, AC Drive Energy Savings

Air Handling Unit 4 (AHU 4) currently has no speed control to control airflow. AHU 4 is going to be retrofitted with an AC drive to control airflow. Payback period is the length of time AHU 4 must run with the AC drive for the energy savings to pay for the drive and the drive installation. Below is the data needed to enter into the Electric Motor Drive Energy Savings spreadsheet.

DATA
- Project Name: AHU 4
- Cost of Electricity: $0.10/kWh
- Motor: 25 HP
- Run Time: 24 hr/day, 365 days/yr
- Drive Cost: $3900.00 plus $985.00 for installation
- The percentage of time that AHU 4 runs at various flow rates is as follows:

FLOW (in %)	TIME (in %)
100%	2%
90%	5%
80%	8%
70%	15%
60%	22%
50%	18%
40%	15%
30%	15%

Use the Electric Motor Energy Savings spreadsheet to calculate the savings and payback period. Save the completed spread sheet as AHU 4 and print a copy to turn in. Answer the questions below based on the results from the Electric Motor Drive Energy Savings spreadsheet.

_____ **1.** How many hours does the motor run per year?

_____ **2.** How many hours does AHU 4 run at a 30% flow rate?

_____ **3.** What is the annual energy cost to operate AHU 4 at a 30% flow rate?

_____ **4.** What is the annual energy cost to operate AHU 4 without any form of speed control?

_____ **5.** What is the annual energy cost to operate AHU 4 with an AC drive?

_____ **6.** What is the payback period when AHU 4 is converted to include an AC drive?

ACTIVITY 14-5—Pump Station, AC Drive Energy Savings

Pump Station 12 is part of a rural irrigation district. Pump Station 12 currently has a 100 HP pump motor. Pump Station 12 is currently operated at 100% speed and 100% flow rate. Pump Station 12 is going to be retrofitted with an AC drive to control water flow. Payback period is the length of time Pump Station 12 must run with the AC drive for the energy savings to pay for the drive and the drive installation. Below is the data needed to enter into the Electric Motor Drive Energy Savings spreadsheet.

DATA
- Project Name: Pump Station 12
- Customer Information: Use school name and location
- Cost of Electricity: $0.13/kWh
- Motor: 100 HP
- Runs: 19 hr/day, 365 days/yr
- Drive cost: $11,000.00 plus $1876.00 for installation
- The percentage of time that Pump Station 12 runs at various flow rates is as follows:

FLOW (in %)	TIME (in %)
100%	3%
90%	5%
80%	7%
70%	15%
60%	23%
50%	18%
40%	16%
30%	13%

Use the Electric Motor Drive Energy Savings spreadsheet to calculate the savings and payback period. Save the completed spreadsheet as Pump Station 12 and print a copy to turn in. Answer the questions below based on the results from the Electric Motor Drive Energy Savings spreadsheet.

_____ 1. How many hours does Pump Station 12 run at a 50% flow rate?

_____ 2. What is the annual energy cost to operate Pump Station 12 at 50% flow rate with an AC drive?

_____ 3. What is the annual energy consumption (kWh) of Pump Station 12 at a 50% flow rate?

_____ 4. What is the annual energy cost to operate Pump Station 12 with no speed control?

_____ 5. What is the annual energy cost to operate Pump Station 12 with an AC drive?

_____ 6. What is the payback period when Pump Station 12 is converted to include an AC drive?

Appendix

RESISTIVE CIRCUIT

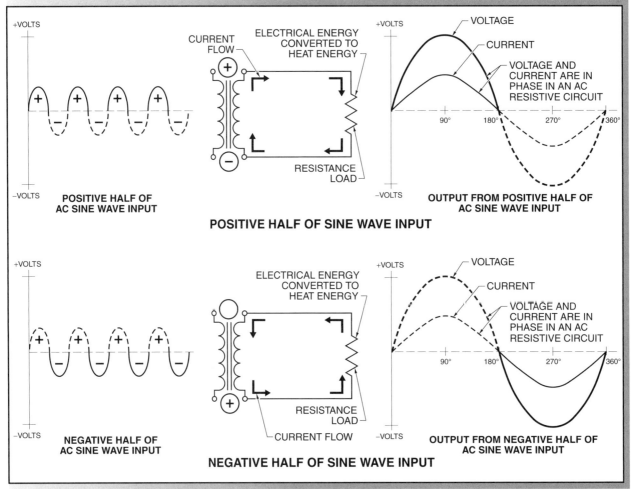

FIGURE 3-15.

CONVENTIONAL DIAGRAM ⊕ *TO* ⊖

INDUCTIVE CIRCUIT

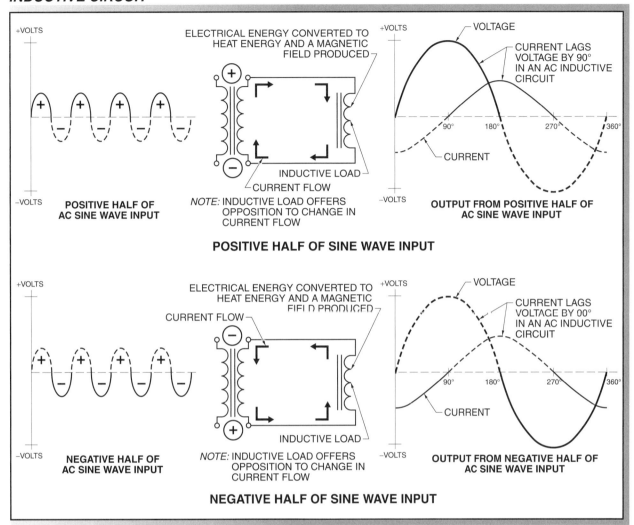

FIGURE 3-16.

CONVENTIONAL DIAGRAM ⊕ TO ⊖

CAPACITIVE CIRCUIT

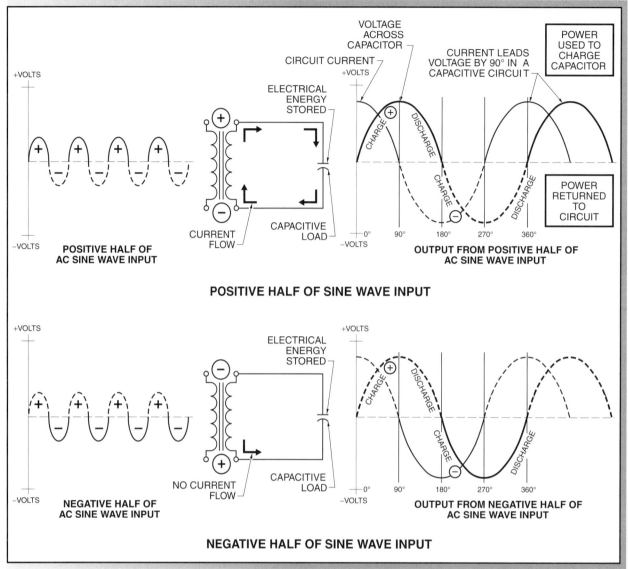

FIGURE 3-17.

CONVENTIONAL DIAGRAM ⊕ TO ⊖

ELECTRIC MOTOR MAGNETIC FIELDS

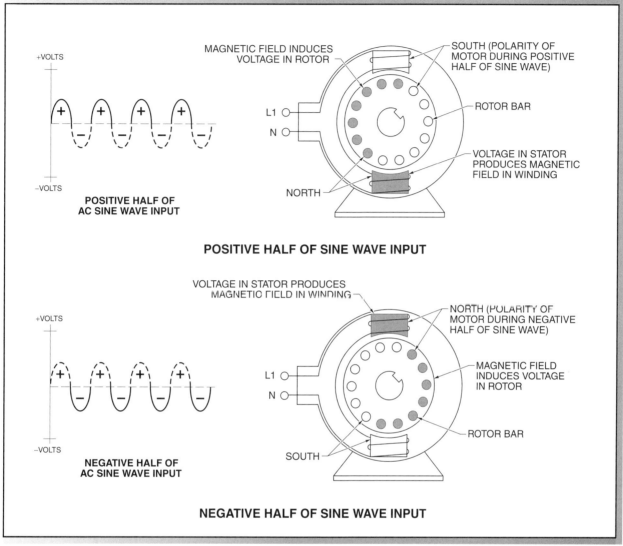

POSITIVE HALF OF SINE WAVE INPUT

NEGATIVE HALF OF SINE WAVE INPUT

FIGURE 4-4.

CONVENTIONAL DIAGRAM ⊕ TO ⊖

DC MOTOR ROTATION

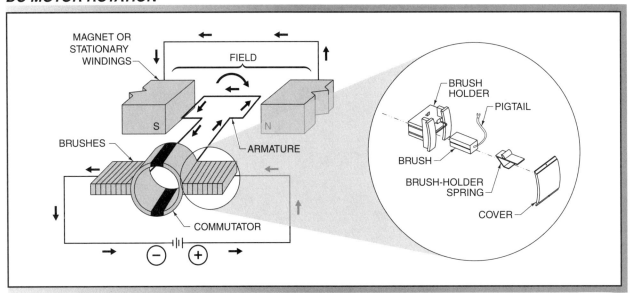

FIGURE 4-20.

CONVENTIONAL DIAGRAM ⊕ TO ⊖

DIODE OPERATION

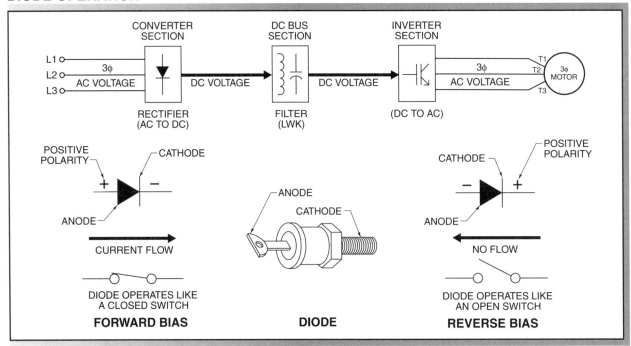

FIGURE 7-5.

CONVENTIONAL DIAGRAM ⊕ TO ⊖

DIODE RATING

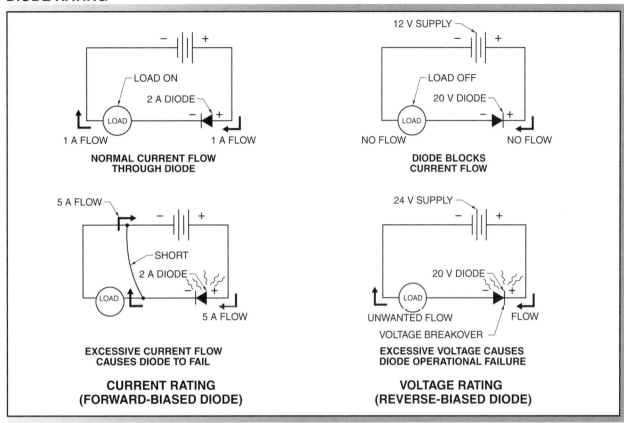

FIGURE 7-6.

HALF-WAVE RECTIFIER

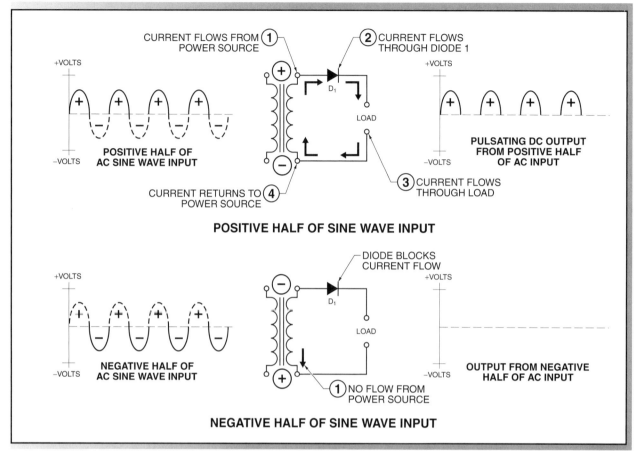

FIGURE 7-7.

CONVENTIONAL DIAGRAM ⊕ TO ⊖

FULL-WAVE RECTIFIER

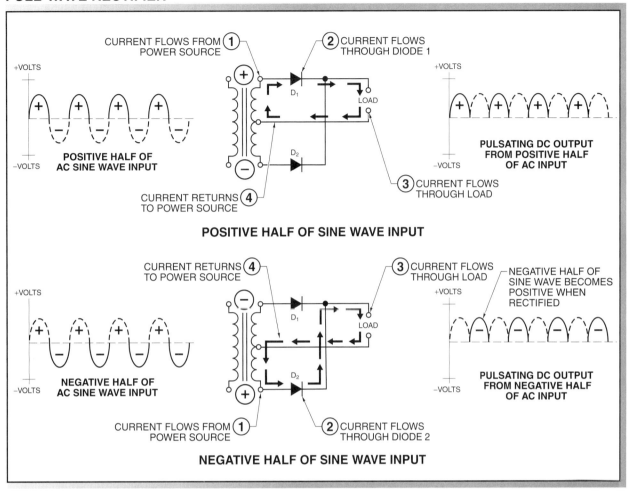

FIGURE 7-8.

CONVENTIONAL DIAGRAM ⊕ TO ⊖

BRIDGE RECTIFIER

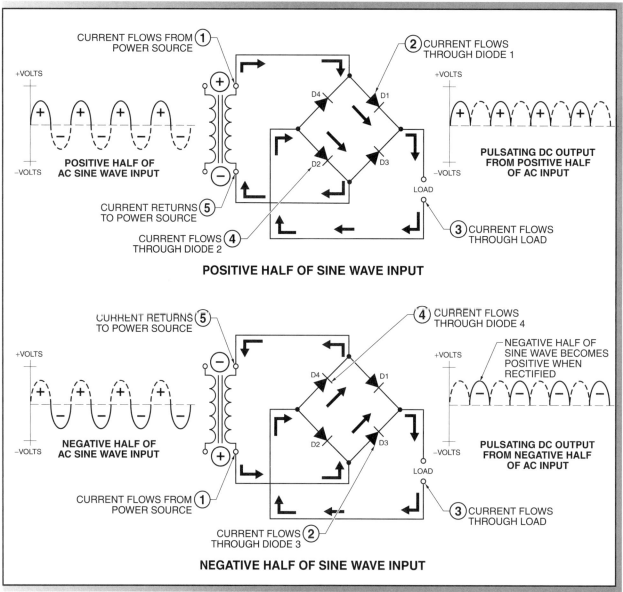

FIGURE 7-9.

SILICON-CONTROLLED RECTIFIER OPERATION

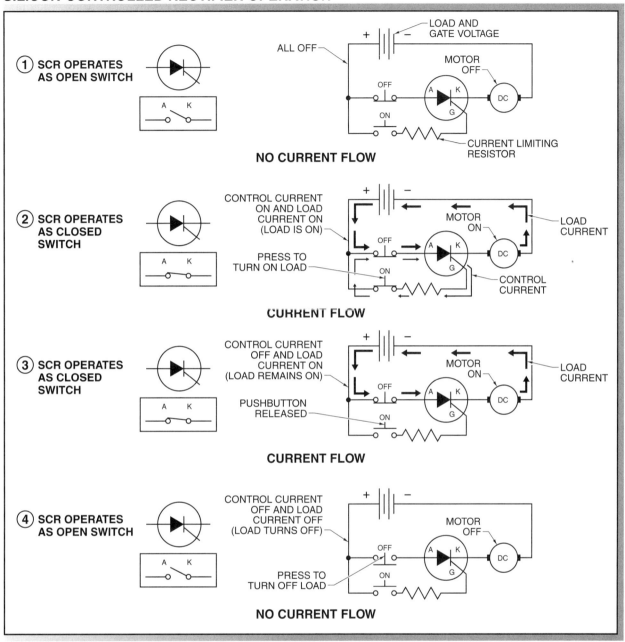

(1) SCR OPERATES AS OPEN SWITCH

A   K

ALL OFF — LOAD AND GATE VOLTAGE

+   −

OFF

ON

MOTOR OFF

A   K   G   DC

CURRENT LIMITING RESISTOR

NO CURRENT FLOW

(2) SCR OPERATES AS CLOSED SWITCH

A   K

CONTROL CURRENT ON AND LOAD CURRENT ON (LOAD IS ON)

PRESS TO TURN ON LOAD

+   −

OFF

ON

MOTOR ON

A   K   G   DC

LOAD CURRENT

CONTROL CURRENT

CURRENT FLOW

(3) SCR OPERATES AS CLOSED SWITCH

A   K

CONTROL CURRENT OFF AND LOAD CURRENT ON (LOAD REMAINS ON)

PUSHBUTTON RELEASED

+   −

OFF

ON

MOTOR ON

A   K   G   DC

LOAD CURRENT

CURRENT FLOW

(4) SCR OPERATES AS OPEN SWITCH

A   K

CONTROL CURRENT OFF AND LOAD CURRENT OFF (LOAD TURNS OFF)

PRESS TO TURN OFF LOAD

+   −

OFF

ON

MOTOR OFF

A   K   G   DC

NO CURRENT FLOW

FIGURE 7-12.

CONVENTIONAL DIAGRAM ⊕ *TO* ⊖

AC SWITCHING USING SILICON-CONTROLLED RECTIFIERS

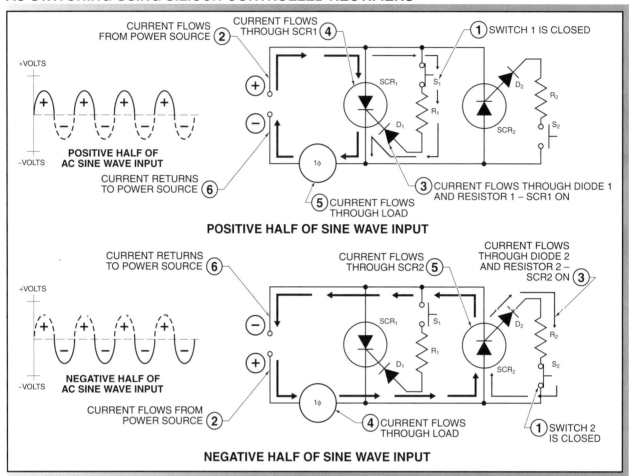

FIGURE 7-15.

CONVENTIONAL DIAGRAM ⊕ TO ⊖

GATE TURN-OFF THYRISTOR OPERATION

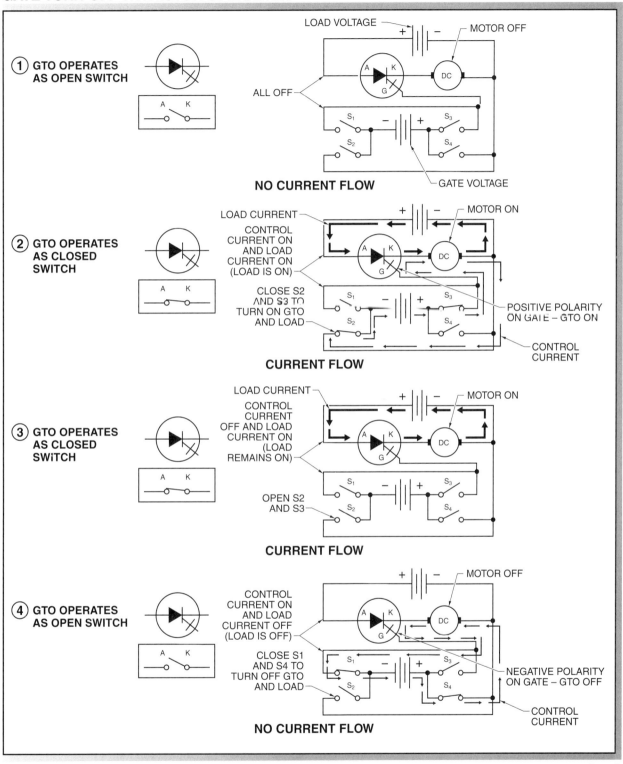

FIGURE 7-16.

TRANSISTOR CIRCUIT CURRENT FLOW

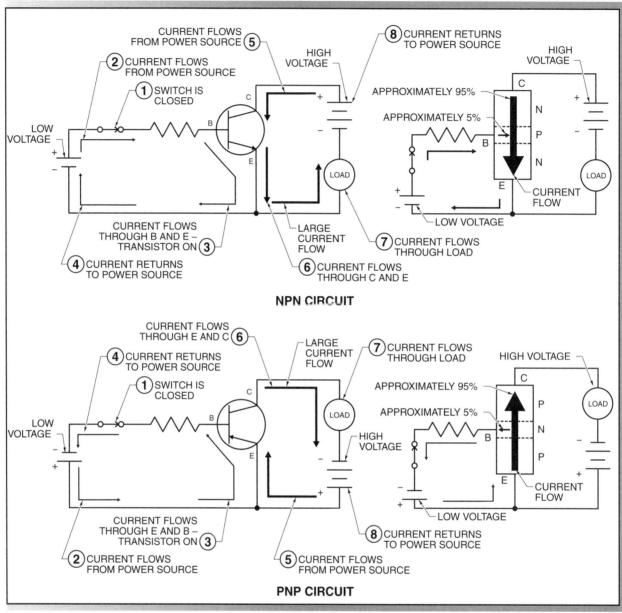

FIGURE 7-18.

CONVENTIONAL DIAGRAM ⊕ TO ⊖

DC MOTOR CONTROL USING TRANSISTORS

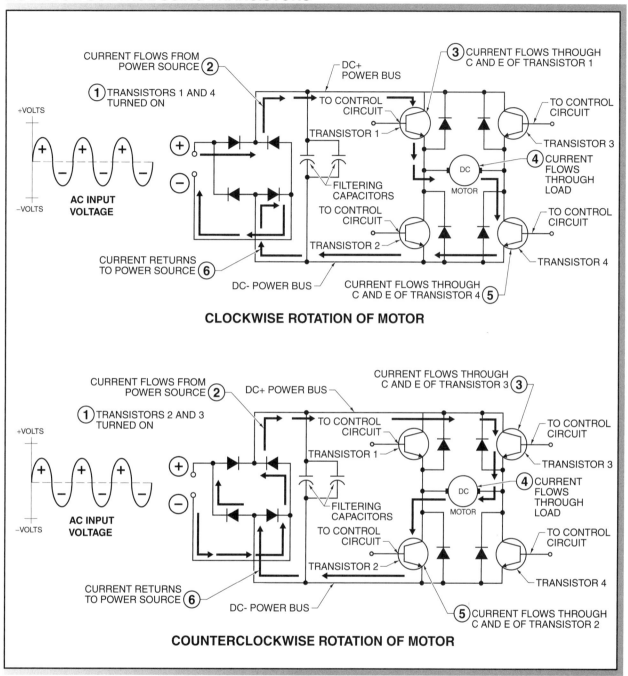

CLOCKWISE ROTATION OF MOTOR

COUNTERCLOCKWISE ROTATION OF MOTOR

FIGURE 7-19.

CONVENTIONAL DIAGRAM ⊕ *TO* ⊖

FLY-BACK DIODE PROTECTION

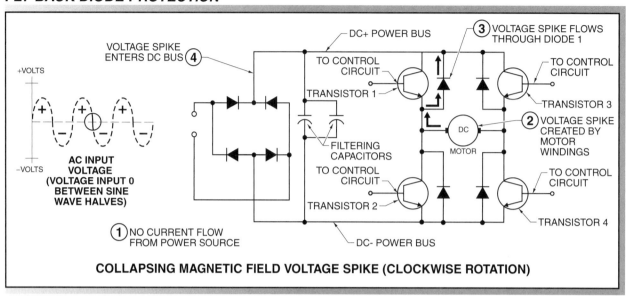

FIGURE 7-20.

IGBTs CONTROLLING MOTOR ROTATION

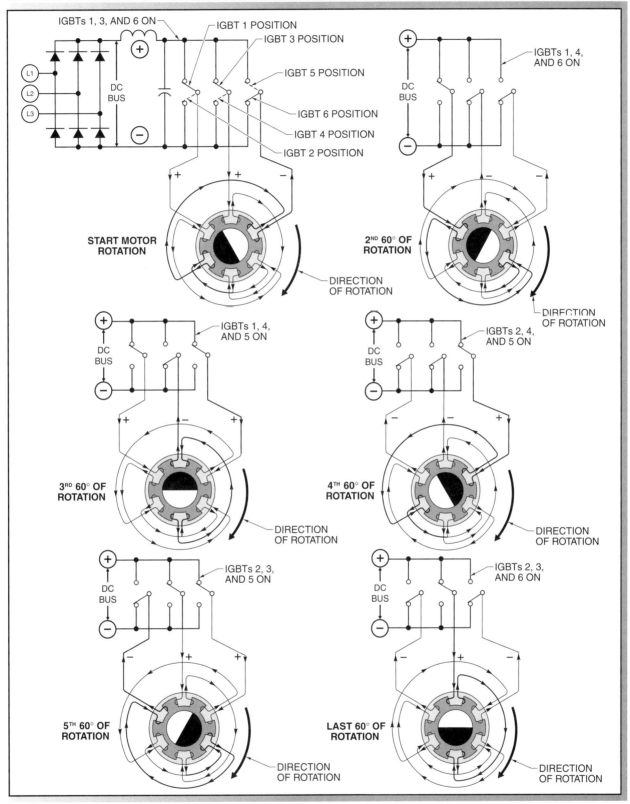

FIGURE 8-13.

INCOMING POWER TROUBLESHOOTING MATRIX			
SYMPTOM/FAULT CODE	PROBLEM	CAUSE	SOLUTION
Electric motor drive overvoltage faults. Blown converter (rectifier) semiconductor	High input voltage/voltage swell	Switching of power factor correction capacitors	Stop switching power factor correction capacitors. Install electric motor drive on another feeder
		Utility switching transformer taps for load adjustment	Install line reactor, or install electric motor drive on another feeder
		Proximity to low-impedance voltage source	Install line reactor
		Transformer secondary voltage is high	Adjust taps on transformer
Electric motor drive undervoltage faults. Difference between no load and full load voltage is greater than 3%	Low input voltage/ voltage sag	Power source for electric motor drive unable to deliver enough current	Increase kVA source rating, or install electric motor drive on another feeder
		Low transformer secondary voltage	Adjust secondary taps on supply transformer
		Large starting load(s)	Increase kVA source rating, or install electric motor drive on another feeder
		Electric motor drive is overloaded	See *Motor and Load Troubleshooting Matrix*
	Voltage sine wave flat-topping at electric motor drive	Power source for electric motor drive unable to deliver enough current	Increase kVA source rating, or install electric motor drive on another feeder
Electric motor drive does not turn on. Blown fuse or tripped circuit breaker	No input voltage	Incorrect fuse or circuit breaker	Install correct fuse or circuit breaker
		Conductors feeding electric motor drive shorted or have ground fault	Repair or replace conductors
		Problem with electric motor drive	See *Electric Motor Drive Troubleshooting Matrix*
Electric motor drive overload faults	Voltage unbalance greater than 2%	Unbalance from utility	Contact utility
		Single-phase loads on the same feeder as the electric motor drive switching ON and OFF	Install electric motor drive on separate feeder
Intermittent electric motor drive faults and/or electric motor drive intermittently does not operate per design	Improper grounding	Undersized grounds or no grounds	Install proper size ground
		Loose ground connections	Tighten ground connections
	Intermittent fault/ erratic operation	Electric noise, EMI/RFI	See *Electric Motor Drive Troubleshooting Matrix*
Harmonics	Harmonics present on electrical distribution system	Electric motor drive or existing nonlinear loads can be source	Install input reactor

ELECTRIC MOTOR DRIVE TROUBLESHOOTING MATRIX			
ELECTRIC MOTOR DRIVE FAULTS . . .			
SYMPTOM/FAULT CODE	PROBLEM	CAUSE	SOLUTION
Electric motor drive overvoltage fault	Electric motor drive overvoltage	Deceleration time is too short	Increase deceleration time
		High input voltage/ voltage swell	See *Incoming Power Troubleshooting Matrix*
		Load is overhauling motor	Add dynamic braking resistors and/or increase deceleration time. Note: also see *Motor and Load Troubleshooting Matrix*
Electric motor drive overcurrent fault	Electric motor drive overcurrent	Motor nameplate data incorrectly programmed	Check motor nameplate and program data correctly
		Acceleration time is too short	Increase acceleration time
		Start boost or continuous boost set too high	Lower start boost or continuous boost
		Short in inverter semiconductor	Replace inverter semiconductor or replace electric motor drive
		Contactor between electric motor drive and motor is changing state while electric motor drive outputs more than 0 Hz	Wire contactor to change state only when electric motor drive outputs 0 Hz
		Motor trying to start in a spinning mode	Enable flying start
Electric motor drive overcurrent fault. Load conductor current readings not equal	Electric motor drive overcurrent	Motor or conductors feeding motor are shorted or have ground fault	See *Motor and Load Troubleshooting Matrix*
Electric motor drive overcurrent fault. Current readings for load conductors are 105% or greater than motor nameplate current when motor powers driven load	Electric motor drive overcurrent	Problem with motor	See *Motor and Load Troubleshooting Matrix*
Electric motor drive overload fault. Current readings for load conductors are 105% or greater than motor nameplate current when motor powers driven load	Electric motor drive overload	Problem with motor and/or load	See *Motor and Load Troubleshooting Matrix*

ELECTRIC MOTOR DRIVE TROUBLESHOOTING MATRIX			
... ELECTRIC MOTOR DRIVE FAULTS			
SYMPTOM/FAULT CODE	PROBLEM	CAUSE	SOLUTION
Electric motor drive undervoltage fault	Electric motor drive undervoltage	Low input voltage/ voltage sag	See *Incoming Power Troubleshooting Matrix*
Electric motor drive overtemperature fault	Electric motor drive overtemperature	High ambient temperature	Add cooling to electric motor drive enclosure, or relocate electric motor drive enclosure
		Drive cooling fan defective	Replace electric motor drive cooling fan, or replace electric motor drive
		Heat sink dirty, or air intake clogged	Clean heatsink or air intake
		Problem with motor and/or load	See *Motor and Load Troubleshooting Matrix*

ELECTRIC MOTOR DRIVE COMPONENT FAILURES . . .			
SYMPTOM/FAULT CODE	PROBLEM	CAUSE	SOLUTION
Electric motor drive does not turn on. Blown fuse or tripped circuit breaker	Defective converter (rectifier) semiconductor	High input voltage/voltage swell	Replace converter semiconductor or replace electric motor drive. See *Incoming Power Troubleshooting Matrix*
		Electric motor drive cooling fan is defective.	Replace converter semiconductor and electric motor drive cooling fan, or replace electric motor drive.
Electric motor drive does not turn on. Blown fuse or tripped circuit breaker. No AC output from electric motor drive. DC bus fuse (if present) is blown. Electric motor drive overcurrent fault	Defective inverter semiconductor	Electric motor drive cooling fan is defective.	Replace inverter semiconductor and electric motor drive cooling fan, or replace electric motor drive.
Electric motor drive overtemperature fault or electric motor drive component failure	Defective electric motor drive cooling fan	Age. Electric motor drive manufacturers recommend replacing electric motor drive cooling fans every 3 to 5 years. *Note: Fan life expectancy is influenced by ambient temperature.*	Replace electric motor drive cooling fan or replace electric motor drive.
DC bus capacitor is swollen and/or pressure relief valve is protruding. DC bus capacitor is destroyed. Unstable DC bus voltage when electric motor drive is running at a constant speed. *Note: Defective capacitor balancing resistors can damage the DC bus capacitors. If there is a problem with a DC bus capacitor, also check the capacitor balancing resistor*	Defective DC bus capacitors	Age. Manufacturers recommend replacing capacitors every 5 to 10 years.	Replace DC bus capacitors or replace electric motor drive.
		Electric motor drive cooling fan is defective.	Replace DC bus capacitors and electric motor drive cooling fan, or replace electric motor drive.
		Capacitor balancing resistors are defective or incorrect value.	Replace DC bus capacitors and capacitor balancing resistors, or replace electric motor drive. Note: Verify that replacement resistors are correct ohm and watt value.

... ELECTRIC MOTOR DRIVE COMPONENT FAILURES			
SYMPTOM/FAULT CODE	PROBLEM	CAUSE	SOLUTION
Resistors and resistor connections are discolored and/or burned. *Note: Defective capacitor balancing resistors can damage the DC bus capacitors. If there is a problem with a capacitor balancing resistor, also check the DC bus capacitor.*	Defective capacitor balancing resistors	Incorrect capacitor balancing resistors	Replace capacitor balancing resistors, or replace electric motor drive. Verify that replacement resistor are correct ohm and watt value.
		Excessive heat	Verify that electric motor drive cooling fan is working, replace if defective. Replace capacitor balancing resistors, or replace electric motor drive. Verify that replacement resistors are correct ohm and watt value. *Note: Also see Drive Overtemperature.*
		Loose connections	Replace capacitor balancing resistors, or replace electric motor drive. Verify that replacement resistors are correct ohm and watt value. Tighten connections.

ELECTRIC MOTOR DRIVE INPUT AND OUTPUT PROBLEMS			
SYMPTOM/FAULT CODE	PROBLEM	CAUSE	SOLUTION
Electric motor drive operates correctly when set to default parameters (excluding motor nameplate data and control mode), input mode is keypad, and motor/load are connected. Electric motor drive does not operate correctly when input mode is other than keypad, motor/load are connected, and electric motor drive is operated per design.	Externally connected inputs and outputs incorrect	Input(s) and/or output(s) incorrectly wired. Input or output devices not correct.	Verify the wiring to input and output devices. Check for loose connections. Verify the devices are correct for the application. Correct as needed
		Problem with separate system that supplies start, stop, reference, or feedback signals to the electric motor drive, e.g., HVAC control system	Verify the operation of separate system. Correct as needed
	Parameters incorrect	Parameter(s) incorrectly programmed.	See *Electric Motor Drive Parameter Problems*

ELECTRIC MOTOR DRIVE PARAMETER PROBLEMS			
SYMPTOM/FAULT CODE	PROBLEM	CAUSE	SOLUTION
Electric motor drive operates correctly when set to default parameters, input mode is keypad, and motor/load are disconnected. Electric motor drive does not operate correctly when set to default parameters (excluding motor nameplate data and control mode), input mode is keypad, and motor/load are connected	Parameters incorrect	Parameter(s) incorrectly programmed	Verify that the motor nameplate data parameter matches the actual motor nameplate data. Verify that the control mode parameter is correct for the type of load, e.g., electric motor drive set to variable torque for a pump or fan load. See *Electric Motor Drive Faults* for other causes. Correct as needed
		Parameter(s) incorrectly programmed	Verify that all parameters that pertain to application are set correctly. It is possible there is a conflict between two parameters, e.g., minimum motor frequency and maximum motor frequency. See *Electric Motor Drive Faults* for other causes. Correct as needed
	Problem with motor and/or load	Problem with motor and/or load	See *Motor and Load Troubleshooting Matrix*
Unusual noises or vibrations when electric motor drive is powering the load.	Parameters incorrect	Parameter(s) incorrectly programmed	Adjust skip frequency parameter
	Problem with motor and/or load	Problem with motor and/or load	See *Motor and Load Troubleshooting Matrix*
Electric motor drive operates correctly when set to default parameters (excluding motor nameplate data and control mode), input mode is keypad, and motor/load are connected. Electric motor drive does not operate correctly when input mode is other than keypad, motor/load are connected, and electric motor drive is operated per design	Parameters incorrect	Parameter(s) incorrectly programmed	Verify that input mode parameter matches the existing input, e.g., serial communication is programmed if electric motor drive is controlled via serial communication. See *Electric Motor Drive Faults* for other causes. Correct as needed
	Externally connected inputs and outputs incorrect	Input(s) and/or output(s) incorrectly wired. Input or ouput devices not correct.	See *Electric Motor Drive Input and Output Problems*

ELECTRIC MOTOR DRIVE OPERATIONAL PROBLEMS			
SYMPTOM/FAULT CODE	PROBLEM	CAUSE	SOLUTION
Intermittent electric motor drive faults and/or intermittently electric motor drive does not operate as per design	Intermittent fault/erratic operation	Incoming power problems	See *Incoming Power Troubleshooting Matrix*
		Electrical noise, EMI/RFI	Verify RC snubbers, MOVs, or flywheel diodes are installed across coils operating near electric motor drive or controlled by electric motor drive. Correct as needed
		Electrical noise, EMI/RFI	Verify that input and output conductors are installed in separate metal raceways or separate shielded cables. Verify proper grounding. Correct as needed
		Electrical noise, EMI/RFI	Verify that proper separation is maintained between power and control conductors, including separate metal raceways. Correct as needed
		Electrical noise, EMI/RFI	Verify that analog signals are run in a separate metal conduit using shielded twisted pair cable, and only grounded at one end. Correct as needed
Motor rotation incorrect when powered by electric motor drive	Incorrect phasing	Wiring	Interchange two of the load conductors at the power terminal strip to reverse the direction of rotation
Motor rotation incorrect when powered through bypass contactors	Incorrect phasing	Wiring	Interchange two of the line conductors that feed the electric motor drive and the bypass. Note: This assumes that the electric motor drive and bypass share a common feed

MOTOR AND LOAD TROUBLESHOOTING MATRIX			
MOTOR AND LOAD PROBLEMS . . .			
SYMPTOM/FAULT CODE	PROBLEM	CAUSE	SOLUTION
Electric motor drive overvoltage fault	Electric motor drive overvoltage	Load is overhauling motor	Contact drive manufacturer for assistance. Note: Also see *Electric Motor Drive Matrix*
Electric motor drive overcurrent fault. Current readings for load conductors/motor not equal. Low megohmmeter readings for conductors feeding motor	Electric motor drive overcurrent	Conductors feeding motor are shorted or have ground fault	Repair or replace conductors
Electric motor drive overcurrent fault. Current readings for load conductors/motor not equal. Low megohmmeter readings for motor	Electric motor drive overcurrent	Motor is shorted or has ground fault	Repair or replace motor
Electric motor drive overcurrent fault, electric motor drive overload fault, or electric motor drive overtemperature fault. Current readings for load conductors/motor greater than 105% of motor nameplate current when motor powers driven load	Electric motor drive overload	Misalignment between motor and driven load	Align motor and load
		Motor and/or driven load not securely fastened	Securely fasten motor and/or driven load
		Defective bearing(s) in motor and/or driven load	Replace defective bearing(s). Replace motor
		Object preventing motor or load from turning	Remove object
		Motor and/or electric motor drive not properly sized for load	Contact drive manufacturer for assistance
Unusual noise or vibration when electric motor drive powers load	Problem with motor and/or load	Misalignment between motor and driven load	Align motor and load
		Motor and/or driven load not securely fastened	Securely fasten motor and/or driven load
		Defective bearings(s) in motor and/or driven load	Replace defective bearing(s). Replace motor
		Motor and/or electric motor drive not properly sized for load	Contact drive manufacturer for assistance
	Parameter(s) incorrect	Parameter(s) incorrectly programmed	See *Electric Motor Drive Troubleshooting Matrix*
Bind in rotation of motor shaft or noise when shaft is rotated by hand	Problem with motor	Defective bearing(s) in motor	Replace defective bearing(s). Replace motor

MOTOR AND LOAD TROUBLESHOOTING MATRIX			
. . . MOTOR AND LOAD PROBLEMS			
SYMPTOM/FAULT CODE	PROBLEM	CAUSE	SOLUTION
Bind in rotation of load or noise when load is rotated by hand	Problem with load	Defective bearing(s) in load. Mechanical problem with load	Replace defective bearing(s). Correct mechanical problem
Current readings equal to or greater than motor nameplate current when motor run with load disconnected	Problem with motor	Defective bearing(s) in motor	Replace defective bearing(s). Replace motor
RELATED PROBLEMS			
Electric motor drive operates correctly when set to default parameters, Input Mode is keypad, and motor/load are disconnected. Drive does not operate correctly when set to default parameters (excluding Motor Nameplate Data, and Control Mode), Input Mode is keypad, and motor/load are connected	Problem with motor and/or load	Problem with motor and/or load	See *Motor and Load Problems.*
	Parameters incorrect	Parameter(s) incorrectly programmed	See *Electric Motor Drive Troubleshooting Matrix*

ELECTRIC MOTOR DRIVE RECORD SHEET

ELECTRIC MOTOR DRIVE INFORMATION			
Drive Name:	DRIVE LOCATION		
Drive Manufacturer:	Building:		Floor:
Drive Model:	System:		
Drive Serial Number:	Machine:		
Drive Horsepower:	Software Version:		

INPUT	OUTPUT	PARAMETERS CHANGED FROM DEFAULT	
Voltage:	Voltage:	P –	P –
Current:	Current:	P –	P –
Frequency:	Frequency:	P –	P –
Phase:	Phase:	P –	P –
KVA:	KVA:	P –	P –

MOTOR INFORMATION	
Horsepower or Kilowatts:	Service Factor:
Voltage:	Current:
Speed:	Frequency:
Magnetizing Current:	Stator Resistance:
NEMA Design:	NEMA Efficiency:
Duty:	Frame:
Motor is NEMA MG-1 Section IV part 31 compliant	☐ YES ☐ NO

DRIVE MAINTENANCE INFORMATION				
LINE VOLTAGE			LINE CURRENT	LOAD CURRENT
L1 to L2:	L1 to L3:	L2 to L3:	L1:	T1:
DC Bus Voltage:		Heat Sink temp:	L2:	T2:
			L3:	T3:
MEGOHMMETER–MOTOR AND LOAD CONDUCTOR TEST				
T1 to Ground:		T2 to Ground:		T3 to Ground:

MAINTENANCE LOG		
DATE	PROBLEM FOUND	SOLUTION

AC DRIVE STARTUP SHEET

DRIVE ITEMS

- ☐ Visual inspection of drive interior.
- ☐ Vaccum interior of drive.
- ☐ Test for unwanted grounds.
- ☐ Test converter section semiconductors.
- ☐ Test inverter section semiconductors.
- ☐ Verify line voltage is correct.

L1 TO L2 _____

L1 TO L3 _____

L2 TO L3 _____

- ☐ Verify line conductors are terminated at correct location and connections are tight.
- ☐ Verify control wires are terminated at correct location and connections are tight.
- ☐ Set DIP switches, jumpers, and overloads if applicable.
- ☐ Apply power to the drive and program parameters.
- ☐ Perform drive test. Run drive with motor disconnected.
- ☐ Perform initial drive and motor test. Verify direction of rotation. Measure load current.

T1 _____

T2 _____

T3 _____

- ☐ Perform final drive and motor test. Verify drive, motor, and load function as per design. Measure line and load currents.

L1 _____

L2 _____

L3 _____

T1 _____

T2 _____

T3 _____

- ☐ Record parameter values that were changed from default.
- ☐ Perform bypass contactor test. Verify line voltage cannot be applied to drive load terminals. Verify direction of rotation in bypass mode.

COMMENTS

AC DRIVE TROUBLESHOOTING SHEET

CUSTOMER INFORMATION

Customer Name:

Customer Location:

Customer Contact: | Contact Phone #: | Contact Email:

Drive Location:

Technician:

Date: | Time:

DRIVE INFORMATION

Drive Manufacturer: | Drive Serial Number:

Drive Model: | Drive Horsepower:

INPUT	OUTPUT
Voltage:	Voltage:
Current:	Current:
Frequency:	Frequency:
Phase:	Phase:
KVA:	KVA:

MOTOR INFORMATION

Horsepower or Kilowatts:	Service Factor:
Magnetizing Current:	Current:
Voltage:	Frequency:
Speed:	Stator Resistance:
Motor is NEMA MG-1 Section IV Part 31 compliant	YES ☐ NO ☐

INFORMATION/INSPECTION

☐ Speak with in-house personnel regarding what the drive was doing when it failed, e.g., running, accelerating, decelerating.

☐ Is failure linked to a time of day, change in load, recent repair?

☐ Gather drive manuals and list of parameter settings. Download parameters to PC, if possible.

☐ Record drive and motor nameplate data. Verify motor and drive power rating correspond.

☐ Verify all power disconnects are on.

☐ Access drive fault history.

☐ Visual inspection of drive and motor for damage, overheating, and inadequate ventilation.

☐ Motor and load aligned and fastened in place. Coupling is secure, and nothing preventing motor or load from rotating.

INCOMING POWER TESTS

☐ Verify line conductors are terminated at correct location and connections are tight.

☐ Verify that line conductors are the proper size, in a separate metal raceway or cable, and properly grounded.

☐ Verify line voltage is present and correct with the drive not running, no load.

L1 to L2 _____ L1 to L3 _____ L2 to L3 _____

☐ Measure line voltage with the drive running, full load. A difference greater than 3% between no-load and full-load voltage indicates a problem.

L1 to L2 _____ L1 to L3 _____ L2 to L3 _____

☐ Calculate voltage unbalance using no load voltages. A value greater than 2% is not acceptable.

AC DRIVE TROUBLESHOOTING SHEET

DRIVE TESTS
☐ Program drive parameters to factory default and input mode to keypad.
☐ Perform drive test. Run drive with motor disconnected.
☐ If no problems occur during the drive test, the drive is not the source of the problem. Reconnect the motor and proceed to drive, motor, and load test.
☐ If problems occur during the drive test, proceed to drive component tests.
☐ Perform drive, motor, and load test. Run drive with motor and load connected.
☐ Measure Load current.
T1 _____
T2 _____
T3 _____
☐ If no problems occur during the drive, motor, and load test, the drive, motor, and load are not the source of the problem. Proceed to drive, motor, load, and I/O test.
☐ If problems occur during the drive, motor, and load test, it is a drive parameter problem, motor problem, or load problem.
☐ Program drive parameters for the specific application.
☐ Perform drive, motor, load, and I/O test.
☐ If no problems occur during the drive, motor, load, and I/O test, the drive, motor, load, and I/O are not the source of the problem. The drive application is ready for actual use.
☐ If problems occur during the drive, motor, load, and I/O test, it is a drive parameter problem or an I/O problem.

DRIVE COMPONENT TESTS
☐ Inspect DC bus capacitors for protruding pressure relief valve(s) or bulging.
☐ Inspect capacitor balancing resistors for damage or overheating.
☐ Inspect cooling fan for proper functioning.
☐ Test converter section semiconductors.
☐ Test inverter section semiconductors.

COMMENTS

MOTOR TROUBLESHOOTING SHEET

MOTOR TESTS

☐ Test motor and load conductors with megohmmeter.

[Test conducted at drive with load conductors disconnected from drive.]

T1 to GROUND
T2 to GROUND
T3 to GROUND

☐ Test motor and load conductors individually with megohmeter, if low readings were obtained above.

MOTOR	LOAD CONDUCTORS
T1 to GROUND _____	T1 to GROUND _____
T2 to GROUND _____	T2 to GROUND _____
T3 to GROUND _____	T3 to GROUND _____

☐ Verify motor and load are aligned. Verify the motor is fastened in place, the coupling is secure, and nothing is preventing motor from rotating.

☐ Disconnect the motor from the driven load. Rotate motor shaft by hand. Verify shaft rotates freely and is not noisy.

☐ Run drive and motor with the load disconnected. Measure load current. [Current should be less than motor nameplate.]

T1 _____
T2 _____
T3 _____

☐ If no problems occurred during the motor tests, the motor is not the source of the problem. The load is the problem.

☐ If problems occurred during the motor tests, the motor is the source of the problem.

LOAD TESTS

☐ Verify the driven load is fastened in place, and nothing preventing motor from rotating.

☐ Rotate the load by hand. Verify shaft rotates freely and is not noisy.

☐ Reconnect the motor with the driven load. Run drive and motor/driven load. Measure load current. [Current greater than 105% of the motor nameplate indicates an overloaded condition.]

T1 _____
T2 _____
T3 _____

☐ If no problems occurred during the load tests, the load is not the source of the problem.

☐ If problems occurred during the load tests, the load is the source of the problem.

COMMENTS

AC DRIVE MAINTENANCE SHEET

CUSTOMER INFORMATION

Customer Name:	
Customer Location:	
Customer Contact:	
Contact Phone #:	Contact Email:
Drive Location:	
Technician:	
Date:	Time:

DRIVE INFORMATION

Drive Manufacturer:
Drive Model:
Drive Serial Number:
Drive Horsepower:

INPUT

Voltage:
Current:
Frequency:
Phase:
KVA:

OUTPUT

Voltage:
Current:
Frequency:
Phase:
KVA:

MOTOR INFORMATION

Horsepower or Kilowatts:
Voltage:
Speed:
Magnetizing Current:
Service Factor:
Current:
Frequency:
Stator Resistance:
Motor is NEMA MG-1 Section IV Part 31 compliant   YES ☐   NO ☐

COMMENTS

AC DRIVE MAINTENANCE SHEET

DRIVE MAINTENANCE DATA

☐ Measure line voltage

L1 TO L2 _____

L1 TO L3 _____

L2 TO L3 _____

☐ Measure DC bus voltage

DC+ to DC- _____

☐ Measure line and load currents

L1 _____   T1 _____

L2 _____   T2 _____

L3 _____   T3 _____

☐ Test motor and load conductors with megohmmeter. [Test conducted at drive with load conductors disconnected from drive.]

T1 _____

T2 _____

T3 _____

☐ Measure heat sink temperature

_____ °F or °C

REPLACEMENT PARTS

☐ Inspect cooling fan for proper functioning.

☐ Date, if fan is replaced _____

☐ Inspect DC bus capacitors for protruding pressure relief valve(s) or bulging.

☐ Date, if capacitors are replaced _____

☐ Inspect capacitor balancing resistors for damage or overheating.

☐ Inspect battery of clean text display.

☐ Date, if battery is replaced _____

VISUAL INSPECTION

☐ Inspect drive for damage or overheating.

☐ Inspect for adequate ventilation near drive.

☐ Inspect heat sink for adequate air flow.

☐ Inspect bypass contactors for damage or overheating.

☐ Tighten bypass contactor connections.

☐ Tighten connections on control and power terminal strip.

☐ Inspect motor for damage or overheating.

☐ Inspect for adequate ventilation near motor.

☐ Motor and load aligned and fastened in place.

☐ Coupling/connection between motor and load is not loose or broken.

COMMENTS

_____

_____

_____

_____

_____

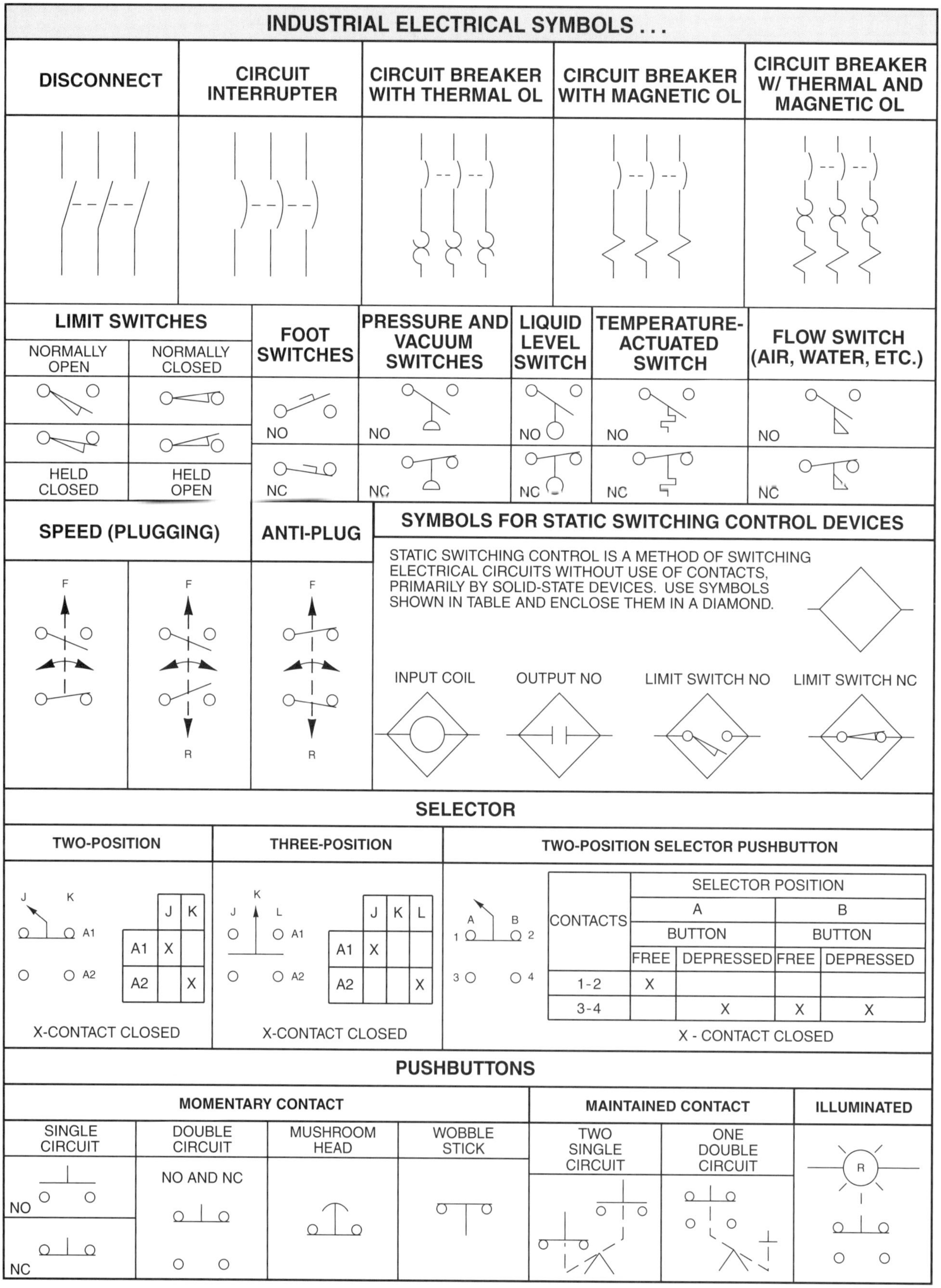

INDUSTRIAL ELECTRICAL SYMBOLS . . .

DISCONNECT	CIRCUIT INTERRUPTER	CIRCUIT BREAKER WITH THERMAL OL	CIRCUIT BREAKER WITH MAGNETIC OL	CIRCUIT BREAKER W/ THERMAL AND MAGNETIC OL

LIMIT SWITCHES

NORMALLY OPEN	NORMALLY CLOSED	FOOT SWITCHES	PRESSURE AND VACUUM SWITCHES	LIQUID LEVEL SWITCH	TEMPERATURE-ACTUATED SWITCH	FLOW SWITCH (AIR, WATER, ETC.)
		NO	NO	NO	NO	NO
HELD CLOSED	HELD OPEN	NC	NC	NC	NC	NC

SPEED (PLUGGING)	ANTI-PLUG	SYMBOLS FOR STATIC SWITCHING CONTROL DEVICES

STATIC SWITCHING CONTROL IS A METHOD OF SWITCHING ELECTRICAL CIRCUITS WITHOUT USE OF CONTACTS, PRIMARILY BY SOLID-STATE DEVICES. USE SYMBOLS SHOWN IN TABLE AND ENCLOSE THEM IN A DIAMOND.

INPUT COIL   OUTPUT NO   LIMIT SWITCH NO   LIMIT SWITCH NC

SELECTOR

TWO-POSITION	THREE-POSITION	TWO-POSITION SELECTOR PUSHBUTTON

TWO-POSITION:

	J	K
A1	X	
A2		X

X-CONTACT CLOSED

THREE-POSITION:

	J	K	L
A1	X		
A2			X

X-CONTACT CLOSED

TWO-POSITION SELECTOR PUSHBUTTON:

CONTACTS	SELECTOR POSITION			
	A		B	
	BUTTON		BUTTON	
	FREE	DEPRESSED	FREE	DEPRESSED
1-2	X			
3-4		X	X	X

X - CONTACT CLOSED

PUSHBUTTONS

MOMENTARY CONTACT				MAINTAINED CONTACT		ILLUMINATED
SINGLE CIRCUIT	DOUBLE CIRCUIT	MUSHROOM HEAD	WOBBLE STICK	TWO SINGLE CIRCUIT	ONE DOUBLE CIRCUIT	
NO	NO AND NC					R
NC						

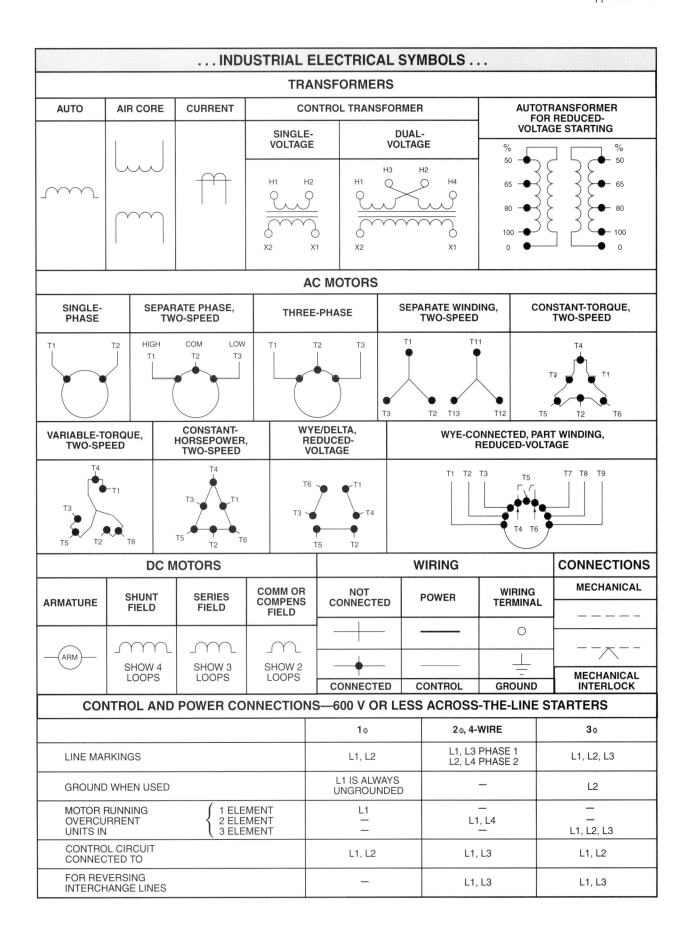

. . . INDUSTRIAL ELECTRICAL SYMBOLS . . .

TRANSFORMERS

AUTO	AIR CORE	CURRENT	CONTROL TRANSFORMER		AUTOTRANSFORMER FOR REDUCED-VOLTAGE STARTING
			SINGLE-VOLTAGE	DUAL-VOLTAGE	

AC MOTORS

SINGLE-PHASE	SEPARATE PHASE, TWO-SPEED	THREE-PHASE	SEPARATE WINDING, TWO-SPEED	CONSTANT-TORQUE, TWO-SPEED

VARIABLE-TORQUE, TWO-SPEED	CONSTANT-HORSEPOWER, TWO-SPEED	WYE/DELTA, REDUCED-VOLTAGE	WYE-CONNECTED, PART WINDING, REDUCED-VOLTAGE

DC MOTORS / WIRING / CONNECTIONS

DC MOTORS				WIRING			CONNECTIONS
ARMATURE	SHUNT FIELD	SERIES FIELD	COMM OR COMPENS FIELD	NOT CONNECTED	POWER	WIRING TERMINAL	MECHANICAL
ARM	SHOW 4 LOOPS	SHOW 3 LOOPS	SHOW 2 LOOPS	CONNECTED	CONTROL	GROUND	MECHANICAL INTERLOCK

CONTROL AND POWER CONNECTIONS—600 V OR LESS ACROSS-THE-LINE STARTERS

		1φ	2φ, 4-WIRE	3φ
LINE MARKINGS		L1, L2	L1, L3 PHASE 1 L2, L4 PHASE 2	L1, L2, L3
GROUND WHEN USED		L1 IS ALWAYS UNGROUNDED	—	L2
MOTOR RUNNING OVERCURRENT UNITS IN	1 ELEMENT 2 ELEMENT 3 ELEMENT	L1 — —	— L1, L4 —	— — L1, L2, L3
CONTROL CIRCUIT CONNECTED TO		L1, L2	L1, L3	L1, L2
FOR REVERSING INTERCHANGE LINES		—	L1, L3	L1, L3

. . . INDUSTRIAL ELECTRICAL SYMBOLS . . .

CONTACTS								OVERLOAD RELAYS	
INSTANT OPERATING				TIMED CONTACTS - CONTACT ACTION RETARDED AFTER COIL IS:				THERMAL	MAGNETIC
WITH BLOWOUT		WITHOUT BLOWOUT		ENERGIZED		DE-ENERGIZED			
NO	NC	NO	NC	NOTC	NCTO	NOTO	NCTC		

SUPPLEMENTARY CONTACT SYMBOLS

SPST NO		SPST NC		SPDT		TERMS
SINGLE BREAK	DOUBLE BREAK	SINGLE BREAK	DOUBLE BREAK	SINGLE BREAK	DOUBLE BREAK	SPST SINGLE-POLE, SINGLE-THROW
DPST, 2NO		DPST, 2NC		DPDT		SPDT SINGLE-POLE, DOUDLE-TIIROW
SINGLE BREAK	DOUBLE BREAK	SINGLE BREAK	DOUBLE BREAK	SINGLE BREAK	DOUBLE BREAK	DPST DOUBLE-POLE, SINGLE-THROW
						DPDT DOUBLE-POLE, DOUBLE-THROW
						NO NORMALLY OPEN
						NC NORMALLY CLOSED

METER (INSTRUMENT)					PILOT LIGHTS	
INDICATE TYPE BY LETTER	TO INDICATE FUNCTION OF METER OR INSTRUMENT, PLACE SPECIFIED LETTER OR LETTERS WITHIN SYMBOL.				INDICATE COLOR BY LETTER	
					NON PUSH-TO-TEST	PUSH-TO-TEST

AM or A	AMMETER	VA	VOLTMETER
AH	AMPERE HOUR	VAR	VARMETER
μA	MICROAMMETER	VARH	VARHOUR METER
mA	MILLAMMETER	W	WATTMETER
PF	POWER FACTOR	WH	WATTHOUR METER
V	VOLTMETER		

INDUCTORS		COILS			
IRON CORE			DUAL-VOLTAGE MAGNET COILS		BLOWOUT COIL
			HIGH-VOLTAGE	LOW-VOLTAGE	
AIR CORE					

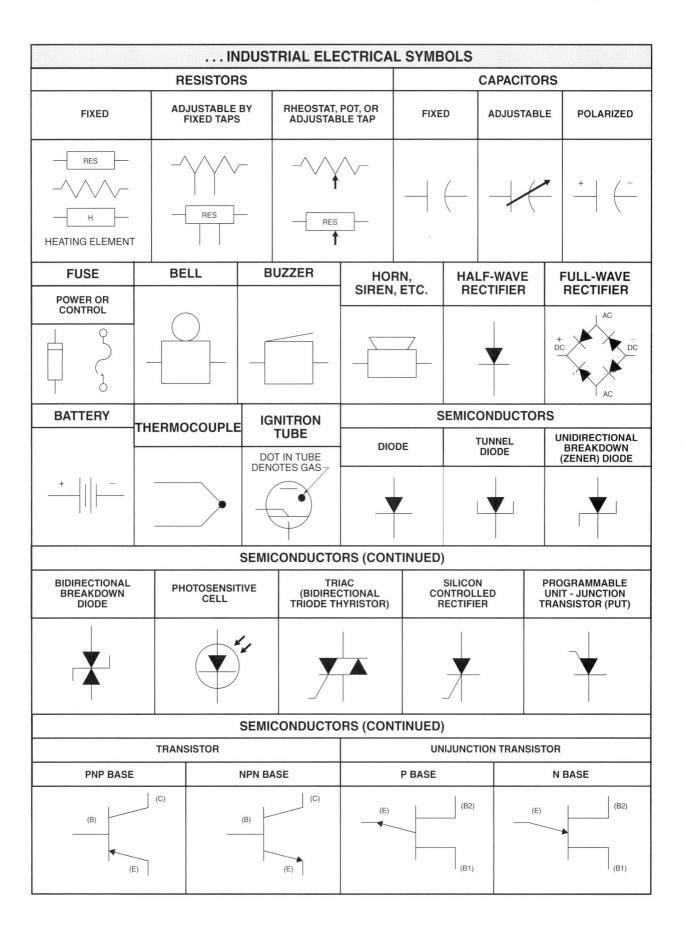

ELECTRICAL/ELECTRONIC ABBREVIATIONS/ACRONYMS

Abbr/ Acronym	Meaning	Abbr/ Acronym	Meaning	Abbr/ Acronym	Meaning
A	Ammeter; Ampere; Anode; Armature	FU	Fuse	PNP	Positive-Negative-Positive
AC	Alternating Current	FWD	Forward	POS	Positive
AC/DC	Alternating Current; Direct Current	G	Gate; Giga; Green; Conductance	POT.	Potentiometer
A/D	Analog to Digital	GEN	Generator	P-P	Peak-to-Peak
AF	Audio Frequency	GRD	Ground	PRI	Primary Switch
AFC	Automatic Frequency Control	GY	Gray	PS	Pressure Switch
Ag	Silver	H	Henry; High Side of Transformer; Magnetic Flux	PSI	Pounds Per Square Inch
ALM	Alarm			PUT	Pull-Up Torque
AM	Ammeter; Amplitude Modulation	HF	High Frequency	Q	Transistor
AM/FM	Amplitude Modulation; Frequency Modulation	HP	Horsepower	R	Radius; Red; Resistance; Reverse
		Hz	Hertz	RAM	Random-Access Memory
ARM.	Armature	I	Current	RC	Resistance-Capacitance
Au	Gold	IC	Integrated Circuit	RCL	Resistance-Inductance-Capacitance
AU	Automatic	INT	Intermediate; Interrupt	REC	Rectifier
AVC	Automatic Volume Control	INTLK	Interlock	RES	Resistor
AWG	American Wire Gauge	IOL	Instantaneous Overload	REV	Reverse
BAT.	Battery (electric)	IR	Infrared	RF	Radio Frequency
BCD	Binary Coded Decimal	ITB	Inverse Time Breaker	RH	Rheostat
BJT	Bipolar Junction Transistor	ITCB	Instantaneous Trip Circuit Breaker	rms	Root Mean Square
BK	Black	JB	Junction Box	ROM	Read-Only Memory
BL	Blue	JFET	Junction Field-Effect Transistor	rpm	Revolutions Per Minute
BR	Brake Relay; Brown	K	Kilo; Cathode	RPS	Revolutions Per Second
C	Celsius; Capacitance; Capacitor	L	Line; Load; Coil; Inductance	S	Series; Slow; South; Switch
CAP.	Capacitor	LB-FT	Pounds Per Foot	SCR	Silicon Controlled Rectifier
CB	Circuit Breaker; Citizen's Band	LB-IN.	Pounds Per Inch	SEC	Secondary
CC	Common-Collector Configuration	LC	Inductance-Capacitance	SF	Service Factor
CCW	Counterclockwise	LCD	Liquid Crystal Display	1 PH; 1φ	Single-Phase
CE	Common-Emitter Configuration	LCR	Inductance-Capacitance-Resistance	SOC	Socket
CEMF	Counter Electromotive Force	LED	Light Emitting Diode	SOL	Solenoid
CKT	Circuit	LRC	Locked Rotor Current	SP	Single-Pole
CONT	Continuous; Control	LS	Limit Switch	SPDT	Single-Pole, Double-Throw
CPS	Cycles Per Second	LT	Lamp	SPST	Single-Pole, Single-Throw
CPU	Central Processing Unit	M	Motor; Motor Starter; Motor Starter Contacts	SS	Selector Switch
CR	Control Relay			SSW	Safety Switch
CRM	Control Relay Master	MAX.	Maximum	SW	Switch
CT	Current Transformer	MB	Magnetic Brake	T	Tera; Terminal; Torque; Transformer
CW	Clockwise	MCS	Motor Circuit Switch	TB	Terminal Board
D	Diameter; Diode; Down	MEM	Memory	3 PH; 3φ	Three-Phase
D/A	Digital to Analog	MED	Medium	TD	Time Delay
DB	Dynamic Braking Contactor; Relay	MIN	Minimum	TDF	Time Delay Fuse
DC	Direct Current	MN	Manual	TEMP	Temperature
DIO	Diode	MOS	Metal-Oxide Semiconductor	THS	Thermostat Switch
DISC.	Disconnect Switch	MOSFET	Metal-Oxide Semiconductor Field-Effect Transistor	TR	Time Delay Relay
DMM	Digital Multimeter			TTL	Transistor-Transistor Logic
DP	Double-Pole	MTR	Motor	U	Up
DPDT	Double-Pole, Double-Throw	N; NEG	North; Negative	UCL	Unclamp
DPST	Double-Pole, Single-Throw	NC	Normally Closed	UHF	Ultrahigh Frequency
DS	Drum Switch	NEUT	Neutral	UJT	Unijunction Transistor
DT	Double-Throw	NO	Normally Open	UV	Ultraviolet; Undervoltage
DVM	Digital Voltmeter	NPN	Negative-Positive-Negative	V	Violet; Volt
EMF	Electromotive Force	NTDF	Nontime-Delay Fuse	VA	Volt Amp
F	Fahrenheit; Fast; Field; Forward; Fuse	O	Orange	VAC	Volts Alternating Current
FET	Field-Effect Transistor	OCPD	Overcurrent Protection Device	VDC	Volts Direct Current
FF	Flip-Flop	OHM	Ohmmeter	VHF	Very High Frequency
FLC	Full-Load Current	OL	Overload Relay	VLF	Very Low Frequency
FLS	Flow Switch	OZ/IN.	Ounces Per Inch	VOM	Volt-Ohm-Milliammeter
FLT	Full-Load Torque	P	Peak; Positive; Power; Power Consumed	W	Watt; White
FM	Frequency Modulation	PB	Pushbutton	w/	With
FREQ	Frequency	PCB	Printed Circuit Board	X	Low Side of Transformer
FS	Float Switch	PH; φ	Phase	Y	Yellow
FTS	Foot Switch	PLS	Plugging Switch	Z	Impedance

THREE-PHASE VOLTAGE VALUES

For 208 V × 1.732, use 360
For 230 V × 1.732, use 398
For 240 V × 1.732, use 416
For 440 V × 1.732, use 762
For 460 V × 1.732, use 797
For 480 V × 1.732, use 831
For 2400 V × 1.732, use 4157
For 4160 V × 1.732, use 7205

POWER FORMULA ABBREVIATIONS AND SYMBOLS

P = Watts	V = Volts
I = Amps	VA = Volt Amps
A = Amps	φ = Phase
R = Ohms	√ = Square Root
E = Volts	

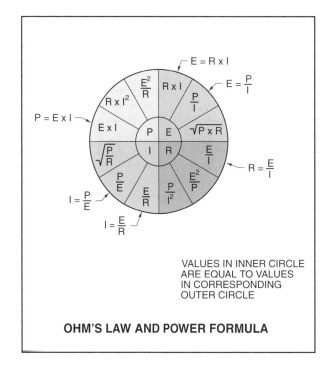

VALUES IN INNER CIRCLE ARE EQUAL TO VALUES IN CORRESPONDING OUTER CIRCLE

OHM'S LAW AND POWER FORMULA

POWER FORMULAS—1φ, 3φ

Phase	To Find	Use Formula	Given	Find	Solution
1φ	I	$I = \dfrac{VA}{V}$	32,000 VA, 240 V	I	$I = \dfrac{VA}{V}$ $I = \dfrac{32,000\ VA}{240\ V}$ $I = $ **133 A**
1φ	VA	$VA = I \times V$	100 A, 240 V	VA	$VA = I \times A$ $VA = 100\ A \times 240\ V$ $VA = $ **24,000 VA**
1φ	V	$V = \dfrac{VA}{I}$	42,000 VA, 350 A	V	$V = \dfrac{VA}{I}$ $V = \dfrac{42,000\ VA}{350\ A}$ $V = $ **120 V**
3φ	I	$I = \dfrac{VA}{V \times \sqrt{3}}$	72,000 VA, 208 V	I	$I = \dfrac{VA}{V \times \sqrt{3}}$ $I = \dfrac{72,000\ VA}{360\ V}$ $I = $ **200 A**
3φ	VA	$VA = I \times V \times \sqrt{3}$	2 A, 240 V	VA	$VA = I \times V \times \sqrt{3}$ $VA = 2 \times 416$ $VA = $ **832 VA**

AC/DC FORMULAS				
To Find	**DC**	**AC**		
		1φ, 115 or 220 V	**1φ, 208 or 230, or 240V**	**3φ − All Voltages**
I, HP known	$\dfrac{HP \times 746}{E \times E_{ff}}$	$\dfrac{HP \times 746}{E \times E_{ff} \times PF}$	$\dfrac{HP \times 746}{E \times E_{ff} \times PF}$	$\dfrac{HP \times 746}{1.73 \times E \times E_{ff} \times PF}$
I, kW known	$\dfrac{kW \times 1000}{E}$	$\dfrac{kW \times 1000}{E \times PF}$	$\dfrac{kW \times 1000}{E \times PF}$	$\dfrac{kW \times 1000}{1.73 \times E \times PF}$
I, kVA known		$\dfrac{kVA \times 1000}{E}$	$\dfrac{kVA \times 1000}{E}$	$\dfrac{kVA \times 1000}{1.763 \times E}$
kW	$\dfrac{I \times E}{1000}$	$\dfrac{I \times E \times PF}{1000}$	$\dfrac{I \times E \times PF}{1000}$	$\dfrac{I \times E \times 1.73 \times PF}{1000}$
kVA		$\dfrac{I \times E}{1000}$	$\dfrac{I \times E}{1000}$	$\dfrac{I \times E \times 1.73}{1000}$
HP (output)	$\dfrac{I \times E \times E_{ff}}{746}$	$\dfrac{I \times E \times E_{ff} \times PF}{746}$	$\dfrac{I \times E \times E_{ff} \times PF}{746}$	$\dfrac{I \times E \times 1.73 \times E_{ff} \times PF}{746}$

E_{ff} = efficiency

HORSEPOWER FORMULAS				
To Find	**Use Formula**	**Example**		
		Given	**Find**	**Solution**
HP	$HP = \dfrac{I \times E \times E_{ff}}{746}$	240 V, 20 A, 85% E_{ff}	HP	$HP = \dfrac{I \times E \times E_{ff}}{746}$ $HP \dfrac{20\,A \times 240\,V \times 85\%}{746}$ $HP = \textbf{5.5}$
I	$I = \dfrac{HP \times 746}{E \times E_{ff} \times PF}$	10 HP, 240 V, 90% E_{ff}, 88% PF	I	$I = \dfrac{HP \times 746}{E \times E_{ff} \times PF}$ $I \dfrac{10\,HP \times 746}{240\,V \times 90\% \times 88\%}$ $I = \textbf{39 A}$

VOLTAGE DROP FORMULAS—1φ, 3φ					
Phase	**To Find**	**Use Formula**	**Example**		
			Given	**Find**	**Solution**
1φ	VD	$VD = \dfrac{2 \times R \times L \times I}{1000}$	240 V, 40 A, 60′ L, .764R	VD	$VD = \dfrac{2 \times R \times L \times I}{1000}$ $VD = \dfrac{2 \times .764 \times 60 \times 40}{1000}$ $VD = \textbf{3.67 V}$
3φ	VD	$VD = \dfrac{2 \times R \times L \times I}{1000} \times .866$	208 V, 110 A, 75′ L, .194 R, .866 multiplier	VD	$VD = \dfrac{2 \times R \times L \times I}{1000} \times .866$ $VD = \dfrac{2 \times .194 \times 75 \times 110}{1000} \times .866$ $VD = \textbf{2.77 V}$

CAPACITORS			
Connected in Series		**Connected in Parallel**	**Connected in Series/Parallel**
Two Capacitors	**Three or More Capacitors**		
$C_T = \dfrac{C_1 \times C_2}{C_1 + C_2}$ where C_T = total capacitance (in μF) C_1 = capacitance of capacitor 1 (in μF) C_2 = capacitance of capacitor 2 (in μF)	$\dfrac{1}{C_T} = \dfrac{1}{C_1} + \dfrac{1}{C_2} + \ldots$	$C_T = C_1 + C_2 + \ldots$	1. Calculate the capacitance of the parallel branch. 2. Calculate the capacitance of the series combination. $C_T = \dfrac{C_1 \times C_2}{C_1 + C_2}$

SINE WAVES		
Frequency	**Period**	**Peak-to-Peak Value**
$f = \dfrac{1}{T}$ where f – frequency (in hertz) 1 = constant T = period (in seconds)	$T = \dfrac{1}{f}$ where T = period (in seconds) 1 = constant f = frequency (in hertz)	$V_{p\text{-}p} = 2 \times V_{max}$ where 2 = constant $V_{p\text{-}p}$ = peak-to-peak value V_{max} = peak value

Average Value	**rms Value**
$V_{avg} = V_{max} \times .637$ where V_{avg} = average value (in volts) V_{max} = peak value (in volts) .637 = constant	$V_{rms} = V_{max} \times .707$ where V_{rms} = rms value (in volts) V_{max} = peak value (in volts) .707 = constant

CONDUCTIVE LEAKAGE CURRENT
$I_L = \dfrac{V_A}{R_I}$ where I_L = leakage current (in microamperes) V_A = applied voltage (in volts) R_I = insulation resistance (in megohms)

TEMPERATURE CONVERSIONS	
Convert C to F	**Convert F to C**
$°F = (1.8 \times °C) + 32$	$°C = \dfrac{°F - 32}{1.8}$

FLOW RATE
$Q = \dfrac{N \times V_d}{231}$ where Q = flow rate (in gpm) N = pump drive speed (in rpm) V_d = pump displacement (in cu in./rev) 231 = constant

BRANCH CIRCUIT VOLTAGE DROP
$\%V_D = \dfrac{V_{NL} - V_{FL}}{V_{FL}} \times 100$ where $\%V_D$ = percent voltage drop (in volts) V_{NL} = no-load voltage drop (in volts) V_{FL} = full-load voltage drop (in volts) 100 = constant

METRIC PREFIXES

Multiples and Submultiples	Prefixes	Symbols	Meaning
$1,000,000,000,000 = 10^{12}$	tera	T	trillion
$1,000,000,000 = 10^{9}$	giga	G	billion
$1,000,000 = 10^{6}$	mega	M	million
$1000 = 10^{3}$	kilo	k	thousand
$100 = 10^{2}$	hecto	h	hundred
$10 = 10^{1}$	deka	d	ten
Unit $1 = 10^{0}$			
$.1 = 10^{-1}$	deci	d	tenth
$.01 = 10^{-2}$	centi	c	hundredth
$.001 = 10^{-3}$	milli	m	thousandth
$.000001 = 10^{-6}$	micro	μ	millionth
$.000000001 = 10^{-9}$	nano	n	billionth
$.000000000001 = 10^{-12}$	pico	p	trillionth

METRIC CONVERSIONS

Initial Units	Final Units											
	giga	mega	kilo	hecto	deka	base unit	deci	centi	milli	micro	nano	pico
giga		3R	6R	7R	8R	9R	10R	11R	12R	15R	18R	21R
mega	3L		3R	4R	5R	6R	7R	8R	9R	12R	15R	18R
kilo	6L	3L		1R	2R	3R	4R	5R	6R	9R	12R	15R
hecto	7L	4L	1L		1R	2R	3R	4R	5R	8R	11R	14R
deka	8L	5L	2L	1L		1R	2R	3R	4R	7R	10R	13R
base unit	9L	6L	3L	2L	1L		1R	2R	3R	6R	9R	12R
deci	10L	7L	4L	3L	2L	1L		1R	2R	5R	8R	11R
centi	11L	8L	5L	4L	3L	2L	1L		1R	4R	7R	10R
milli	12L	9L	6L	5L	4L	3L	2L	1L		3R	6R	9R
micro	15L	12L	9L	8L	7L	6L	5L	4L	3L		3R	6R
nano	18L	15L	12L	11L	10L	9L	8L	7L	6L	3L		3R
pico	21L	18L	15L	14L	13L	12L	11L	10L	9L	6L	3L	

COMMON PREFIXES

Symbol	Prefix	Equivalent
G	giga	1,000,000,000
M	mega	1,000,000
k	kilo	1000
base unit	—	1
m	milli	0.001
μ	micro	0.000001
n	nano	0.000000001
p	pico	0.000000000001
Z	impedance	ohms — Ω

POWERS OF 10				
1×10^4	=	10,000	=	$10 \times 10 \times 10 \times 10$ · Read ten to the fourth power
1×10^3	=	1000	=	$10 \times 10 \times 10$ · Read ten to the third power or ten cubed
1×10^2	=	100	=	10×10 · Read ten to the second power or ten squared
1×10^1	=	10	=	10 · Read ten to the first power
1×10^0	=	1	=	1 · Read ten to the zero power
1×10^{-1}	=	.1	=	1/10 · Read ten to the minus first power
1×10^{-2}	=	.01	=	1/(10 × 10) or 1/100 · Read ten to the minus second power
1×10^{-3}	=	.001	=	1/(10 × 10 × 10) or 1/1000 · Read ten to the minus third power
1×10^{-4}	=	.0001	=	1/(10 × 10 × 10 × 10) or 1/10,000 · Read ten to the minus fourth power

UNITS OF ENERGY					
Energy	**Btu**	**ft lb**	**J**	**kcal**	**kWh**
British thermal unit	1	777.9	1.056	0.252	2.930×10^{-4}
Foot-pound	1.285×10^0	1	1.356	3.240×10^{-4}	3.766×10^{-7}
Joule	9.481×10^{-4}	0.7376	1	2.390×10^{-4}	2.778×10^{-7}
Kilocalorie	3.968	3.086	4.184	1	1.163×10^{-3}
Kilowatt-hour	3.413	2.655×10^6	3.6×10^6	860.2	1

UNITS OF POWER				
Power	**W**	**ft lb/s**	**HP**	**kW**
Watt	1	0.7376	$.341 \times 10^{-3}$	0.001
Foot-pound/sec	1.356	1	$.818 \times 10^{-3}$	1.356×10^{-3}
Horsepower	745.7	550	1	0.7457
Kilowatt	1000	736.6	1.341	1

STANDARD SIZES OF FUSES AND CBs

NEC® 240.6(a) lists standard ampere ratings of fuses and fixed-trip CBs as follows:
15, 20, 25, 30, 35, 40, 45,
50, 60, 70, 80, 90, 100, 110,
125, 150, 175, 200, 225,
250, 300, 350, 400, 450,
500, 600, 700, 800,
1000, 1200, 1600,
2000, 2500, 3000, 4000, 5000, 6000

VOLTAGE CONVERSIONS		
To Convert	**To**	**Multiply By**
rms	Average	.9
rms	Peak	1.414
Average	rms	1.111
Average	Peak	1.567
Peak	rms	.707
Peak	Average	.637
Peak	Peak-to-peak	2

COMPARISON OF NUMBERING SYSTEMS

Decimal	Binary	Octal	Hexadecimal	BCD
0	000	0	0	0000 0000
1	001	1	1	0000 0001
2	010	2	2	0000 0010
3	011	3	3	0000 0011
4	100	4	4	0000 0100
5	101	5	5	0000 0101
6	110	6	6	0000 0110
7	111	7	7	0000 0111
8	1000	10	8	0000 1000
9	1001	11	9	0000 1001
10	1010	12	A	0001 1000
11	1011	13	B	0001 0001
12	1100	14	C	0001 0010
13	1101	18	D	0001 0011
14	1110	16	E	0001 0100
15	1111	17	F	0001 0101

BINARY NUMBERING SYSTEM

Binary				Decimal	Binary				Decimal
8s	4s	2s	1s		8s	4s	2s	1s	
0	0	0	0	0	1	0	0	0	8
0	0	0	1	1	1	0	0	1	9
0	0	1	0	2	1	0	1	0	10
0	0	1	1	3	1	0	1	1	11
0	1	0	0	4	1	1	0	0	12
0	1	0	1	5	1	1	0	1	13
0	1	1	0	6	1	1	1	0	14
0	1	1	1	7	1	1	1	1	15

STANDARD NEMA RATINGS OF AC CONTACTORS, 60 HZ

Size	8 Hr Open Rating (A)	Power*				
		3φ			1φ	
		200	230	230/460	115	230
00	9	1½	1½	2	⅓	1
0	18	3	3	5	1	2
1	27	7½	7½	10	2	3
2	45	10	15	25	3	7½
3	90	25	30	50	–	–
4	135	40	50	100	–	–
5	270	75	100	200	–	–
6	540	150	200	400	–	–
7	810	–	300	600	–	–
8	1215	–	450	900	–	–
9	2250	–	800	1600	–	–

* in volts

3φ, 230 V MOTORS AND CIRCUITS – 240 V SYSTEM

1		2		3	4	5				6	
Size of motor		Motor overload protection				Controller termination temperature rating				Minimum size of copper wire and trade conduit	
		Low-peak or Fusetron®				60°C		75°C			
HP	Amp	Motor less than 40°C or greater than 1.15 SF (Max fuse 125%)	All other motors (Max fuse 115%)	Switch 115% minimum or HP rated or fuse holder size	Minimum size of starter	TW	THW	TW	THW	Wire size (AWG or kcmil)	Conduit (inches)
½	2	2½	2¼	30	00	•	•	•	•	14	½
¾	2.8	3½	3²⁄₁₀	30	00	•	•	•	•	14	½
1	3.6	4½	4	30	00	•	•	•	•	14	½
1½	5.2	6¼	5⁶⁄₁₀	30	00	•	•	•	•	14	½
2	6.8	8	7½	30	0	•	•	•	•	14	½
3	9.6	12	10	30	0	•	•	•	•	14	½
5	15.2	17½	17½	30	1	•	•	•	•	14	½
7½	22	25	25	30	1	•	•	•	•	10	½
10	28	35	30	60	2	•	•	•		8	¾
									•	10	½
15	42	50	45	60	2	•	•	•	•	6	1
										6	¾
20	54	60	60	100	3	•	•	•	•	4	1
25	68	80	75	100	3	•	•			3	1¼
								•		3	1
									•	4	1
30	80	100	90	100	3	•	•	•		1	1¼
									•	3	1¼
40	104	125	110	200	4	•	•	•		2/0	1½
									•	1	1¼
50	130	150	150	200	4	•	•	•		3/0	2
									•	2/0	1½
75	192	225	200	400	5	•	•	•		300	2½
									•	250	2½
100	248	300	250	400	5	•	•	•		500	3
									•	350	2½
150	360	450	400	600	6	•	•	•		300-2/φ*	2-2½*
									•	4/0-2/φ*	2-2*

* two sets of multiple conductors and two runs of conduit required

3φ, 460 V MOTORS AND CIRCUITS – 480 V SYSTEM											
1	2		3	4	5				6		
Size of motor	Motor overload protection				Controller termination temperature rating				Minimum size of copper wire and trade conduit		
	Low-peak or Fusetron®				60°C		75°C				
	Motor less than 40°C or greater than 1.15 SF (Max fuse 125%)	All other motors (Max fuse 115%)	Switch 115% minimum or HP rated or fuse holder size	Minimum size of starter					Wire size (AWG or kcmil)	Conduit (inches)	
HP	Amp					TW	THW	TW	THW		
½	1	1¼	1⅛	30	00	•	•	•	•	14	½
¾	1.4	1⁶⁄₁₀	1⁶⁄₁₀	30	00	•	•	•	•	14	½
1	1.8	2¼	2	30	00	•	•	•	•	14	½
1½	2.6	3²⁄₁₀	2⁶⁄₁₀	30	00	•	•	•	•	14	½
2	3.4	4	3½	30	00	•	•	•	•	14	½
3	4.8	5⁶⁄₁₀	5	30	0	•	•	•	•	14	½
5	7.6	9	8	30	0	•	•	•	•	14	½
7½	11	12	12	30	1	•	•	•	•	14	½
10	14	17½	15	30	1	•	•	•	•	14	½
15	21	25	20	30	2	•	•	•	•	10	½
20	27	30	30	60	2	•	•	•		8	¾
									•	10	½
25	34	40	35	60	2	•	•	•		6	1
									•	8	¾
30	40	50	45	60	3	•	•	•		6	1
									•	8	¾
40	52	60	60	100	3	•	•	•		4	1
									•	6	1
50	65	80	70	100	3	•	•	•		3	1¼
									•	4	1
60	77	90	80	100	4	•	•	•		1	1¼
									•	3	1¼
75	96	110	110	200	4	•	•	•		1/0	1½
									•	1	1¼
100	124	150	125	200	4	•	•	•		3/0	2
									•	2/0	1½
125	156	175	175	200	5	•	•	•		4/0	2
									•	3/0	2
150	180	225	200	400	5	•	•	•		300	2½
									•	4/0	2
200	240	300	250	400	5	•	•	•		500	3
									•	350	2½
250	302	350	325	400	6	•	•	•		4/0-2/φ*	2-2*
									•	3/0-2/φ*	2-2*
300	361	450	400	600	6	•	•	•		300-2/φ*	2-1½ *
									•	4/0-2/φ*	2-2*

* two sets of multiple conductors and two runs of conduit required

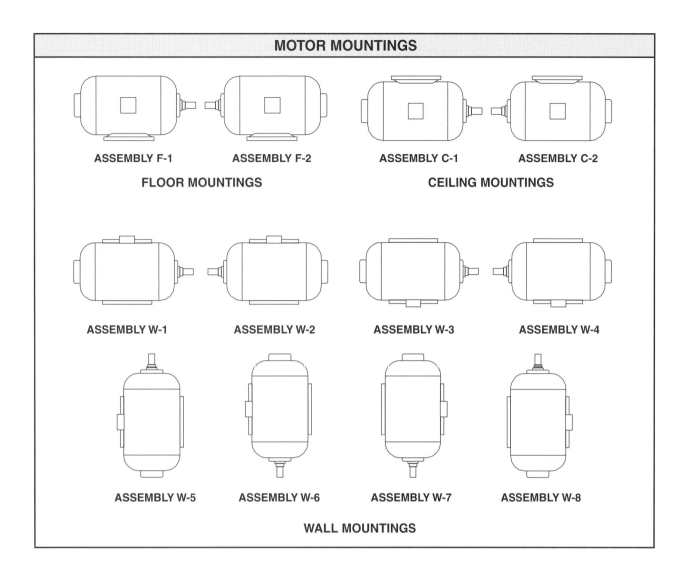

MOTOR MOUNTINGS

ASSEMBLY F-1 ASSEMBLY F-2

ASSEMBLY C-1 ASSEMBLY C-2

FLOOR MOUNTINGS **CEILING MOUNTINGS**

ASSEMBLY W-1 ASSEMBLY W-2 ASSEMBLY W-3 ASSEMBLY W-4

ASSEMBLY W-5 ASSEMBLY W-6 ASSEMBLY W-7 ASSEMBLY W-8

WALL MOUNTINGS

IEC—REFERENCE CHART

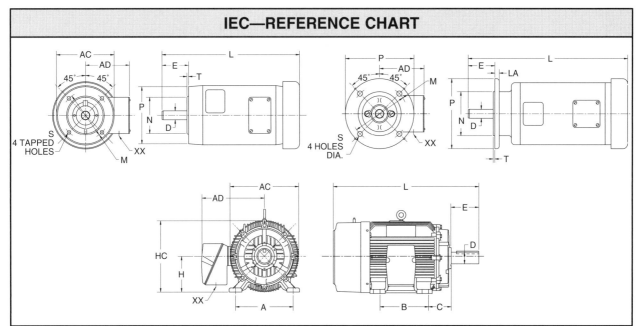

IEC Frame	Type	Foot Mounting*				Shaft*		B5 Flange*						B14 Face*					General*				
		A	B	C	H	D	E	LA	M	N	P	S	T	M	N	P	S	T	L	AC	AD	HC	XX
63	300	3.397	3.150	1.570	2.480	.433	.906	.313	4.528	3.740	5.512	.354	.118	2.953	2.362	3.540	M5	.098	†	4.690	4 / 4.567‡	4.760 / 5.375‡	.500 / .880‡
71	300 / 400	4.409	3.543	1.770	2.800	.551	1.181	.313	5.118	4.331	6.299	.393	.138	3.347	2.756	4.130	M6	.098	†	4.690 / 5.690‡	4	5.140 / 5.880‡	.690 / .844‡
80	400 / 500	4.921	3.937	1.969	3.150	.748	1.575	.500	6.496	5.118	7.874	.430	.138	3.937	3.150	4.724	M6	.118	†	5.690 / 6.614‡	4.510 / 5.120	6 / 6.380‡	.880 / .844‡
90	S / L	5.511	3.937 / 4.921	2.205	3.543	.945	1.969	.500	6.496	5.118	7.874	.472	.138	4.530	3.740	5.512	M8	.118	†	6.614 / 5.687‡	5.120 / 4.250‡	6.810 / 6.531‡	.880 / .844‡
100	S / L	6.300	4.409 / 5.512	2.480	3.937	1.102	2.362	.562	8.465	7.087	9.840	.560	.160	5.108	4.331	6.299	M8	1.38	†	7.875	5.875 / 6.060‡	7.906 / 9.440‡	1.062
112	S / M	7.480	4.488 / 5.512	2.760	4.409	1.102	2.362	.562	8.465	7.087	9.840	.560	.160	5.108	4.331	6.299	M8	.138	†	7.875	5.875	8.437	1.062
132	S / M	8.504	5.512 / 7.008	3.504	5.197	1.496	3.150	.562	10.433	9.055	11.811	.560	.160	6.496	5.118	7.874	M8	.138	†	9.562	7.375	10.062	1.062
160	M / L	10	8.268 / 10	4.252	6.299	1.654	4.331	.787	11.811	9.842	13.780	.748	.200	8.465	7.087	9.840	M12	.160	†	12.940	9.510	12.940	1.375
180	M / L	10.984	9.488 / 10.984	4.764	7.087	1.890	4.331		11.811	9.842	13.780	.748	.200						†	15.560	13.120	14.640	2.008
200	L / M	12.520	10.512 / 12.008	5.236	7.874	2.165	4.331		13.780	11.811	15.748	.748							†	17.375	14.125	16.375	2.500
225	S / M	14.016	11.260 / 12.244	5.866	8.858	2.362	5.512		15.748	13.780	17.716	.748							†	19.488	15.079	19.016	2.500
250	S / M	15.984	12.244 / 13.740	6.614	9.843	2.756	5.512												†	20.472	17.992	20.197	2.500
280	S / M	17.992	14.488 / 16.496	7.485	11.025	3.150	6.693												†	24.252	19.567	22.874	2.500
315	S / M	20	16 / 18	8.500	12.400	3.346	6.693												†	29.900	26.880	28.840	4
355	S / L	24	19.690 / 24.800	10	13.980	3.346	6.693												†	29.900	26.880	28.320	4

* In in.
† Contact manufacturer for "L" dimensions
‡ DC motor

MOTOR FRAME DIMENSIONS . . .

Frame No.	Shaft U	Shaft V	Key W	Key T	Key L	A	B	D	E	F	BA
48	$\frac{1}{2}$	$1\frac{1}{2}$*	flat	$\frac{3}{64}$	—	$5\frac{5}{8}$*	$3\frac{1}{2}$*	3	$2\frac{1}{8}$	$1\frac{3}{8}$	$2\frac{1}{2}$
56	$\frac{5}{8}$	$1\frac{7}{8}$*	$\frac{3}{16}$	$\frac{3}{16}$	$1\frac{3}{8}$	$6\frac{1}{2}$*	$4\frac{1}{4}$*	$3\frac{1}{2}$	$2\frac{7}{16}$	$1\frac{1}{2}$	$2\frac{3}{4}$
143T	$\frac{7}{8}$	2	$\frac{3}{16}$	$\frac{3}{16}$	$1\frac{3}{8}$	7	6	$3\frac{1}{2}$	$2\frac{3}{4}$	2	$2\frac{1}{4}$
145T	$\frac{7}{8}$	2	$\frac{3}{16}$	$\frac{3}{16}$	$1\frac{3}{8}$	7	6	$3\frac{1}{2}$	$2\frac{3}{4}$	$2\frac{1}{2}$	$2\frac{1}{4}$
182	$\frac{7}{8}$	2	$\frac{3}{16}$	$\frac{3}{16}$	$1\frac{3}{8}$	9	$6\frac{1}{2}$	$4\frac{1}{2}$	$3\frac{3}{4}$	$2\frac{1}{4}$	$2\frac{3}{4}$
182T	$1\frac{1}{8}$	$2\frac{1}{2}$	$\frac{1}{4}$	$\frac{1}{4}$	$1\frac{3}{4}$	9	$6\frac{1}{2}$	$4\frac{1}{2}$	$3\frac{3}{4}$	$2\frac{1}{4}$	$2\frac{3}{4}$
184	$\frac{7}{8}$	2	$\frac{3}{16}$	$\frac{3}{16}$	$1\frac{3}{8}$	9	$7\frac{1}{2}$	$4\frac{1}{2}$	$3\frac{3}{4}$	$2\frac{3}{4}$	$2\frac{3}{4}$
184T	$1\frac{1}{8}$	$2\frac{1}{2}$	$\frac{1}{4}$	$\frac{1}{4}$	$1\frac{3}{4}$	9	$7\frac{1}{2}$	$4\frac{1}{2}$	$3\frac{3}{4}$	$2\frac{3}{4}$	$2\frac{3}{4}$
203	$\frac{3}{4}$	2	$\frac{3}{16}$	$\frac{3}{16}$	$1\frac{3}{8}$	10	$7\frac{1}{2}$	5	4	$2\frac{3}{4}$	$3\frac{1}{8}$
204	$\frac{3}{4}$	2	$\frac{3}{16}$	$\frac{3}{16}$	$1\frac{3}{8}$	10	$8\frac{1}{2}$	5	4	$3\frac{1}{4}$	$3\frac{1}{8}$
213	$1\frac{1}{8}$	$2\frac{3}{4}$	$\frac{1}{4}$	$\frac{1}{4}$	2	$10\frac{1}{2}$	$7\frac{1}{2}$	$5\frac{1}{4}$	$4\frac{1}{4}$	$2\frac{3}{4}$	$3\frac{1}{2}$
213T	$1\frac{3}{8}$	$3\frac{1}{8}$	$\frac{5}{16}$	$\frac{5}{16}$	$2\frac{3}{8}$	$10\frac{1}{2}$	$7\frac{1}{2}$	$5\frac{1}{4}$	$4\frac{1}{4}$	$2\frac{3}{4}$	$3\frac{1}{2}$
215	$1\frac{1}{8}$	$2\frac{3}{4}$	$\frac{1}{4}$	$\frac{1}{4}$	2	$10\frac{1}{2}$	9	$5\frac{1}{4}$	$4\frac{1}{4}$	$3\frac{1}{2}$	$3\frac{1}{2}$
215T	$1\frac{3}{8}$	$3\frac{1}{8}$	$\frac{5}{16}$	$\frac{5}{16}$	$2\frac{3}{8}$	$10\frac{1}{2}$	9	$5\frac{1}{4}$	$4\frac{1}{4}$	$3\frac{1}{2}$	$3\frac{1}{2}$
224	1	$2\frac{3}{4}$	$\frac{1}{4}$	$\frac{1}{4}$	2	11	$8\frac{3}{4}$	$5\frac{1}{2}$	$4\frac{1}{2}$	$3\frac{3}{8}$	$3\frac{1}{2}$
225	1	$2\frac{3}{4}$	$\frac{1}{4}$	$\frac{1}{4}$	2	11	$9\frac{1}{2}$	$5\frac{1}{2}$	$4\frac{1}{2}$	$3\frac{3}{4}$	$3\frac{1}{2}$
254	$1\frac{1}{8}$	$3\frac{1}{8}$	$\frac{1}{4}$	$\frac{1}{4}$	$2\frac{3}{8}$	$12\frac{1}{2}$	$10\frac{3}{4}$	$6\frac{1}{4}$	5	$4\frac{1}{8}$	$4\frac{1}{4}$
254U	$1\frac{3}{8}$	$3\frac{1}{2}$	$\frac{5}{16}$	$\frac{5}{16}$	$2\frac{3}{4}$	$12\frac{1}{2}$	$10\frac{3}{4}$	$6\frac{1}{4}$	5	$4\frac{1}{8}$	$4\frac{1}{4}$
254T	$1\frac{5}{8}$	$3\frac{3}{4}$	$\frac{3}{8}$	$\frac{3}{8}$	$2\frac{7}{8}$	$12\frac{1}{2}$	$10\frac{3}{4}$	$6\frac{1}{4}$	5	$4\frac{1}{8}$	$4\frac{1}{4}$
256U	$1\frac{3}{8}$	$3\frac{1}{2}$	$\frac{5}{16}$	$\frac{5}{16}$	$2\frac{3}{4}$	$12\frac{1}{2}$	$12\frac{1}{2}$	$6\frac{1}{4}$	5	5	$4\frac{1}{4}$
256T	$1\frac{5}{8}$	$3\frac{3}{4}$	$\frac{3}{8}$	$\frac{3}{8}$	$2\frac{7}{8}$	$12\frac{1}{2}$	$12\frac{1}{2}$	$6\frac{1}{4}$	5	5	$4\frac{1}{4}$
284	$1\frac{1}{4}$	$3\frac{1}{2}$	$\frac{1}{4}$	$\frac{1}{4}$	$2\frac{3}{4}$	14	$12\frac{1}{2}$	7	$5\frac{1}{2}$	$4\frac{3}{4}$	$4\frac{3}{4}$
284U	$1\frac{5}{8}$	$4\frac{5}{8}$	$\frac{3}{8}$	$\frac{3}{8}$	$3\frac{3}{4}$	14	$12\frac{1}{2}$	7	$5\frac{1}{2}$	$4\frac{3}{4}$	$4\frac{3}{4}$
284T	$1\frac{7}{8}$	$4\frac{3}{8}$	$\frac{1}{2}$	$\frac{1}{2}$	$3\frac{1}{4}$	14	$12\frac{1}{2}$	7	$5\frac{1}{2}$	$4\frac{3}{4}$	$4\frac{3}{4}$
284TS	$1\frac{5}{8}$	3	$\frac{3}{8}$	$\frac{3}{8}$	$1\frac{7}{8}$	14	$12\frac{1}{2}$	7	$5\frac{1}{2}$	$4\frac{3}{4}$	$4\frac{3}{4}$
286U	$1\frac{5}{8}$	$4\frac{5}{8}$	$\frac{3}{8}$	$\frac{3}{8}$	$3\frac{3}{4}$	14	14	7	$5\frac{1}{2}$	$5\frac{1}{2}$	$4\frac{3}{4}$
286T	$1\frac{7}{8}$	$4\frac{3}{8}$	$\frac{1}{2}$	$\frac{1}{2}$	$3\frac{1}{4}$	14	14	7	$5\frac{1}{2}$	$5\frac{1}{2}$	$4\frac{3}{4}$
286TS	$1\frac{5}{8}$	3	$\frac{3}{8}$	$\frac{3}{8}$	$1\frac{7}{8}$	14	14	7	$5\frac{1}{2}$	$5\frac{1}{2}$	$4\frac{3}{4}$
324	$1\frac{5}{8}$	$4\frac{5}{8}$	$\frac{3}{8}$	$\frac{3}{8}$	$3\frac{3}{4}$	16	14	8	$6\frac{1}{4}$	$5\frac{1}{4}$	$5\frac{1}{4}$
324U	$1\frac{7}{8}$	$5\frac{3}{8}$	$\frac{1}{2}$	$\frac{1}{2}$	$4\frac{1}{4}$	16	14	8	$6\frac{1}{4}$	$5\frac{1}{4}$	$5\frac{1}{4}$
324S	$1\frac{5}{8}$	3	$\frac{3}{8}$	$\frac{3}{8}$	$1\frac{7}{8}$	16	14	8	$6\frac{1}{4}$	$5\frac{1}{4}$	$5\frac{1}{4}$
324T	$2\frac{1}{8}$	5	$\frac{1}{2}$	$\frac{1}{2}$	$3\frac{7}{8}$	16	14	8	$6\frac{1}{4}$	$5\frac{1}{4}$	$5\frac{1}{4}$
324TS	$1\frac{7}{8}$	$3\frac{1}{2}$	$\frac{1}{2}$	$\frac{1}{2}$	2	16	14	8	$6\frac{1}{4}$	$5\frac{1}{4}$	$5\frac{1}{4}$
326	$1\frac{5}{8}$	$4\frac{5}{8}$	$\frac{3}{8}$	$\frac{3}{8}$	$3\frac{3}{4}$	16	$15\frac{1}{2}$	8	$6\frac{1}{4}$	6	$5\frac{1}{4}$
326U	$1\frac{7}{8}$	$5\frac{3}{8}$	$\frac{1}{2}$	$\frac{1}{2}$	$4\frac{1}{4}$	16	$15\frac{1}{2}$	8	$6\frac{1}{4}$	6	$5\frac{1}{4}$
326S	$1\frac{5}{8}$	3	$\frac{3}{8}$	$\frac{3}{8}$	$1\frac{7}{8}$	16	$15\frac{1}{2}$	8	$6\frac{1}{4}$	6	$5\frac{1}{4}$
326T	$2\frac{1}{8}$	5	$\frac{1}{2}$	$\frac{1}{2}$	$3\frac{7}{8}$	16	$15\frac{1}{2}$	8	$6\frac{1}{4}$	6	$5\frac{1}{4}$
326TS	$1\frac{7}{8}$	$3\frac{1}{2}$	$\frac{1}{2}$	$\frac{1}{2}$	2	16	$15\frac{1}{2}$	8	$6\frac{1}{4}$	6	$5\frac{1}{4}$
364	$1\frac{7}{8}$	$5\frac{3}{8}$	$\frac{1}{2}$	$\frac{1}{2}$	$4\frac{1}{4}$	18	$15\frac{1}{4}$	9	7	$5\frac{5}{8}$	$5\frac{7}{8}$
364S	$1\frac{5}{8}$	3	$\frac{3}{8}$	$\frac{3}{8}$	$1\frac{7}{8}$	18	$15\frac{1}{4}$	9	7	$5\frac{5}{8}$	$5\frac{7}{8}$
364U	$2\frac{1}{8}$	$6\frac{1}{8}$	$\frac{1}{2}$	$\frac{1}{2}$	5	18	$15\frac{1}{4}$	9	7	$5\frac{5}{8}$	$5\frac{7}{8}$

. . . MOTOR FRAME DIMENSIONS

Frame No.	Shaft U	Shaft V	Key W	Key T	Key L	Dimensions – Inches A	B	D	E	F	BA
364US	$1\frac{7}{8}$	$3\frac{1}{2}$	$\frac{1}{2}$	$\frac{1}{2}$	2	18	$15\frac{1}{4}$	9	7	$5\frac{5}{8}$	$5\frac{7}{8}$
405	$2\frac{1}{8}$	$6\frac{1}{8}$	$\frac{1}{2}$	$\frac{1}{2}$	5	20	$17\frac{3}{4}$	10	8	$6\frac{7}{8}$	$6\frac{5}{8}$
405S	$1\frac{7}{8}$	$3\frac{1}{2}$	$\frac{1}{2}$	$\frac{1}{2}$	2	20	$17\frac{3}{4}$	10	8	$6\frac{7}{8}$	$6\frac{5}{8}$
405U	$2\frac{3}{8}$	$6\frac{7}{8}$	$\frac{5}{8}$	$\frac{5}{8}$	$5\frac{1}{2}$	20	$17\frac{3}{4}$	10	8	$6\frac{7}{8}$	$6\frac{5}{8}$
405US	$2\frac{1}{8}$	4	$\frac{1}{2}$	$\frac{1}{2}$	$2\frac{3}{4}$	20	$17\frac{3}{4}$	10	8	$6\frac{7}{8}$	$6\frac{5}{8}$
405T	$2\frac{7}{8}$	7	$\frac{3}{4}$	$\frac{3}{4}$	$5\frac{5}{8}$	20	$17\frac{3}{4}$	10	8	$6\frac{7}{8}$	$6\frac{5}{8}$
405TS	$2\frac{1}{8}$	4	$\frac{1}{2}$	$\frac{1}{2}$	$2\frac{3}{4}$	20	$17\frac{3}{4}$	10	8	$6\frac{7}{8}$	$6\frac{5}{8}$
444	$2\frac{3}{8}$	$6\frac{7}{8}$	$\frac{5}{8}$	$\frac{5}{8}$	$5\frac{1}{2}$	22	$18\frac{1}{2}$	11	9	$7\frac{1}{4}$	$7\frac{1}{2}$
444S	$2\frac{1}{8}$	4	$\frac{1}{2}$	$\frac{1}{2}$	$2\frac{3}{4}$	22	$18\frac{1}{2}$	11	9	$7\frac{1}{4}$	$7\frac{1}{2}$
444U	$2\frac{7}{8}$	$8\frac{3}{8}$	$\frac{3}{4}$	$\frac{3}{4}$	7	22	$18\frac{1}{2}$	11	9	$7\frac{1}{4}$	$7\frac{1}{2}$
444US	$2\frac{1}{8}$	4	$\frac{1}{2}$	$\frac{1}{2}$	$2\frac{3}{4}$	22	$18\frac{1}{2}$	11	9	$7\frac{1}{4}$	$7\frac{1}{2}$
444T	$3\frac{3}{8}$	$8\frac{1}{4}$	$\frac{7}{8}$	$\frac{7}{8}$	$6\frac{7}{8}$	22	$18\frac{1}{2}$	11	9	$7\frac{1}{4}$	$7\frac{1}{2}$
444TS	$2\frac{3}{8}$	$4\frac{1}{2}$	$\frac{5}{8}$	$\frac{5}{8}$	3	22	$18\frac{1}{2}$	11	9	$7\frac{1}{4}$	$7\frac{1}{2}$
445	$2\frac{3}{8}$	$6\frac{7}{8}$	$\frac{5}{8}$	$\frac{5}{8}$	$5\frac{1}{2}$	22	$20\frac{1}{2}$	11	9	$8\frac{1}{4}$	$7\frac{1}{2}$
445S	$2\frac{1}{8}$	4	$\frac{1}{2}$	$\frac{1}{2}$	$2\frac{3}{4}$	22	$20\frac{1}{2}$	11	9	$8\frac{1}{4}$	$7\frac{1}{2}$
445U	$2\frac{7}{8}$	$8\frac{3}{8}$	$\frac{3}{4}$	$\frac{3}{4}$	7	22	$20\frac{1}{2}$	11	9	$8\frac{1}{4}$	$7\frac{1}{2}$
445US	$2\frac{1}{8}$	4	$\frac{1}{2}$	$\frac{1}{2}$	$2\frac{3}{4}$	22	$20\frac{1}{2}$	11	9	$8\frac{1}{4}$	$7\frac{1}{2}$
445T	$3\frac{3}{8}$	$8\frac{1}{4}$	$\frac{7}{8}$	$\frac{7}{8}$	$6\frac{7}{8}$	22	$20\frac{1}{2}$	11	9	$8\frac{1}{4}$	$7\frac{1}{2}$
445TS	$2\frac{3}{8}$	$4\frac{1}{2}$	$\frac{5}{8}$	$\frac{5}{8}$	3	22	$20\frac{1}{2}$	11	9	$8\frac{1}{4}$	$7\frac{1}{2}$
504U	$2\frac{7}{8}$	$8\frac{3}{8}$	$\frac{3}{4}$	$\frac{3}{4}$	$7\frac{1}{4}$	25	21	$12\frac{1}{2}$	10	8	$8\frac{1}{2}$
504S	$2\frac{1}{8}$	4	$\frac{1}{2}$	$\frac{1}{2}$	$2\frac{3}{4}$	25	21	$12\frac{1}{2}$	10	8	$8\frac{1}{2}$
505	$2\frac{7}{8}$	$8\frac{3}{8}$	$\frac{3}{4}$	$\frac{3}{4}$	$7\frac{1}{4}$	25	23	$12\frac{1}{2}$	10	9	$8\frac{1}{2}$
505S	$2\frac{1}{8}$	4	$\frac{1}{2}$	$\frac{1}{2}$	$2\frac{3}{4}$	25	23	$12\frac{1}{2}$	10	9	$8\frac{1}{2}$

* not NEMA standard dimensions

MOTOR FRAME LETTERS

LETTER	DESIGNATION
G	Gasoline pump motor
K	Sump pump motor
M and N	Oil burner motor
S	Standard short shaft for direct connection
T	Standard dimensions established
U	Previously used as frame designation for which standard dimensions are established
Y	Special mounting dimensions required from manufacturer
Z	Standard mounting dimensions except shaft extension

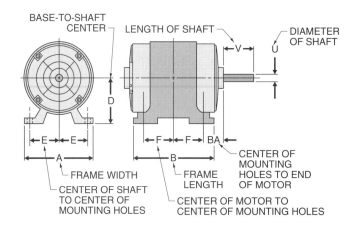

MOTOR FRAME TABLE

Frame No. Series	Third/Fourth Digit of Frame No.							
	D	1	2	3	4	5	6	7
140	3.50	3.00	3.50	4.00	4.50	5.00	5.50	6.25
160	4.00	3.50	4.00	4.50	5.00	5.50	6.25	7.00
180	4.50	4.00	4.50	5.00	5.50	6.25	7.00	8.00
200	5.00	4.50	5.00	5.50	6.50	7.00	8.00	9.00
210	5.25	4.50	5.00	5.50	6.25	7.00	8.00	9.00
220	5.50	5.00	5.50	6.25	6.75	7.50	9.00	10.00
250	6.25	5.50	6.25	7.00	8.25	9.00	10.00	11.00
280	7.00	6.25	7.00	8.00	9.50	10.00	11.00	12.50
320	8.00	7.00	8.00	9.00	10.50	11.00	12.00	14.00
360	9.00	8.00	9.00	10.00	11.25	12.25	14.00	16.00
400	10.00	9.00	10.00	11.00	12.25	13.75	16.00	18.00
440	11.00	10.00	11.00	12.50	14.50	16.50	18.00	20.00
500	12.50	11.00	12.50	14.00	16.00	18.00	20.00	22.00
580	14.50	12.50	14.00	16.00	18.00	20.00	22.00	25.00
680	17.00	16.00	18.00	20.00	22.00	25.00	28.00	32.00

Frame No. Series	Third/Fourth Digit of Frame No.							
	D	8	9	10	11	12	13	14	15
140	3.50	7.00	8.00	9.00	10.00	11.00	12.50	14.00	16.00
160	4.00	8.00	9.00	10.00	11.00	12.50	14.00	16.00	18.00
180	4.50	9.00	10.00	11.00	12.50	14.00	16.00	18.00	20.00
200	5.00	10.00	11.00	—	—	—	—	—	—
210	5.25	10.00	11.00	12.50	14.00	16.00	18.00	20.00	22.00
220	5.50	11.00	12.50	—	—	—	—	—	—
250	6.25	12.50	14.00	16.00	18.00	20.00	22.00	25.00	28.00
280	7.00	14.00	16.00	18.00	20.00	22.00	25.00	28.00	32.00
320	8.00	16.00	18.00	20.00	22.00	25.00	28.00	32.00	36.00
360	9.00	18.00	20.00	22.00	25.00	28.00	32.00	36.00	40.00
400	10.00	20.00	22.00	25.00	28.00	32.00	36.00	40.00	45.00
440	11.00	22.00	25.00	28.00	32.00	36.00	40.00	45.00	50.00
500	12.50	25.00	28.00	32.00	36.00	40.00	45.00	50.00	56.00
580	14.50	28.00	32.00	36.00	40.00	45.00	50.00	56.00	63.00
680	17.00	36.00	40.00	45.00	50.00	56.00	63.00	71.00	80.00

MOTOR RATINGS

Classification	Rating	Size
Milli	W	1, 1.5, 2, 3, 5, 7.5, 10, 15, 25, 35
Fractional	HP	$\frac{1}{20}$, $\frac{1}{12}$, $\frac{1}{8}$, $\frac{1}{6}$, $\frac{1}{4}$, $\frac{1}{3}$, $\frac{1}{2}$, $\frac{3}{4}$
Full	HP	1, $1\frac{1}{2}$, 2, 3, 5, $7\frac{1}{2}$, 10, 15, 20, 25, 30, 40, 50, 60, 75, 100, 125, 150, 200, 250, 300
Full-Special Order	HP	350, 400, 450, 500, 600, 700, 800, 900, 1000, 1250, 1500, 1750, 2000, 2250, 2500, 3000, 3500, 4000, 4500, 5000, 5500, 6000, 7000, 8000, 9000, 10,000, 11,000, 12,000, 13,000, 14,000, 15,000, 16,000, 17,000, 18,000, 19,000, 20,000, 22,500, 30,000, 32,500, 35,000, 37,500, 40,000, 45,000, 50,000

AC MOTOR CHARACTERISTICS

Motor Type 1φ	Typical Voltage	Starting Ability (Torque)	Size (HP)	Speed Range (rpm)	Cost*	Typical Uses
Shaded-pole	115 V, 230 V	Very low 50% to 100% of full load	Fractional ½ HP to ⅓ HP	Fixed 900, 1200, 1800, 3600	Very low 75% to 85%	Light-duty applications such as small fans, hair dryers, blowers, and computers
Split-phase	115 V, 230 V	Low 75% to 200% of full load	Fractional ⅓ HP or less	Fixed 900, 1200, 1800, 3600	Low 85% to 95%	Low-torque applications such as pumps, blowers, fans, and machine tools
Capacitor-start	115 V, 230 V	High 200% to 350% of full load	Fractional to 3 HP	Fixed 900, 1200, 1800	Low 90% to 110%	Hard-to-start loads such as refrigerators, air compressors, and power tools
Capacitor-run	115 V, 230 V	Very low 50% to 100% of full load	Fractional to 5 HP	Fixed 900, 1200, 1800	Low 90% to 110%	Applications that require a high running torque such as pumps and conveyors
Capacitor start-and-run	115 V, 230 V	Very high 350% to 450% of full load	Fractional to 10 HP	Fixed 900, 1200, 1800	Low 100% to 115%	Applications that require both a high starting and running torque such as loaded conveyors
3φ Induction	230 V, 460 V	Low 100% to 175% of full load	Fractional to over 500 HP	Fixed 900, 1200, 3600	Low 100%	Most industrial applications
Wound rotor	230 V, 460 V	High 200% to 300% of full load	½ HP to 200 HP	Varies by changing resistance in rotor	Very high 250% to 350%	Applications that require high torque at different speeds such as cranes and elevators
Synchronous	230 V, 460 V	Very low 40% to 100% of full load	Fractional to 250 HP	Exact constant speed	High 200% to 250%	Applications that require very slow speeds and correct power factors

* based on standard 3φ induction motor

DC AND UNIVERSAL MOTOR CHARACTERISTICS

Motor Type	Typical Voltage	Starting Ability (Torque)	Size (HP)	Speed Range (rpm)	Cost*	Typical Uses
DC Series	12 V, 90 V, 120 V, 180 V	Very high 400% to 450% of full load	Fractional to 100 HP	Varies 0 to full speed	High 175% to 225%	Applications that require very high torque such as hoists and bridges
Shunt	12 V, 90 V, 120 V, 180 V	Low 125% to 250% of full load	Fractional to 100 HP	Fixed or adjustable below full speed	High 175% to 225%	Applications that require better speed control than a series motor such as woodworking machines
Compound	12 V, 90 V, 120 V, 180 V	High 300% to 400% of full load	Fractional to 100 HP	Fixed or adjustable	High 175% to 225%	Applications that require high torque and speed control such as printing presses, conveyors, and hoists
Permanent-magnet	12 V, 24 V, 36 V, 120 V	Low 100% to 200% of full load	Fractional	Varies from 0 to full speed	High 150% to 200%	Applications that require small DC-operated equipment such as automobile power windows, seats, and sun roofs
Stepping	5 V, 12 V, 24 V	Very low** .5 to 5000 oz/in.	Size rating is given as holding torque and number of steps	Rated in number of steps per sec (maximum)	Varies based on number of steps and rated torque	Applications that require low torque and precise control such as indexing tables and printers
AC/DC Universal	115 VAC, 230 VAC, 12 VDC, 24 VDC, 36 VDC, 120 VDC	High 300% to 400% of full load	Fractional	Varies 0 to full speed	High 175% to 225%	Most portable tools such as drills, routers, mixers, and vacuum cleaners

* based on standard 3φ induction motor
** torque is rated as holding torque

OVERCURRENT PROTECTION DEVICES

Motor Type	Code Letter	Motor Size	TDF	NTDF	ITB	ITCB
AC*	—	—	175	300	150	700
AC*	A	—	150	150	150	700
AC*	B – E	—	175	250	200	700
AC*	F – V	—	175	300	250	700
DC	—	⅛ to 50 HP	150	150	150	150
DC	—	Over 50 HP	150	150	150	175

* full voltage and resistor starting

FULL-LOAD CURRENTS – DC MOTORS		
Motor rating (HP)	Current (A)	
	120 V	240 V
1/4	3.1	1.6
1/3	4.1	2.0
1/2	5.4	2.7
3/4	7.6	3.8
1	9.5	4.7
1 1/2	13.2	6.6
2	17	8.5
3	25	12.2
5	40	20
7 1/2	48	29
10	76	38

FULL-LOAD CURRENTS – 1φ, AC MOTORS		
Motor rating (HP)	Current (A)	
	115 V	230 V
1/6	4.4	2.2
1/4	5.8	2.9
1/3	7.2	3.6
1/2	9.8	4.9
3/4	13.8	6.9
1	16	8
1 1/2	20	10
2	24	12
3	34	17
5	56	28
7 1/2	80	40

FULL-LOAD – 3φ, AC INDUCTION MOTORS				
Motor rating (HP)	Current (A)			
	208 V	230 V	460 V	575 V
1/4	1.11	.96	.48	.38
1/3	1.34	1.18	.59	.47
1/2	2.2	2.0	1.0	.8
3/4	3.1	2.8	1.4	1.1
1	4.0	3.6	1.8	1.4
1 1/2	5.7	5.2	2.6	2.1
2	7.5	6.8	3.4	2.7
3	10.6	9.6	4.8	3.9
5	16.7	15.2	7.6	6.1
7 1/2	24.0	22.0	11.0	9.0
10	31.0	28.0	14.0	11.0
15	46.0	42.0	21.0	17.0
20	59	54	27	22
25	75	68	34	27
30	88	80	40	32
40	114	104	52	41
50	143	130	65	52
60	169	154	77	62
75	211	192	96	77
100	273	248	124	99
125	343	312	156	125
150	396	360	180	144
200	—	480	240	192
250	—	602	301	242
300			362	288
350			413	337
400			477	382
500			590	472

TYPICAL MOTOR EFFICIENCIES					
HP	Standard Motor (%)	Energy-Effiecient Motor (%)	HP	Standard Motor (%)	Energy-Efficient Motor (%)
1	76.5	84.0	30	88.1	93.1
1.5	78.5	85.5	40	89.3	93.6
2	79.9	86.5	50	90.4	93.7
3	80.8	88.5	75	90.8	95.0
5	83.1	88.6	100	91.6	95.4
7.5	83.8	90.2	125	91.8	95.8
10	85.0	90.3	150	92.3	96.0
15	86.5	91.7	200	93.3	96.1
20	87.5	92.4	250	93.6	96.2
25	88.0	93.0	300	93.8	96.5

ENCLOSURES				
Type	Use	Service Conditions	Tests	Comments
1	Indoor	No unusual	Rod entry, rust resistance	
3	Outdoor	Windblown dust, rain, sleet, and ice on enclosure	Rain, external icing, dust, and rust resistance	Do not provide protection against internal condensation or internal icing
3R	Outdoor	Falling rain and ice on enclosure	Rod entry, rain, external icing, and rust resistance	Do not provide protection against dust, internal condensation, or internal icing
4	Indoor/outdoor	Windblown dust and rain, splashing water, hose-directed water, and ice on enclosure	Hosedown, external icing, and rust resistance	Do not provide protection against internal condensation or internal icing
4X	Indoor/outdoor	Corrosion, windblown dust and rain, splashing water, hose-directed water, and ice on enclosure	Hosedown, external icing, and corrosion resistance	Do not provide protection against internal condensation or internal icing
6	Indoor/outdoor	Occasional temporary submersion at a limited depth		
6P	Indoor/outdoor	Prolonged submersion at a limited depth		
7	Indoor locations classified as Class I, Groups A, B, C, or D, as defined in the NEC®	Withstand and contain an internal explosion of specified gases, contain an explosion sufficiently so an explosive gas-air mixture in the atmosphere is not ignited	Explosion, hydrostatic, and temperature	Enclosed heat-generating devices shall not cause external surfaces to reach temperatures capable of igniting explosive gas-air mixtures in the atmosphere
9	Indoor locations classified as Class II, Groups E or G, as defined in the NEC®	Dust	Dust penetration, temperature, and gasket aging	Enclosed heat-generating devices shall not cause external surfaces to reach temperatures capable of igniting explosive gas-air mixtures in the atmosphere
12	Indoor	Dust, falling dirt, and dripping noncorrosive liquids	Drip, dust, and rust resistance	Do not provide protection against internal condensation
13	Indoor	Dust, spraying water, oil, and noncorrosive coolant	Oil explosion and rust resistance	Do not provide protection against internal condensation

HAZARDOUS LOCATIONS			
Class	Division	Group	Material
I	1 or 2	A	Acetylene
	1 or 2	B	Hydrogen, butadiene, ethylene oxide, propylene oxide
	1 or 2	C	Carbon monoxide, ether, ethylene, hydrogen sulfide, morpholine, cyclopropane
	1 or 2	D	Gasoline, benzene, butane, propane, alcohol, acetone, ammonia, vinyl chloride
II	1 or 2	E	Metal dusts
	1 or 2	F	Carbon black, coke dust, coal
	1 or 2	G	Grain dust, flour, starch, sugar, plastics
III	1 or 2	No groups	Wood chips, cotton, flax, and nylon

IEC ENCLOSURE CLASSIFICATION

IEC Publication 529 describes standard degrees of protection that enclosures of a product must provide when properly installed. The degree of protection is indicated by two letters, IP, and two numerals. International Standard IEC 529 contains descriptions and associated test requirements to define the degee of protection that each numeral specifies. The following table indicates the general degrees of protection. For complete test requirements refer to IEC 529.

First Numeral*†	Second Numeral*†
Protection of persons against access to hazardous parts and protection against penetration of solid foreign objects.	*Protection against liquids‡ under test conditions specified in IEC 529.*
0 Not protected	**0** Not protected
1 Protection against objects greater than 50 mm in diameter (hands)	**1** Protection against vertically falling drops of water (condensation)
2 Protection against objects greater than 12.5 mm in diameter (fingers)	**2** Protection against falling water with enclosure tilted 15°
3 Protection against objects greater than 2.5 mm in diameter (tools, wires)	**3** Protection against spraying or falling water with enclosure tilted 60°
4 Protection against objects greater than 1.0 mm in diameter (tools, small wires)	**4** Protection against splashing water
5 Protection against dust (dust may enter during test but must not interfere with equipment operation or impair safety)	**5** Protection against low-pressure water jets
6 Dusttight (no dust observable inside enclosure at end of test)	**6** Protection against powerful water jets
	7 Protection against temporary submersion
	8 Protection against continuous submersion

Example: IP41 describes an enclosure that is designed to protect against the entry of tools or objects greater than 1 mm in diameter and to protect against vertically dripping water under specified test conditions

* All first and second numerals up to and including numeral 6 imply compliance with the requirements of all preceeding numerals in their respective series. Second numerals 7 and 8 do not imply suitability for exposure to water jets unless dual coded e.g., IP 5 IP 7
† The IEC permits use of certain supplementary leters with the characteristic numerals. If such letters are used, refer to IEC 529 for an explanation
‡ The IEC test requirements for degrees of protection against liquid ingress refer only to water

HORSEPOWER AND TORQUE UNITS OF MEASURE

To Convert	To	Multiply By
oz-in.	lb-in.	.0625
oz-in.	lb-ft	.0052
oz-in.	N-m	.0071
lb-in.	oz-in.	16
lb-in.	lb-ft	.0833
lb-in.	N-m	.113
lb-ft	oz-in.	192
lb-ft	lb-in.	12
lb-ft	N-m	1.356
N-m	oz-in.	141.612
N-m	lb-in.	8.851
N-m	lb-ft	.7376

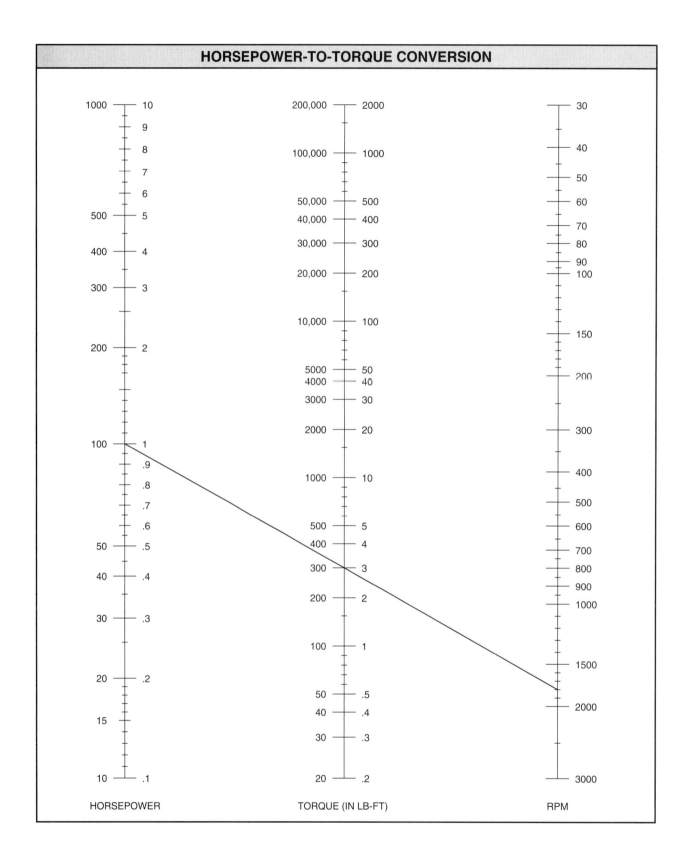

HORSEPOWER-TO-TORQUE CONVERSION

HORSEPOWER

TORQUE (IN LB-FT)

RPM

HAZARDOUS LOCATIONS – 500

Hazardous Location – A location where there is an increased risk of fire or explosion due to the presence of flammable gases, vapors, liquids, combustible dusts, or easily-ignitable fibers or flyings.

Location – A position or site.

Flammable – Capable of being easily ignited and of burning quickly.

Gas – A fluid (such as air) that has no independent shape or volume but tends to expand indefinitely.

Vapor – A substance in the gaseous state as distinguished from the solid or liquid state.

Liquid – A fluid (such as water) that has no independent shape but has a definite volume. A liquid does not expand indefinitely and is only slightly compressible.

Combustible – Capable of burning.

Ignitable – Capable of being set on fire.

Fiber – A thread or piece of material.

Flyings – Small particles of material.

Dust – Fine particles of matter.

Classes	Likelihood that a flammable or combustible concentration is present
I	Sufficient quantities of flammable gases and vapors present in air to cause an explosion or ignite hazardous materials
II	Sufficient quantities of combustible dust are present in air to cause an explosion or ignite hazardous materials
III	Easily ignitable fibers or flyings are present in air, but not in a sufficient quantity to cause an explosion of ignite hazardous materials.

Divisions	Location containing hazardous substances
1	Hazardous location in which hazardous substance is normally present in air in sufficient quantities to cause an explosion or ignite hazardous materials
2	Hazardous location in which hazardous substance is not normally present in air in sufficient quantities to cause an explosion or ignite hazardous materials

Groups	Atmosphere containing flammable gases or vapors or combustible dust		
	Class I	Class II	Class III
	A B C D	E F G	none

DIVISION I EXAMPLES

Class I:
Spray booth interiors

Areas adjacent to spraying or painting operations using volatile flammable solvents

Open tanks or vats of volatile flammable liquids

Drying or evaporation rooms for flammable vents

Areas where fats and oil extraction equipment using flammable solvents are operated

Cleaning and dyeing plant rooms that use flammable liquids that do not contain adequate ventilation

Refrigeration or freezer interiors that store flammable materials

All other locations where sufficient ignitable quantities of flammable gases or vapors are likely to occur during routine operations

Class II:
Grain and grain products

Pulverized sugar and cocoa

Dried egg and milk powders

Pulverized spices

Starch and pastes

Potato and wood flour

Oil meal from beans and seeds

Dried hay

Any other organic material that may produce combustible dusts during their use or handling

Class III:
Portions of rayon, cotton, or other textile mills

Manufacturing and processing plants for combustible fibers, cotton gins, and cotton seed mills

Flax processing plants

Clothing manufacturing plants

Woodworking plants

Other establishments involving similar hazardous processes or conditions

THREE-PHASE VOLTAGE VALUES

For 208 V × 1.732, use 360
For 230 V × 1.732, use 398
For 240 V × 1.732, use 416
For 440 V × 1.732, use 762
For 460 V × 1.732, use 797
For 480 V × 1.732, use 831

CAPACITOR RATINGS

110–125 VAC, 50/60 Hz, Starting Capacitors

| Typical Ratings* | Dimensions** | | Model Number*** |
	Diameter	Length	
88 – 106	$1^7/_{16}$	$2^3/_4$	EC8815
108 – 130	$1^7/_{16}$	$2^3/_4$	EC10815
130 – 156	$1^7/_{16}$	$2^3/_4$	EC13015
145 – 174	$1^7/_{16}$	$2^3/_4$	EC14515
161 – 193	$1^7/_{16}$	$2^3/_4$	EC16115
189 – 227	$1^7/_{16}$	$2^3/_4$	EC18915A
216 – 259	$1^7/_{16}$	$3^3/_8$	EC21615
233 – 280	$1^7/_{16}$	$3^3/_8$	EC23315A
243 – 292	$1^7/_{16}$	$3^3/_8$	EC24315A
270 – 324	$1^7/_{16}$	$3^3/_8$	EC27015A
324 – 389	$1^7/_{16}$	$3^3/_8$	EC2R10324N
340 – 408	$1^{13}/_{16}$	$3^3/_8$	EC34015
378 – 454	$1^{13}/_{16}$	$3^3/_8$	EC37815
400 – 480	$1^{13}/_{16}$	$3^3/_8$	EC40015
430 – 516	$1^{13}/_{16}$	$3^3/_8$	EC43015A
460 – 553	$1^{13}/_{16}$	$4^3/_8$	EC5R10460
540 – 648	$1^{13}/_{16}$	$4^3/_8$	EC54015B
590 – 708	$1^{13}/_{16}$	$4^3/_8$	EC59015A
708 – 850	$1^{13}/_{16}$	$4^3/_8$	EC70815
815 – 978	$1^{13}/_{16}$	$4^3/_8$	EC81515
1000 – 1200	$2^1/_{16}$	$4^3/_8$	EC100015A

220–250 VAC, 50/60 Hz, Starting Capacitors

| Typical Ratings* | Dimensions** | | Model Number*** |
	Diameter	Length	
53 – 64	$1^7/_{16}$	$3^3/_8$	EC5335
64 – 77	$1^7/_{16}$	$3^3/_8$	EC6435
88 – 106	$1^{13}/_{16}$	$3^3/_8$	EC8835
108 – 130	$1^{13}/_{16}$	$3^3/_8$	EC10835A
124 – 149	$1^{13}/_{16}$	$4^3/_8$	EC12435
130 – 154	$1^{13}/_{16}$	$4^3/_8$	EC13035
145 – 174	$2^1/_{16}$	$3^3/_8$	EC6R22145
161 – 193	$2^1/_{16}$	$3^3/_8$	EC6R2216N
216 – 259	$2^1/_{16}$	$4^3/_8$	EC21635A
233 – 280	$2^1/_{16}$	$4^3/_8$	EC23335A
270 – 324	$2^1/_{16}$	$4^3/_8$	EC27035A

* in µF
** in inches
*** Model numbers vary by manufacturer.

CAPACITOR RATINGS

270 VAC, 50/60 Hz, Running Capacitors

| Typical Ratings* | Dimensions** | | Model Number*** |
	Oval	Length	
2		$2^1/_8$	VH5502
3		$2^1/_8$	VH5503
4	$1^5/_{16} × 2^5/_{32}$	$2^1/_8$	VH5704
5		$2^1/_8$	VH5705
6		$2^5/_8$	VH5706
7.5		$2^7/_8$	VH9001
10	$1^5/_{16} × 2^5/_{32}$	$2^7/_8$	VH9002
12.5		$3^7/_8$	VH9003
15	$1^{29}/_{32} × 2^{29}/_{32}$	$2^1/_8$	VH9121
17.5		$2^7/_8$	VH9123
20		$2^7/_8$	VH5463
25	$1^{29}/_{32} × 2^{29}/_{32}$	$3^7/_8$	VH9069
30		$3^7/_8$	VH5465
35	$1^{29}/_{32} × 2^{29}/_{32}$	$3^7/_8$	VH9071
40		$3^7/_8$	VH9073
45	$1^{31}/_{32} × 3^{21}/_{32}$	$3^7/_8$	VH9115
50		$3^7/_8$	VH9075

440 VAC, 50/60 Hz, Running Capacitors

10	$1^5/_{16} × 2^5/_{32}$	$3^7/_8$	VH5300
15	$1^{29}/_{32} × 2^{29}/_{32}$	$2^7/_8$	VH5304
17.5	$1^{29}/_{32} × 2^{29}/_{32}$	$3^7/_8$	VH9141
20	$1^{29}/_{32} × 2^{29}/_{32}$	$3^7/_8$	VH9082
25	$1^{29}/_{32} × 2^{29}/_{32}$	$3^7/_8$	VH5310
30		$4^3/_4$	VH9086
35	$1^{29}/_{32} × 2^{29}/_{32}$	$4^3/_4$	VH9088
40		$4^3/_4$	VH9641
45		$3^7/_8$	VH5351
50	$1^{31}/_{32} × 3^{21}/_{32}$	$3^7/_8$	VH5320
55		$4^3/_4$	VH9084

* in µF
** in inches
*** Model numbers vary by manufacturer.

RESISTOR COLOR CODES

| Color | Number | | Multiplier | Tolerance (%) |
	1st	2nd		
Black (BK)	0	0	1	0
Brown (BR)	1	1	10	—
Red (R)	2	2	100	—
Orange (O)	3	3	1000	—
Yellow (Y)	4	4	10,000	—
Green (G)	5	5	100,000	—
Blue (BL)	6	6	1,000,000	—
Violet (V)	7	7	10,000,000	—
Gray (GY)	8	8	100,000,000	—
White (W)	9	9	1,000,000,000	—
Gold (Au)	—	—	0.1	5
Silver (Ag)	—	—	0.01	10
None	—	—	0	20

DMM TERMINOLOGY . . .

TERM	SYMBOL	DEFINITION
AC		Continually changing current that reverses direction at regular intervals. Standard U.S. frequency is 60 Hz
AC COUPLING		Signal that passes an AC signal and blocks a DC signal. Used to measure AC signals that are riding on a DC signal
ACCURACY ANALOG METER		Largest allowable error (in percent of full scale) made under normal operating conditions. The reading of a DMM set on the 250 V range with an accuracy rating of ±2% could vary ±5 V. Analog DMMs have greater accuracy when readings are taken on the upper half of the scale
ACCURACY DIGITAL METER		Largest allowable error (in percent of reading) made under normal operating conditions. A reading of 100.0 V on a DMM with an accuracy of ±2% is between 98.0 V and 102.0 V. Accuracy may also include a specified amount of digits (counts) that are added to the basic accuracy rating. For example, an accuracy of ±2% (±2 digits) means that a display reading of 100.0 V on the DMM is between 97.8 V and 102.2 V
AC/DC		Indicates ability to read or operate on alternating and direct current
AC FREQUENCY RESPONSE		Frequency range over which AC voltage measurements are accurate
ALLIGATOR CLIP		Long-jawed, spring-loaded clamp connected to the end of a test lead. Used to make temporary electrical connections
AMBIENT TEMPERATURE		Temperature of air surrounding a DMM or equipment to which the DMM is connected
AMMETER		DMM that measures electric current
AMMETER SHUNT		Low-resistance conductor that is connected in parallel with the terminals of an ammeter to extend the range of current values measured by the ammeter
AMPLITUDE		Measure of AC signal alternation expressed in values such as peak or peak-to-peak
ATTENUATION		Decrease in amplitude of a signal
AUDIBLE		Sound that can be heard
AUTOMATIC TOUCH HOLD® MODE		Function that captures a measurement, beeps, and locks the measurement on the digital display for later viewing
AUTORANGE MODE		Function that automatically selects a DMM's range based on signals received
AVERAGE VALUE		Value equal to .637 times the amplitude of a measured value

... DMM TERMINOLOGY ...

TERM	SYMBOL	DEFINITION
BACKLIGHT		Light that brightens the DMM display
BANANA JACK	⊙	DMM jack that accepts a banana plug
BANANA PLUG		Long, thick terminal connection on one end of a test lead used to make a connection to a DMM
BATTERY SAVE		Feature that enables a DMM to shut down when battery level is too low or no key is pressed within a set time
BNC		Coaxial-type input connector used on some DMMs
CAPTURE		Function that records and displays measured values
CELSIUS	$^\circ$C	Temperature measured on a scale for which the freezing point of water is 0° and the boiling point is 100°
CLOSED CIRCUIT		Circuit in which two or more points allow a predesigned current to flow
CONTINUITY CAPTURE		Function used to detect intermittent open and short circuits as brief as 250 μs
COUNTS		Unit of measure of DMM resolution. A 1999 count DMM cannot display a measurement of $1/10$ of a volt when measuring 200 V or more. A 3200 count DMM can display a measurement of $1/10$ of a volt up to 320 V
CREST FACTOR		Ratio of peak value to the rms value. The higher the DMM crest factor, the wider the range of waveforms it can measure. In a pure sine wave, the crest factor is 1.41
db READOUT		Decibels (dB) unit of measure used to express the ratio between two quantities such as the gain or loss of amplifiers, filters, or attenuators in telecommunications or audio applications
DC	——— — — —	Current that constantly flows in one direction
DECIBEL (dB)		Measurement that indicates voltage or power comparison in a logarithmic scale
DIGITS		Indication of the resolution of a DMM. A $3\frac{1}{2}$ digit DMM can display three full digits and one half digit. The three full digits display a number from 0 to 9. The half digit displays a 1 or is left blank. A $3\frac{1}{2}$ digit DMM displays readings up to 1999 counts of resolution. A $4\frac{1}{2}$ digit DMM displays readings up to 19,999 counts of resolution
DIODE	▶⊢	Semiconductor that allows current to flow in only one direction
DISCHARGE		Removal of an electric charge
DUAL TRACE		Feature that allows two separate waveforms to be displayed simultaneously
EARTH GROUND		Reference point that is directly connected to ground

. . . DMM TERMINOLOGY . . .

TERM	SYMBOL	DEFINITION
EFFECTIVE VALUE		Value equal to .707 of the peak in a sine wave
FAHRENHEIT	°F	Temperature measured on a scale for which the freezing point of water is 32° and the boiling point is 212°
FREEZE		Function that holds a waveform (or measurement) for closer examination
FREQUENCY		Number of complete cycles occurring per unit of time
FUNCTION SWITCH		Switch that selects the function (AC voltage, DC voltage, etc.) that a DMM is to measure
GLITCH		Momentary spike in a waveform
GLITCH DETECT		Function that increases the DMM sampling rate to maximize the detection of the glitch(es)
GROUND		Common connection to a point in a circuit whose potential is taken as zero
HARD COPY		Function that allows a printed copy of the displayed measurement
HARMONICS		Currents generated by electronic devices (nonlinear loads) which draw current in short pulses, not as a smooth sine wave. Harmonic currents are whole-number multiples of the fundamental (typically 60 Hz) current
HOLD BUTTON	HOLD H	Button that allows a DMM to capture and hold a stable measurement
INPUT ALERT		Function that provides an audible warning if test leads are in current input jacks but the function switch is not in amps position
LIQUID CRYSTAL DISPLAY (LCD)		Display that uses liquid crystals to display waveforms, measurements, and text on its screen
MEASURING RANGE		Minimum and maximum quantity that a DMM can safely and accurately measure
MIN MAX		Function that captures and stores the highest and lowest measurements for later viewing. Function can be used with any DMM measurement function such as volts, amps, etc.
MIN MAX INSTANTANEOUS PEAK		High speed, 1 ms response time used to capture MIN MAX readings of a waveform peak value. Can be used for crest factor calculations or to capture transient voltage or momentary voltage surge measurements
NOISE		Unwanted extraneous electrical signals
OPEN CIRCUIT		Circuit in which two (or more) points do not provide a path for current flow
OVERFLOW		Condition of a DMM that occurs when a quantity to be measured is greater than the quantity the DMM can display
OVERLOAD	OL	Condition of a DMM that occurs when a quantity to be measured is greater than the quantity the DMM can safely handle for the DMM range setting
PEAK		Maximum value of positive or negative alternation in a sine wave
PEAK-TO-PEAK		Value measured from the maximum negative to the maximum positive alternation in a sine wave

TERM	SYMBOL	DEFINITION
	. . . DMM TERMINOLOGY	
POLARITY		Orientation of the positive (+) and negative (–) side of direct current or voltage
PROBE		Pointed metal tip of a test lead used to make contact with the circuit under test
PULSE		Waveform that increases from a constant value, then decreases to its original value
PULSE TRAIN		Repetitive series of pulses
RANGE		Quantities between two points or levels
RECALL		Function that allows stored information (or measurements) to be displayed
RESOLUTION		Sensitivity of a DMM. A DMM may have a resolution of current flow
RISING SLOPE		Part of a waveform displaying a rise in voltage
ROOT-MEAN-SQUARE		Value equal to .707 of the amplitude of a measured value
SAMPLE		Momentary reading taken from an input signal
SAMPLING RATE		Number of readings taken on a signal over time
SHORT CIRCUIT		Two or more points in a circuit that allow an unplanned current flow
TERMINAL		Point to which DMM test leads are connected
TERMINAL VOLTAGE		Voltage level that DMM terminals can safely handle
TRACE		Displayed waveform that shows the voltage variations of the input signal as a function of time
TRIGGER		Device which determines the beginning point of a wavelength
WAVEFORM		Pattern defined by an electrical signal
ZOOM		Function that allows a waveform (or part of waveform) to be magnified

SELECTED DMM ABBREVIATIONS

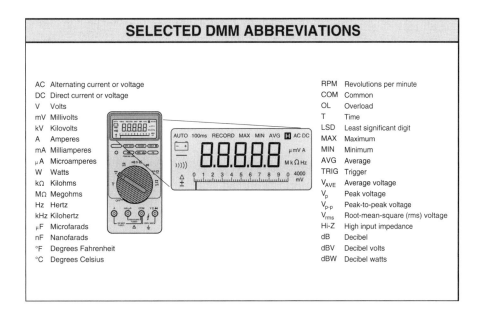

AC	Alternating current or voltage	RPM	Revolutions per minute
DC	Direct current or voltage	COM	Common
V	Volts	OL	Overload
mV	Millivolts	T	Time
kV	Kilovolts	LSD	Least significant digit
A	Amperes	MAX	Maximum
mA	Milliamperes	MIN	Minimum
μA	Microamperes	AVG	Average
W	Watts	TRIG	Trigger
kΩ	Kilohms	V_{AVE}	Average voltage
MΩ	Megohms	V_p	Peak voltage
Hz	Hertz	V_{p-p}	Peak-to-peak voltage
kHz	Kilohertz	V_{rms}	Root-mean-square (rms) voltage
μF	Microfarads	Hi-Z	High input impedance
nF	Nanofarads	dB	Decibel
°F	Degrees Fahrenheit	dBV	Decibel volts
°C	Degrees Celsius	dBW	Decibel watts

SELECTED DMM SYMBOLS

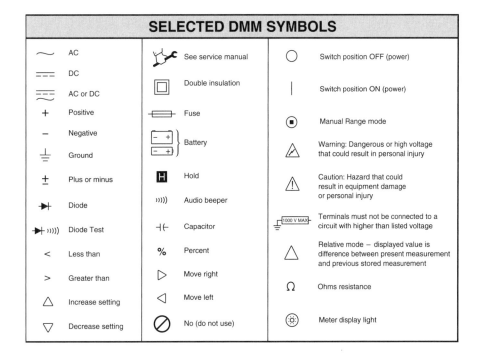

～	AC		See service manual	○	Switch position OFF (power)
===	DC		Double insulation	I	Switch position ON (power)
===	AC or DC		Fuse	■	Manual Range mode
+	Positive		Battery		Warning: Dangerous or high voltage that could result in personal injury
−	Negative				
	Ground	H	Hold		Caution: Hazard that could result in equipment damage or personal injury
±	Plus or minus	)))))	Audio beeper		
	Diode	−|(−	Capacitor	1000 V MAX	Terminals must not be connected to a circuit with higher than listed voltage
)))))	Diode Test	%	Percent		Relative mode − displayed value is difference between present measurement and previous stored measurement
<	Less than	▷	Move right		
>	Greater than	◁	Move left	Ω	Ohms resistance
△	Increase setting	⊘	No (do not use)		
▽	Decrease setting				Meter display light

Glossary

A

acceleration time (ramp-up time): A motor control parameter that determines the length of time an electric motor drive takes to accelerate a motor from a standstill (0 rpm) to maximum motor speed (max rpm).

AC drive motor: A motor specifically designed for use with an electric motor drive.

Affinity Laws: Physics laws that cover the relationships between speed, flow, pressure, and horsepower for variable torque loads.

alternation: One-half of a cycle.

ambient temperature: The temperature of the air surrounding an object.

amplification: The process of taking a small signal and increasing the signal size.

analog signal: A type of input signal to an electric motor drive that can be either varying voltage or varying current.

angular misalignment: Misalignment caused by two shafts that are not parallel.

anode: The positive lead of a diode.

apparent power: The total power delivered.

arc blast: An explosion that occurs when the surrounding air becomes ionized and conductive.

armature: The rotating part of a DC motor.

automatic calibration (automatic tuning): A parameter that fine tunes an electric motor drive to the characteristics of a motor for optimum performance.

automatic restart after a fault: A parameter that allows an electric motor drive to start automatically, provided the start inputs are still closed and the fault has cleared.

automatic restart after a power outage: A parameter that allows an electric motor drive to start automatically when power is once again present on the supply lines.

autoranging DMM: A meter that automatically adjusts to a higher range setting if the range is not high enough.

autotransformer starting: A motor starting method that uses a tapped 3φ autotransformer to provide reduced-voltage starting.

B

bandwidth: The range of frequencies to which the meter can respond. Average-responding and true-rms multimeters and clamp-on ammeters have a bandwidth rating.

bar graph: A graph composed of segments that function as an analog pointer.

base speed: The nameplate speed (rpm) at which a motor develops rated horsepower at rated load and voltage.

basic motor starter: A starter that has a manually operated switch (contactor) and includes motor overload protection.

bearing current: The result of induced voltage in the motor rotor created by the electric motor drive.

bipolar junction transistor (BJT): A transistor that controls the flow of current through the emitter (E) and collector (C) with a properly biased base (B).

branch circuit: The portion of a distribution system between the final overcurrent protection device and the outlet (receptacle) or load.

breakdown torque (BDT): The maximum torque a motor can provide without an abrupt reduction in motor speed.

bridge rectifier: A circuit containing four diodes that permits both halves of the input AC sine wave to pass.

brush: The sliding contact that makes the connection between the rotating armature and the external circuit (power supply) to the DC motor.

bypass contactor: A contactor that allows line power to a motor that is normally controlled by an electric motor drive.

bypass contactor test: A test that verifies that the bypass contactor works and that the motor rotates in the correct direction when the bypass contactor is energized.

C

capacitance (C): The ability of a component or circuit to store energy in the form of an electrical charge.

capacitive load: A load that contains only electrical capacitance.

capacitive reactance (XC): The opposition to current flow by a capacitor.

capacitor: An electrical device designed to store a voltage charge of electrical energy.

capacitor motor: A 1φ motor with a capacitor connected in series with the stator windings to produce phase displacement in the starting or running winding and add higher starting and/or running torque.

carrier frequency: The frequency that controls the number of times the solid state switches in the inverter section of a PWM electric motor drive turn ON and turn OFF.

cathode: The negative lead of a diode.

caution signal word: A word used to indicate a potentially hazardous situation which, if not avoided, may result in minor or moderate injury.

centrifugal switch: A switch that opens to disconnect the starting winding when the rotor reaches a preset speed and reconnects the starting winding when the speed falls below the preset value.

clamp-on meter: A meter that measures current in a circuit by measuring the strength of the magnetic field around a single conductor.

closed loop drive: An electric motor drive that operates using a feedback sensor such as an encoder or tachometer connected to the shaft of the motor to send information about motor speed back to the drive.

closed loop system: A system with feedback from the motor sensors to the electric motor drive.

closed-loop vector control (vector control): A control mode that allows an AC motor to have torque characteristics identical to a DC motor.

coast stop: A method of stopping a motor in which the electric motor drive shuts OFF the voltage to the motor, allowing the motor to coast to a stop.

commutator: The part of the armature that connects each armature coil to the brushes by using copper bars (segments) that are insulated from each other with pieces of mica.

constant horsepower (CH) load: A load that requires high torque at low speeds and low torque at high speeds.

constant torque (CT) load: A load that requires the torque to remain constant.

continuity: The presence of a complete path for current flow.

control mode (volts-per-hertz pattern): A motor control parameter that determines the relationship between the voltage and frequency an electric motor drive outputs to a motor.

control wiring: All external wiring connected to an electric motor drive, excluding the line and load conductors.

converter: An electronic device that changes AC voltage into DC voltage.

converter and inverter test: A test used to verify correct operation of semiconductor components in the converter and inverter sections of an electric motor drive.

cover gloves: Gloves worn over latex electrical gloves to prevent penetration of the latex gloves and provide added protection against electrical shock.

current: The quantity of electrons flowing through an electrical circuit.

current unbalance: The unbalance that occurs when the current is not equal at the leads of a 3φ motor or other 3φ load.

cutoff region: The point at which the transistor is turned OFF and no current flows.

cycle: One complete wave of alternating voltage or current.

D

danger signal word: A word used to indicate an imminently hazardous situation, which if not avoided, results in death or serious injury.

DC compound motor: A DC motor with the field connected in series and shunt with the armature.

DC injection braking: A deceleration method which brings a motor to a smooth, quick stop and can be used to hold a motor shaft stationary for brief periods of time.

DC permanent-magnet motor: A motor that uses magnets, not a coil of wire, for the field winding.

DC series motor: A motor with the field connected in series with the armature.

DC shunt motor: A motor with the field connected in shunt (parallel) with the armature.

deceleration time (ramp-down time): A motor control parameter that determines how long the length of time an electric motor drive takes to decelerate a motor from maximum motor frequency speed to a standstill.

decibel (dB): A unit of measure used to express the relative intensity of sound.

digital display: A display that shows numerical values using light-emitting diodes (LEDs) or liquid crystal displays (LCDs).

digital multimeter (DMM): A test tool used to measure two or more electrical values.

digital signals: Signals that have only two states, ON or OFF like a momentary pushbutton.

diode: An electronic device that allows current to pass through in only one direction.

direct current: Current that flows in one direction only.

direct current (DC) motor: A motor that uses direct current connected to the field and armature to produce shaft rotation.

disconnect: A device that isolates an electric motor drive and/or motor from the voltage source to allow safe access for maintenance or repair.

dual-function switch: A switch that performs two different switching functions such as forward or reverse.

dual-voltage motor: A motor that operates at more than one voltage level.

dynamic braking: A motor deceleration method that brings a motor to a smooth, quick stop, similar to DC injection braking.

E

earmuff: An ear protection device worn over the ears.

earplug: An ear protection device made of moldable rubber, foam, or plastic, and inserted into the ear canal.

electrical energy: Energy made available by the flow of electric charge.

electrical gloves: Gloves made of latex that are used to provide maximum insulation from electrical shock.

electrical power system: A system that generates, transmits, distributes, and/or delivers electrical power to satisfactorily operate electrical loads designed for connection to the system.

electrical safety standard: A document that provides information to reduce safety hazards that occur when using electrical test equipment such as DMMs, clamp-on meters, and ohmmeters.

electrical warning signal word: A word used to indicate a high-voltage location and conditions that could result in death or serious personal injury from an electrical shock if proper precautions are not taken.

electric motor: A rotating output device that converts electrical energy into rotating mechanical energy.

electric motor drive: An electronic device that controls the direction, speed, torque, and other operating functions of an electric motor in addition to providing motor protection and monitoring functions.

electric motor drive component test: A test that identifies which components in a drive are defective.

electric motor drive feature: A function and/or accessory of a drive that is required for a specific application.

electric motor drive information parameter: A parameter that provides information used for periodic maintenance and troubleshooting.

electric motor drive input and output test: A test used to verify that the inputs and outputs of a drive function properly when operated as designed

electric motor drive, motor, and load test: A test used to verify that a drive and motor function together properly to rotate the driven load.

electric shock: A shock that results any time a body becomes part of an electrical circuit.

electromagnetic compatibility (EMC): A comparison of how different pieces of equipment work together with varying levels of interference.

electromagnetic interference (EMI): Unwanted electrical noise generated by electrical and electronic equipment.

electrostatic discharge (ESD): The movement of electrons from a source to an object.

encoder: A sensor (transducer) that produces discrete electrical pulses during each increment of shaft rotation.

energy: The capacity to do work.

engineered application: An electric motor drive application that requires a licensed engineer to be safely implemented.

equipment grounding: An equal potential between all metal components of an installation and a low-impedance path for fault currents to operate overcurrent protection devices.

explosion warning signal word: A word used to indicate locations and conditions where exploding parts may cause death or serious personal injury if proper precautions and procedures are not followed.

F

face shield: An eye and face protection device that covers the entire face with a plastic shield, and is used for protection from flying objects.

fan curve: A graph that shows the relationship between flow and pressure for a particular fan at a single speed.

field: The stationary windings, or magnets, of a DC motor.

field-effect transistor (FET): A transistor that controls the flow of current through the drain (D) and source (S) with a properly biased gate (G).

final checks: Checks used to limit the source of potential problems during start-up of an electric motor drive.

fixed capacitor: A capacitor with one capacitance value.

fixed resistor: A resistor with a set value.

flat-topping: The covering of the peaks of the voltage sine wave.

fly-back (freewheeling) diode: A diode that handles regenerated voltage when an AC motor decelerates.

flying start: A parameter that allows an electric motor drive to lock in on the speed of a motor and ramp the motor up from that speed to setpoint.

foot protection: Shoes worn to prevent foot injuries that are typically caused by objects falling less than 4″ and having an average weight less than 65 lb. Safety shoes with reinforced steel toes protect against injuries caused by compression and impact.

force: A form of energy that changes the position, motion, direction, or shape of an object.

foreign control voltage: A voltage that originates outside the electric motor drive, such as voltage from a control panel or sensor that interfaces with the drive.

forward bias: The condition of a diode when the diode allows current flow.

forward breakover voltage: The forward bias voltage necessary for a semiconductor to go into conduction mode.

frequency setpoint: A parameter that sets the frequency at which an electric motor drive operates when speed reference is internal.

full-load torque (FLT): The torque required to produce the rated power at full speed of the motor.

full-wave rectifier: A circuit containing two diodes and a center-tapped transformer that permits both halves of the input AC sine wave to pass.

fully loaded motor: A motor that has an amperage reading of 95% to 105% of nameplate current.

fundamental frequency: The frequency of the voltage used to control motor speed.

G

gate turn-off thyristor (GTO): A solid state device that allows for a controlled turn ON and controlled turn OFF of current using the gate.

ghost voltage: A voltage reading on a DMM that is not connected to an energized circuit.

goggles: An eye protection device with a flexible frame that is secured on the face with an elastic headband.

ground fault circuit interrupter (GFCI): A device that protects against electrical shock by detecting an imbalance of current in the normal conductor pathways and opening the circuit.

grounding: The connection of all exposed non-current-carrying metal parts to the earth. Grounding provides a direct path for unwanted fault current to the earth without causing harm to persons or equipment.

grounding conductor: A conductor that does not normally carry current, except during a fault (short circuit).

H

half-wave rectifier: A circuit containing one diode that allows only half of the input AC sine wave to pass.

harmonics: Frequencies that are whole-number multiples (second, third, fourth, fifth, etc.) of the fundamental frequency.

hasp: A multiple lockout/tagout device.

heat sink: A device that conducts and dissipates heat away from an electrical component.

hertz: The international unit of frequency, equal to one cycle per second.

horsepower (HP): A unit of power equal to 746 W, 550 lb-ft/sec, or 33,000 lb-ft/min.

human interface module (HIM): A manually operated input control unit that includes programming keys, system operating keys, and normally a status display.

I

impedance (Z): The total opposition of resistance, inductive reactance, and capacitive reactance offered to the flow of alternating current.

improper phase sequence: The changing of the sequence of any two phases (phase reversal) in a 3φ motor control circuit.

inductance (L): The property of a circuit that causes it to oppose a change in current due to energy stored in a magnetic field of a coil.

induction motor: A motor that has no physical electrical connection to the rotor because there are no brushes.

inductive load: A load that contains only electrical inductance.

inductive reactance (XL): An inductor's opposition to alternating current.

initial test: A test that verifies if an electric motor drive is operational.

in phase: The state when voltage and current reach their maximum amplitude and zero level simultaneously.

input current: The current required by an electric motor drive to avoid sine wave flat-topping.

input mode (operating mode): A display mode that determines how an electric motor drive is controlled during starting and stopping.

input voltage: The voltage supplied to an electric motor drive. It must be the same as the nameplate voltage of the motor.

insulated-gate bipolar transistor (IGBT): A high power switching device that can switch high currents and high voltages.

insulation spot-test: A test that checks the insulation integrity of the stator windings and load conductors of a motor.

integrated power module: A grouping of separate parts designed to perform a set function or task, such as providing a time delay, power gain, or filtering of a signal.

International Electrotechnical Commission (IEC): An organization that develops international safety standards for electrical equipment.

inverter: An electronic device that changes DC voltage into AC voltage.

inverter duty motor: An electric motor specifically designed to work with electric motor drives.

K

keypad: An input control device for electric motor drives using LED or LCD displays.

keypad control: A parameter that disables some or all of the buttons on an electric motor drive keypad.

knee pad: A rubber, leather, or plastic pad strapped onto the knees for protection.

L

lead length : The length of the conductors (motor leads) between the electric motor drive and the motor.

line frequency: The number of complete electrical cycles per second of a power source.

load mechanical test: A test that checks the mechanical operation of a load.

locked in step: A motor condition that occurs when the field of the stator and the field of the rotor are parallel to one another, not allowing the shaft to rotate.

locked rotor: The condition in which a motor is loaded so heavily that the motor shaft cannot turn.

locked rotor torque (LRT): The torque a motor produces when the shaft (rotor) is stationary and full power is applied to the motor.

lockout: The process of removing the source of electrical power and installing a lock, which prevents the power from being turned ON.

M

macro: A parameter that contains predefined values for a group of parameters.

magnetic motor starter: A starter that has an electrically operated switch (contactor) and includes motor overload protection.

magnetic motor starter contactor: An electrical control device that uses a small control current to energize or de-energize 3φ power to a load.

manual control circuit: A circuit that requires technicians or operators to initiate an action in order for the circuit and motor to operate.

manual motor starter: A control device used to control a motor by having technicians or operators control the motor directly at the location of the starter.

megohmmeter: A high-resistance-range meter that is used to measure the insulation integrity of individual conductors, motor windings, transformer windings, etc.

memory: A control function that keeps a motor running after the start pushbutton is released.

minimum holding current: The minimum amount of current required to keep a device operating.

model number: A number that identifies the design of a drive and how the drive is to be used.

motor control circuit: A circuit that provides control functions such as starting, stopping, jogging, direction of rotation control, speed control, and motor protection.

motor current test: A test used to find hidden motor problems not found with the motor mechanical test.

motor mechanical test: A test that checks the mechanical operation of a motor.

motor efficiency: The measure of the effectiveness with which a motor converts electrical energy to mechanical energy.

motor nameplate: A metal plate attached to a motor that lists the technical specifications of a motor.

motor nameplate data: Data that consists of parameters related to motor specifications.

motor power: The strength of a motor measured in watts or horsepower.

motor torque: The force that produces or tends to produce rotation of a motor shaft.

N

nonlinear load: Any load where the instantaneous load current is not proportional to the instantaneous voltage.

nonsinusoidal waveform: A waveform that has a distorted appearance when compared with a pure sine waveform.

O

open loop drive: An electric motor drive that operates without any feedback to the drive about motor speed.

open-loop vector control (sensorless control): A control mode that uses complex mathematical formulas to control the flux-producing and torque-producing currents to an AC motor.

open loop vector drive: An electric motor drive that has no feedback method.

open motor enclosure : A motor enclosure with openings to allow passage of air to cool the windings.

output current: The current sent to a motor to cause rotation; output current must be equal to or greater than the motor nameplate current.

overloaded motor: A motor that has a current reading greater than 105% of nameplate rating.

P

parallel misalignment: Misalignment caused by two shafts that are parallel but on the same axis.

parameter: A property of an electric motor drive that is programmed or adjusted.

parameter protection: A parameter that limits access to electric motor drive parameters.

parameter protection and keypad controls: A group of parameters that control the functionality of a keypad and limit access to parameters along with resetting the drive to factory default settings.

part-winding starting: A motor starting method that first applies power to part of the coil windings of a motor for starting and then applies power to the remaining coil windings for normal operation.

period (T): The time required to produce one complete cycle of a waveform.

permissible temperature rise : The difference between the ambient temperature and the listed (nameplate) ambient rating of a motor.

personal computer (PC): A desktop or laptop computer intended for personal use in a home, office, or factory.

personal protective equipment (PPE): Gear worn by a technician to reduce the possibility of injury in the work area.

phase unbalance (imbalance): The unbalance that occurs when the 3ϕ power lines are more than or less than 120° out of phase.

pigtail: An extended, flexible connection or a braided copper conductor.

polarity: The positive (+) or negative (−) electrical state of an object.

potentiometer: An input control device that sends various resistance values to an electric motor drive.

power: The rate of work (lb-ft) produced per unit of time (sec). Power is work divided by time.

power distribution system: Wires, wire ways, switchgear, and panels used to deliver the required type (DC, 1ϕ AC, or 3ϕ AC) and level (120 V, 230 V, 460 V, etc.) of electricity to loads connected to the system.

power factor: The ratio of true power to apparent power.

power factor correction capacitor: A capacitor used to improve a facility's power factor by improving voltage levels, increasing system capacity, and reducing line losses.

power reactor: A device used to condition the supplied power to an electric motor drive and/or motor and protect the diodes, transistors, and other electronic components inside the drive from damage caused by transient voltages.

primary resistor starting: A motor starting method that uses resistors connected in the motor conductors to produce a voltage drop.

programmable logic controller (PLC): A solid state control device that is programmed to automatically control an industrial process or machine.

protective clothing: Clothing that provides protection from contact with sharp objects, hot equipment, and harmful materials.

protective helmet: A hard hat that is used in the workplace to prevent injury from the impact of falling and flying objects, and from electrical shock.

pull-up torque (PUT): The torque required to bring a load up to its rated speed.

pulse width modulation (PWM): A method of controlling the amount of voltage sent to a motor.

pump curve: A graph that shows the relationship between flow and pressure for a particular pump at a single speed.

PWM frequency parameter: A parameter that allows the PWM frequency (carrier frequency) of an electric motor drive to be adjusted.

Q

quadratic volts-per-hertz: A control mode that provides a nonlinear voltage to the frequency ratio.

qualified person: A person who is trained and has special knowledge of the construction and operation of electrical equipment or a specific task, and is trained to recognize and avoid electrical hazards that might be present with respect to the equipment or specific task.

R

ramp stop: A method of stopping a motor in which the frequency applied to a motor is reduced, which decelerates the motor to a stop.

read-only parameter: A parameter value that can be displayed, but not set or changed.

rectification: The changing of AC voltage into DC voltage. Rectifiers can be half-wave, full-wave, or bridge rectifiers.

rectifier: A device that changes AC voltage into DC voltage. Alternating current (AC) power is used because it is more efficiently generated and transmitted over long distances than direct current (DC) power.

repulsion motor: A motor with the rotor connected to the power supply through brushes that ride on a commutator.

resistance: 1. Mechanical resistance is any force that tends to hinder the movement of an object. **2.** Electrical resistance is the opposition to the flow of electrons.

resistive load: A load that contains only electrical resistance.

resistor: An electrical device used in the DC bus section to limit the charging current to capacitors, to discharge capacitors, and to absorb unwanted voltages.

reverse bias: The condition of a diode when it does not allow current flow and acts as an insulator.

reverse inhibit: A parameter that prevents an electric motor drive from running a motor in reverse.

rotor: The rotating part of an AC motor.

rubber insulating matting: A floor covering that provides technicians protection from electrical shock when working on live electrical circuits.

S

safety glasses: An eye protection device with special impact-resistant glass or plastic lenses, reinforced frames, and side shields.

safety label: A label that indicates areas or tasks that can pose a hazard to personnel and/or equipment.

saturation region: The maximum current that can flow in the transistor load circuit.

S-curve (smoothing): A parameter that changes the acceleration and deceleration profile from a ramp to an S-curve slope.

secondary check: A check that is performed between the initial checks and final checks of an electric motor drive.

secondary electric motor drive test: A test that verifies the basic functionality of a drive.

self-excited shunt field: A shunt field connected to the same power supply as the armature.

semiconductor device: An electronic device that has electrical conductivity between that of a conductor (high conductivity) and that of an insulator (low conductivity).

separately-excited shunt field: A shunt field connected to a different power supply than the armature.

serial communication: A communications port that uses a D-shell connector to connect an electric motor drive with other drives, PLCs, or PCs.

serial communication signals: Digital data signals from an external source such as a PLC or PC.

serial number: A number that identifies when a drive was manufactured, what modifications were made to the design, and what software is used.

service factor (SF): A multiplier that represents the percentage of extra load that can be placed on a motor for short periods of time without damaging the motor.

shaded-pole motor: An AC motor that uses a shaded pole for starting.

shaft grounding system: A system that connects (shorts) the rotor voltage to ground via a brush or other device in contact with the shaft to discharge unwanted voltage.

shim: A thin piece of metal that is thousandths of an inch thick and made of steel or brass.

shock (impact) load: A load which varies from a fraction of rating (horsepower and torque) to several hundred percent of rating.

silicon-controlled rectifier (SCR): A solid state device with the ability to rapidly switch high currents.

single-phasing: The operation of a 3ϕ load on two phases because one phase is lost.

single-voltage motor: A motor that operates at only one voltage level.

single-function switch: A switch that performs only one switching function such as START or STOP.

skip frequency: A parameter that prevents an electric motor drive from operating within a particular frequency range.

slip: The difference between the synchronous speed and the actual speed of a motor.

soft starter: A motor control device that provides a gradual voltage increase (ramp up) during AC motor starting and a gradual decrease (ramp down) during stopping.

solid state starter: A motor starting device that uses solid state switches to control the voltage to a motor.

solid state switch: An electronic switching device that opens and closes circuits at a precise point in time to control current flow and voltage levels.

speed range: The minimum speed and maximum speed at which an electric motor drive or motor can operate under constant torque or variable torque conditions.

speed reference (frequency source): A signal that informs an electric motor drive of the speed at which to operate.

speed regulation: The numerical measure of how accurately an electric motor drive can maintain the speed of a motor when the load changes.

split-phase motor: An AC motor that has running and starting windings.

static electricity: An electrical charge at rest.

stator: The stationary part of an AC motor that produces a rotating magnetic field.

stop mode: A parameter that determines how an electric motor drive stops when it receives a stop command.

surge suppressor: An electrical device that provides protection from transient voltages by limiting the level of voltage allowed downstream from the surge suppressor.

synchronous speed: The theoretical speed of a motor based on the number of poles of the motor and line frequency.

system curve: A graph that shows the relationship between the pressure and flow required by a system.

T

tachometer: A sensor that monitors the speed of a rotating shaft.

tagout: The process of placing a danger tag on the source of electrical power, which indicates that the equipment may not be operated until the danger tag is removed.

tapped resistor: A resistor that contains fixed tap points of different resistances.

temperature rise: The difference between the winding temperature of a running motor and ambient temperature.

three-phase manual motor starter: An electrical control device that manually energizes or de-energizes 3ϕ power to a load.

three-wire control: An input control for an electric motor drive requiring three conductors to complete a circuit.

torque: The force that produces rotation.

totally enclosed motor enclosure: A motor enclosure that prevents air from entering the motor.

transient voltage: A high-energy, high-voltage, short-duration spike in an electrical system.

transistor: A solid state device that controls current according to the amount of voltage applied to the base.

troubleshooting: The systematic elimination of various parts of a system to locate a malfunctioning part.

two-wire control: An input control for an electric motor drive requiring two conductors to complete a circuit.

U

underloaded motor: A motor that has an amperage reading of 0% to 95% of nameplate current.

utility rebates: Monetary benefits received after retrofitting an application with an electric motor drive.

V

variable capacitor: A capacitor that can be varied in capacitance value.

variable (adjustable) resistor: A resistor with a set range of values.

variable torque (VT) load: A load that requires varying torque and horsepower at different speeds.

vector or inverter duty-rated motor: A motor made with wire and insulation that resist voltage spikes and high temperatures to extend the life expectancy of the motor.

voltage drop: The amount of voltage consumed by a device or component as current passes through it.

voltage sag: A drop in voltage of not more than 10% below the normal rated line voltage lasting from 8 ms to 1 min.

voltage surge: A higher-than-normal voltage that temporarily exists in one or more power lines.

voltage swell: An increase in voltage of not more than 10% above the normal rated line voltage lasting from 8 ms to 1 min.

voltage unbalance: The unbalance that occurs when the voltages at the three motor terminals or other 3ϕ loads are not equal.

volts-per-hertz (V/Hz): A control mode that provides a linear voltage ratio to the frequency of a motor from 0 rpm to base speed.

W

warning signal word: A word used to indicate a potentially hazardous situation which, if not avoided, could result in death or serious injury.

watt (W): An electrical measurement equal to the power produced by 1 A of current across a potential difference of 1 V.

work: A force (measured in lb) applied at a distance (measured in ft).

wrap-around bar graph: A bar graph that displays a fraction of the full range on the graph.

Index

Using the Electric Motor Drive Installation and Troubleshooting CD-ROM

Before removing the CD-ROM, please note that the book cannot be returned if the CD-ROM sleeve seal is broken.

System Requirements

The *Electric Motor Drive Troubleshooting and Installation CD-ROM* is designed to work best on a computer with a processor speed of 200 MHz or faster, running Microsoft® Windows® 95, 98, 2000, Me, NT®, or XP™. Adobe® Acrobat® Reader™ software is required for opening many resources provided on this CD-ROM. If necessary, Adobe® Acrobat® Reader™ can be installed from the CD-ROM. Microsoft® Windows® 2000, NT™, or XP™ users that are connected to a server-based network may be required to log on with administrative rights to allow installation of the application. See your Information Systems group for further information. Additional information is available from the Adobe® web site at www.adobe.com. Adobe® Acrobat® Reader™ provides the user with the ability to enlarge images for greater clarity as well as provides other navigational functions. The Energy Savings Spreadsheet requires Microsoft® Excel® software. The Internet links require Microsoft® Internet Explorer™ 3.0 or Netscape® 3.0 or later browser software and an Internet connection.

Opening Files

Insert the CD-ROM into the computer CD-ROM drive. Within a few seconds, the start screen will be displayed. Click on START to open the home screen. Information about the usage of the CD-ROM can be accessed by clicking on USING THE CD-ROM. The Chapter Quick Quizzes®, Illustrated Glossary, Electrical Power Data Sheets, Energy Savings Spreadsheet, Test Tool Connections, Media Clips and ATPeResources.com can be accessed by clicking on the appropriate button located on the home screen. Clicking on the American Tech logo accesses the American Tech web site (www.go2atp.com) for information on related educational products. Unauthorized reproduction of the material on the CD-ROM is strictly prohibited.

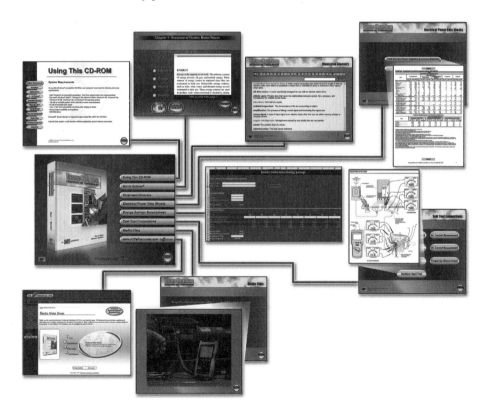